Intermediate Algebra
Concepts and Graphs

THIRD EDITION

Intermediate Algebra
Concepts and Graphs

Charles P. McKeague
Cuesta College

Saunders College Publishing
HARCOURT BRACE COLLEGE PUBLISHERS
Fort Worth   Philadelphia   San Diego   New York   Orlando   Austin
San Antonio   Toronto   Montreal   London   Sydney   Tokyo

Vice President/Publisher: Emily Barrosse
Acquisitions Editor: Bill Hoffman
Product Manager: Nick Agnew
Developmental Editor: Carol Loyd
Project Editor: Phyllis Niklas
Production Manager: Alicia Jackson
Art Director: Lisa Caro
Cover Designer: Lisa Caro

Cover Credit: Private Collection/Superstock

Printed in the United States of America

INTERMEDIATE ALGEBRA: Concepts and Graphs

ISBN: 0-03-019468-7

Library of Congress Catalog Card Number: 97-61727

7890123456   039   10 987654321

Basic Definitions and Properties   1

Equations and Inequalities in One Variable   73

Equations and Inequalities in Two Variables   135

Exponential and Logarithmic Functions   453

Sequences and Series   491

Conic Sections   527

If you have a difficult time covering all the topics in intermediate algebra, you will like the sequence of topics in this book: **The review of elementary algebra has been condensed considerably.**

Chapter 1 reviews real numbers, exponents, and polynomials. It can be covered in detail or simply left for reference, depending on the background of your students.

With polynomials and factoring covered in Chapter 1, the sequence and variety of topics in Chapter 2 becomes more interesting. For example, Section 2.1 covers linear equations in one variable, as well as quadratic equations that can be solved by factoring. Covering these topics together is a pleasant deviation from the usual sequence of intermediate algebra topics. It also allows for a wider variety of problems in the sections on applications and formulas that make up the next part of the chapter.

You will find other interesting changes in both the scope and the order of topics in the rest of the book, which result from the condensed review of elementary algebra. In general, both the level and content of the course are enhanced by spending less time on elementary algebra topics and more time on new topics.

Other Features of the Book

Emphasis on Graphing Graphing starts early in the book and is contained in almost every chapter. The chart below illustrates.

CHAPTER	TYPE OF GRAPHING PROBLEM
3	Bar charts, linear equations, parabolas, inverse variation
4	Simple rational functions
5	Square and cube root equations
6	Parabolas
7	Systems of equations
8	Exponential and logarithmic functions
10	Conic sections

Early Coverage of Functions and Graphs The material on functions and graphs has been rewritten for this edition. It starts in Section 3.1 with coverage of paired data, bar charts, and line graphs. Functions are introduced in Section 3.5 and function notation in Section 3.6. The concept of a function is then carried through to the remaining chapters in the book.

Using Technology   Scattered throughout the book is new material that shows how graphing calculators, spreadsheet programs, and computer graphing programs can be used to enhance the topics being covered. This material is easy to find because it appears under the heading "Using Technology." I recommend that you look at this material, even if you do not use technology in your classroom, because it is designed to provide insight into the standard algebraic procedures and problem solving it accompanies.

Blueprint for Problem Solving   New to this edition, the Blueprint for Problem Solving (in Section 2.3) is a detailed outline of the steps needed to be successful with application problems. It is intended as a guide to problem solving in general, and it overlays the solution process to all the application problems in the first few chapters of the book. As students become more familiar with problem solving, the steps in the Blueprint are streamlined.

Facts from Geometry   New to this edition, many important facts from geometry are listed under the heading "Facts from Geometry." In most cases, an example or two accompanies each fact to give students a chance to see how topics from geometry are related to the algebra they are learning.

Unit Analysis   Chapter 4 contains new problems requiring students to convert from one unit of measure to another. The conversion method we use is the same as the method students will use if they take a chemistry class. Since it is similar to the method used to multiply rational expressions, we introduce it in Section 4.7 as an application of multiplication and division of rational expressions.

Chapter Openings   Also new to this edition are the chapter openings. They contain the following items:

1. **Introduction** Each chapter opens with an introduction in which a real-world application, historical example, or a link between topics is used to stimulate interest in the chapter. Many of these introductions are reexamined later in the chapter and then carried through to topics found further on in the book.
2. **Overview** A general overview of the chapter follows the chapter introduction. The overview lists the important topics that will be covered in the chapter, along with their connection to one another and to topics covered previously in the text.
3. **Study skills** Found in the first six chapter openings are a list of study skills intended to help students become organized and efficient with their time. These lists are more detailed than the general study skills listed in the Preface to the Student.

Organization of the Problem Sets   Six main ideas are incorporated into the problem sets.

1. **Drill** There are enough problems in each set to ensure student proficiency in the material.
2. **Progressive difficulty** The problems increase in difficulty as the problem set progresses.
3. **Odd–even similarities** Each pair of consecutive problems is similar. Since the answers to the odd-numbered problems are listed in the back of the book, the similarity of the odd–even pairs of problems allows your students to check their work on an odd-numbered problem and then try the similar even-numbered problem.
4. **Application problems** I have found that students are more likely to put some time and effort into trying application problems if they do not have to work an overwhelming number of them at one time, and if they work on them every day. For these reasons, I have placed a few application problems toward the end of almost every problem set in the book.
5. **Review problems** Each problem set, beginning with Chapter 2, contains a few review problems. Where appropriate, the review problems cover material that will be needed in the next section. Otherwise, they cover material from the previous chapter. That is, the review problems in Chapter 5 cover the important points covered in Chapter 4. Likewise, the review problems in Chapter 6 review the important material from Chapter 5. If you give tests on two chapters at a time, you will find this to be a time-saving feature. Your students will review one chapter as they study the next chapter.
6. **One step further problems** Most of the problem sets end with a few problems under the heading "One Step Further." These problems are more challenging than those in the problem set, or they are problems that extend some of the topics covered in the section.

Chapter Summaries Each chapter summary lists the new properties and definitions found in the chapter. The margins in the chapter summaries contain examples that illustrate the topics being reviewed.

Chapter Reviews Each chapter ends with a set of review problems that cover all the different types of problems found in the chapter. The chapter reviews are longer and more extensive than the chapter tests.

Chapter Tests Each chapter test contains a representative sample of the problems covered in the chapter.

Additional Changes in the Third Edition

Increased Visualization of Topics This edition contains many more diagrams, charts, and graphs than the previous edition. The purpose of this is to give students additional information, in visual form, to help them understand the topics we cover.

Content Changes The section on solving quadratic inequalities has been moved from Chapter 2 to Chapter 6 so that we can include how technology can be used in

the solution process. The material in Chapter 2 on equations and inequalities involving absolute value has been split into two sections instead of one. The section
at the end of Chapter 6 on graphing parabolas that open left and right has been
removed from this edition. Some new material on addition, subtraction, and multiplication of matrices has been added to Section 7.3. Also in Chapter 7, the section
on matrix solutions to systems of equations has been deleted. The material on conic
sections has been moved to the end of the book. Exponential functions and inverse
functions have been combined with the material on logarithms, which make up
Chapter 8 of this edition.

Supplements to the Textbook

This third edition of *Intermediate Algebra: Concepts and Graphs* is accompanied
by a number of useful supplements.

For the Instructor

■ **Printed test bank**   The test bank consists of multiple-choice and short-answer
test items organized by chapter, section, and difficulty level. Answers for every
test item are provided. In addition, a list of answers to even-numbered problems
in the text is included in this supplement.

■ *ExaMaster+*™ **computerized test bank**   A flexible, powerful, computerized
testing system, *ExaMaster+*™ offers teachers a wide range of integrated testing
options and features. Available in IBM, Macintosh, or Windows format, it offers
teachers the ability to select, edit, and create not only test items but algorithms
for test items as well. Teachers can tailor tests according to a variety of criteria,
scramble the order of test items, and administer tests on-line. *ExaMaster+*™ also
includes full-function gradebook and graphing features.

■ **Research projects on the web**   My web site, mckeague.com, contains a list of
research projects for intermediate algebra that can be used as group projects or
for individual extra credit. These projects fit in very nicely with the material in
the book and are very easy to access from the web site. The complete internet
address is

http://www.mckeague.com

For the Student

■ **Videotape package**   Free to adopters, the videotape package consists of 10
VHS videotapes, one for each chapter of the book. Each chapter tape is an hour
to an hour and a half in length and is divided into lessons that correspond to each
section of the chapter.

■ **Core concept video**   This single videotape is over 4 hours in length and contains
more than 50 problem-solving sessions, covering most sections of the text. Tailor-
made as a take-home tutorial for students with a demanding schedule, this video
can be used as a preview of what is to be covered in class, as an aid to completing
homework assignments, or as a tool to review for a test.

■ *Student's Solutions Manual*   This manual contains complete annotated solutions to every other odd-numbered problem in the problem sets and all chapter review and chapter test exercises.

■ *MathCue Tutorial*   This computer software package of tutorials contains problems that correspond to every section in the series. The software presents problems to solve and tutors students by displaying annotated, step-by-step solutions. Students may view partial solutions to get started on a problem, see a continuous record of progress, and back up to review missed problems. Student scores can also be printed. Available for Windows and Macintosh.

■ *MathCue Solution Finder*   This software allows students to enter their own problems into the computer and get annotated, step-by-step solutions in return. This unique program simulates working with a tutor, tracks student progress, refers students to specific sections in the text when appropriate, and prints student scores. Available for Windows and Macintosh.

■ *MathCue Practice*   This algorithm-based software allows students to generate large numbers of practice problems keyed to problem types from each section of the book. *MathCue Practice* scores students' performance, and saves students' scores session to session. Available for Windows and Macintosh.

■ *MathCue F/C Graph: The Functions and Conics Grapher*   For use with *Intermediate Algebra: Concepts and Graphs,* this computer program allows students to graph and analyze any polynomial, logarithmic, exponential, or trigonometric function or conic equation they choose. *F/C Graph* can zoom, trace, display function values or coordinates of selected points, graph up to four functions simultaneously, and save and retrieve setups. Students can use *F/C Graph* to relate algebraic and visual forms of functions and conic sections, and to explore how changing parameters affect the graph. A set of computer lab exercises accompanies the software to direct student investigations of a variety of topics. Available for Windows and Macintosh.

■ *Intermediate Algebra and the Graphing Calculator: A Learning Resource*   This workbook helps both instructors and students understand how to use graphing calculators, and presents exercises and exploratory investigations into graphing functions, linear equations and inequalities, polynomials and factoring, lines and inequalities, quadratic equations, and systems of linear equations.

Acknowledgments

A project of this size cannot be completed without help from many people. In particular, Bill Hoffman got things off to a good start by finding excellent reviewers, suggesting that we focus on a few key things to incorporate into the revision, and lining up the right people to handle the production process. Phyllis Niklas, the project editor, did an outstanding job of producing the book. I am very lucky to have someone with her experience and capabilities work on this project. My son, Patrick, assisted me with this revision, and his influence on the use of technology has made this a better book. Kate Pawlik and Martha Brown did all the accuracy checking and proofreading. Their attention to detail and ability to get work done on time are unmatched. Thanks also to Chaudra Hallock, Hali Hallock, and Mike Rosenborg for their help with proofreading and problem checking. My thanks to these people; this book would not have been possible without them.

Thanks also to Diane McKeague and Amy McKeague for their encouragement with all my writing endeavors.

Finally, I am grateful to the following instructors for their suggestions and comments on this revision. Some reviewed the entire manuscript, while others were asked to evaluate the development of specific topics or the overall sequence of topics. My thanks go to the people listed below:

Carole Bauer, Triton College

Marsha J. Driskill, Aims Community College

Sharon Edgmon, Bakersfield College

Maria Kelly, Kings River Community College

Diane Murphy, Howard College

Ted Ostrander, Fresno City College

Mary Jane Turner, Birmingham Southern College

Charles P. McKeague

Many of my algebra students are apprehensive at first because they are worried that they will not understand the topics we cover. When I present a new topic that they do not grasp completely, they think something is wrong with them for not understanding it.

On the other hand, some students are excited about the course from the beginning. They are not worried about understanding algebra and, in fact, *expect* to find some topics difficult.

What is the difference between these two types of students?

Those who are excited about the course know from experience (as you do) that a certain amount of confusion is associated with most new topics in mathematics. They don't worry about it because they also know that the confusion gives way to understanding in the process of reading the textbook, working the problems, and getting questions answered. If they find a topic they are having difficulty with, they work as many problems as necessary to grasp the subject. They don't wait for the understanding to come to them; they go out and get it by working lots of problems. In contrast, the students who lack confidence tend to give up when they become confused. Instead of working more problems, they sometimes stop working problems altogether—and that, of course, guarantees that they will remain confused.

If you are worried about this course because you lack confidence in your ability to understand algebra, and you want to change the way you feel about mathematics, then look forward to the first topic that causes you some confusion. As soon as that topic comes along, make it your goal to master it, in spite of your apprehension. You will see that each and every topic covered in this course is one you can eventually master, even if your initial introduction to it is accompanied by some confusion. As long as you have passed a college-level beginning algebra course (or its equivalent), you are ready to take this course.

If you have decided to do well in algebra, the following list will be important to you:

How To Be Successful in Algebra

1. **Attend all class sessions on time** You cannot know exactly what goes on in class unless you are there. Missing class and then expecting to find out what went on from someone else is not the same as being there yourself.

2. **Read the book** It is best to read the section that will be covered in class beforehand. Reading in advance, even if you do not understand everything you read, is still better than going to class with no idea of what will be discussed.

3. **Work problems every day and check your answers** The key to success in mathematics is working problems. The more problems you work, the better you will become at working them. The answers to the odd-numbered problems are given in the back of the book. When you have finished an assignment, be sure to compare your answers with those in the book. If you have made a mistake, find out what it is, and correct it.

4. **Do it on your own** Don't be misled into thinking someone else's work is your own. Having someone else show you how to work a problem is not the same as working the same problem yourself. It is okay to get help when you are stuck. As a matter of fact, it is a good idea. Just be sure you do the work yourself.

5. **Review every day** After you have finished the problems your instructor has assigned, take another fifteen minutes and review a section you have already completed. The more you review, the longer you will retain the material you have learned.

6. **Don't expect to understand every new topic the first time you see it** Sometimes you will understand everything you are doing, and sometimes you won't. That's just the way things are in mathematics. Expecting to understand each new topic the first time you see it can lead to disappointment and frustration. The process of understanding algebra takes time. It requires that you read the book, work problems, and get your questions answered.

7. **Spend as much time as it takes for you to master the material** No set formula exists for the exact amount of time you need to spend on algebra to master it. You will find out as you go along what is or isn't enough time for you. If you end up spending two or more hours on each section in order to master the material there, then that's how much time it takes; trying to get by with less will not work.

8. **Relax** It's probably not as difficult as you think.

Intermediate Algebra
Concepts and Graphs

Basic Definitions and Properties

Contents

Introduction

If you major in business or economics, you will see equations and charts like the ones below many times during your college career:

$$P = R - C$$

where

P = Profit; the difference between revenue and cost

R = Revenue; the amount of money a company brings in by selling its product

C = Cost; the amount of money a company pays out to produce its product

Revenue and Cost to Duplicate a 30 Minute Video

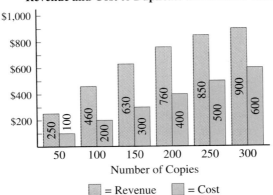

$\square$ = Revenue $\square$ = Cost

The formula for profit tells us that we can find the profits in the chart by finding the difference between the heights of each pair of connected columns.

The material we cover in this chapter will help you understand equations and charts like the ones above, and get you started on your way to working with some of the more detailed applications in business and economics.

Overview

The material in Chapter 1 is some of the most important material in the book. Be sure that you master it. Your success in the following chapters is directly related to how well you understand this chapter.

Here is a list of the essential concepts from Chapter 1 that you will need to be successful in the succeeding chapters:

1. You must know how to add, subtract, multiply, and divide both positive and negative numbers.
2. You must understand and recognize the commutative, associative, and distributive properties—the three most important properties of real numbers.
3. You must know the difference between whole numbers, integers, rational numbers, and real numbers.
4. You must be able to use the properties of exponents to simplify expressions involving exponents.
5. You must have a working knowledge of scientific notation.
6. You must be able to add, subtract, multiply, and factor polynomials.

Study Skills

At the beginning of each of the first few chapters of this book you will find a section like this in which we list the skills that are necessary for success in algebra. If you have just completed an introductory algebra class successfully, you have acquired most of these skills. If it has been some time since you have taken a math class, you must pay attention to the sections on study skills.

Here is a list of things you can do to begin to develop effective study skills.

1. **Put yourself on a schedule** The general rule is that you spend 2 hours on homework for every hour you are in class. Make a schedule for yourself, setting aside 2 hours each day to work on algebra. Once you make the schedule, stick to it. Don't just complete your assignments and stop. Use all the time you have set aside. If you complete an assignment and have time left over, read the next section in the book, and work more problems. As the course progresses you may find that 2 hours a day is not enough time for you to master the material in this course. If it takes you longer to reach your goals for this course, then that's how much time you need to spend. Trying to get by with less will not work.
2. **Find your mistakes and correct them** There is more to studying algebra than just working problems. You must always check your answers with the answers in the back of the book. When you have made a mistake, find out what it is, and correct it. Making mistakes is part of the process of learning mathematics. The key to discovering what you do not understand can be found by correcting your mistakes.

3. **Imitate success** Your work should look like the work you see in this book and the work your instructor shows. The steps shown in solving problems in this book were written by someone who has been successful in mathematics. The same is true of your instructor. Your work should imitate the work of people who have been successful in mathematics.

4. **Don't let your intuition fool you** As you become more experienced and more successful in mathematics you will be able to trust your mathematical intuition. For now, though, it can get in the way of your success. For example, if students are asked to "subtract 3 from -5," many will answer -2 or 2. Both answers are incorrect, even though they may seem intuitively true. Likewise, some students will expand $(a + b)^2$ and arrive at $a^2 + b^2$, which is incorrect. In both cases, intuition leads directly to the wrong answer.

SECTION

1.1

Fundamental Definitions and Notation

The diagram below is called a *bar chart*. This one shows the net price of a popular intermediate algebra textbook. (The net price is the price the bookstore pays for the book.)

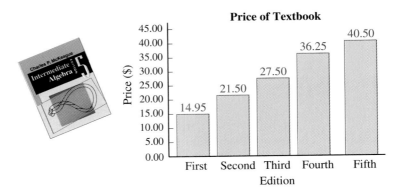

From the chart, we can find many relationships between numbers. We may notice that the price of the third edition was less than the price of the fourth edition. In mathematics we use symbols to represent relationships between quantities. If we let P represent the price of the book, then the relationship just mentioned, between the price of the third edition and the price of the fourth edition, can be written this way:

$$P(3) < P(4)$$

You will be on your way to understanding expressions like the one above by studying the material in this section.

This section is, for the most part, simply a list of many of the basic symbols and definitions we will be using throughout the book.

Comparison Symbols

IN SYMBOLS	IN WORDS
$a = b$	a is equal to b
$a \neq b$	a is not equal to b
$a < b$	a is less than b
$a \leq b$	a is less than or equal to b
$a \nless b$	a is not less than b
$a > b$	a is greater than b
$a \geq b$	a is greater than or equal to b
$a \ngtr b$	a is not greater than b
$a \Leftrightarrow b$	a is equivalent to b

Operation Symbols

OPERATION	IN SYMBOLS	IN WORDS
Addition	$a + b$	The sum of a and b
Subtraction	$a - b$	The difference of a and b
Multiplication	ab, $a \cdot b$, $a(b)$, $(a)b$, or $(a)(b)$	The product of a and b
Division	$a \div b$, a/b, or $\frac{a}{b}$	The quotient of a and b

The key words are **sum, difference, product,** and **quotient.** They are used frequently in mathematics. For instance, we may say "the product of 3 and 4 is 12." We mean that both the statements "3 · 4" and "12" are called the product of 3 and 4. The important idea here is that the word *product* implies multiplication, regardless of whether it is written 3 · 4, 12, 3(4), or (3)4.

The following example shows how we translate sentences written in English into expressions written in symbols.

EXAMPLE 1

IN ENGLISH	IN SYMBOLS
The sum of x and 5 is less than 2.	$x + 5 < 2$
The product of 3 and x is 21.	$3x = 21$
The quotient of y and 6 is 4.	$\frac{y}{6} = 4$
Twice the difference of b and 7 is greater than 5.	$2(b - 7) > 5$
The difference of twice b and 7 is greater than 5.	$2b - 7 > 5$

Exponents

Consider the expression 3^4. The 3 is called the **base** and the 4 is called the **exponent.** The exponent 4 tells us the number of times the base appears in the product. That is,

$$3^4 = 3 \cdot 3 \cdot 3 \cdot 3 = 81$$

The expression 3^4 is said to be in *exponential form,* while $3 \cdot 3 \cdot 3 \cdot 3$ is said to be in *expanded form.*

EXAMPLES Expand and multiply.

2. $5^2 = 5 \cdot 5 = 25$ Base 5, exponent 2

3. $2^5 = 2 \cdot 2 \cdot 2 \cdot 2 \cdot 2 = 32$ Base 2, exponent 5

4. $4^3 = 4 \cdot 4 \cdot 4 = 64$ Base 4, exponent 3

Order of Operations

It is important when evaluating arithmetic expressions in mathematics that each expression have only one answer in reduced form. Consider the expression

$$3 \cdot 7 + 2$$

If we find the product of 3 and 7 first, then add 2, the answer is 23. On the other hand, if we first combine the 7 and 2, then multiply by 3, we have 27. The problem seems to have two distinct answers depending on whether we multiply first or add first. To avoid this situation, we follow the rule that multiplication in a situation like this will always be done before addition. In this case, only the first answer, 23, is correct.

Here is the complete set of rules for evaluating expressions.

RULE: ORDER OF OPERATIONS

When evaluating a mathematical expression, we will perform the operations in the following order, beginning with the expression in the innermost parentheses or brackets and working our way out.

Step 1 Simplify all numbers with exponents, working from left to right if more than one of these expressions is present.

Step 2 Then, do all multiplications and divisions left to right.

Step 3 Perform all additions and subtractions left to right.

Here are some examples that illustrate the use of this rule.

EXAMPLES Simplify each expression using the rule for order of operations.

5. $5 + 3(2 + 4) = 5 + 3(6)$ Simplify inside parentheses.

$\qquad\qquad\qquad = 5 + 18$ Then, multiply.

$\qquad\qquad\qquad = 23$ Add.

6. $5 \cdot 2^3 - 4 \cdot 3^2 = 5 \cdot 8 - 4 \cdot 9$ Simplify exponentials left to right.
$\qquad\qquad\quad = 40 - 36$ Multiply left to right.
$\qquad\qquad\quad = 4$ Subtract.

7. $20 - (2 \cdot 5^2 - 30) = 20 - (2 \cdot 25 - 30)$ ⎱ Simplify inside
$\qquad\qquad\qquad\quad = 20 - (50 - 30)$ ⎰ parentheses,
$\qquad\qquad\qquad\quad = 20 - (20)$ evaluating
$\qquad\qquad\qquad\quad = 0$ exponents first,
then multiplying,
and finally subtracting.

Sets

A **set** is a collection of objects or things. The objects in the set are called **elements** or **members** of the set.

Sets are usually denoted by capital letters and elements of sets by lowercase letters. We use braces, { }, to enclose the elements of a set.

To show that an element is contained in a set we use the symbol $\in$. That is,

$x \in A$ is read   "x is an element (member) of set A"

For example, if A is the set $\{1, 2, 3\}$, then $2 \in A$. On the other hand, $5 \notin A$, means 5 is not an element of set A.

Set A is a **subset** of set B, written $A \subset B$, if every element in A is also an element of B. That is,

$A \subset B$ if and only if A is contained in B

EXAMPLES

8. The set of numbers used to count things is $\{1, 2, 3, \ldots\}$. The dots mean the set continues indefinitely in the same manner. This is an example of an **infinite** set.

9. The set of all numbers represented by the dots on the faces of a regular die is $\{1, 2, 3, 4, 5, 6\}$. This set is a subset of the set in Example 8. It is an example of a **finite** set, since it has a limited number of elements.

The set with no members is called the **empty** or **null set.** It is denoted by the symbol $\varnothing$. The empty set is considered a subset of every set.

Operations with Sets

These diagrams are called *Venn diagrams* after John Venn (1834–1923). They can be used to visualize operations with sets. The region inside the circle labeled A is set A, while the region inside the circle labeled B is set B. The shaded region in Figure 1 is the union of A and B. The shaded region in Figure 2 is the intersection of A and B.

Two basic operations are used to combine sets: union and intersection.

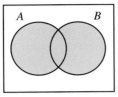

$A \cup B$

FIGURE I
The *union* of two sets

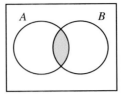

$A \cap B$

FIGURE 2
The *intersection* of two sets

DEFINITION

The **union** of two sets A and B, written $A \cup B$, is the set of all elements that are either in A or in B, or in both A and B. The key word here is *or*. For an element to be in $A \cup B$ it must be in A or B. In symbols, the definition looks like this:

$$x \in A \cup B \qquad \text{if and only if} \qquad x \in A \text{ or } x \in B$$

DEFINITION

The **intersection** of two sets A and B, written $A \cap B$, is the set of elements in both A and B. The key word in this definition is the word *and*. For an element to be in $A \cap B$ it must be in both A and B. In symbols,

$$x \in A \cap B \qquad \text{if and only if} \qquad x \in A \text{ and } x \in B$$

EXAMPLES   Let $A = \{1, 3, 5\}$, $B = \{0, 2, 4\}$, and $C = \{1, 2, 3, \ldots\}$. Then:

10. $A \cup B = \{0, 1, 2, 3, 4, 5\}$

11. $A \cap B = \varnothing$ A and B have no elements in common.

12. $A \cap C = \{1, 3, 5\} = A$

13. $B \cup C = \{0, 1, 2, 3, \ldots\}$

Up to this point we have described the sets encountered by listing all the elements and then enclosing them with braces { }. There is another notation we can use to describe sets. It is called **set-builder notation.** Here is how we would write our definition for the union of two sets A and B using set-builder notation:

$$A \cup B = \{x \mid x \in A \text{ or } x \in B\}$$

The right side of this statement is read "the set of all x such that x is a member of A or x is a member of B." As you can see, the vertical line after the first x is read "such that."

EXAMPLE 14 If $A = \{1, 2, 3, 4, 5, 6\}$, find $C = \{x \mid x \in A \text{ and } x \geq 4\}$.

SOLUTION We are looking for all the elements of A that are also greater than or equal to 4. They are 4, 5, and 6. Using set notation, we have

$$C = \{4, 5, 6\}$$

The set we will work with most often in this book is the set of real numbers. To develop the real numbers, we start with the real number line.

The Real Numbers

The **real number line** is constructed by drawing a straight line and labeling a convenient point with the number 0. Positive numbers are in increasing order to the right of 0; negative numbers are in decreasing order to the left of 0. The point on the line corresponding to 0 is called the **origin.**

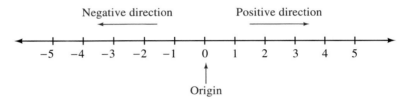

FIGURE 3
Constructing a number line

The numbers on the number line increase in size as we move to the right. When we compare the size of two numbers on the number line, the number on the left is always the smaller number.

The numbers associated with the points on the line are called **coordinates** of those points. Every point on the line has a number associated with it. The set of all these numbers makes up the set of real numbers.

> **DEFINITION**
>
> A **real number** is any number that is the coordinate of a point on the real number line.

EXAMPLE 15 Locate the numbers -4.5, -0.75, $\frac{1}{2}$, $\sqrt{2}$, π, and 4.1 on the real number line.

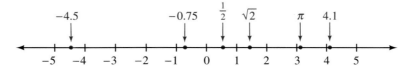

Note: In this book we will refer to real numbers as being on the real number line. Actually, real numbers are *not* on the line; only the points they represent are on the line. We can save some writing, however, if we simply refer to real numbers as being on the number line.

Subsets of the Real Numbers

Next, we consider some of the more important subsets of the real numbers. Each set listed here is a subset of the real numbers:

$$\text{\textbf{Counting} (or \textbf{Natural}) \textbf{numbers}} = \{1, 2, 3, \ldots\}$$
$$\text{\textbf{Whole numbers}} = \{0, 1, 2, 3, \ldots\}$$
$$\text{\textbf{Integers}} = \{\ldots, -3, -2, -1, 0, 1, 2, 3, \ldots\}$$
$$\text{\textbf{Rational numbers}} = \left\{\left.\frac{a}{b}\right| a \text{ and } b \text{ are integers}, b \neq 0\right\}$$

Any number that can be written in the form

$$\frac{\text{Integer}}{\text{Integer}}$$

is a rational number. Rational numbers are numbers that can be written as the ratio of two integers. Each of the following is a rational number:

$\dfrac{3}{4}$ Because it is the ratio of the integers 3 and 4

-8 Because it can be written as the ratio of -8 to 1: $\dfrac{-8}{1}$

0.75 Because it is the ratio of 75 to 100 (or 3 to 4 if you reduce to lowest terms)

$0.333\ldots$ Because it can be written as the ratio of 1 to 3

There are still other numbers on the number line that are not members of the subsets we have listed so far. They are real numbers, but they cannot be written as the ratio of two integers. That is, they are not rational numbers. For that reason, we call them irrational numbers.

$$\text{\textbf{Irrational numbers}} = \{x \mid x \text{ is real, but not rational}\}$$

The following are irrational numbers:

$$\sqrt{2} \qquad -\sqrt{3} \qquad 4 + 2\sqrt{3} \qquad \pi \qquad \pi + 5\sqrt{6}$$

EXAMPLE 16   For the set $\{-5, -3.5, 0, \frac{3}{4}, \sqrt{3}, \sqrt{5}, 9\}$, list the numbers that are:
(a) Whole numbers   (b) Integers   (c) Rational numbers
(d) Irrational numbers   (e) Real numbers

SOLUTION
(a) Whole numbers $= \{0, 9\}$
(b) Integers $= \{-5, 0, 9\}$ (c) Rational numbers $= \{-5, -3.5, 0, \frac{3}{4}, 9\}$
(d) Irrational numbers $= \{\sqrt{3}, \sqrt{5}\}$ (e) They are all real numbers.

The diagram below gives a visual representation of the relationships among subsets of the real numbers.

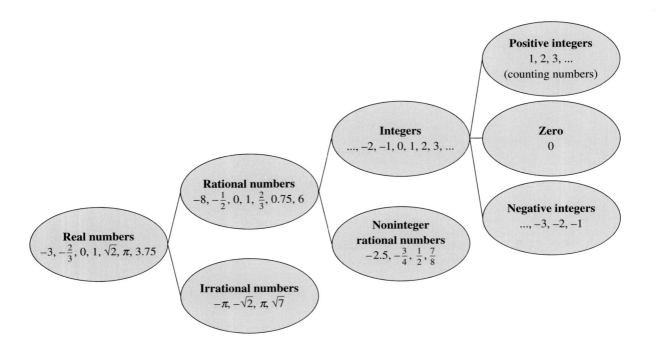

Prime Numbers and Factoring

The following diagram shows the relationship between multiplication and factoring:

Multiplication

Factors → $3 \cdot 4 = 12$ ← Product

Factoring

When we read the problem from left to right, we say the product of 3 and 4 is 12. Or we multiply 3 and 4 to get 12. When we read the problem in the other direction, from right to left, we say we have factored 12 into 3 times 4, or 3 and 4 are **factors** of 12.

The number 12 can be factored still further:

$$12 = 4 \cdot 3$$
$$= 2 \cdot 2 \cdot 3$$
$$= 2^2 \cdot 3$$

The numbers 2 and 3 are called **prime** factors of 12 because neither of them can be factored any further.

Here is a list of the first few prime numbers:

Prime numbers = {2, 3, 5, 7, 11, 13, 17, 19, 23, 29, 31, 37, 41, . . .}

When a number is not prime, we can factor it into the product of prime numbers.

EXAMPLE 17   Factor 525 into the product of primes.

SOLUTION   Since 525 ends in 5, it is divisible by 5:

$$525 = 5 \cdot 105$$
$$= 5 \cdot 5 \cdot 21$$
$$= 5 \cdot 5 \cdot 3 \cdot 7$$
$$= 3 \cdot 5^2 \cdot 7$$

EXAMPLE 18   Reduce $\dfrac{210}{231}$ to lowest terms.

SOLUTION   First we factor 210 and 231 into the product of prime factors. Then we reduce to lowest terms by dividing the numerator and denominator by any factors they have in common.

$$\frac{210}{231} = \frac{2 \cdot 3 \cdot 5 \cdot 7}{3 \cdot 7 \cdot 11}$$ Factor the numerator and denominator completely.

$$= \frac{2 \cdot 3 \cdot 5 \cdot 7}{3 \cdot 7 \cdot 11}$$ Divide the numerator and denominator by $3 \cdot 7$.

$$= \frac{2 \cdot 5}{11}$$

$$= \frac{10}{11}$$

The small lines we have drawn through the factors that are common to the numerator and denominator are used to indicate that we have divided the numerator and denominator by those factors.

Problem Set

1.1

NOTE: If you are like my students, most of the time you spend with this book will be spent here, in the problem sets. Since that is the case, I am going to write some notes to you, right here in the problem sets, that will assist you in working your way through the course.

Translate each of the following sentences into symbols.

1. The sum of x and 5 is 2.

2. The sum of y and -3 is 9.

3. The difference of 6 and x is y.

4. The difference of x and 6 is $-y$.

5. The product of t and 2 is less than y.

6. The product of $5x$ and y is equal to z.

7. The sum of x and y is less than the difference of x and y.

8. Twice the sum of a and b is 15.

9. The rectangle below is made up of two smaller rectangles, which are numbered 1 and 2. The area of a rectangle is the product of the length and the width. (In symbols, this is written $A = lw$.)
  (a) Find the area of the two smaller rectangles and add them together to find the area of the large rectangle.
  (b) Find the area of the large rectangle by multiplying its length by its width.

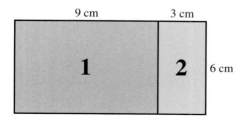

10. The large rectangle below is made up of four smaller rectangles, which have been numbered 1–4. To find the area of a rectangle, we multiply the length by the width.
  (a) Find the area of the four smaller rectangles and add them together to find the area of the large rectangle.
  (b) Find the area of the large rectangle by multiplying its length by its width.

Simplify each expression using the rule for order of operations.

11. $2 + 8 \cdot 5$

12. $12 - 3 \cdot 3$

13. $(2 + 8)5$

14. $(12 - 3)3$

15. $6 + 3 \cdot 4 - 2$

16. $8 + 2 \cdot 7 - 3$

17. $6 + 3(4 - 2)$

18. $8 + 2(7 - 3)$

19. $(6 + 3)(4 - 2)$

20. $(8 + 2)(7 - 3)$

21. $(5 + 3)^2$

22. $(8 - 3)^2$

23. $5^2 + 3^2$

24. $8^2 - 3^2$

25. $5^2 + 2(5)(3) + 3^2$

26. $8^2 - 2(8)(3) + 3^2$

27. $(7 - 4)(7 + 4)$

28. $(8 - 5)(8 + 5)$

29. $7^2 - 4^2$

30. $8^2 - 5^2$

31. $5 \cdot 10^3 + 4 \cdot 10^2 + 3 \cdot 10 + 1$

32. $6 \cdot 10^3 + 5 \cdot 10^2 + 4 \cdot 10 + 3$

33. $40 - [10 - (4 - 2)]$

34. $50 - [17 - (8 - 3)]$

35. $40 - 10 - 4 - 2$

36. $50 - 17 - 8 - 3$

37. $3 + 2(2 \cdot 3^2 + 1)$

38. $4 + 5(3 \cdot 2^2 - 5)$

39. $(3 + 2)(2 \cdot 3^2 + 1)$

40. $(4 + 5)(3 \cdot 2^2 - 5)$

41. $3[2 + 4(5 + 2 \cdot 3)]$
42. $2[4 + 2(6 + 3 \cdot 5)]$
43. $6[3 + 2(5 \cdot 3 - 10)]$
44. $8[7 + 2(6 \cdot 9 - 14)]$
45. $5(7 \cdot 4 - 3 \cdot 4) + 8(5 \cdot 9 - 4 \cdot 9)$
46. $4(3 \cdot 9 - 2 \cdot 9) + 5(6 \cdot 8 - 5 \cdot 8)$

NOTE: Are you checking your answers with the answers in the back of the book? It is very important that you do. That is how you will find out what you know and what you don't know.

Median Incomes

The bar chart below gives the top five median incomes for women age 30 and older in the United States, according to the field of study in which they earned a bachelor's degree. Use the chart to answer questions 47–50.

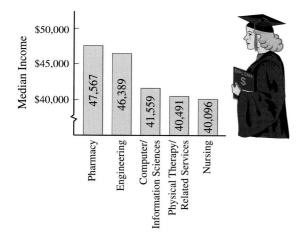

47. What is the difference, in dollars, between the highest median wage and the lowest median wage?

48. Find the difference in median wage between women studying engineering and women studying nursing.

49. What does the word *median,* as used in the chart above, mean? (*Hint:* Use a dictionary.)

50. For the numbers in the set below, which one is the median?

$$\{20, 25, 27, 29, 35, 40, 41\}$$

The bar chart below gives the top five median incomes for men age 30 and older in the United States, according to the field of study in which they earned a bachelor's degree. Use the chart to answer questions 51–54.

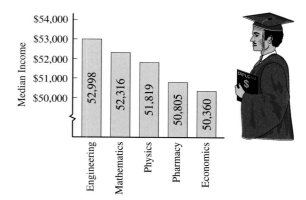

51. What is the difference, in dollars, between the highest median wage and the lowest median wage?

52. Find the difference in median wage between men studying engineering and men studying physics.

53. Using the information in both of the median income charts, find the difference between the median incomes for men studying engineering and women studying engineering.

54. What is the difference, in dollars, between the median incomes of men studying pharmacy and women studying pharmacy?

Sets and Real Numbers

Let $A = \{0, 2, 4, 6\}$, $B = \{1, 2, 3, 4, 5\}$, and $C = \{1, 3, 5, 7\}$. List the elements of each of the following sets.

55. $A \cup B$
56. $A \cup C$
57. $A \cap B$
58. $A \cap C$
59. $B \cap C$
60. $B \cup C$
61. $A \cup (B \cap C)$
62. $C \cup (A \cap B)$
63. $\{x \mid x \in A \text{ and } x < 4\}$
64. $\{x \mid x \in B \text{ and } x > 3\}$
65. $\{x \mid x \in A \text{ and } x \notin B\}$
66. $\{x \mid x \in B \text{ and } x \notin C\}$

67. $\{x \mid x \in A \text{ or } x \in C\}$
68. $\{x \mid x \in A \text{ or } x \in B\}$

 When your answer is different from the answer in the back of the book, do you work the problem again to find your mistake? It is very important that you do so. Your mistakes are the keys to your success in this course. Find your mistakes and correct them.

For the set

$$\left\{-6, -5.2, -\sqrt{7}, -\pi, 0, 1, 2, 2.3, \tfrac{9}{2}, \sqrt{17}\right\}$$

list all the elements that are named in each of the following problems.

69. Counting numbers   **70.** Whole numbers

71. Rational numbers   **72.** Integers

73. Irrational numbers   **74.** Real numbers

Label the following true or false.

75. The number 39 is a prime number.

76. Some irrational numbers are also rational numbers.

77. All whole numbers are integers.

78. Every real number is a rational number.

79. All integers are rational numbers.

80. Zero is not considered a real number.

81. An irrational number is also a real number.

82. The number 0.25 is a rational number.

Factor each number into the product of prime factors.

83. 266   **84.** 385
85. 111   **86.** 735
87. 369   **88.** 1,155

Reduce each fraction to lowest terms.

89. $\dfrac{165}{385}$ **90.** $\dfrac{550}{735}$

91. $\dfrac{385}{735}$ **92.** $\dfrac{266}{285}$

93. $\dfrac{111}{185}$ **94.** $\dfrac{279}{310}$

One Step Further

Most of the problem sets in this book end with a few problems like those that follow. These problems challenge you to extend your knowledge of the material in the problem set. In most cases, there are no examples in the text similar to these problems. You should approach these problems with a positive point of view, because even though you may not complete them correctly, just the process of attempting them will increase your knowledge and ability in algebra.

The expression n! is read "n factorial" and is the product of all the consecutive integers from n down to 1. For example,

$$1! = 1$$
$$2! = 2 \cdot 1 = 2$$
$$3! = 3 \cdot 2 \cdot 1 = 6$$
$$4! = 4 \cdot 3 \cdot 2 \cdot 1 = 24$$

95. Calculate 5!

96. Calculate 6!

97. Show that this statement is true: $6! = 6 \cdot 5!$

98. Show that this statement is false: $(2 + 3)! = 2! + 3!$

Properties of Real Numbers

The area of the large rectangle shown at the top of the next page can be found in two ways: we can multiply its length a by its width $b + c$, or we can find the areas of the two smaller rectangles and add those areas to find the total area.

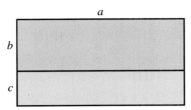

Area of large rectangle: $a(b + c)$

Sum of the areas of two smaller rectangles: $ab + ac$

Since the area of the large rectangle is the sum of the areas of the two smaller rectangles, we can write

$$a(b + c) = ab + ac$$

This expression is called the *distributive property.* It is one of the properties we will be discussing in this section. Before we arrive at the distributive property, we need to review some basic definitions and vocabulary.

Opposites and Reciprocals

> **DEFINITION**
>
> Any two real numbers that are the same distance from 0, but in opposite directions from 0 on the number line, are called **opposites** or *additive inverses.*

EXAMPLE 1 The numbers -3 and 3 are opposites. So are π and $-\pi$, $\frac{3}{4}$ and $-\frac{3}{4}$, and $\sqrt{2}$ and $-\sqrt{2}$.

The negative sign in front of a number can be read in a number of different ways. It can be read as "negative" or "the opposite of." We say -4 is the opposite of 4 or negative 4. The one we use will depend on the situation. For instance, the expression $-(-3)$ is best read "the opposite of negative 3." Since the opposite of -3 is 3, we have $-(-3) = 3$. In general, if a is any positive real number, then

$$-(-a) = a \quad \text{The opposite of a negative is positive.}$$

Review of Multiplication with Fractions

Before we go further with our study of the number line, we need to review multiplication with fractions. Recall that for the fraction $\frac{a}{b}$, a is called the numerator and b is called the denominator. To multiply two fractions we simply multiply numerators and multiply denominators.

EXAMPLES Multiply.

2. $\dfrac{3}{5} \cdot \dfrac{7}{8} = \dfrac{3 \cdot 7}{5 \cdot 8} = \dfrac{21}{40}$

3. $8 \cdot \dfrac{1}{5} = \dfrac{8}{1} \cdot \dfrac{1}{5} = \dfrac{8 \cdot 1}{1 \cdot 5} = \dfrac{8}{5}$

4. $\left(\dfrac{2}{3}\right)^4 = \dfrac{2}{3} \cdot \dfrac{2}{3} \cdot \dfrac{2}{3} \cdot \dfrac{2}{3} = \dfrac{16}{81}$ ◢

The idea of multiplication of fractions is useful in understanding the concept of the reciprocal of a number. Here is the definition.

DEFINITION

Any two real numbers whose product is 1 are called **reciprocals** or *multiplicative inverses*.

EXAMPLES Give the reciprocal of each number.

	NUMBER	RECIPROCAL	
5.	3	$\frac{1}{3}$	Because $3 \cdot \frac{1}{3} = \frac{3}{1} \cdot \frac{1}{3} = \frac{3}{3} = 1$
6.	$\frac{1}{6}$	6	Because $\frac{1}{6} \cdot 6 = \frac{1}{6} \cdot \frac{6}{1} = \frac{6}{6} = 1$
7.	$\frac{4}{5}$	$\frac{5}{4}$	Because $\frac{4}{5} \cdot \frac{5}{4} = \frac{20}{20} = 1$
8.	a	$\dfrac{1}{a}$	Because $a \cdot \dfrac{1}{a} = \dfrac{a}{1} \cdot \dfrac{1}{a} = \dfrac{a}{a} = 1 \quad (a \neq 0)$

◢

Although we will not develop multiplication with negative numbers until later in this chapter, you should know that the reciprocal of a negative number is also a negative number. For example, the reciprocal of -5 is $-\frac{1}{5}$.

The Absolute Value of a Real Number

DEFINITION

The **absolute value** of a number (also called its *magnitude*) is the distance the number is from 0 on the number line. If x represents a real number, then the absolute value of x is written $|x|$.

This definition of absolute value is geometric in form since it defines absolute value in terms of the number line. Here is an alternative definition of absolute value that is algebraic in form since it involves only symbols.

If x represents a real number, then the **absolute value** of x is written $|x|$, and is given by

$$|x| = \begin{cases} x & \text{if } x \geq 0 \\ -x & \text{if } x < 0 \end{cases}$$

If the original number is positive or 0, then its absolute value is the number itself. If the number is negative, its absolute value is its opposite (which must be positive).

Note: It is important to recognize that if x is a real number, $-x$ is not necessarily negative. For example, if x is 5, then $-x$ is -5. On the other hand, if x were -5, then $-x$ would be $-(-5)$, which is 5.

EXAMPLES

Write each expression without absolute value symbols.

Note that there is a negative sign *outside* the absolute value bars in Examples 12–14.

9. $|5| = 5$ **10.** $|-2| = 2$

11. $\left|-\frac{1}{2}\right| = \frac{1}{2}$ **12.** $-|-3| = -3$

13. $-|5| = -5$ **14.** $-|-\sqrt{2}| = -\sqrt{2}$

Properties of Real Numbers

We know that adding 3 and 7 gives the same answer as adding 7 and 3. The order of two numbers in an addition problem can be changed without changing the result. This fact about numbers and addition is called the **commutative property of addition.**

For all the properties listed in this section, a, b, and c represent real numbers.

In Symbols: $a + b = b + a$
In Words: The *order* of the numbers in a sum does not affect the result.

In Symbols: $a \cdot b = b \cdot a$
In Words: The *order* of the numbers in a product does not affect the result.

EXAMPLES

15. The statement $3 + 7 = 7 + 3$ is an example of the commutative property of addition.

16. The statement $3 \cdot x = x \cdot 3$ is an example of the commutative property of multiplication.

The other two basic operations (subtraction and division) are not commutative. If we change the order in which we are subtracting or dividing two numbers, we will usually change the result.

Another property of numbers you have used many times has to do with grouping. When adding $3 + 5 + 7$, we can add the 3 and 5 first and then the 7, or we can add the 5 and 7 first and then the 3. Mathematically, it looks like this: $(3 + 5) + 7 = 3 + (5 + 7)$. Operations that behave in this manner are called **associative** operations.

ASSOCIATIVE PROPERTY OF ADDITION

In Symbols: $a + (b + c) = (a + b) + c$
In Words: The *grouping* of the numbers in a sum does not affect the result.

ASSOCIATIVE PROPERTY OF MULTIPLICATION

In Symbols: $a(bc) = (ab)c$
In Words: The *grouping* of the numbers in a product does not affect the result.

The following examples illustrate how the associative properties can be used to simplify expressions that involve both numbers and variables.

EXAMPLES   Simplify by using the associative property.

17. $2 + (3 + y) = (2 + 3) + y$ Associative property
$ = 5 + y$ Addition

18. $5(4x) = (5 \cdot 4)x$ Associative property
$ = 20x$ Multiplication

19. $\dfrac{1}{4}(4a) = \left(\dfrac{1}{4} \cdot 4\right)a$ Associative property

$\phantom{\dfrac{1}{4}(4a)} = 1a$ Multiplication
$\phantom{\dfrac{1}{4}(4a)} = a$

Our next property involves both addition and multiplication. It is called the **distributive property** and is stated as follows.

DISTRIBUTIVE PROPERTY

In Symbols: $a(b + c) = ab + ac$
In Words: Multiplication *distributes* over addition.

You will see as we progress through the book that the distributive property is used very frequently in algebra. To see that the distributive property works, compare the following:

$$3(4 + 5) \qquad 3(4) + 3(5)$$
$$= 3(9) \qquad = 12 + 15$$
$$= 27 \qquad = 27$$

In both cases the result is 27. Since the results are the same, the original two expressions must be equal; that is, $3(4 + 5) = 3(4) + 3(5)$.

EXAMPLES Apply the distributive property to each expression and then simplify the result.

20. $5(4x + 3) = 5(4x) + 5(3)$ Distributive property
           $= 20x + 15$ Multiplication

21. $6(3x + 2y) = 6(3x) + 6(2y)$ Distributive property
            $= 18x + 12y$ Multiplication

22. $\dfrac{1}{2}(3x + 6) = \dfrac{1}{2}(3x) + \dfrac{1}{2}(6)$ Distributive property

            $= \dfrac{3}{2}x + 3$ Multiplication

23. $2(3y + 4) + 2 = 2(3y) + 2(4) + 2$ Distributive property
            $= 6y + 8 + 2$ Multiplication
            $= 6y + 10$ Addition

Note: Although the properties we are listing are stated for only two or three real numbers, they hold for as many numbers as needed. For example, the distributive property holds for expressions like $3(x + y + z + 2)$. That is,

$$3(x + y + z + 2) = 3x + 3y + 3z + 6$$

Combining Similar Terms The distributive property can also be used to combine similar terms. (For now, we define a *term* as the product of a number with one or more variables.) **Similar terms** are terms with the same variable part. The terms $3x$ and $5x$ are similar, as are $2y$, $7y$, and $-3y$, because the variable parts are the same.

EXAMPLES Use the distributive property to combine similar terms.

24. $3x + 5x = (3 + 5)x$ Distributive property
        $= 8x$ Addition

25. $3y + y = (3 + 1)y$ Distributive property
        $= 4y$ Addition

Review of Addition with Fractions

To add fractions, each fraction must have the same denominator.

20

> **DEFINITION**
>
> The **least common denominator (LCD)** for a set of denominators is the smallest number divisible by *all* the denominators.

The first step in adding fractions is to find a common denominator for all the denominators. We then rewrite each fraction (if necessary) as an equivalent fraction with the common denominator. Finally, we add the numerators and reduce to lowest terms if necessary.

EXAMPLE 26 Add: $\dfrac{5}{12} + \dfrac{7}{18}$.

SOLUTION The least common denominator for the denominators 12 and 18 must be the smallest number divisible by both 12 and 18. We can factor 12 and 18 completely and then build the LCD from these factors.

12 divides the LCD

$$\left.\begin{array}{l}12 = 2 \cdot 2 \cdot 3\\18 = 2 \cdot 3 \cdot 3\end{array}\right\} \text{LCD} = 2 \cdot 2 \cdot 3 \cdot 3 = 36$$

18 divides the LCD

Next we rewrite our original fractions as equivalent fractions with denominators of 36. To do so, we multiply each original fraction by an appropriate form of the number 1:

$$\frac{5}{12} + \frac{7}{18} = \frac{5}{12} \cdot \frac{3}{3} + \frac{7}{18} \cdot \frac{2}{2}$$
$$= \frac{15}{36} + \frac{14}{36}$$

Finally, we add numerators and place the result over the common denominator, 36:

$$\frac{15}{36} + \frac{14}{36} = \frac{15 + 14}{36} = \frac{29}{36}$$

Simplifying Expressions

We can use the commutative, associative, and distributive properties together to simplify expressions.

EXAMPLE 27 Simplify: $7x + 4 + 6x + 3$.

SOLUTION We begin by applying the commutative and associative properties to group similar terms:

$$7x + 4 + 6x + 3$$

$$
\begin{aligned}
&= (7x + 6x) + (4 + 3) && \text{Commutative and associative} \\
& && \text{properties} \\
&= (7 + 6)x + (4 + 3) && \text{Distributive property} \\
&= 13x + 7 && \text{Addition}
\end{aligned}
$$

EXAMPLE 28   Simplify: $4 + 3(2y + 5) + 8y$.

SOLUTION   Since our rule for order of operations indicates that we are to multiply before adding, we must distribute the 3 across $2y + 5$ first:

$$4 + 3(2y + 5) + 8y$$

$$
\begin{aligned}
&= 4 + 6y + 15 + 8y && \text{Distributive property} \\
&= (6y + 8y) + (4 + 15) && \text{Commutative and associative} \\
& && \text{properties} \\
&= (6 + 8)y + (4 + 15) && \text{Distributive property} \\
&= 14y + 19 && \text{Addition}
\end{aligned}
$$

The remaining properties of real numbers have to do with the numbers 0 and 1.

ADDITIVE IDENTITY PROPERTY

There exists a unique number 0 such that

$$a + 0 = a \quad \text{and} \quad 0 + a = a$$

MULTIPLICATIVE IDENTITY PROPERTY

There exists a unique number 1 such that

$$a(1) = a \quad \text{and} \quad 1(a) = a$$

The numbers 0 and 1 are called the **additive identity** and **multiplicative identity,** respectively. Combining 0 with a number under addition does not change the identity of the number. Likewise, combining 1 with a number under multiplication does not alter the identity of the number. So we see that 0 is to addition what 1 is to multiplication.

ADDITIVE INVERSE PROPERTY

For each real number a, there exists a unique real number $-a$ such that:

In Symbols: $a + (-a) = 0$
In Words: Opposites add to 0.

> ### MULTIPLICATIVE INVERSE PROPERTY
>
> For every real number a, except 0, there exists a unique real number $\dfrac{1}{a}$ such that:
>
> **In Symbols:** $\quad a\left(\dfrac{1}{a}\right) = 1$
>
> **In Words:** $\quad$ Reciprocals multiply to 1.

The following examples illustrate how we use these properties.

EXAMPLES

29. $7(1) = 7$ $\qquad\qquad$ Multiplicative identity property

30. $4 + (-4) = 0$ $\qquad$ Additive inverse property

31. $6\left(\dfrac{1}{6}\right) = 1$ $\qquad$ Multiplicative inverse property

32. $(5 + 0) + 2 = 5 + 2$ $\quad$ Additive identity property

Problem Set
1.2

NOTE: The problems in the problem sets are arranged in roughly the same order as the examples in the section. So if you are stuck halfway through the problem set, look to the middle of the section for an example similar to the problem you are stuck on.

Use the associative property to rewrite each of the following expressions and then simplify the result.

1. $4 + (2 + x)$
2. $6 + (5 + 3x)$
3. $(a + 3) + 5$
4. $(4a + 5) + 7$
5. $5(3y)$
6. $7(4y)$
7. $\frac{1}{3}(3x)$
8. $\frac{1}{5}(5x)$
9. $4(\frac{1}{4}\,a)$
10. $7(\frac{1}{7}\,a)$
11. $\frac{2}{3}(\frac{3}{2}\,x)$
12. $\frac{4}{3}(\frac{3}{4}\,x)$

Apply the distributive property to each expression. Simplify when possible.

13. $3(x + 6)$
14. $5(x + 9)$
15. $2(6x + 4)$
16. $3(7x + 8)$
17. $5(3a + 2b)$
18. $7(2a + 3b)$
19. $\frac{1}{3}(4x + 6)$
20. $\frac{1}{2}(3x + 8)$
21. $\frac{1}{5}(10 + 5y)$
22. $\frac{1}{6}(12 + 6y)$
23. $3(5x + 2) + 4$
24. $4(3x + 2) + 5$
25. $5(1 + 3t) + 4$
26. $2(1 + 5t) + 6$

Add the following fractions.

27. $\frac{2}{5} + \frac{1}{15}$
28. $\frac{5}{8} + \frac{1}{4}$
29. $\frac{17}{30} + \frac{11}{42}$
30. $\frac{19}{42} + \frac{13}{70}$
31. $\frac{9}{48} + \frac{3}{54}$
32. $\frac{6}{28} + \frac{5}{42}$
33. $\frac{25}{84} + \frac{41}{90}$
34. $\frac{23}{70} + \frac{29}{84}$

Use the commutative, associative, and distributive properties to simplify the following.

35. $5a + 7 + 8a + a$
36. $6a + 4 + a + 4a$
37. $3y + y + 5 + 2y + 1$
38. $4y + 2y + 3 + y + 7$
39. $2(5x + 1) + 2x$
40. $3(4x + 1) + 9x$

41. $7 + 2(4y + 2)$ **42.** $6 + 3(5y + 2)$
43. $3 + 4(5a + 3) + 4a$
44. $8 + 2(4a + 2) + 5a$
45. $5x + 2(3x + 8) + 4$
46. $7x + 3(4x + 1) + 7$
47. $2t + 3(1 + 6t) + 2$
48. $3t + 2(4 + 2t) + 6$

NOTE: Once again we want to remind you that checking answers is very important. If you are working problems without checking your answers, or if you are just going to the back of the book and writing answers without working problems, you are not getting what you should from your homework.

Find the area of each rectangle below in two ways: first, by multiplying length and width, and then by adding the areas of the two smaller rectangles together.

49.
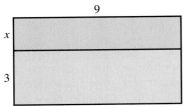

50.
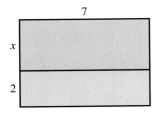

51.
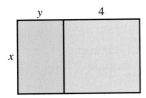

52.
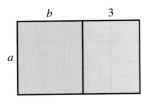

Identify the property (or properties) of real numbers that justifies each of the following.

53. $3 + 2 = 2 + 3$ **54.** $3(ab) = (3a)b$
55. $5x = x5$ **56.** $2 + 0 = 2$
57. $4 + (-4) = 0$ **58.** $1(6) = 6$
59. $x + (y + 2) = (y + 2) + x$
60. $(a + 3) + 4 = a + (3 + 4)$
61. $4(5 \cdot 7) = 5(4 \cdot 7)$
62. $6(xy) = (xy)6$
63. $4 + (x + y) = (4 + y) + x$
64. $(r + 7) + s = (r + s) + 7$
65. $3(4x + 2) = 12x + 6$
66. $5(\frac{1}{5}) = 1$

Write each of the following without absolute value symbols.

67. $|-2|$ **68.** $|-7|$
69. $|-\frac{3}{4}|$ **70.** $|\frac{5}{6}|$
71. $|\pi|$ **72.** $|-\sqrt{2}|$
73. $-|4|$ **74.** $-|5|$
75. $-|-2|$ **76.** $-|-10|$

Multiply the following.

77. $\frac{3}{5} \cdot \frac{7}{8}$ **78.** $\frac{6}{7} \cdot \frac{9}{5}$
79. $\frac{1}{3} \cdot 6$ **80.** $\frac{1}{4} \cdot 8$
81. $(\frac{2}{3})^3$ **82.** $(\frac{4}{5})^2$
83. $(\frac{1}{10})^4$ **84.** $(\frac{1}{2})^5$
85. $\frac{3}{5} \cdot \frac{4}{7} \cdot \frac{6}{11}$ **86.** $\frac{4}{5} \cdot \frac{6}{7} \cdot \frac{3}{11}$
87. $9(\frac{1}{3})^2$ **88.** $25(\frac{1}{5})^2$

Applying the Concepts

89. Name two numbers that are their own reciprocals.

90. Give the number that has no reciprocal.

91. Name the number that is its own opposite.

92. The reciprocal of a negative number is negative—true or false?

93. Show that the statement $5x - 5 = x$ is not correct by replacing x with 4 and simplifying both sides.

94. Show that the statement $8x - x = 8$ is not correct by replacing x with 5 and simplifying both sides.

95. Simplify the expressions $15 - (8 - 2)$ and $(15 - 8) - 2$ to show that subtraction is not an associative operation.

96. Simplify the expressions $(48 \div 6) \div 2$ and $48 \div (6 \div 2)$ to show that division is not an associative operation.

Arithmetic with Real Numbers

The temperature at the airport is 70°F. A plane takes off and reaches its cruising altitude of 28,000 feet, where the temperature is -40°F. Find the difference in the temperatures at takeoff and at cruising altitude.

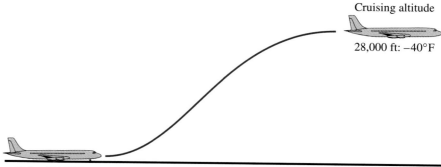

Cruising altitude

28,000 ft: $-40°$F

Takeoff: 70°F

We know intuitively that the difference in temperature is 110°F. If we write this problem using symbols, we have

$$70 - (-40) = 110$$

In this section we review the rules for arithmetic with real numbers, which will include problems such as the one above.

Adding Real Numbers

The purpose of this section is to review the rules for arithmetic with real numbers and the justification for those rules. We can justify the rules for addition of real numbers geometrically by use of the real number line. Consider the sum of -5 and 3:

$$-5 + 3$$

We can interpret this expression as meaning "start at the origin and move 5 units in the negative direction and then 3 units in the positive direction." With the aid of a number line we can visualize the process:

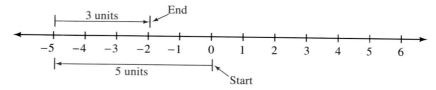

Since the process ends at -2, we say the sum of -5 and 3 is -2:

$$-5 + 3 = -2$$

We can eliminate actually drawing a number line by simply visualizing it mentally. The following example gives the results of all possible sums of positive and negative 5 and 7.

EXAMPLE 1 Add all combinations of positive and negative 5 and 7.

SOLUTION

$5 + 7 = 12$	$5 + (-7) = -2$
$-5 + 7 = 2$	$-5 + (-7) = -12$

Looking closely at the relationships in Example 1 (and trying other similar examples if necessary), we can arrive at the following rule for adding two real numbers.

> **Strategy for Adding Two Real Numbers**
>
> **With the *Same* Sign:**
>
> **Step 1** Add their absolute values.
> **Step 2** Attach their common sign. If both numbers are positive, their sum is positive; if both numbers are negative, their sum is negative.
>
> **With *Opposite* Signs:**
>
> **Step 1** Subtract the smaller absolute value from the larger.
> **Step 2** Attach the sign of the number whose absolute value is larger.

Subtracting Real Numbers

In order to have as few rules as possible, we will not attempt to list new rules for the *difference* of two real numbers. We will instead define it in terms of addition and apply the rule for addition.

> **DEFINITION**
>
> If a and b are any two real numbers, then the **difference** of a and b, written $a - b$, is given by
>
> $$\underbrace{a - b}_{\text{To subtract } b,} = \underbrace{a + (-b)}_{\text{add the opposite of } b.}$$

We define the process of subtracting b from a as being equivalent to adding the opposite of b to a. In short, we say, "subtraction is addition of the opposite."

EXAMPLES Subtract.

2. $5 - 3 = 5 + (-3)$ Subtracting 3 is equivalent to adding -3.
    $= 2$

3. $-7 - 6 = -7 + (-6)$ Subtracting 6 is equivalent to adding -6.

        $= -13$

4. $9 - (-2) = 9 + 2$ Subtracting -2 is equivalent to adding 2.

        $= 11$

5. $-6 - (-5) = -6 + 5$ Subtracting -5 is equivalent to adding 5.

        $= -1$

EXAMPLE 6   Subtract -3 from -9.

SOLUTION   Since subtraction is not commutative, we must be sure to write the numbers in the correct order. Because we are subtracting -3, the problem looks like this when translated into symbols:

$$-9 - (-3) = -9 + 3 \quad \text{Change to addition of the opposite.}$$
$$= -6 \quad \text{Add.}$$

EXAMPLE 7   Add -4 to the difference of -2 and 5.

SOLUTION   The difference of -2 and 5 is written $-2 - 5$. Adding -4 to that difference gives us

$$(-2 - 5) + (-4) = -7 + (-4) \quad \text{Simplify inside parentheses.}$$
$$= -11 \quad \text{Add.}$$

Multiplying Real Numbers

Multiplication with whole numbers is simply a shorthand way of writing repeated addition. For example, $3(-2)$ can be evaluated as follows:

$$3(-2) = -2 + (-2) + (-2) = -6$$

We can evaluate the product $-3(2)$ in a similar manner if we first apply the commutative property of multiplication:

$$-3(2) = 2(-3) = -3 + (-3) = -6$$

From these results it seems reasonable to say that the product of a positive number and a negative number is a negative number.

The last case we must consider is the product of two negative numbers, such as $-3(-2)$. In order to evaluate this product we will first look at the expression $-3[2 + (-2)]$ in two different ways. First, since $2 + (-2) = 0$, we know the expression $-3[2 + (-2)]$ is equal to 0. On the other hand, we can apply the distributive property to get

$$-3[2 + (-2)] = -3(2) + (-3)(-2) = -6 + ?$$

We know the expression is equal to 0, so it must be true that our ? is 6, since 6 is the only number we can add to -6 to get 0. Therefore, we have

$$-3(-2) = 6$$

Here is a summary of what we know so far:

Original Numbers Have:		The Answer Is:
The same sign	$3(2) = 6$	Positive
Different signs	$3(-2) = -6$	Negative
Different signs	$-3(2) = -6$	Negative
The same sign	$-3(-2) = 6$	Positive

Strategy for Multiplying Two Real Numbers

To multiply two real numbers, simply:

Step 1 Multiply their absolute values.
Step 2 If the two numbers have the *same* sign, the product is positive. If the two numbers have *opposite* signs, the product is negative.

EXAMPLE 8 Multiply all combinations of positive and negative 7 and 3.

SOLUTION

$$7(3) = 21 \qquad -7(3) = -21$$
$$7(-3) = -21 \qquad -7(-3) = 21$$

Dividing Real Numbers

DEFINITION

If a and b are any two real numbers, where $b \neq 0$, then the **quotient** of a and b, written $\dfrac{a}{b}$, is given by

$$\frac{a}{b} = a \cdot \left(\frac{1}{b}\right)$$

Dividing a by b is equivalent to multiplying a by the reciprocal of b. In short, we say, "division is multiplication by the reciprocal."

Since division is defined in terms of multiplication, the same rules hold for assigning the correct sign to a quotient as held for assigning the correct sign to a product. That is, *the quotient of two numbers with like signs is positive, while the quotient of two numbers with unlike signs is negative.*

EXAMPLES Divide.

9. $\dfrac{6}{3} = 6 \cdot \left(\dfrac{1}{3}\right) = 2$

Note: These examples indicate that if a and b are positive real numbers, then

10. $\dfrac{6}{-3} = 6 \cdot \left(-\dfrac{1}{3}\right) = -2$ $\qquad \dfrac{-a}{b} = \dfrac{a}{-b} = -\dfrac{a}{b}$

11. $\dfrac{-6}{3} = -6 \cdot \left(\dfrac{1}{3}\right) = -2$ and

12. $\dfrac{-6}{-3} = -6 \cdot \left(-\dfrac{1}{3}\right) = 2$ $\qquad \dfrac{-a}{-b} = \dfrac{a}{b}$

The second step in the preceding examples is written only to show that each quotient can be written as a product. It is not actually necessary to show this step when working problems.

In the examples that follow, we find a combination of operations. In each case we use the rule for order of operations.

EXAMPLES   Simplify each expression as much as possible.

13. $(-2 - 3)(5 - 9) = (-5)(-4)$ Simplify inside parentheses.
$\qquad\qquad\qquad\quad = 20$ Multiply.

14. $2 - 5(7 - 4) - 6 = 2 - 5(3) - 6$ Simplify inside parentheses.
$\qquad\qquad\qquad\quad = 2 - 15 - 6$ Then, multiply.
$\qquad\qquad\qquad\quad = -19$ Finally, subtract, left to right.

15. $2(4 - 7)^3 + 3(-2 - 3)^2$
$\qquad = 2(-3)^3 + 3(-5)^2$ Simplify inside parentheses.
$\qquad = 2(-27) + 3(25)$ Evaluate numbers with exponents.
$\qquad = -54 + 75$ Multiply.
$\qquad = 21$ Add.

16. $\dfrac{-5(-4) + 2(-3)}{2(-1) - 5} = \dfrac{20 - 6}{-2 - 5}$ **17.** $\dfrac{2^3 + 3^3}{2^2 - 3^2} = \dfrac{8 + 27}{4 - 9}$

$\qquad\qquad\qquad\quad = \dfrac{14}{-7}$ $\qquad\qquad\qquad\quad = \dfrac{35}{-5}$

$\qquad\qquad\qquad\quad = -2$ $\qquad\qquad\qquad\quad = -7$

Remember, since subtraction is defined in terms of addition, we can restate the distributive property in terms of subtraction. That is, if a, b, and c are real numbers, then $a(b - c) = ab - ac$.

EXAMPLE 18   Simplify: $3(2y - 1) + y$.

SOLUTION   We begin by multiplying the 3 and $2y - 1$. Then, we combine similar terms:

$$3(2y - 1) + y = 6y - 3 + y \qquad \text{Distributive property}$$
$$= 7y - 3 \qquad \text{Combine similar terms.}$$

EXAMPLE 19   Simplify: $8 - 3(4x - 2) + 5x$.

SOLUTION   First we distribute the -3 across the $4x - 2$.

$$8 - 3(4x - 2) + 5x = 8 - 12x + 6 + 5x$$
$$= -7x + 14$$

EXAMPLE 20   Simplify: $5(2a + 3) - (6a - 4)$.

SOLUTION   We begin by applying the distributive property to remove the parentheses. The expression $-(6a - 4)$ can be thought of as $-1(6a - 4)$. Thinking of it in this way allows us to apply the distributive property:

$$-1(6a - 4) = -1(6a) - (-1)(4) = -6a + 4$$

Here is the complete problem:

$$5(2a + 3) - (6a - 4) = 10a + 15 - 6a + 4 \quad \text{Distributive property}$$
$$= 4a + 19 \quad\quad\quad\quad \text{Combine similar terms.}$$

Dividing Fractions

We end this section by reviewing division with fractions and division with the number 0.

EXAMPLES   Divide and reduce to lowest terms.

21. $\dfrac{3}{4} \div \dfrac{6}{11} = \dfrac{3}{4} \cdot \dfrac{11}{6}$ Definition of division

$\quad\quad\quad = \dfrac{33}{24}$ Multiply numerators, multiply denominators.

$\quad\quad\quad = \dfrac{11}{8}$ Divide numerator and denominator by 3.

22. $10 \div \dfrac{5}{6} = \dfrac{10}{1} \cdot \dfrac{6}{5}$ Definition of division

$\quad\quad\quad = \dfrac{60}{5}$ Multiply numerators, multiply denominators.

$\quad\quad\quad = 12$ Divide.

23. $-\dfrac{3}{8} \div 6 = -\dfrac{3}{8} \cdot \dfrac{1}{6}$ Definition of division

$\quad\quad\quad = -\dfrac{3}{48}$ Multiply numerators, multiply denominators.

$\quad\quad\quad = -\dfrac{1}{16}$ Divide numerator and denominator by 3.

Division with the Number 0

For every division problem there is an associated multiplication problem involving the same numbers. For example, the following two problems say the same thing about the numbers 2, 3, and 6:

DIVISION   MULTIPLICATION

$$\frac{6}{3} = 2 \qquad 6 = 2(3)$$

We can use this relationship between division and multiplication to clarify division involving the number 0.

First of all, dividing 0 by a number other than 0 is allowed and always results in 0. To see this, consider dividing 0 by 5. We know the answer is 0 because of the relationship between multiplication and division. This is how we write it:

$$\frac{0}{5} = 0 \qquad \text{because} \qquad 0 = 0(5)$$

On the other hand, dividing a nonzero number by 0 is not allowed in the real numbers. Suppose we were attempting to divide 5 by 0. We don't know whether there is an answer to this problem, but if there is, let's say the answer is a number that we can represent with the letter n. If 5 divided by 0 is a number n, then

$$\frac{5}{0} = n \qquad \text{and} \qquad 5 = n(0)$$

But this is impossible, because no matter what number n is, when we multiply it by 0 the answer must be 0. It can never be 5. In algebra, we say expressions like $\frac{5}{0}$ are undefined, because there is no answer to them. That is, division by 0 is not allowed in the real numbers.

Problem Set

1.3

Are you reading the section that will be covered in class before going to class? If you are, you know that the lecture you attend afterward is more understandable, and you have an easier time with your homework than if you don't read ahead.

Find each of the following sums.

1. $6 + (-2)$

2. $11 + (-5)$

3. $-6 + 2$

4. $-11 + 5$

5. $-\frac{1}{2} + (-\frac{1}{6}) + (-\frac{1}{18})$

6. $-\frac{1}{2} + (-\frac{1}{4}) + (-\frac{1}{10})$

Find each of the following differences.

7. $-7 - 3$

8. $-6 - 9$

9. $-7 - (-3)$

10. $-6 - (-9)$

11. $\frac{3}{4} - (-\frac{5}{6})$

12. $\frac{2}{3} - (-\frac{7}{5})$

13. $\frac{11}{42} - \frac{17}{30}$

14. $\frac{13}{70} - \frac{19}{42}$

Simplify as much as possible.

15. $6 - (-2) + 11$

16. $8 - (-3) + 12$

17. $-\frac{4}{3} - (-\frac{1}{2}) - \frac{3}{2}$

18. $-\frac{1}{6} - (-\frac{1}{3}) - \frac{1}{2}$

19. $-5 - (2 - 6) - 3$

20. $-4 - (5 - 9) - 2$

21. $-(2 - 5) - (7 - 3)$

22. $-(8 - 10) - (6 - 1)$

23. Subtract 5 from -3.

24. Subtract -3 from 5.

25. Find the difference of -4 and 8.

26. Find the difference of 8 and -4.

27. Subtract $4x$ from $-3x$.

28. Subtract $-5x$ from $7x$.

29. What number do you subtract from 5 to get -8?

30. What number do you subtract from -3 to get 9?

31. Add -7 to the difference of 2 and 9.

32. Add -3 to the difference of 9 and 2.

33. Subtract $3a$ from the sum of $8a$ and a.

34. Subtract $-3a$ from the sum of $3a$ and $5a$.

Find the following products.

35. $3(-5)$

36. $-3(5)$

37. $-3(-5)$

38. $4(-6)$

39. $-8(3)$

40. $-7(-6)$

41. $-2(-1)(-6)$

42. $-3(-2)(5)$

43. $2(-3)(4)$

44. $-2(3)(-4)$

45. $-2(5x)$

46. $-5(4x)$

47. $-\frac{1}{3}(-3x)$

48. $-\frac{1}{6}(-6x)$

49. $-\frac{2}{3}(-\frac{3}{2}y)$

50. $-\frac{2}{5}(-\frac{5}{2}y)$

51. $-2(4x - 3)$

52. $-6(2x - 1)$

53. $-4(-3t + 7)$

54. $-2(-5t + 6)$

55. $-\frac{1}{2}(6a - 8)$

56. $-\frac{1}{3}(6a - 9)$

57. $-\frac{1}{2}(-3x - 4)$

58. $-\frac{1}{2}(-5x - 8)$

Simplify each expression as much as possible.

59. $3(-4) - 2$

60. $-3(-4) - 2$

61. $4(-3) - 6(-5)$

62. $-6(-3) - 5(-7)$

63. $2 - 5(-4) - 6$

64. $3 - 8(-1) - 7$

65. $2 - 5(-4 - 6)$

66. $3 - 8(-1 - 7)$

67. $(2 - 5)(-4 - 6)$

68. $(3 - 8)(-1 - 7)$

69. $4 - 3(7 - 1) - 5$

70. $8 - 5(6 - 3) - 7$

71. $2(-3)^2 - 4(-2)^3$

72. $5(-2)^2 - 2(-3)^3$

73. $(2 - 8)^2 - (3 - 7)^2$

74. $(5 - 8)^2 - (4 - 8)^2$

75. $7(3 - 5)^3 - 2(4 - 7)^3$

76. $3(-7 + 9)^3 - 5(-2 + 4)^3$

77. $-3(2 - 9) - 4(6 - 1)$

78. $-5(5 - 6) - 7(2 - 8)$

79. $-5(-8 - 2) - 3(-2 - 8)$

80. $-3(-5 - 15) - 4(-12 - 8)$

81. $2 - 4[3 - 5(-1)]$

82. $6 - 5[2 - 4(-8)]$

83. $(8 - 7)[4 - 7(-2)]$

84. $(6 - 9)[15 - 3(-4)]$

85. $-3 + 4[6 - 8(-3 - 5)]$

86. $-2 + 7[2 - 6(-3 - 4)]$

87. $5 - 6[-3(2 - 9) - 4(8 - 6)]$

88. $9 - 4[-2(4 - 8) - 5(3 - 1)]$

 There are videotapes that accompany this text. On the tapes, I work some of the problems from this problem set. You may want to look into these tapes as another resource to get you to the place you want to be in this course.

Simplify each expression.

89. $3(5x + 4) - x$

90. $4(7x + 3) - x$

91. $6 - 7(3 - m)$

92. $3 - 5(5 - m)$

93. $7 - 2(3x - 1) + 4x$

94. $8 - 5(2x - 3) + 4x$

95. $5(3y + 1) - (8y - 5)$

96. $4(6y + 3) - (6y - 6)$

97. $4(2 - 6x) - (3 - 4x)$

98. $7(1 - 2x) - (4 - 10x)$

99. $10 - 4(2x + 1) - (3x - 4)$

100. $7 - 2(3x + 5) - (2x - 3)$

Use the definition of division to write each division problem as a multiplication problem; then simplify.

101. $\dfrac{8}{-4}$

102. $\dfrac{-8}{4}$

103. $\dfrac{-8}{-4}$

104. $\dfrac{-12}{-4}$

105. $\dfrac{4}{0}$

106. $\dfrac{-7}{0}$

107. $\dfrac{0}{-3}$

108. $\dfrac{0}{5}$

109. $-\frac{3}{4} \div \frac{9}{8}$

110. $-\frac{2}{3} \div \frac{4}{9}$

111. $-8 \div \left(-\frac{1}{4}\right)$

112. $-12 \div \left(-\frac{2}{3}\right)$

113. $-40 \div \left(-\frac{5}{8}\right)$

114. $-30 \div \left(-\frac{5}{6}\right)$

115. $\frac{4}{9} \div (-8)$

116. $\frac{3}{7} \div (-6)$

Simplify as much as possible.

117. $\dfrac{6(-2) - 8}{-15 - (-10)}$

118. $\dfrac{8(-3) - 6}{-7 - (-2)}$

119. $\dfrac{3(-1) - 4(-2)}{8 - 5}$

120. $\dfrac{6(-4) - 5(-2)}{7 - 6}$

121. $8 - (-6)\left[\dfrac{2(-3) - 5(4)}{-8(6) - 4}\right]$

122. $-9 - 5\left[\dfrac{11(-1) - 9}{4(-3) + 2(5)}\right]$

123. $6 - (-3)\left[\dfrac{2 - 4(3 - 8)}{1 - 5(1 - 3)}\right]$

124. $8 - (-7)\left[\dfrac{6 - 1(6 - 10)}{4 - 3(5 - 7)}\right]$

125. Subtract -5 from the product of 12 and $-\frac{2}{3}$.

126. Subtract -3 from the product of -12 and $\frac{3}{4}$.

127. Add -5 to the quotient of -3 and $\frac{1}{2}$.

128. Add -7 to the quotient of 6 and $-\frac{1}{2}$.

129. Add $8x$ to the product of -2 and $3x$.

130. Add $7x$ to the product of -5 and $-2x$.

SECTION
1.4

Exponents and Scientific Notation

Below are a square and a cube, each with a side of length 1.5 centimeters. To find the area of the square, we raise 1.5 to the second power: 1.5^2. To find the volume of the cube, we raise 1.5 to the third power: 1.5^3.

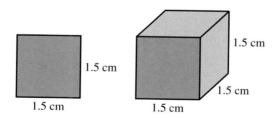

1.5 cm

1.5 cm

1.5 cm

1.5 cm

1.5 cm

1.5 cm

Because the area of the square is 1.5^2, we say second powers are *squares;* that is, x^2 is read "*x* squared." Likewise, since the volume of the cube is 1.5^3, we say third powers are *cubes;* that is, x^3 is read "*x* cubed." Exponents and the vocabulary associated with them are one of the topics we will study in this section.

Properties of Exponents

In this section, we will be concerned with the simplification of expressions that involve exponents. We begin by making some generalizations about exponents, based on specific examples.

EXAMPLE 1 Write the product $x^3 \cdot x^4$ with a single exponent.

SOLUTION
$$
\begin{aligned}
x^3 \cdot x^4 &= (x \cdot x \cdot x)(x \cdot x \cdot x \cdot x) \\
&= (x \cdot x \cdot x \cdot x \cdot x \cdot x \cdot x) \\
&= x^7 \qquad\qquad \textit{Notice: } 3 + 4 = 7
\end{aligned}
$$

We can generalize this result into the first property of exponents.

> **PROPERTY 1 FOR EXPONENTS**
>
> If a is a real number, and r and s are integers, then
> $$a^r \cdot a^s = a^{r+s}$$

Note: We are stating the properties of exponents for integer exponents instead of just positive integer exponents. As you will see, the definition for negative integer exponents is stated in such a way that we can change any expression with a negative exponent to an equivalent expression with a positive exponent. This allows us to state the properties for all integers, not just for positive ones.

EXAMPLE 2 Write $(5^3)^2$ with a single exponent.

SOLUTION
$$
\begin{aligned}
(5^3)^2 &= 5^3 \cdot 5^3 \\
&= 5^6 \qquad \textit{Notice: } 3 \cdot 2 = 6
\end{aligned}
$$

Generalizing this result, we have a second property of exponents.

> **PROPERTY 2 FOR EXPONENTS**
>
> If a is a real number, and r and s are integers, then
> $$(a^r)^s = a^{r \cdot s}$$

A third property of exponents arises when we have the product of two or more numbers raised to an integer power.

EXAMPLE 3 Expand $(3x)^4$ and then multiply.

SOLUTION
$$
\begin{aligned}
(3x)^4 &= (3x)(3x)(3x)(3x) \\
&= (3 \cdot 3 \cdot 3 \cdot 3)(x \cdot x \cdot x \cdot x) \\
&= 3^4 \cdot x^4 \\
&= 81x^4
\end{aligned}
$$

Notice: The exponent 4 distributes over the product $3x$.

Generalizing Example 3, we have property 3 for exponents.

> ### PROPERTY 3 FOR EXPONENTS
>
> If a and b are any two real numbers, and r is an integer, then
> $$(ab)^r = a^r \cdot b^r$$

The next property of exponents deals with negative integer exponents.

> ### PROPERTY 4 FOR EXPONENTS
>
> If a is any nonzero real number, and r is a positive integer, then
> $$a^{-r} = \frac{1}{a^r}$$

Note: This property is actually a definition. That is, we are defining negative integer exponents as indicating reciprocals. Doing so gives us a way to write an expression with a negative exponent as an equivalent expression with a positive exponent.

EXAMPLES Write with positive exponents; then simplify.

4. $5^{-2} = \dfrac{1}{5^2} = \dfrac{1}{25}$

5. $(-2)^{-3} = \dfrac{1}{(-2)^3} = \dfrac{1}{-8} = -\dfrac{1}{8}$

6. $\left(\dfrac{3}{4}\right)^{-2} = \dfrac{1}{(\frac{3}{4})^2} = \dfrac{1}{\frac{9}{16}} = \dfrac{16}{9}$

If we generalize the result in Example 6, we have the following extension of property 4:
$$\left(\frac{a}{b}\right)^{-r} = \left(\frac{b}{a}\right)^r$$

which indicates that raising a fraction to a negative power is equivalent to raising the reciprocal of the fraction to the positive power.

Property 3 indicated that exponents distribute over products. Since division is defined in terms of multiplication, we can expect that exponents will distribute over quotients as well. Property 5 is the formal statement of this fact.

> ### PROPERTY 5 FOR EXPONENTS
>
> If a and b are any two real numbers with $b \neq 0$, and r is an integer, then
> $$\left(\frac{a}{b}\right)^r = \frac{a^r}{b^r}$$

Proof of Property 5

$$\left(\frac{a}{b}\right)^r = \underbrace{\left(\frac{a}{b}\right)\left(\frac{a}{b}\right)\left(\frac{a}{b}\right)\cdots\left(\frac{a}{b}\right)}_{r \text{ factors}}$$

$$= \frac{a \cdot a \cdot a \cdots a}{b \cdot b \cdot b \cdots b} \quad \begin{array}{l} \leftarrow r \text{ factors} \\ \leftarrow r \text{ factors} \end{array}$$

$$= \frac{a^r}{b^r}$$

Since multiplication with the same base resulted in addition of exponents, it seems reasonable to expect division with the same base to result in subtraction of exponents.

PROPERTY 6 FOR EXPONENTS

If a is any nonzero real number, and r and s are any two integers, then

$$\frac{a^r}{a^s} = a^{r-s}$$

Notice again that we have specified r and s to be any integers. Our definition of negative exponents is such that the properties of exponents hold for all integer exponents, whether positive or negative integers. Here is a proof of property 6.

Proof of Property 6
Our proof is centered around the fact that division by a number is equivalent to multiplication by the reciprocal of the number.

$$\frac{a^r}{a^s} = a^r \cdot \frac{1}{a^s} \quad \text{Dividing by } a^s \text{ is equivalent to multiplying by } \frac{1}{a^s}.$$

$$= a^r a^{-s} \quad \text{Property 4}$$
$$= a^{r+(-s)} \quad \text{Property 1}$$
$$= a^{r-s} \quad \text{Definition of subtraction}$$

EXAMPLES Apply property 6 to each expression and then simplify the result. All answers that contain exponents should contain positive exponents only.

7. $\dfrac{2^8}{2^3} = 2^{8-3} = 2^5 = 32$

8. $\dfrac{x^2}{x^{18}} = x^{2-18} = x^{-16} = \dfrac{1}{x^{16}}$

9. $\dfrac{a^6}{a^{-8}} = a^{6-(-8)} = a^{14}$

10. $\dfrac{m^{-5}}{m^{-7}} = m^{-5-(-7)} = m^2$

Let's complete our list of properties by looking at how the numbers 0 and 1 behave when used as exponents.

We can use the original definition for exponents when the number 1 is used as an exponent:

$$a^1 = \underbrace{a}_{1 \text{ factor}}$$

For 0 as an exponent, consider the expression $\dfrac{3^4}{3^4}$. Since $3^4 = 81$, we have

$$\frac{3^4}{3^4} = \frac{81}{81} = 1$$

On the other hand, since we have the quotient of two expressions with the same base, we can subtract exponents:

$$\frac{3^4}{3^4} = 3^{4-4} = 3^0$$

Hence, 3^0 must be the same as 1.

Summarizing these results, we have our last property for exponents.

PROPERTY 7 FOR EXPONENTS

If a is any real number, then

$$a^1 = a$$

and

$$a^0 = 1 \qquad \text{(as long as } a \neq 0)$$

EXAMPLES Simplify.

11. $(2x^2y^4)^0 = 1$ **12.** $(2x^2y^4)^1 = 2x^2y^4$

Here are some examples that use many of the properties of exponents. There are a number of different ways to proceed on problems like these. You should use the method that works best for you.

EXAMPLES Simplify.

13.
$$\frac{(x^3)^{-2}(x^4)^5}{(x^{-2})^7} = \frac{x^{-6}x^{20}}{x^{-14}} \qquad \text{Property 2}$$

$$= \frac{x^{14}}{x^{-14}} \qquad \text{Property 1}$$

$$= x^{28} \qquad \text{Property 6: } x^{14-(-14)} = x^{28}$$

Note: This last answer can also be written as $\dfrac{a^2b^3}{2}$. Either answer is correct.

14.
$$\frac{6a^5b^{-6}}{12a^3b^{-9}} = \frac{6}{12} \cdot \frac{a^5}{a^3} \cdot \frac{b^{-6}}{b^{-9}} \qquad \text{Write as separate fractions.}$$

$$= \frac{1}{2}a^2b^3 \qquad \text{Property 6}$$

15. $\dfrac{(4x^{-5}y^3)^2}{(x^4y^{-6})^{-3}} = \dfrac{16x^{-10}y^6}{x^{-12}y^{18}}$ Properties 2, 3

$\qquad\qquad = 16x^2y^{-12}$ Property 6

$\qquad\qquad = 16x^2 \cdot \dfrac{1}{y^{12}}$ Property 4

$\qquad\qquad = \dfrac{16x^2}{y^{12}}$ Multiplication

16. $\left(\dfrac{rs^4t^{-3}}{r^5s^{-3}t^{-2}}\right)^{-2} = (r^{-4}s^7t^{-1})^{-2}$ Property 6

$\qquad\qquad = r^8s^{-14}t^2$ Properties 2, 3

$\qquad\qquad = r^8 \cdot \dfrac{1}{s^{14}} \cdot t^2$ Property 4

$\qquad\qquad = \dfrac{r^8t^2}{s^{14}}$ Multiplication

Scientific Notation

The last topic we will cover in this section is scientific notation. Scientific notation is a way to write very large or very small numbers in a more manageable form. Here is the definition.

> **DEFINITION**
>
> A number is written in **scientific notation** if it is written as the product of a number between 1 and 10 and an integer power of 10. A number written in scientific notation has the form
>
> $$n \times 10^r$$
>
> where $1 \le n < 10$ and r is an integer.

EXAMPLE 17   Write 376,000 in scientific notation.

SOLUTION   We must rewrite 376,000 as the product of a number between 1 and 10 and a power of 10. To do so, we move the decimal point 5 places to the left so that it appears between the 3 and the 7. Then, we multiply this number by 10^5. The number that results has the same value as our original number and is written in scientific notation:

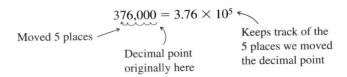

$$376{,}000 = 3.76 \times 10^5$$

Moved 5 places

Decimal point originally here

Keeps track of the 5 places we moved the decimal point

If a number is greater than or equal to 10, then when the number is written in scientific notation the exponent on 10 will be positive. A number that is less than 1 will have a negative exponent when written in scientific notation.

EXAMPLE 18   Write 4.52×10^3 in expanded form—that is, as a regular number.

SOLUTION   Since 10^3 is 1,000, we can think of this as simply a multiplication problem. That is,

$$4.52 \times 10^3 = 4.52 \times 1,000 = 4,520$$

On the other hand, we can think of the exponent 3 as indicating the number of places we need to move the decimal point in order to write the number in expanded form. Since our exponent is positive 3, we move the decimal point 3 places to the right:

$$4.52 \times 10^3 = 4,520$$

The table that follows lists some additional examples of numbers written in expanded form and in scientific notation. In each case, note the relationship between the number of places the decimal point is moved and the exponent on 10.

Number Written in Expanded Form		Number Written in Scientific Notation
376,000	=	3.76×10^5
49,500	=	4.95×10^4
3,200	=	3.2×10^3
591	=	5.91×10^2
46	=	4.6×10^1
8	=	8×10^0
0.47	=	4.7×10^{-1}
0.093	=	9.3×10^{-2}
0.00688	=	6.88×10^{-3}
0.0002	=	2×10^{-4}
0.000098	=	9.8×10^{-5}

Calculator Note: Some scientific calculators have a key that allows you to enter numbers in scientific notation. The key is labeled

| EXP | or | EE | or | SCI |

To enter the number 3.45×10^6 you would first enter the decimal number, then press the scientific notation key, and finally, enter the exponent:

3.45 | EXP | 6

To enter 6.2×10^{-27} you would use the following sequence:

6.2 | EXP | 27 | +/− |

We can use the properties of exponents to do arithmetic with numbers written in scientific notation. Here are some examples.

EXAMPLES Simplify each expression and write all answers in scientific notation.

19. $(2 \times 10^8)(3 \times 10^{-3}) = (2)(3) \times (10^8)(10^{-3})$
$$= 6 \times 10^5$$

20. $\dfrac{4.8 \times 10^9}{2.4 \times 10^{-3}} = \dfrac{4.8}{2.4} \times \dfrac{10^9}{10^{-3}}$
$$= 2 \times 10^{9-(-3)}$$
$$= 2 \times 10^{12}$$

21. $\dfrac{(6.8 \times 10^5)(3.9 \times 10^{-7})}{7.8 \times 10^{-4}} = \dfrac{(6.8)(3.9)}{7.8} \times \dfrac{(10^5)(10^{-7})}{10^{-4}}$
$$= 3.4 \times 10^2$$

Calculator Note: If you have a scientific calculator with a scientific notation key, then the sequence of keys you would use to do Example 20 would look like this:

4.8 [EXP] 9 [÷] 2.4 [EXP] 3 [+/−] [=]

Problem Set

1.4

NOTE: Do you study with other people in your class? If not, you should consider forming a study group. There are many benefits to this, one of which is finding out how your classmates are doing in the course. If you feel as if you are the only one in class who is having difficulty, you will find out quickly that you are not alone.

Evaluate each of the following.

1. 4^2

2. $(-4)^2$

3. -4^2

4. $-(-4)^2$

5. -0.3^3

6. $(-0.3)^3$

7. 2^5

8. 2^4

9. $(\frac{1}{2})^3$

10. $(\frac{3}{4})^2$

11. $(-\frac{5}{6})^2$

12. $(-\frac{7}{8})^2$

Use the properties of exponents to simplify each of the following as much as possible.

13. $x^5 \cdot x^4$

14. $x^6 \cdot x^3$

15. $(2^3)^2$

16. $(3^2)^2$

17. $(-\frac{2}{3}x^2)^3$

18. $(-\frac{3}{5}x^4)^3$

19. $-3a^2(2a^4)$

20. $5a^7(-4a^6)$

Write each of the following with positive exponents. Then simplify as much as possible.

21. 3^{-2}

22. $(-5)^{-2}$

23. $(-2)^{-5}$

24. 2^{-5}

25. $(\frac{3}{4})^{-2}$

26. $(\frac{3}{5})^{-2}$

27. $(\frac{1}{3})^{-2} + (\frac{1}{2})^{-3}$

28. $(\frac{1}{2})^{-2} + (\frac{1}{3})^{-3}$

Simplify each expression. Write all answers with positive exponents only. (Assume all variables are nonzero.)

29. $x^{-4}x^7$

30. $x^{-3}x^8$

31. $(a^2b^{-5})^3$

32. $(a^4b^{-3})^3$

33. $(5y^4)^{-3}(2y^{-2})^3$

34. $(3y^5)^{-2}(2y^{-4})^3$

35. $(\frac{1}{2}x^3)(\frac{2}{3}x^4)(\frac{3}{5}x^{-7})$

36. $(\frac{1}{7}x^{-3})(\frac{7}{8}x^{-5})(\frac{8}{9}x^8)$

37. $(4a^5b^2)(2b^{-5}c^2)(3a^7c^4)$

38. $(3a^{-2}c^3)(5b^{-6}c^5)(4a^6b^{-2})$

39. $(2x^2y^{-5})^3(3x^{-4}y^2)^{-4}$
40. $(4x^{-4}y^9)^{-2}(5x^4y^{-3})^2$

Use the properties of exponents to simplify each expression. Write all answers with positive exponents only. (Assume all variables are nonzero.)

41. $\dfrac{x^{-1}}{x^9}$

42. $\dfrac{x^{-3}}{x^5}$

43. $\dfrac{a^4}{a^{-6}}$

44. $\dfrac{a^5}{a^{-2}}$

45. $\dfrac{t^{-10}}{t^{-4}}$

46. $\dfrac{t^{-8}}{t^{-5}}$

47. $\left(\dfrac{x^5}{x^3}\right)^6$

48. $\left(\dfrac{x^7}{x^4}\right)^5$

49. $\dfrac{(x^5)^6}{(x^3)^4}$

50. $\dfrac{(x^7)^3}{(x^4)^5}$

51. $\dfrac{(x^{-2})^3(x^3)^{-2}}{x^{10}}$

52. $\dfrac{(x^{-4})^3(x^3)^{-4}}{x^{10}}$

53. $\dfrac{5a^8b^3}{20a^5b^{-4}}$

54. $\dfrac{7a^6b^{-2}}{21a^2b^{-5}}$

55. $\dfrac{(3x^{-2}y^8)^4}{(9x^4y^{-3})^2}$

56. $\dfrac{(6x^{-3}y^{-5})^2}{(3x^{-4}y^{-3})^4}$

57. $\left(\dfrac{8x^2y}{4x^4y^{-3}}\right)^4$

58. $\left(\dfrac{5x^4y^5}{10xy^{-2}}\right)^3$

59. $\left(\dfrac{x^{-5}y^2}{x^{-3}y^5}\right)^{-2}$

60. $\left(\dfrac{x^{-8}y^{-3}}{x^{-5}y^6}\right)^{-1}$

61. $\left(\dfrac{ab^{-3}c^{-2}}{a^{-3}b^0c^{-5}}\right)^{-1}$

62. $\left(\dfrac{a^3b^2c^1}{a^{-1}b^{-2}c^{-3}}\right)^{-2}$

Write each number in scientific notation.

63. 378,000
64. 3,780,000
65. 4,900
66. 490
67. 0.00037
68. 0.000037
69. 0.00495
70. 0.0495

Write each number in expanded form.

71. 5.34×10^3
72. 5.34×10^2
73. 7.8×10^6
74. 7.8×10^4
75. 3.44×10^{-3}
76. 3.44×10^{-5}
77. 4.9×10^{-1}
78. 4.9×10^{-2}

Use the properties of exponents to simplify each of the following expressions. Write all answers in scientific notation.

79. $(4 \times 10^{10})(2 \times 10^{-6})$
80. $(3 \times 10^{-12})(3 \times 10^4)$

81. $\dfrac{8 \times 10^{14}}{4 \times 10^5}$

82. $\dfrac{6 \times 10^8}{2 \times 10^3}$

83. $\dfrac{(5 \times 10^6)(4 \times 10^{-8})}{8 \times 10^4}$

84. $\dfrac{(6 \times 10^{-7})(3 \times 10^9)}{5 \times 10^6}$

85. $\dfrac{(2.4 \times 10^{-3})(3.6 \times 10^{-7})}{(4.8 \times 10^6)(1 \times 10^{-9})}$

86. $\dfrac{(7.5 \times 10^{-6})(1.5 \times 10^9)}{(1.8 \times 10^4)(2.5 \times 10^{-2})}$

Applying the Concepts

87. The number 237×10^4 is not written in scientific notation because 237 is larger than 10. Write 237×10^4 in scientific notation.

88. Write 46.2×10^{-3} in scientific notation.

89. **Large Numbers** If you are 20 years old, you have been alive for more than 630,000,000 seconds. Write this last number in scientific notation.

90. **Large Numbers** Use the information from Problem 89 to give the approximate number of seconds you have lived if you are 40 years old. Write your answer in scientific notation.

91. **Very Large Numbers** The mass of the earth is approximately 5.98×10^{24} kilograms. If this number were written in expanded form, how many zeros would it contain?

92. Very Small Numbers The mass of a single hydrogen atom is approximately 1.67×10^{-27} kilogram. If this number were written in expanded form, how many digits would there be to the right of the decimal point?

93. Very Large Numbers A light-year, the distance light travels in 1 year, is approximately 5.9×10^{12} miles. The Andromeda galaxy is approximately 1.7×10^{6} light-years from our galaxy. Find the distance in miles between our galaxy and the Andromeda galaxy.

94. Very Large Numbers The distance from the earth to the sun is approximately 9.3×10^{7} miles. If light travels 1.2×10^{7} miles in 1 minute, how many minutes does it take the light from the sun to reach the earth?

One Step Further

Assume all variable exponents represent positive integers and simplify each expression.

95. $x^{m+2} \cdot x^{-2m} \cdot x^{m-5}$

96. $x^{m-4} x^{m+9} x^{-2m}$

97. $(y^{m})^{2}(y^{-3})^{m}(y^{m+3})$

98. $(y^{m})^{-4}(y^{3})^{m}(y^{m+6})$

99. $\dfrac{x^{n+2}}{x^{n-3}}$

100. $\dfrac{x^{n-3}}{x^{n-7}}$

Polynomials: Sums, Differences, and Products

The chart below is from a company that duplicates videotapes. It shows the revenue and cost to duplicate a 30 minute video. From the chart you can see that 300 copies will bring in $900 in revenue, with a cost of $600. The profit is the difference between revenue and cost, or $300.

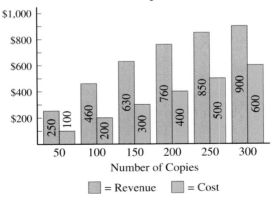

Revenue and Cost to Duplicate a 30 Minute Video

The relationship between profit, revenue, and cost is one application of the polynomials we will study in this section. Let's begin with a definition that we will use to build polynomials.

> **DEFINITION**
>
> A **term** (or **monomial**) is a constant or the product of a constant and one or more variables raised to whole number exponents.

The following are monomials (or terms):

$$-16 \qquad 3x^2y \qquad -\frac{2}{5}a^3b^2c \qquad xy^2z$$

The numerical part of each monomial is called the **numerical coefficient,** or just **coefficient** for short. For the preceding terms, the coefficients are -16, 3, $-\frac{2}{5}$, and 1. Notice that the coefficient for xy^2z is understood to be 1.

> **DEFINITION**
>
> A **polynomial** is any finite sum of terms. Since subtraction can be written in terms of addition, finite differences are also included in this definition.

The following are polynomials:

$$2x^2 - 6x + 3 \qquad -5x^2y + 2xy^2 \qquad 4a - 5b + 6c + 7d$$

Polynomials can be classified further according to the number of terms present. If a polynomial consists of two terms, it is said to be a **binomial.** If it has three terms, it is called a **trinomial.** And, as stated above, a polynomial with only one term is said to be a **monomial.**

> **DEFINITION**
>
> The **degree** of a polynomial with one variable is the highest power to which the variable is raised in any one term.

EXAMPLES Identify the type and degree of each polynomial.

1. $6x^2 + 2x - 1$ A trinomial of degree 2

2. $5x - 3$ A binomial of degree 1

3. $7x^6 - 5x^3 + 2x - 4$ A polynomial of degree 6

4. $-7x^4$ A monomial of degree 4

5. 15 A monomial of degree 0

Polynomials in one variable are usually written in decreasing powers of the variable. When this is the case, the coefficient of the first term is called the **leading coefficient.** In Example 1, the leading coefficient is 6. In Example 2, it is 5. The leading coefficient in Example 3 is 7.

DEFINITION

Two or more terms that differ only in their numerical coefficients are called **similar** or **like terms.** Since similar terms differ only in their coefficients, they have identical variable parts.

As mentioned earlier (Section 1.2), we can use the distributive property to combine the similar terms $6x^2$ and $9x^2$ as follows:

$$6x^2 + 9x^2 = (6 + 9)x^2 \quad \text{Distributive property}$$
$$= 15x^2 \quad \text{The sum of 6 and 9 is 15.}$$

Adding Polynomials

To add two polynomials, we simply apply the commutative and associative properties to group similar terms together and then use the distributive property as we did in the preceding example.

EXAMPLE 6   Add $5x^2 - 4x + 2$ and $3x^2 + 9x - 6$.

SOLUTION

$$(5x^2 - 4x + 2) + (3x^2 + 9x - 6)$$
$$= (5x^2 + 3x^2) + (-4x + 9x) + (2 - 6) \quad \text{Commutative and associative properties}$$
$$= (5 + 3)x^2 + (-4 + 9)x + (2 - 6) \quad \text{Distributive property}$$
$$= 8x^2 + 5x + (-4)$$
$$= 8x^2 + 5x - 4$$

EXAMPLE 7   Find the sum of $-8x^3 + 7x^2 - 6x + 5$ and $10x^3 + 3x^2 - 2x - 6$.

SOLUTION   We can add the two polynomials using the method of Example 6, or we can arrange similar terms in columns and add vertically. Using the column method, we have

$$-8x^3 + 7x^2 - 6x + 5$$
$$\underline{10x^3 + 3x^2 - 2x - 6}$$
$$2x^3 + 10x^2 - 8x - 1$$

Subtracting Polynomials

To find the difference of two polynomials, we need to use the fact that the opposite of a sum is the sum of the opposites. That is,

$$-(a + b) = -a + (-b)$$

One way to remember this fact is to observe that $-(a + b)$ is equivalent to $-1(a + b) = (-1)a + (-1)b = -a + (-b)$.

If there is a negative sign directly preceding the parentheses surrounding a polynomial, we may remove the parentheses and preceding negative sign by changing the sign of each term within the parentheses. For example:

$$-(3x + 4) = -3x + (-4) = -3x - 4$$
$$-(5x^2 - 6x + 9) = -5x^2 + 6x - 9$$
$$-(-x^2 + 7x - 3) = x^2 - 7x + 3$$

EXAMPLE 8   Subtract $4x^2 + 2x - 3$ from $9x^2 - 3x + 5$.

SOLUTION   First, we write the problem in terms of subtraction. Then, we subtract by adding the opposite of each term in the polynomial that follows the subtraction sign:

$$(9x^2 - 3x + 5) - (4x^2 + 2x - 3)$$

$= 9x^2 - 3x + 5 + (-4x^2) + (-2x) + 3$ The opposite of a sum is the sum of the opposites.

$= (9x^2 - 4x^2) + (-3x - 2x) + (5 + 3)$ Commutative and associative properties

$= 5x^2 - 5x + 8$ Combine similar terms.

When one set of grouping symbols is contained within another, it is best to begin the process of simplification within the innermost grouping symbols and work outward from there.

EXAMPLE 9   Simplify: $4x - 3[2 - (3x + 4)]$.

SOLUTION   Removing the innermost parentheses first, we have

$$4x - 3[2 - (3x + 4)] = 4x - 3(2 - 3x - 4)$$
$$= 4x - 3(-3x - 2)$$
$$= 4x + 9x + 6$$
$$= 13x + 6$$

In the example that follows we will find the value of a polynomial for a given value of the variable.

EXAMPLE 10   Find the value of $5x^3 - 3x^2 + 4x - 5$ when x is 2.

SOLUTION   We begin by substituting 2 for x in the original polynomial:

When $x = 2$:

the polynomial $5x^3 - 3x^2 + 4x - 5$

becomes $5 \cdot 2^3 - 3 \cdot 2^2 + 4 \cdot 2 - 5$
$$= 5 \cdot 8 - 3 \cdot 4 + 4 \cdot 2 - 5$$
$$= 40 - 12 + 8 - 5$$
$$= 31$$

Three quantities that occur very frequently in business and economics are profit, revenue, and cost. If a company manufactures and sells x items, then the revenue R is the total amount of money obtained by selling all x items. The cost C is the total amount of money it costs the company to manufacture the x items. The profit P obtained by selling all x items is the difference between the revenue and the cost and is given by the equation

$$P = R - C$$

EXAMPLE 11 A company produces and sells copies of an accounting program for home computers. The total weekly cost (in dollars) for the company to produce x copies of the program is $C = 8x + 500$. Find the weekly profit if the total revenue obtained from selling all x programs is $R = 35x - 0.1x^2$. How much profit will the company make if they produce and sell 100 programs a week?

SOLUTION Using the equation $P = R - C$ and the information given in the problem, we have

$$\begin{aligned} P &= R - C \\ &= 35x - 0.1x^2 - (8x + 500) \\ &= 35x - 0.1x^2 - 8x - 500 \\ &= -500 + 27x - 0.1x^2 \qquad \text{Weekly profit} \end{aligned}$$

If the company produces and sells 100 copies of the program, the weekly profit will be

$$\begin{aligned} P &= -500 + 27(100) - 0.1(100)^2 \\ &= -500 + 27(100) - 0.1(10,000) \\ &= -500 + 2,700 - 1,000 \\ &= 1,200 \end{aligned}$$

The weekly profit is $1,200 when 100 copies are sold.

Multiplying Polynomials

The distributive property is the key to multiplying polynomials. The simplest type of multiplication occurs when we multiply a polynomial by a monomial.

EXAMPLE 12 Find the product of $4x^3$ and $5x^2 - 3x + 1$.

SOLUTION To multiply, we apply the distributive property:

Notice that we multiply coefficients and add exponents.

$$\begin{aligned} &4x^3(5x^2 - 3x + 1) \\ &= 4x^3(5x^2) + 4x^3(-3x) + 4x^3(1) \qquad \text{Distributive property} \\ &= 20x^5 - 12x^4 + 4x^3 \end{aligned}$$

The distributive property can also be applied to multiply a polynomial by a polynomial. Let's consider the case where both polynomials have two terms.

EXAMPLE 13   Multiply $2x - 3$ and $x + 5$.

SOLUTION   Distributing the $2x - 3$ across the sum $x + 5$ gives us

$$\begin{aligned}
(2x - 3)(x + 5) \\
= (2x - 3)x + (2x - 3)5 \qquad &\text{Distributive property} \\
= 2x(x) + (-3)x + 2x(5) + (-3)5 \qquad &\text{Distributive property} \\
= 2x^2 - 3x + 10x - 15 \\
= 2x^2 + 7x - 15 \qquad &\text{Combine like terms.}
\end{aligned}$$

Notice the third line in Example 13. It consists of all possible products of terms in the first binomial and those of the second binomial. We can generalize this into a rule for multiplying two polynomials.

> **RULE: MULTIPLYING TWO POLYNOMIALS**
>
> Multiply each term in the first polynomial by each term in the second polynomial.

Multiplying polynomials can be accomplished by a method that looks very similar to long multiplication with whole numbers. Reworking Example 13, we have

$$\begin{array}{rr}
2x & -\ \ 3 \\
x & +\ \ 5 \\
\hline
2x^2 & -\ \ 3x \\
& 10x - 15 \\
\hline
2x^2 & +\ \ 7x - 15
\end{array}$$

Multiply x by $2x - 3$.
Multiply $+5$ by $2x - 3$.
Add in columns.

EXAMPLE 14   Multiply $2x - 3y$ and $3x^2 - xy + 4y^2$ vertically.

SOLUTION

$$\begin{array}{r}
3x^2 - \quad xy + \ 4y^2 \\
2x - \quad 3y \\
\hline
6x^3 - \ 2x^2y + \ 8xy^2 \\
- \ 9x^2y + \ 3xy^2 - 12y^3 \\
\hline
6x^3 - 11x^2y + 11xy^2 - 12y^3
\end{array}$$

Multiply $3x^2 - xy + 4y^2$ by $2x$.
Multiply $3x^2 - xy + 4y^2$ by $-3y$.
Add similar terms.

Multiplying Binomials—The FOIL Method

The product of two binomials occurs very frequently in algebra. Since this type of product is so common, we have a special method of multiplication that applies only to products of binomials.

Consider the product of $2x - 5$ and $3x - 2$. Distributing $3x - 2$ over $2x$ and -5, we have

$$\begin{aligned}
(2x - 5)(3x - 2) &= (2x)(3x - 2) + (-5)(3x - 2) \\
&= (2x)(3x) + (2x)(-2) + (-5)(3x) + (-5)(-2) \\
&= 6x^2 - 4x - 15x + 10 \\
&= 6x^2 - 19x + 10
\end{aligned}$$

Looking closely at the second and third lines, we notice the following relationships:

1. $6x^2$ comes from multiplying the *first* terms in each binomial:

$$(2x - 5)(3x - 2) \qquad 2x(3x) = 6x^2 \qquad \textit{First } \text{terms}$$

2. $-4x$ comes from multiplying the *outside* terms in the product:

$$(2x - 5)(3x - 2) \qquad 2x(-2) = -4x \qquad \textit{Outside } \text{terms}$$

3. $-15x$ comes from multiplying the *inside* terms in the product:

$$(2x - 5)(3x - 2) \qquad -5(3x) = -15x \qquad \textit{Inside } \text{terms}$$

4. 10 comes from multiplying the *last* two terms in the product:

$$(2x - 5)(3x - 2) \qquad -5(-2) = 10 \qquad \textit{Last } \text{terms}$$

Once we know where the terms in the answer come from, we can reduce the number of steps used in finding the product:

$$(2x - 5)(3x - 2) = \underset{\text{First}}{6x^2} - \underset{\text{Outside}}{4x} - \underset{\text{Inside}}{15x} + \underset{\text{Last}}{10}$$
$$= 6x^2 - 19x + 10$$

This is called the **FOIL method.**

EXAMPLES   Multiply using the FOIL method.

15. $(4a - 5b)(3a + 2b) = \underset{\text{F}}{12a^2} + \underset{\text{O}}{8ab} - \underset{\text{I}}{15ab} - \underset{\text{L}}{10b^2}$
$$= 12a^2 - 7ab - 10b^2$$

16. $(3 - 2t)(4 + 7t) = \underset{\text{F}}{12} + \underset{\text{O}}{21t} - \underset{\text{I}}{8t} - \underset{\text{L}}{14t^2}$
$$= 12 + 13t - 14t^2$$

17. $(2x + \frac{1}{2})(4x - \frac{1}{2}) = \underset{\text{F}}{8x^2} - \underset{\text{O}}{x} + \underset{\text{I}}{2x} - \underset{\text{L}}{\frac{1}{4}}$
$$= 8x^2 + x - \frac{1}{4}$$

18. $(a^5 + 3)(a^5 - 7) = \underset{\text{F}}{a^{10}} - \underset{\text{O}}{7a^5} + \underset{\text{I}}{3a^5} - \underset{\text{L}}{21}$
$$= a^{10} - 4a^5 - 21$$

19. $(2x + 3)(5y - 4) = \underset{\text{F}}{10xy} - \underset{\text{O}}{8x} + \underset{\text{I}}{15y} - \underset{\text{L}}{12}$

The Square of a Binomial

EXAMPLE 20   Find $(4x - 6)^2$.

SOLUTION   Applying the definition of exponents and then the FOIL method, we have

$$
\begin{aligned}
(4x - 6)^2 &= (4x - 6)(4x - 6) \\
&= 16x^2 - 24x - 24x + 36 \\
&\qquad\quad \text{F} \qquad \text{O} \qquad \text{I} \qquad \text{L} \\
&= 16x^2 - 48x + 36
\end{aligned}
$$

This example is the square of a binomial. This type of product occurs frequently enough in algebra that we have special formulas for squares of binomials:

$$(a + b)^2 = (a + b)(a + b) = a^2 + ab + ab + b^2 = a^2 + 2ab + b^2$$
$$(a - b)^2 = (a - b)(a - b) = a^2 - ab - ab + b^2 = a^2 - 2ab + b^2$$

Observing the results in both cases, we have the following rule.

The Square of a Binomial

The square of a binomial is the sum of the square of the first term, twice the product of the two terms, and the square of the last term. That is:

$$(a + b)^2 = \quad a^2 \quad + \quad 2ab \quad + \quad b^2$$

| | Square of first term | Twice the product of the two terms | Square of last term |

$$(a - b)^2 = \quad a^2 \quad - \quad 2ab \quad + \quad b^2$$

EXAMPLES   Use the preceding formulas to expand each binomial square.

21. $(x + 7)^2 = x^2 + 2(x)(7) + 7^2 = x^2 + 14x + 49$

22. $(3t - 5)^2 = (3t)^2 - 2(3t)(5) + 5^2 = 9t^2 - 30t + 25$

23. $(4x + 2y)^2 = (4x)^2 + 2(4x)(2y) + (2y)^2 = 16x^2 + 16xy + 4y^2$

24. $(5 - a^3)^2 = 5^2 - 2(5)(a^3) + (a^3)^2 = 25 - 10a^3 + a^6$

25. $(x + \frac{1}{3})^2 = x^2 + 2(x)(\frac{1}{3}) + (\frac{1}{3})^2 = x^2 + \frac{2}{3}x + \frac{1}{9}$

Products That Are the Difference of Two Squares

Another frequently occurring kind of product is found when multiplying two binomials that differ only in the sign between their terms.

EXAMPLE 26   Multiply $(3x - 5)$ and $(3x + 5)$.

SOLUTION   Applying the FOIL method, we have

$$(3x - 5)(3x + 5) = 9x^2 + 15x - 15x - 25 \quad \text{Two middle terms}$$
$$ \text{F} \quad\ \text{O} \qquad \text{I} \qquad \text{L} \qquad \text{add to 0.}$$
$$= 9x^2 - 25$$

The outside and inside products in Example 26 are opposites and therefore add to 0.

Here it is in general:

$$(a - b)(a + b) = a^2 + ab - ab - b^2 \quad \text{Two middle terms add to 0.}$$
$$= a^2 - b^2$$

> ### The Product of the Sum and Difference of Two Terms
>
> To multiply two binomials that differ only in the sign between their two terms, simply subtract the square of the second term from the square of the first term:
>
> $$(a - b)(a + b) = a^2 - b^2$$

The expression $a^2 - b^2$ is called the **difference of two squares.**

Once we memorize and understand this rule, we can multiply binomials of this form with a minimum of work.

EXAMPLES   Find the following products.

27. $(x - 5)(x + 5) = x^2 - 25$

28. $(2a - 3)(2a + 3) = 4a^2 - 9$

29. $(x^2 + 4)(x^2 - 4) = x^4 - 16$

30. $(x^3 - 2a)(x^3 + 2a) = x^6 - 4a^2$

Problem Set

1.5

NOTE: Working problems every day is the most important thing you can do to be successful in this course. If you are working hard one day and then taking three days off, you should readjust your schedule. You will be better off if you spread your work over all the days of the week— not just two or three.

Identify those of the following that are monomials, binomials, or trinomials. Give the degree of each and name the leading coefficient.

1. $5x^2 - 3x + 2$

2. $2x^2 + 4x - 1$

3. $3x - 5$

4. $5y + 3$

5. $8a^2 + 3a - 5$

6. $9a^2 - 8a - 4$

7. $4x^3 - 6x^2 + 5x - 3$

8. $9x^4 + 4x^3 - 2x^2 + x$

9. $-\frac{3}{4}$ **10.** -16

11. $4x - 5 + 6x^3$ **12.** $9x + 2 + 3x^3$

Simplify each of the following by combining similar terms.

13. $(4x + 2) + (3x - 1)$

14. $(8x - 5) + (-5x + 4)$

15. $2x^2 - 3x + 10x - 15$

16. $6x^2 - 4x - 15x + 10$

17. $12a^2 + 8ab - 15ab - 10b^2$

18. $28a^2 - 8ab + 7ab - 2b^2$

19. $(5x^2 - 6x + 1) - (4x^2 + 7x - 2)$

20. $(11x^2 - 8x) - (4x^2 - 2x - 7)$

21. $(\frac{1}{2}x^2 - \frac{1}{3}x - \frac{1}{6}) - (\frac{1}{4}x^2 + \frac{7}{12}x) + (\frac{1}{3}x - \frac{1}{12})$

22. $(\frac{2}{3}x^2 - \frac{1}{2}x) - (\frac{1}{4}x^2 + \frac{1}{6}x + \frac{1}{12}) - (\frac{1}{2}x^2 + \frac{1}{4})$

23. Subtract $2x^2 - 4x$ from $2x^2 - 7x$.

24. Subtract $-3x + 6$ from $-3x + 9$.

Simplify each of the following. Begin by working on the innermost parentheses first.

25. $-5[-(x - 3) - (x + 2)]$

26. $-6[(2x - 5) - 3(8x - 2)]$

27. $4x - 5[3 - (x - 4)]$

28. $x - 7[3x - (2 - x)]$

29. Find the value of $2x^2 - 3x - 4$ when x is 2.

30. Find the value of $4x^2 + 3x - 2$ when x is -1.

31. Find the value of $\frac{3}{2}x^2 - \frac{3}{4}x + 1$ when x is 12.

32. Find the value of $\frac{2}{5}x^2 - \frac{1}{10}x + 2$ when x is 10.

33. Find the value of $x^3 - x^2 + x - 1$ when x is -2.

34. Find the value of $x^3 + x^2 + x + 1$ when x is -2.

Multiply the following by applying the distributive property.

35. $2x(6x^2 - 5x + 4)$

36. $-3x(5x^2 - 6x - 4)$

37. $2a^2b(a^3 - ab + b^3)$

38. $5a^2b^2(8a^2 - 2ab + b^2)$

Multiply the following vertically.

39. $(3a + 5)(2a^3 - 3a^2 + a)$

40. $(2a - 3)(3a^2 - 5a + 1)$

41. $(a - b)(a^2 + ab + b^2)$

42. $(a + b)(a^2 - ab + b^2)$

43. $(2x + y)(4x^2 - 2xy + y^2)$

44. $(x - 3y)(x^2 + 3xy + 9y^2)$

Multiply the following using the FOIL method.

45. $(2a + 3)(3a + 2)$ **46.** $(5a - 4)(2a + 1)$

47. $(5 - 3t)(4 + 2t)$ **48.** $(7 - t)(6 - 3t)$

49. $(x^3 + 3)(x^3 - 5)$ **50.** $(x^3 + 4)(x^3 - 7)$

51. $(3t + \frac{1}{3})(6t - \frac{2}{3})$ **52.** $(5t - \frac{1}{5})(10t + \frac{3}{5})$

53. $(b - 4a^2)(b + 3a^2)$

54. $(b + 5a^2)(b - 2a^2)$

Find the following special products.

55. $(2a - 3)^2$ **56.** $(3a + 2)^2$

57. $(5x + 2y)^2$ **58.** $(3x - 4y)^2$

59. $(5 - 3t^3)^2$ **60.** $(7 - 2t^4)^2$

61. $(2a + 3b)(2a - 3b)$

62. $(6a - 1)(6a + 1)$

63. $(3r^2 + 7s)(3r^2 - 7s)$

64. $(5r^2 - 2s)(5r^2 + 2s)$

65. $(\frac{1}{3}x - \frac{2}{5})(\frac{1}{3}x + \frac{2}{5})$

66. $(\frac{3}{4}x - \frac{1}{7})(\frac{3}{4}x + \frac{1}{7})$

Expand and simplify.

67. $(x - 2)^3$ **68.** $(x + 4)^3$

69. $(2x - 1)^3$ **70.** $(4x + 1)^3$

71. $(b^2 + 8)(a^2 + 1)$ **72.** $(b^2 + 1)(a^4 - 5)$

73. $(x - 2)(3y^2 + 4)$ **74.** $(x - 4)(2y^3 + 1)$

75. $(x + 1)^2 + (x + 2)^2 + (x + 3)^2$

76. $(x - 1)^2 + (x - 2)^2 + (x - 3)^2$

77. $(2x + 3)^2 - (2x - 3)^2$

78. $(5x - 4)^2 - (5x + 4)^2$

79. $(x - 1)^3 - (x + 1)^3$

80. $(x - 3)^3 - (x + 3)^3$

In Words

81. **Profits** The total cost (in dollars) for a company to manufacture and sell x items per week is $C = 60x + 300$. If the revenue brought in by selling all x items is $R = 100x - 0.5x^2$, find the weekly profit. How much profit will be made by producing and selling 60 items each week?

82. **Profits** The total cost (in dollars) for a company to produce and sell x items per week is $C = 200x + 1{,}600$. If the revenue brought in by selling all x items is $R = 300x - 0.6x^2$, find the weekly profit. How much profit will be made by producing and selling 50 items each week?

83. Profits Suppose it costs a company selling patterns $C = 800 + 6.5x$ dollars to produce and sell x patterns a month. If the revenue obtained by selling x patterns is $R = 10x - 0.002x^2$, what is the profit equation? How much profit will be made if 1,000 patterns are produced and sold in May?

84. Profits Suppose a company manufactures and sells x picture frames each month with a total cost of $C = 1,200 + 3.5x$ dollars. If the revenue obtained by selling x frames is $R = 9x - 0.003x^2$, find the profit equation. How much profit will be made if 1,000 frames are manufactured and sold in June?

85. Height of an Object If an object is thrown straight up into the air with velocity of 128 feet/second, then its height h above the ground t seconds later is given by the formula

$$h = -16t^2 + 128t$$

Find the height after 3 seconds and after 5 seconds.

86. Height of an Object The formula for the height of an object that has been thrown

straight up with a velocity of 64 feet/second is

$$h = -16t^2 + 64t$$

Find the height after 1 second and after 3 seconds.

87. Compound Interest If you deposit $100 in an account with an interest rate r that is compounded annually, then the amount of money in that account at the end of 4 years is given by the formula $A = 100(1 + r)^4$. Expand the right side of this formula.

88. Compound Interest If you deposit P dollars in an account with an annual interest rate r that is compounded twice a year, then at the end of a year the amount of money in that account is given by the formula

$$A = P\left(1 + \frac{r}{2}\right)^2$$

Expand the right side of this formula.

One Step Further

Assume n is a positive integer and multiply.

89. $(x^n - 2)(x^n - 3)$
90. $(x^n + 4)(x^n - 1)$
91. $(x^{2n} + 3)(x^{2n} - 3)$
92. $(x^{3n} + 4)(x^{3n} - 4)$
93. $(2x^n + 3)(5x^n - 1)$
94. $(4x^n - 3)(7x^n + 2)$
95. $(x^n + 5)^2$
96. $(x^n - 3)^2$

SECTION 1.6

Factoring

In general, as mentioned in Section 1.1, factoring is the reverse of multiplication. The following diagram illustrates the relationship between factoring and multiplication:

Multiplication

Factors → $3 \cdot 7 = 21$ ← Product

Factoring

Reading from left to right, we say the product of 3 and 7 is 21. Reading in the other direction, from right to left, we say 21 factors into 3 times 7. Or, 3 and 7 are factors of 21.

> **DEFINITION**
>
> The **greatest common factor** for a polynomial is the largest monomial that divides (is a factor of) each term of the polynomial.

The greatest common factor for the polynomial $25x^5 + 20x^4 - 30x^3$ is $5x^3$ since it is the largest monomial that is a factor of each term. We can apply the distributive property and write

$$25x^5 + 20x^4 - 30x^3 = 5x^3(5x^2) + 5x^3(4x) - 5x^3(6)$$
$$= 5x^3(5x^2 + 4x - 6)$$

The last line is written in factored form.

EXAMPLE 1 Factor the greatest common factor from $8a^3 - 8a^2 - 48a$.

SOLUTION The greatest common factor is $8a$. It is the largest monomial that divides each term of the polynomial. We can write each term in the polynomial as the product of $8a$ and another monomial. Then, we apply the distributive property to factor $8a$ from each term:

$$8a^3 - 8a^2 - 48a = 8a(a^2) - 8a(a) - 8a(6)$$
$$= 8a(a^2 - a - 6)$$

Note: The phrase *largest monomial,* as used here, refers to the monomial with the largest integer exponent and whose coefficient has the greatest absolute value. We could have factored the polynomial in Example 1 correctly by taking out $-8a$, instead of $8a$. However, when factoring like this, it is usually better to have a positive coefficient on the greatest common factor, even though it is not incorrect to have a negative coefficient.

EXAMPLE 2 Factor the greatest common factor from $5x^2(a + b) - 6x(a + b) - 7(a + b)$.

SOLUTION The greatest common factor is $a + b$. Factoring it from each term, we have

$$5x^2(a + b) - 6x(a + b) - 7(a + b) = (a + b)(5x^2 - 6x - 7)$$

Factoring by Grouping

Many polynomials have no greatest common factor other than the number 1. Some of these can be factored using the distributive property if those terms with a common factor are grouped together.

For example, the polynomial $5x + 5y + x^2 + xy$ can be factored by noticing that the first two terms have a 5 in common, whereas the last two have an x in common. Applying the distributive property, we have

$$5x + 5y + x^2 + xy = 5(x + y) + x(x + y)$$

This last expression can be thought of as having two terms, $5(x + y)$ and $x(x + y)$, each of which has a common factor $(x + y)$. We apply the distributive property again to factor $(x + y)$ from each term:

$$5(x + y) + x(x + y)$$

$$= (x + y)(5 + x)$$

EXAMPLE 3   Factor: $a^2b^2 + b^2 + 8a^2 + 8$.

SOLUTION   The first two terms have b^2 in common; the last two have 8 in common:

$$a^2b^2 + b^2 + 8a^2 + 8 = b^2(a^2 + 1) + 8(a^2 + 1)$$
$$= (a^2 + 1)(b^2 + 8)$$

Factoring Trinomials with Leading Coefficient 1

In Section 1.5 we multiplied binomials:

$$(x - 2)(x + 3) = x^2 + x - 6$$
$$(x + 5)(x + 2) = x^2 + 7x + 10$$

In each case the product of two binomials is a trinomial. The first term in the resulting trinomial is obtained by multiplying the first term in each binomial. The middle term comes from adding the products of the two inside terms and the two outside terms. The last term is the product of the last term in each binomial.
    In general,

$$(x + a)(x + b) = x^2 + ax + bx + ab$$
$$= x^2 + (a + b)x + ab$$

Writing this as a factoring problem, we have

$$x^2 + (a + b)x + ab = (x + a)(x + b)$$

To factor a trinomial with a leading coefficient of 1, we simply find the two numbers a and b whose sum is the coefficient of the middle term and whose product is the constant term.

EXAMPLE 4   Factor: $x^2 + 2x - 15$.

SOLUTION   The leading coefficient is 1. We need two integers whose product is -15 and whose sum is $+2$. The integers are $+5$ and -3:

$$x^2 + 2x - 15 = (x + 5)(x - 3)$$

If a trinomial is factorable, then its factors are unique. For instance, in Example 4 we found factors of $x + 5$ and $x - 3$. These are the only two factors for $x^2 + 2x - 15$. There is no other pair of binomials whose product is $x^2 + 2x - 15$.
    If a trinomial is not factorable, we say that it is a **prime polynomial.**

EXAMPLE 5   Factor: $x^2 - xy - 12y^2$.

SOLUTION   We need two numbers whose product is $-12y^2$ and whose sum is $-y$. The numbers are $-4y$ and $3y$:

$$x^2 - xy - 12y^2 = (x - 4y)(x + 3y)$$

Check   Checking this result gives

$$(x - 4y)(x + 3y) = x^2 + 3xy - 4xy - 12y^2$$
$$= x^2 - xy - 12y^2$$

EXAMPLE 6   Factor: $x^2 - 8x + 6$.

SOLUTION   Since there is no pair of integers whose product is 6 and whose sum is -8, the trinomial $x^2 - 8x + 6$ is not factorable. It is a prime polynomial.

Factoring Other Trinomials by Trial and Error

We want to turn our attention now to trinomials with leading coefficients other than 1 and with no greatest common factor other than 1.

Suppose we want to factor $3x^2 - x - 2$. The factors will be a pair of binomials. The product of the first terms will be $3x^2$ and the product of the last terms will be -2. We can list all the possible factors along with their products as follows:

Possible Factors	First Term	Middle Term	Last Term
$(x + 2)(3x - 1)$	$3x^2$	$+5x$	-2
$(x - 2)(3x + 1)$	$3x^2$	$-5x$	-2
$(x + 1)(3x - 2)$	$3x^2$	$+x$	-2
$(x - 1)(3x + 2)$	$3x^2$	$-x$	-2

From the last line we see that the factors of $3x^2 - x - 2$ are $(x - 1)(3x + 2)$. That is,

$$3x^2 - x - 2 = (x - 1)(3x + 2)$$

To factor trinomials with leading coefficients other than 1, when the greatest common factor is 1, we must use trial and error or list all the possible factors. In either case, the idea is this: Look only at pairs of binomials whose products give the correct first and last terms, then look for a combination that will give the correct middle term.

EXAMPLE 7 Factor: $2x^2 + 13xy + 15y^2$.

SOLUTION Listing all possible factors that give $2x^2$ as the product of the first terms and $+15y^2$ as the product of the last terms yields the table at the right.

The last line has the correct middle term:

$$2x^2 + 13xy + 15y^2$$
$$= (2x + 3y)(x + 5y)$$

Possible Factors	Middle Term of Product
$(2x + y)(x + 15y)$	$31xy$
$(2x - y)(x - 15y)$	$-31xy$
$(2x + 15y)(x + y)$	$+17xy$
$(2x - 15y)(x - y)$	$-17xy$
$(2x - 5y)(x - 3y)$	$-11xy$
$(2x + 5y)(x + 3y)$	$+11xy$
$(2x - 3y)(x - 5y)$	$-13xy$
$(2x + 3y)(x + 5y)$	$+13xy$

Actually, we did not need to check four of the pairs of possible factors in the preceding list. All the signs in the trinomial $2x^2 + 13xy + 15y^2$ are positive. The binomial factors must then be of the form $(ax + b)(cx + d)$, where a, b, c, and d are all positive.

There are other ways to reduce the number of possible factors under consideration. For example, if we were to factor the trinomial $2x^2 - 11x + 12$, we would not have to consider the pair of possible factors $(2x - 4)(x - 3)$. If the original trinomial has no greatest common factor other than 1, then neither of its binomial factors will either. The trinomial $2x^2 - 11x + 12$ has a greatest common factor of 1, but the possible factor $2x - 4$ has a greatest common factor of 2: $2x - 4 = 2(x - 2)$. Therefore, we do not need to consider $2x - 4$ as a possible factor.

EXAMPLE 8 Factor $18x^3y + 3x^2y^2 - 36xy^3$.

SOLUTION First, factor out the greatest common factor $3xy$. Then, factor the remaining trinomial:

$$18x^3y + 3x^2y^2 - 36xy^3 = 3xy(6x^2 + xy - 12y^2)$$
$$= 3xy(3x - 4y)(2x + 3y)$$

EXAMPLE 9 Factor: $12x^4 + 17x^2 + 6$.

SOLUTION This is a trinomial in x^2:

$$12x^4 + 17x^2 + 6 = (4x^2 + 3)(3x^2 + 2)$$

EXAMPLE 10 Factor: $2x^2(x - 3) - 5x(x - 3) - 3(x - 3)$.

SOLUTION We begin by factoring out the greatest common factor $(x - 3)$. Then, we factor the trinomial that remains:

$$2x^2(x - 3) - 5x(x - 3) - 3(x - 3) = (x - 3)(2x^2 - 5x - 3)$$
$$= (x - 3)(2x + 1)(x - 3)$$
$$= (x - 3)^2(2x + 1)$$

Another Method of Factoring Trinomials

As an alternative to the trial-and-error method of factoring trinomials, we present the following method, which does not require as much trial and error. To use this new method, we must rewrite the original trinomial in such a way that the method of factoring by grouping can be applied.

To Factor $ax^2 + bx + c$

Step 1 Form the product ac.
Step 2 Find a pair of numbers whose product is ac and whose sum is b.
Step 3 Rewrite the polynomial to be factored so that the middle term bx is written as the sum of two terms whose coefficients are the two numbers found in Step 2.
Step 4 Factor by grouping.

EXAMPLE 11 Factor $3x^2 - 10x - 8$ using these steps.

SOLUTION The trinomial $3x^2 - 10x - 8$ has the form $ax^2 + bx + c$, where $a = 3$, $b = -10$, and $c = -8$.

Step 1 The product ac is $3(-8) = -24$.

Step 2 We need to find two numbers whose product is -24 and whose sum is -10. Let's list all the pairs of numbers whose product is -24 to find the pair whose sum is -10:

Product	Sum
$1(-24) = -24$	$1 + (-24) = -23$
$-1(24) = -24$	$-1 + 24 = 23$
$2(-12) = -24$	$2 + (-12) = -10$
$-2(12) = -24$	$-2 + 12 = 10$
$3(-8) = -24$	$3 + (-8) = -5$
$-3(8) = -24$	$-3 + 8 = 5$
$4(-6) = -24$	$4 + (-6) = -2$
$-4(6) = -24$	$-4 + 6 = 2$

As you can see, of all the pairs of numbers whose product is -24, only 2 and -12 have a sum of -10.

Step 3 We now rewrite our original trinomial so the middle term $-10x$ is written as the sum of $-12x$ and $2x$:

$$3x^2 - 10x - 8 = 3x^2 - 12x + 2x - 8$$

Step 4 Factoring by grouping, we have

$$3x^2 - 12x + 2x - 8 = 3x(x - 4) + 2(x - 4)$$
$$= (x - 4)(3x + 2)$$

You can see this method works by multiplying $x - 4$ and $3x + 2$ to get

$$3x^2 - 10x - 8$$

EXAMPLE 12   Factor: $9x^2 + 15x + 4$.

SOLUTION   In this case, $a = 9$, $b = 15$, and $c = 4$. The product ac is $9 \cdot 4 = 36$. Listing all the pairs of positive numbers whose product is 36 with their corresponding sums, we have:

Product	Sum
$1(36) = 36$	$1 + 36 = 37$
$2(18) = 36$	$2 + 18 = 20$
$3(12) = 36$	$3 + 12 = 15$
$4(9) = 36$	$4 + 9 = 13$
$6(6) = 36$	$6 + 6 = 12$

Notice that we list only positive numbers since both the sum and product we are looking for are positive. The numbers 3 and 12 are the numbers we are looking for. Their product is 36 and their sum is 15. We now rewrite the original polynomial $9x^2 + 15x + 4$ with the middle term written as $3x + 12x$. We then factor by grouping:

$$9x^2 + 15x + 4 = 9x^2 + 3x + 12x + 4$$
$$= 3x(3x + 1) + 4(3x + 1)$$
$$= (3x + 1)(3x + 4)$$

The polynomial $9x^2 + 15x + 4$ factors into the product $(3x + 1)(3x + 4)$.

EXAMPLE 13   Factor: $8x^2 - 2x - 15$.

SOLUTION   The product ac is $8(-15) = -120$. There are many pairs of numbers whose product is -120. We are looking for the pair whose sum is also -2. The numbers are -12 and 10. Writing $-2x$ as $-12x + 10x$ and then factoring by grouping, we have

$$8x^2 - 2x - 15 = 8x^2 - 12x + 10x - 15$$
$$= 4x(2x - 3) + 5(2x - 3)$$
$$= (2x - 3)(4x + 5)$$

Problem Set 1.6

Factor the greatest common factor from each of the following. (The answers in the back of the book all show greatest common factors whose coefficients are positive.)

1. $10x^3 - 15x^2$
2. $12x^5 + 18x^7$
3. $9y^6 + 18y^3$
4. $24y^4 - 8y^2$
5. $9a^2b - 6ab^2$
6. $30a^3b^4 + 20a^4b^3$
7. $21xy^4 + 7x^2y^2$
8. $14x^6y^3 - 6x^2y^4$
9. $3a^2 - 21a + 30$
10. $3a^2 - 3a - 6$
11. $4x^3 - 16x^2 - 20x$
12. $2x^3 - 14x^2 + 20x$
13. $5x(a - 2b) - 3y(a - 2b)$
14. $3a(x - y) - 7b(x - y)$
15. $3x^2(x + y)^2 - 6y^2(x + y)^2$
16. $10x^3(2x - 3y) - 15x^2(2x - 3y)$

17. $2x^2(x + 5) + 7x(x + 5) + 6(x + 5)$
18. $2x^2(x + 2) + 13x(x + 2) + 15(x + 2)$

Factor each of the following by grouping.

19. $3xy + 3y + 2ax + 2a$
20. $5xy^2 + 5y^2 + 3ax + 3a$
21. $x^2y + x + 3xy + 3$
22. $x^3y^3 + 2x^3 + 5x^2y^3 + 10x^2$
23. $x^2 - ax - bx + ab$
24. $ax - x^2 - bx + ab$
25. $ab + 5a - b - 5$
26. $x^2 - xy - ax + ay$

NOTE: Remember, confusion is part of the process of learning mathematics. If you are confused about some of the problems in this section, it is still important that you work through the assignment and not use confusion as an excuse for stopping before your assignment is finished.

Factor each of the following trinomials.

27. $x^2 + 7x + 12$	**28.** $x^2 - 7x + 12$
29. $x^2 - x - 12$	**30.** $x^2 + x - 12$
31. $y^2 + y - 6$	**32.** $y^2 - y - 6$
33. $16 - 6x - x^2$	**34.** $3 + 2x - x^2$
35. $12 + 8x + x^2$	**36.** $15 - 2x - x^2$
37. $x^2 + 3xy + 2y^2$	**38.** $x^2 - 5xy - 24y^2$
39. $a^2 + 3ab - 18b^2$	**40.** $a^2 - 8ab - 9b^2$
41. $x^2 - 2xa - 48a^2$	**42.** $x^2 + 14xa + 48a^2$
43. $x^2 - 12xb + 36b^2$	**44.** $x^2 + 10xb + 25b^2$

Factor completely, if possible. Be sure to factor out the greatest common factor first if it is other than 1.

45. $2x^2 + 7x - 15$	**46.** $2x^2 - 7x - 15$
47. $2x^2 + x - 15$	**48.** $2x^2 - x - 15$
49. $2x^2 - 13x + 15$	**50.** $2x^2 + 13x + 15$
51. $2x^2 - 11x + 15$	**52.** $2x^2 + 11x + 15$
53. $2x^2 + 7x + 15$	**54.** $2x^2 + x - 15$
55. $2 + 7a + 6a^2$	**56.** $2 - 7a + 6a^2$
57. $60y^2 - 15y - 45$	**58.** $72y^2 + 60y - 72$
59. $6x^4 - x^3 - 2x^2$	**60.** $3x^4 + 2x^3 - 5x^2$
61. $40r^3 - 120r^2 + 90r$	
62. $40r^3 + 200r^2 + 250r$	
63. $4x^2 - 11xy - 3y^2$	
64. $3x^2 + 19xy - 14y^2$	

65. $10x^2 - 3xa - 18a^2$
66. $9x^2 + 9xa - 10a^2$
67. $18a^2 + 3ab - 28b^2$
68. $6a^2 - 7ab - 5b^2$
69. $600 + 800t - 800t^2$
70. $200 - 600t - 350t^2$
71. $9y^4 + 9y^3 - 10y^2$
72. $4y^5 + 7y^4 - 2y^3$
73. $24a^2 - 2a^3 - 12a^4$
74. $60a^2 + 65a^3 - 20a^4$
75. $8x^4y^2 - 2x^3y^3 - 6x^2y^4$
76. $8x^4y^2 - 47x^3y^3 - 6x^2y^4$
77. $300x^4 + 1,000x^2 + 300$
78. $600x^4 - 100x^2 - 200$

Factor each of the following by first factoring out the greatest common factor, and then factoring the trinomial that remains.

79. $2x^2(x + 5) + 7x(x + 5) + 6(x + 5)$
80. $2x^2(x + 2) + 13x(x + 2) + 15(x + 2)$
81. $x^2(2x + 3) + 7x(2x + 3) + 10(2x + 3)$
82. $2x^2(x + 1) + 7x(x + 1) + 6(x + 1)$

In Words

83. One factor of the trinomial $a^2 + 260a + 2,500$ is $a + 10$. What is the other factor?

84. One factor of the trinomial $a^2 - 75a - 2,500$ is $a + 25$. What is the other factor?

85. **Compound Interest** If P dollars are placed in a savings account in which the rate of interest r is compounded yearly, then at the end of 1 year the amount of money in the account can be written as $P + Pr$. At the end of 2 years the amount of money in the account is

$$P + Pr + (P + Pr)r$$

Use factoring by grouping to show that this last expression can be written as $P(1 + r)^2$.

86. **Compound Interest** At the end of 3 years, the amount of money in the savings account in Problem 85 will be

$$P(1 + r)^2 + P(1 + r)^2r$$

Use factoring to show that this last expression can be written as $P(1 + r)^3$.

Special Factoring

To find the area of the large square in the margin, we can square the length of its side, giving us $(a + b)^2$. On the other hand, we can add the areas of the four smaller figures to arrive at the same result.

Since the area of the large square is the same whether we find it by squaring a side, or by adding the four smaller areas, we can write the following relationship:

$$(a + b)^2 = a^2 + 2ab + b^2$$

This is the formula for the square of a binomial that we found in Section 1.5. The figure gives us a geometric interpretation for one of the special multiplication formulas. We begin this section by looking at the special multiplication formulas from a factoring perspective.

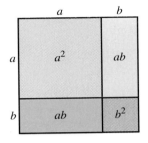

Perfect Square Trinomials

We previously listed some special products found in multiplying polynomials. Two of the formulas looked like this:

$$(a + b)^2 = a^2 + 2ab + b^2$$
$$(a - b)^2 = a^2 - 2ab + b^2$$

If we exchange the left and right sides of each formula, we have two special formulas for factoring:

$$a^2 + 2ab + b^2 = (a + b)^2$$
$$a^2 - 2ab + b^2 = (a - b)^2$$

The left side of each formula is called a **perfect square trinomial.** The right sides are binomial squares. Perfect square trinomials can always be factored using the usual methods for factoring trinomials. However, if we notice that the first and last terms of a trinomial are perfect squares, it is wise to see whether the trinomial factors as a binomial square before attempting to factor by the usual method.

EXAMPLE I   Factor: $x^2 - 6x + 9$.

SOLUTION   Since the first and last terms are perfect squares, we attempt to factor according to the preceding formulas:

$$x^2 - 6x + 9 = (x - 3)^2$$

If we expand $(x - 3)^2$, we have $x^2 - 6x + 9$, indicating that we have factored correctly.

EXAMPLES   Factor each of the following perfect square trinomials.

2. $16a^2 + 40ab + 25b^2 = (4a + 5b)^2$

3. $49 - 14t + t^2 = (7 - t)^2$

4. $9x^4 - 12x^2 + 4 = (3x^2 - 2)^2$

5. $(y + 3)^2 + 10(y + 3) + 25 = [(y + 3) + 5]^2 = (y + 8)^2$

EXAMPLE 6   Factor: $8x^2 - 24xy + 18y^2$.

SOLUTION   We begin by factoring the greatest common factor 2 from each term:

$$8x^2 - 24xy + 18y^2 = 2(4x^2 - 12xy + 9y^2)$$
$$= 2(2x - 3y)^2$$

The Difference of Two Squares

Recall the formula that results in the difference of two squares:

$$(a - b)(a + b) = a^2 - b^2$$

Writing this as a factoring formula, we have

$$a^2 - b^2 = (a - b)(a + b)$$

EXAMPLES   Each of the following is the difference of two squares. Use the formula $a^2 - b^2 = (a - b)(a + b)$ to factor each one.

7. $x^2 - 25 = x^2 - 5^2 = (x - 5)(x + 5)$

8. $49 - t^2 = 7^2 - t^2 = (7 - t)(7 + t)$

9. $81a^2 - 25b^2 = (9a)^2 - (5b)^2 = (9a - 5b)(9a + 5b)$

10. $4x^6 - 1 = (2x^3)^2 - 1^2 = (2x^3 - 1)(2x^3 + 1)$

11. $x^2 - \frac{4}{9} = x^2 - (\frac{2}{3})^2 = (x - \frac{2}{3})(x + \frac{2}{3})$

As our next example shows, the difference of two fourth powers can be factored as the difference of two squares.

EXAMPLE 12   Factor: $16x^4 - 81y^4$.

SOLUTION   The first and last terms are perfect squares. We factor according to the preceding formula:

$$16x^4 - 81y^4 = (4x^2)^2 - (9y^2)^2$$
$$= (4x^2 - 9y^2)(4x^2 + 9y^2)$$

Notice that the first factor is also the difference of two squares. Factoring completely, we have

$$16x^4 - 81y^4 = (2x - 3y)(2x + 3y)(4x^2 + 9y^2)$$

Here is another example of the difference of two squares.

EXAMPLE 13 Factor: $(x - 3)^2 - 25$.

SOLUTION This example has the form $a^2 - b^2$ where a is $x - 3$ and b is 5. We factor it according to the formula for the difference of two squares:

$$
\begin{aligned}
(x - 3)^2 - 25 &= (x - 3)^2 - 5^2 && \text{Write 25 as } 5^2. \\
&= [(x - 3) - 5][(x - 3) + 5] && \text{Factor.} \\
&= (x - 8)(x + 2) && \text{Simplify.}
\end{aligned}
$$

Notice in this example that we could have expanded $(x - 3)^2$, subtracted 25, and then factored to obtain the same result:

$$
\begin{aligned}
(x - 3)^2 - 25 &= x^2 - 6x + 9 - 25 && \text{Expand } (x - 3)^2. \\
&= x^2 - 6x - 16 && \text{Simplify.} \\
&= (x - 8)(x + 2) && \text{Factor.}
\end{aligned}
$$

EXAMPLE 14 Factor: $x^2 - 10x + 25 - y^2$.

SOLUTION Notice that the first three terms form a perfect square trinomial. That is, $x^2 - 10x + 25 = (x - 5)^2$. If we replace the first three terms with $(x - 5)^2$, the expression that results has the form $a^2 - b^2$. We can then factor as we did in Example 13:

$$
\begin{aligned}
x^2 - 10x + 25 - y^2 \\
= (x^2 - 10x + 25) - y^2 && \text{Group first three terms together.} \\
= (x - 5)^2 - y^2 && \text{This has the form } a^2 - b^2. \\
= [(x - 5) - y][(x - 5) + y] && \text{Factor according to the formula} \\
&& a^2 - b^2 = (a - b)(a + b) \\
= (x - 5 - y)(x - 5 + y) && \text{Simplify.}
\end{aligned}
$$

We could check this result by multiplying the two factors together. (You may want to do that to convince yourself that we have the correct result.)

EXAMPLE 15 Factor completely: $x^3 + 2x^2 - 9x - 18$.

SOLUTION We use factoring by grouping to begin and then factor the difference of two squares:

$$
\begin{aligned}
x^3 + 2x^2 - 9x - 18 &= x^2(x + 2) - 9(x + 2) \\
&= (x + 2)(x^2 - 9) \\
&= (x + 2)(x - 3)(x + 3)
\end{aligned}
$$

The Sum and Difference of Two Cubes

Here are the formulas for factoring the sum and difference of two cubes:

$$a^3 + b^3 = (a + b)(a^2 - ab + b^2)$$
$$a^3 - b^3 = (a - b)(a^2 + ab + b^2)$$

EXAMPLE 16   Verify the two formulas.

SOLUTION   We verify the formulas by multiplying the right sides and comparing the results with the left sides:

$$
\begin{array}{r}
a^2 - ab + b^2 \\
a \quad + b \\
\hline
a^3 - a^2b + ab^2 \\
a^2b - ab^2 + b^3 \\
\hline
a^3 \qquad\qquad + b^3
\end{array}
\qquad
\begin{array}{r}
a^2 + ab + b^2 \\
a \quad - b \\
\hline
a^3 + a^2b + ab^2 \\
- a^2b - ab^2 - b^3 \\
\hline
a^3 \qquad\qquad - b^3
\end{array}
$$

Here are some examples using the formulas for factoring the sum and difference of two cubes.

EXAMPLE 17   Factor: $64 + t^3$.

SOLUTION   The first term is the cube of 4 and the second term is the cube of t. Therefore,

$$
\begin{aligned}
64 + t^3 &= 4^3 + t^3 \\
&= (4 + t)(16 - 4t + t^2)
\end{aligned}
$$

EXAMPLE 18   Factor: $27x^3 + 125y^3$.

SOLUTION   Writing both terms as perfect cubes, we have

$$
\begin{aligned}
27x^3 + 125y^3 &= (3x)^3 + (5y)^3 \\
&= (3x + 5y)(9x^2 - 15xy + 25y^2)
\end{aligned}
$$

EXAMPLE 19   Factor: $a^3 - \frac{1}{8}$.

SOLUTION   The first term is the cube of a, while the second term is the cube of $\frac{1}{2}$:

$$
\begin{aligned}
a^3 - \tfrac{1}{8} &= a^3 - (\tfrac{1}{2})^3 \\
&= (a - \tfrac{1}{2})(a^2 + \tfrac{1}{2}a + \tfrac{1}{4})
\end{aligned}
$$

EXAMPLE 20   Factor: $x^6 - y^6$.

SOLUTION   We have a choice of how we want to write the two terms to begin. We can write the expression as the difference of two squares, $(x^3)^2 - (y^3)^2$, or as the difference of two cubes, $(x^2)^3 - (y^2)^3$. It is better to start with the difference of two squares if we have a choice:

$$x^6 - y^6 = (x^3)^2 - {}^{\backprime}(y^3)^2$$
$$= (x^3 - y^3)(x^3 + y^3)$$
$$= (x - y)(x^2 + xy + y^2)(x + y)(x^2 - xy + y^2)$$

Factor completely, using the sum and difference of two cubes.

Try this example again, writing the first line as the difference of two cubes instead of the difference of two squares. It will become apparent why it is better to use the difference of two squares.

Factoring: A General Review

We end this section by reviewing all the different methods of factoring we have covered. To begin, here is a list of the steps that can be used to factor polynomials of any type.

To Factor a Polynomial

Step 1 If the polynomial has a greatest common factor other than 1, then factor out the greatest common factor.

Step 2 If the polynomial has two terms (a binomial), then see if it is the difference of two squares, or the sum or difference of two cubes, and then factor accordingly. (*Note:* If it is the *sum* of two squares, it will not factor.)

Step 3 If the polynomial has three terms (a trinomial), then either it is a perfect square trinomial, which will factor into the square of a binomial, or it is not a perfect square trinomial, in which case you must use the trial-and-error method developed in Section 1.6.

Step 4 If the polynomial has more than three terms, try to factor it by grouping.

Step 5 As a final check, see if any of the factors you have written can be factored further. If you have overlooked a common factor, you can catch it here.

Here are some examples illustrating how we use these five steps. There are no new factoring problems here. The problems are all similar to the problems you have seen before, but they are not grouped according to type.

EXAMPLE 21 Factor: $2x^5 - 8x^3$.

SOLUTION First we check to see if the greatest common factor is other than 1. Since the greatest common factor is $2x^3$, we begin by factoring it out. Once we have done this, we notice that the binomial that remains is the difference of two squares, which we factor according to the formula $a^2 - b^2 = (a + b)(a - b)$.

$$2x^5 - 8x^3 = 2x^3(x^2 - 4)$$
Factor out the greatest common factor, $2x^3$.

$$= 2x^3(x + 2)(x - 2)$$
Factor the difference of two squares.

EXAMPLE 22   Factor: $3x^4 - 18x^3 + 27x^2$.

SOLUTION   Step 1 is to factor out the greatest common factor, $3x^2$. After we have done this, we notice that the trinomial that remains is a perfect square trinomial, which will factor as the square of a binomial:

$$3x^4 - 18x^3 + 27x^2 = 3x^2(x^2 - 6x + 9) \quad \text{Factor out } 3x^2.$$
$$= 3x^2(x - 3)^2 \quad \begin{array}{l} x^2 - 6x + 9 \text{ is the} \\ \text{square of } x - 3. \end{array}$$

EXAMPLE 23   Factor: $y^3 + 25y$.

SOLUTION   We begin by factoring out the y that is common to both terms. The binomial that remains after we have done this is the sum of two squares, which does not factor. So, after the first step, we are finished:

$$y^3 + 25y = y(y^2 + 25)$$

EXAMPLE 24   Factor: $6a^2 - 11a + 4$.

SOLUTION   Here we have a trinomial that does not have a greatest common factor other than 1. Since it is not a perfect square trinomial, we factor it by trial and error. Without showing all the different possibilities, here is the answer:

$$6a^2 - 11a + 4 = (3a - 4)(2a - 1)$$

EXAMPLE 25   Factor: $2x^4 + 16x$.

SOLUTION   This binomial has a greatest common factor of $2x$. The binomial that remains after the $2x$ has been factored from each term is the sum of two cubes, which we factor according to the formula $a^3 + b^3 = (a + b)(a^2 - ab + b^2)$.

$$2x^4 + 16x = 2x(x^3 + 8) \quad \text{Factor } 2x \text{ from each term.}$$
$$= 2x(x + 2)(x^2 - 2x + 4) \quad \text{The sum of two cubes}$$

EXAMPLE 26   Factor: $2ab^5 + 8ab^4 + 2ab^3$.

SOLUTION   The greatest common factor is $2ab^3$. We begin by factoring it from each term. After that, we find that the trinomial that remains cannot be factored further:

$$2ab^5 + 8ab^4 + 2ab^3 = 2ab^3(b^2 + 4b + 1)$$

EXAMPLE 27   Factor: $4x^2 - 6x + 2ax - 3a$.

SOLUTION   This polynomial has four terms, so we factor by grouping:

$$4x^2 - 6x + 2ax - 3a = 2x(2x - 3) + a(2x - 3)$$
$$= (2x - 3)(2x + a)$$

Problem Set

1.7

As you go through the assignment for this problem set, mark the problems you have difficulty with. That way you will know where to go when you study for an exam on this material.

Factor each perfect square trinomial.

1. $x^2 - 6x + 9$
2. $x^2 + 10x + 25$
3. $a^2 - 12a + 36$
4. $36 - 12a + a^2$
5. $25 - 10t + t^2$
6. $64 + 16t + t^2$
7. $4y^4 - 12y^2 + 9$
8. $9y^4 + 12y^2 + 4$
9. $16a^2 + 40ab + 25b^2$
10. $25a^2 - 40ab + 16b^2$
11. $\frac{1}{25} + \frac{1}{10}t^2 + \frac{1}{16}t^4$
12. $\frac{1}{9} - \frac{1}{3}t^3 + \frac{1}{4}t^6$
13. $(x + 2)^2 + 6(x + 2) + 9$
14. $(x + 5)^2 + 4(x + 5) + 4$

Factor completely.

15. $49x^2 - 64y^2$
16. $81x^2 - 49y^2$
17. $4a^2 - \frac{1}{4}$
18. $25a^2 - \frac{1}{25}$
19. $x^2 - \frac{9}{25}$
20. $x^2 - \frac{25}{36}$
21. $25 - t^2$
22. $64 - t^2$
23. $16a^4 - 81$
24. $81a^4 - 16b^4$
25. $x^2 - 10x + 25 - y^2$
26. $x^2 - 6x + 9 - y^2$
27. $a^2 + 8a + 16 - b^2$
28. $a^2 + 12a + 36 - b^2$
29. $x^3 + 2x^2 - 25x - 50$
30. $x^3 + 4x^2 - 9x - 36$
31. $2x^3 + 3x^2 - 8x - 12$
32. $3x^3 + 2x^2 - 27x - 18$
33. $4x^3 + 12x^2 - 9x - 27$
34. $9x^3 + 18x^2 - 4x - 8$

Factor each of the following as the sum or difference of two cubes.

35. $x^3 - y^3$
36. $x^3 + y^3$
37. $a^3 + 8$
38. $a^3 - 8$
39. $y^3 - 1$
40. $y^3 + 1$
41. $10r^3 - 1,250$
42. $10r^3 + 1,250$
43. $64 + 27a^3$
44. $27 - 64a^3$
45. $t^3 + \frac{1}{27}$
46. $t^3 - \frac{1}{27}$

The even-numbered problems are not necessarily similar to the odd-numbered problems that precede them in this problem set.

Factor each of the following polynomials completely, if possible. That is, once you are finished factoring, none of the factors you obtain should be factorable.

47. $x^2 - 81$
48. $x^2 - 18x + 81$
49. $x^2 + 2x - 15$
50. $15x^2 + 13x - 6$
51. $x^2y^2 + 2y^2 + x^2 + 2$
52. $21y^2 - 25y - 4$
53. $2a^3b + 6a^2b + 2ab$
54. $6a^2 - ab - 15b^2$
55. $x^2 + x + 1$
56. $x^2y + 3y + 2x^2 + 6$
57. $12a^2 - 75$
58. $18a^2 - 50$
59. $25 - 10t + t^2$
60. $t^2 + 4t + 4 - y^2$
61. $4x^3 + 16xy^2$
62. $16x^2 + 49y^2$
63. $x^3 + 5x^2 - 9x - 45$
64. $x^3 + 5x^2 - 16x - 80$
65. $x^2 + 49$
66. $16 - x^4$
67. $x^2(x - 3) - 14x(x - 3) + 49(x - 3)$
68. $x^2 + 3ax - 2bx - 6ab$
69. $8 - 14x - 15x^2$
70. $5x^4 + 14x^2 - 3$
71. $r^2 - \frac{1}{25}$
72. $27 - r^3$
73. $49x^2 + 9y^2$
74. $12x^4 - 62x^3 + 70x^2$
75. $100x^2 - 100x - 600$
76. $100x^2 - 100x - 1,200$
77. $3x^4 - 14x^2 - 5$
78. $8 - 2x - 15x^2$
79. $24a^5b - 3a^2b$
80. $18a^4b^2 - 24a^3b^3 + 8a^2b^4$
81. $64 - r^3$
82. $r^2 - \frac{1}{9}$
83. $20x^4 - 45x^2$
84. $16x^3 + 16x^2 + 3x$
85. $16x^5 - 44x^4 + 30x^3$
86. $16x^2 + 16x - 1$
87. $y^6 - 1$
88. $25y^7 - 16y^5$
89. $50 - 2a^2$
90. $4a^2 + 2a + \frac{1}{4}$
91. $x^2 - 4x + 4 - y^2$
92. $x^2 - 12x + 36 - b^2$

One Step Further

Assume m and n are positive integers and factor completely.

93. $x^{2n} - y^{2n}$
94. $x^{4n} - y^{4n}$
95. $x^{3n} - 8$
96. $27x^{3n} - 1$
97. $x^{3n} - y^{3n}$
98. $x^{3n} + y^{3m}$

REVIEW FOR CHAPTER 1

Chapter 1 Summary

NOTE: The numbers in brackets refer to the section(s) in which the topic can be found. In the margin we give brief examples of the topics being reviewed.

Examples

1. $10 + (2 \cdot 3^2 - 4 \cdot 2)$
$= 10 + (2 \cdot 9 - 4 \cdot 2)$
$= 10 + (18 - 8)$
$= 10 + 10$
$= 20$

To Use the Order of Operations Rule [1.1]

When evaluating a mathematical expression, we will perform the operations in the following order, beginning with the expression in the innermost parentheses or brackets and working our way out.

Step 1 Simplify all numbers with exponents, working from left to right if more than one of these numbers is present.
Step 2 Perform all multiplications and divisions left to right.
Step 3 Perform all additions and subtractions left to right.

2. If $A = \{0, 1, 2\}$ and
$B = \{2, 3\}$, then
$A \cup B = \{0, 1, 2, 3\}$
and $A \cap B = \{2\}$.

Sets [1.1]

A **set** is a collection of objects or things.

The **union** of two sets A and B, written $A \cup B$, is all the elements that are in A *or* in B.

The **intersection** of two sets A and B, written $A \cap B$, is the set consisting of all elements common to both A *and* B.

Set A is a **subset** of set B, written $A \subset B$, if all elements in set A are also in set B.

3. 5 is a counting number, a whole number, an integer, a rational number, and a real number.
$\frac{3}{4}$ is a rational number and a real number.
$\sqrt{2}$ is an irrational number and a real number.

$$\textbf{Counting numbers} = \{1, 2, 3, \ldots\}$$
$$\textbf{Whole numbers} = \{0, 1, 2, 3, \ldots\}$$
$$\textbf{Integers} = \{\ldots, -3, -2, -1, 0, 1, 2, 3, \ldots\}$$
$$\textbf{Rational numbers} = \left\{ \frac{a}{b} \,\middle|\, a \text{ and } b \text{ are integers, } b \neq 0 \right\}$$
$$\textbf{Irrational numbers} = \{x \,|\, x \text{ is real, but not rational}\}$$
$$\textbf{Real numbers} = \{x \,|\, x \text{ is rational or } x \text{ is irrational}\}$$
$$\textbf{Prime numbers} = \{2, 3, 5, 7, 11, \ldots\}$$
$$= \{x \,|\, x \text{ is a positive integer greater than 1 whose only positive divisors are itself and 1}\}$$

Properties of Real Numbers [1.2]

	FOR ADDITION	FOR MULTIPLICATION
Commutative	$a + b = b + a$	$ab = ba$
Associative	$a + (b + c) = (a + b) + c$	$a(bc) = (ab)c$

	FOR ADDITION	**FOR MULTIPLICATION**
Identity	$a + 0 = a$	$a \cdot 1 = a$
Inverse	$a + (-a) = 0$	$a\left(\dfrac{1}{a}\right) = 1$
Distributive		$a(b + c) = ab + ac$

4.
$$5 + 3 = 8$$
$$5 + (-3) = 2$$
$$-5 + 3 = -2$$
$$-5 + (-3) = -8$$

Addition [1.3]

To add two real numbers:

1. **With the same sign** Add their absolute values and attach their common sign.
2. **With opposite signs** Subtract the smaller absolute value from the larger. Attach the sign of the number whose absolute value is larger.

5.
$$6 - 2 = 6 + (-2) = 4$$
$$6 - (-2) = 6 + 2 = 8$$

Subtraction [1.3]

If a and b are any two real numbers, the **difference** of a and b, written $a - b$, is given by

$$a - b = a + (-b)$$

In other words, to subtract b, add the opposite of b.

6.
$$5(4) = 20$$
$$5(-4) = -20$$
$$-5(4) = -20$$
$$-5(-4) = 20$$

Multiplication [1.3]

To multiply two real numbers, multiply their absolute values.

1. If the two numbers have the *same* sign, attach a positive sign.
2. If the two numbers have *opposite* signs, attach a negative sign.

7.
$$\frac{12}{-3} = -4$$
$$\frac{-12}{-3} = 4$$

Division [1.3]

If a and b are any two real numbers, where $b \neq 0$, then the **quotient** of a and b, written $\dfrac{a}{b}$, is given by

$$\frac{a}{b} = a \cdot \left(\frac{1}{b}\right)$$

In other words, to divide by b, multiply by the reciprocal of b.

8. These statements demonstrate properties of exponents:
(a) $x^2 \cdot x^3 = x^{2+3} = x^5$
(b) $(x^2)^3 = x^{2 \cdot 3} = x^6$
(c) $(3x)^2 = 3^2 \cdot x^2 = 9x^2$
(d) $2^{-3} = \dfrac{1}{2^3} = \dfrac{1}{8}$

Properties of Exponents [1.4]

If a and b are real numbers and r and s are integers, then

Property 1 $a^r \cdot a^s = a^{r+s}$
Property 2 $(a^r)^s = a^{r \cdot s}$
Property 3 $(ab)^r = a^r \cdot b^r$
Property 4 $a^{-r} = \dfrac{1}{a^r}$ $\qquad (a \neq 0)$

REVIEW FOR CHAPTER I

(e) $\left(\dfrac{x}{5}\right)^2 = \dfrac{x^2}{5^2} = \dfrac{x^2}{25}$

Property 5 $\left(\dfrac{a}{b}\right)^r = \dfrac{a^r}{b^r}$ $(b \neq 0)$

(f) $\dfrac{x^7}{x^5} = x^{7-5} = x^2$

Property 6 $\dfrac{a^r}{a^s} = a^{r-s}$ $(a \neq 0)$

(g) $3^1 = 3$
  $3^0 = 1$

Property 7 $a^1 = a$
  $a^0 = 1$ $(a \neq 0)$

9. 49,800,000
  $= 4.98 \times 10^7$
  0.00462
  $= 4.62 \times 10^{-3}$

Scientific Notation [1.4]

A number is written in **scientific notation** if it is written as the product of a number between 1 and 10 and an integer power of 10. A number in scientific notation has the form

$$n \times 10^r$$

where $1 \leq n < 10$ and r is an integer.

10. $(3x^2 + 2x - 5)$
  $+ (4x^2 - 7x + 2)$
  $= 7x^2 - 5x - 3$

Operations with Polynomials [1.5]

To add two polynomials, combine the coefficients of similar terms.

11. $(3x - 5)(x + 2)$
  $= 3x^2 + 6x - 5x - 10$
  $= 3x^2 + x - 10$

To multiply two polynomials, multiply each term in the first polynomial by each term in the second polynomial.

12. The following are examples of the three special products:

$(x + 3)^2 = x^2 + 6x + 9$

$(5 - x)^2 = 25 - 10x + x^2$

$(x + 7)(x - 7) = x^2 - 49$

Special Products [1.5]

$$(a + b)^2 = a^2 + 2ab + b^2$$
$$(a - b)^2 = a^2 - 2ab + b^2$$
$$(a + b)(a - b) = a^2 - b^2$$

13. Factor completely:
(a) $3x^3 - 6x^2$
  $= 3x^2(x - 2)$
(b) $x^2 - 9$
  $= (x + 3)(x - 3)$
$x^3 - 8$
  $= (x - 2)(x^2 + 2x + 4)$
$x^3 + 27$
  $= (x + 3)(x^2 - 3x + 9)$
(c) $x^2 - 6x + 9$
  $= (x - 3)^2$
$6x^2 - 7x - 5$
  $= (2x + 1)(3x - 5)$

To Factor Polynomials in General [1.6, 1.7]

Step 1 If the polynomial has a greatest common factor other than 1, then factor out the greatest common factor.

Step 2 If the polynomial has two terms (a binomial), then see if it is the difference of two squares, or the sum or difference of two cubes, and then factor accordingly. Remember, if it is the sum of two squares, it will not factor.

Step 3 If the polynomial has three terms (a trinomial), then either it is a perfect square trinomial, which will factor into the square of a binomial, or it is not a perfect square trinomial, in which case you must use the trial-and-error method.

Step 4 If the polynomial has more than three terms, then try to factor it by grouping.

(d) $x^2 + ax + bx + ab$
  $= x(x + a) + b(x + a)$
  $= (x + a)(x + b)$

Step 5 As a final check, look to see if any of the factors you have written can be factored further. If you have overlooked a common factor, you can catch it here.

> **NOTE:** These common mistakes are not meant to confuse you. They are listed here because they are mistakes that many students make. If they look correct to you, then you may be one of the students that they are intended for. You must convince yourself that they are mistakes.

Common Mistakes

1. When we subtract one polynomial from another, it is common to forget to add the *opposite* of each term in the second polynomial. For example:

$$(6x - 5) - (3x + 4) = 6x - 5 - 3x + 4 \quad \text{Mistake}$$
$$= 3x - 1$$

This mistake occurs if the negative sign outside the second set of parentheses is not distributed over all terms inside the parentheses. To avoid this mistake, remember: The opposite of a sum is the sum of the opposites, or

$$-(3x + 4) = -3x + (-4)$$

2. Interpreting the square of a sum to be the sum of the squares. That is,

$$(x + y)^2 = x^2 + y^2 \quad \text{Mistake}$$

This can easily be shown as false by trying a couple of numbers for x and y. If $x = 4$ and $y = 3$, we have

$$(4 + 3)^2 = 4^2 + 3^2 \quad \text{Mistake}$$
$$7^2 = 16 + 9$$
$$49 = 25$$

There has obviously been a mistake. The correct formula for $(a + b)^2$ is

$$(a + b)^2 = a^2 + 2ab + b^2$$

Chapter 1 Review Problems

> **NOTE:** The numbers in brackets indicate the sections to which the problems correspond. If you are working these problems to prepare for a test, be sure to check your answers with the answers in the back of the book. Don't wait until you have worked all the problems, check them as you go. If you have made a mistake, find out what it is and correct it.

Translate into symbols. [1.1]

1. The sum of x and 2.

2. The difference of x and 2.

3. The quotient of x and 2.

4. Twice the sum of x and y.

REVIEW FOR CHAPTER I

Simplify each expression. [1.1]

5. $2 + 3 \cdot 5$

6. $9 \cdot 8 - 8 \cdot 7$

7. $3 + 2(5 - 2)$

8. $3 \cdot 4^2 - 2 \cdot 3^2$

Let $A = \{1, 3, 5\}$, $B = \{2, 4, 6\}$, $C = \{0, 1, 2, 3, 4\}$, and find each of the following. [1.1]

9. $A \cup B$

10. $\{x | x \in A \text{ and } x \notin C\}$

11. $B \cap C$

12. $A \cap B$

13. $\{x | x \in C \text{ and } x < 3\}$

14. $\{x | x \in A \text{ or } x \in B\}$

Factor each of the following completely. [1.1]

15. 231

16. 4,356

Reduce to lowest terms. [1.1]

17. $\dfrac{231}{275}$

18. $\dfrac{4,356}{5,148}$

For the set $\{-7, -4.2, -\sqrt{3}, 0, \frac{3}{4}, \pi, 5\}$ list all the elements that are in the following sets. [1.2]

19. Whole numbers

20. Integers

21. Rational numbers

22. Irrational numbers

Match each expression with the letter (or letters) of the appropriate property (or properties) in the list that follows. [1.2]

(a) Commutative property of addition

(b) Commutative property of multiplication

(c) Associative property of addition

(d) Associative property of multiplication

(e) Additive identity

(f) Multiplicative identity

(g) Additive inverse

(h) Multiplicative inverse

23. $x + 3 = 3 + x$

24. $(x + 2) + 3 = x + (2 + 3)$

25. $3(x + 4) = 3(4 + x)$

26. $(5x)y = x(5y)$

27. $(x + 2) + y = (x + y) + 2$

28. $3(1) = 3$

Write each expression without absolute value symbols; then simplify. [1.2]

29. $|-3|$

30. $-|-5|$

31. $|-7| - |-3|$

32. $2|-8| - 5|-2|$

Multiply. [1.2]

33. $\frac{3}{4} \cdot \frac{8}{5} \cdot \frac{5}{6}$

34. $(\frac{3}{4})^3$

35. $\frac{1}{4} \cdot 8$

36. $36(\frac{1}{6})^2$

Find the following sums and differences. [1.3]

37. $5 - 3$

38. $\frac{1}{3} - \frac{1}{4} - \frac{1}{6} - \frac{1}{12}$

39. $|-4| - |-3| + |-2|$

40. $6 - (-3) - 2 - 5$

Find the following products. [1.3]

41. $6(-7)$

42. $-3(5)(-2)$

43. $7(3x)$

44. $-3(2x)$

Apply the distributive property. [1.2, 1.3]

45. $-2(3x - 5)$

46. $-3(2x - 7)$

47. $-\frac{1}{2}(2x - 6)$

48. $-3(5x - 1)$

Divide. [1.3]

49. $-\frac{5}{8} \div \frac{3}{4}$

50. $-12 \div \frac{1}{3}$

51. $\frac{3}{5} \div 6$

52. $\frac{4}{7} \div (-2)$

Simplify each expression as much as possible. [1.3]

53. $6 + 3(-2)$

54. $-3(2) - 5(6)$

55. $8 - 2(6 - 10)$

56. $\dfrac{3(-4) - 8}{-5 - 5}$

57. $\dfrac{2(-3) - 5(4)}{6 - 8}$

58. $6 - (-2)\left[\dfrac{3(-4) - 8}{2(-5) + 6}\right]$

Simplify. [1.3]

59. $2(3x + 1) - 5$

60. $7 - 2(3y - 1) + 4y$

61. $4(3x - 1) - 5(6x + 2)$

62. $4(2a - 5) - (3a + 2)$

Simplify each of the following. [1.4]

63. 5^2

64. -5^2

65. $(\frac{3}{4})^2$

66. $(-1)^8$

67. 2^4

68. $x^3 \cdot x^7$

69. $(5x^3)^2$

70. $(2x^3y)^2(-2x^4y^2)^3$

Write with positive exponents and then simplify. [1.4]

71. 2^{-3}

72. $(-2)^{-3}$

73. $(\frac{2}{3})^{-2}$

74. $2^{-2} + 4^{-1}$

Write in scientific notation. [1.4]

75. 34,500,000

76. 0.0000529

Write in expanded form. [1.4]

77. 4.45×10^4

78. 4.45×10^{-4}

Simplify each expression. All answers should contain positive exponents only. [1.4]

79. $\dfrac{a^{-4}}{a^5}$

80. $\dfrac{2x^{-3}}{x^{-5}}$

81. $2^8 \cdot 2^{-5}$

82. $\dfrac{(2x^3)^2(-4x^5)}{8x^{-4}}$

83. $\dfrac{(4x^2)(-3x^3)^2}{(12x^{-2})^2}$

84. $\dfrac{x^n x^{3n}}{x^{4n-2}}$

Simplify each expression as much as possible. Write all answers in scientific notation. [1.4]

85. $(2 \times 10^3)(4 \times 10^{-5})$

86. $\dfrac{(600,000)(0.000008)}{(4,000)(3,000,000)}$

Simplify by combining similar terms. [1.5]

87. $(3x - 1) + (2x - 4) - (5x + 1)$

88. $(6x^2 - 3x + 2) - (4x^2 + 2x - 5)$

89. $(3x^2 - 4xy + 2y^2) - (4x^2 + 3xy + y^2)$

90. $(x^3 - x) - (x^2 + x) + (x^3 - 3) - (x^2 + 1)$

91. Subtract $3x + 1$ from $5x - 2$.

92. Subtract $2x^2 - 3x + 1$ from $3x^2 - 5x - 2$.

Simplify each expression. [1.5]

93. $-3[2x - 4(3x + 1)]$

94. $x - 6[2x + 4(x - 5)]$

Find the value of each polynomial when x is -2. [1.5]

95. $2x^2 - 3x + 1$

96. $x^3 - 2x^2 + 3x + 1$

Multiply. [1.5]

97. $3x(4x^2 - 2x + 1)$

98. $2a^2b^3(a^2 + 2ab + b^2)$

99. $(6 - y)(3 - y)$

100. $(2x^2 - 1)(3x^2 + 4)$

101. $2t(t + 1)(t - 3)$

102. $(x + 3)(x^2 - 3x + 9)$

103. $(2x - 3)(4x^2 + 6x + 9)$

104. $(x + 3)^2$

105. $(a^2 - 2)^2$

106. $(3x + 5)^2$

107. $(2a + 3b)^2$

108. $(x - \frac{1}{3})(x + \frac{1}{3})$

109. $(x - 1)^3$

110. $(x^m + 2)(x^m - 2)$

Factor out the greatest common factor. [1.6]

111. $6x^4y - 9xy^4 + 18x^3y^3$

112. $4x^2(x + y)^2 - 8y^2(x + y)^2$

Factor by grouping. [1.6]

113. $x^3y^3 + 5x^2y^3 + 2x^3 + 10x^2$

114. $ab - bx - x^2 + ax$

Factor completely. [1.6]

115. $x^2 - 5x + 6$

116. $x^2 - x - 6$

117. $2x^3 + 4x^2 - 30x$

118. $20a^2 - 41ab + 20b^2$

119. $6x^4 - 11x^3 - 10x^2$

120. $20a^2 + 37a + 15$

121. $24x^2y - 6xy - 45y$

122. $6y^4 - 11y^3 - 10y^2$

Factor completely. [1.7]

123. $x^2 - 10x + 25$

124. $9y^2 - 49$

125. $x^4 - 16$
126. $3a^4 + 18a^2 + 27$
127. $a^3 - 8$
128. $5x^3 + 30x^2y + 45xy^2$

129. $3a^3b - 27ab^3$
130. $x^2 - 10x + 25 - y^2$
131. $x^3 + 4x^2 - 9x - 36$
132. $x^3 + 5x^2 - 4x - 20$

Chapter I Test

NOTE: The numbers in brackets indicate the sections in which similar problems can be found. If you are working these problems as a practice test, give yourself 30–45 minutes to work all the problems. Don't check your answers until you are finished. When you do check your answers, divide the number of problems you worked correctly by the number of problems on the test, and then multiply the result by 100. This will give you a percent score for the practice test. Pay attention to the problems where you made mistakes. These are the problems that you need to study.

1. Write the following in symbols. [1.1]

> The difference of $2a$ and
> $3b$ is less than their sum.

2. If $A = \{1, 2, 3, 4\}$ and $B = \{2, 4, 6\}$, find $A \cup B$. [1.1]

3. Factor 770 into the product of prime factors. [1.1]

State the property or properties that justify each of the following. [1.2]

4. $4 + x = x + 4$ **5.** $5(1) = 5$

Simplify each of the following as much as possible. [1.3]

6. $5(-4) + 1$
7. $3(2 - 4)^3 - 5(2 - 7)^2$
8. $\dfrac{-4(-1) - (-10)}{5 - (-2)}$
9. $-\dfrac{3}{8} + \dfrac{5}{12} - \left(\dfrac{-7}{9}\right)$

10. $-\frac{1}{2}(8x)$
11. $-4(3x + 2) + 7x$
12. Subtract $\frac{3}{4}$ from the product of -4 and $\frac{7}{16}$, and simplify. [1.3]

Simplify. (Assume all variables are nonzero.) [1.4]

13. $x^4 \cdot x^7 \cdot x^{-3}$ **14.** 2^{-5}

15. $\dfrac{a^{-5}}{a^{-7}}$ **16.** $\dfrac{(2ab^3)^{-2}(a^4b^{-3})}{(a^{-4}b^3)^4(2a^{-2}b^2)^{-3}}$

17. Write 6,530,000 in scientific notation. [1.4]

18. Perform the indicated operations and write your answer in scientific notation. [1.4]
$$\frac{(6 \times 10^{-4})(4 \times 10^9)}{8 \times 10^{-3}}$$

Simplify. [1.5]
19. $(\frac{3}{4}x^3 - x^2 - \frac{3}{2}) - (\frac{1}{4}x^2 + 2x - \frac{1}{2})$
20. $3 - 4[2x - 3(x + 6)]$

Multiply. [1.5]
21. $(3y - 7)(2y + 5)$
22. $(2x - 5)(x^2 + 4x - 3)$
23. $(8 - 3t^3)^2$
24. $(1 - 6y)(1 + 6y)$

Factor completely. [1.6, 1.7]
25. $12x^4 + 26x^2 - 10$
26. $16a^4 - 81y^4$
27. $7ax^2 - 14ay - b^2x^2 + 2b^2y$
28. $t^3 + \frac{1}{8}$
29. $x^2 - 10x + 25 - b^2$
30. $81 - x^4$

Equations and Inequalities in One Variable

Contents

Introduction

The topics we cover in this course are interdependent. Each topic is linked to other topics, which are linked to still more topics. The links form chains that hold all of mathematics together. The diagram below shows the relationships that exist among some of the major topics we will cover. Some topics are linked to other topics, and many topics are linked to the real world by applications.

The World of Mathematics

Topic

Arithmetic with real numbers

Link
Variables x and y

Topic

Equations in one and two unknowns

Link
Rectangular coordinate system

Topic

Geometric objects: points, lines, conic sections, and other curves

Application

You have $20 in your checkbook, and you write a check for $30. What is your new balance?

Application

Find the sales tax on a total purchase of $3,205.00 if the sales tax rate is 7.25%.

Application

Find the largest box that can be built from a rectangular piece of cardboard 11 inches wide and 17 inches long.

Overview

In this chapter you will learn the basic steps used to solve linear equations and inequalities in one variable. You will also learn to solve equations by factoring. The methods we develop here will be used again and again throughout the rest of the book. Here is a list of the more important concepts needed to begin this chapter:

1. You must know how to add, subtract, multiply, and divide positive and negative numbers.
2. You should know the commutative, associative, and distributive properties.
3. You should understand that opposites add to 0 and reciprocals multiply to 1.
4. You must know the definition of absolute value.
5. You should know that the least common denominator for a set of fractions is divisible by each of the denominators used to find it.
6. You must be able to simplify expressions containing exponents.
7. You must be able to add, subtract, multiply, and factor polynomials.

Study Skills

If you have successfully completed Chapter 1, then you have made a good start at developing the study skills necessary to succeed in all math classes. Some of the study skills for this chapter are a continuation of the skills from Chapter 1, while others are new to this chapter.

1. **Continue to set and keep a schedule** Sometimes I find students do well in Chapter 1 and then become overconfident. They will begin to put in less time with their homework. Don't fall into this trap. Keep to the same schedule.
2. **Increase effectiveness** You want to become more and more effective with the time you spend on your homework. Increase those activities that are the most beneficial and decrease those that have not given you the results you want.
3. **List difficult problems** Begin to make lists of problems that give you the most difficulty. These are the problems in which you are repeatedly making mistakes.
4. **Begin to develop confidence with word problems** It seems that the main difference between people who are good at working word problems and those who are not is confidence. People with confidence know that no matter how long it takes them, they will eventually be able to solve the problem. Those without confidence begin by saying to themselves, "I'll never be able to work this problem." If you are in this second category, then instead of telling yourself that you can't do word problems, decide to do whatever it takes to master them. The more word problems you work, the better you will become at it.

Many of my students keep a notebook that contains everything they need for the course: class notes, homework, quizzes, tests, and research projects. A three-ring binder with tabs is ideal. Organize your notebook so that you can easily get to any item you wish to look at.

SECTION 2.1

Linear and Quadratic Equations in One Variable

A **linear equation in one variable** is any equation that can be put in the form

$$ax + b = c$$

where a, b, and c are constants and $a \neq 0$. For example, each of the equations

$$5x + 3 = 2 \qquad 2x = 7 \qquad 2x + 5 = 0$$

is linear because each can be put in the form $ax + b = c$. In the first equation, $5x$, 3, and 2 are called **terms** of the equation; $5x$ is a variable term; 3 and 2 are constant terms.

> **DEFINITION**
>
> The **solution set** for an equation is the set of all numbers that, when used in place of the variable, make the equation a true statement.

EXAMPLE 1   The solution set for $2x - 3 = 9$ is $\{6\}$, since replacing x with 6 makes the equation a true statement.

$$\text{If} \quad x = 6:$$
$$\text{then} \qquad 2x - 3 = 9$$
$$\text{becomes} \quad 2(\mathbf{6}) - 3 = 9$$
$$12 - 3 = 9$$
$$9 = 9 \quad \text{A true statement}$$

> **DEFINITION**
>
> Two or more equations with the same solution set are called **equivalent equations.**

EXAMPLE 2   The equations $2x - 5 = 9$, $x - 1 = 6$, and $x = 7$ are all equivalent equations since the solution set for each is $\{7\}$.

Properties of Equality

The first property states that adding the same quantity to both sides of an equation preserves equality. Or, more importantly, adding the same amount to both sides of an equation *never changes* the solution set. This property is called the **addition property of equality** and is stated in symbols as follows.

ADDITION PROPERTY OF EQUALITY

For any three algebraic expressions A, B, and C,

$$\text{if} \qquad A = B$$
$$\text{then} \quad A + C = B + C$$

In Words: Adding the same quantity to both sides of an equation will not change the solution.

The second new property is called the **multiplication property of equality** and is stated like this.

MULTIPLICATION PROPERTY OF EQUALITY

For any three algebraic expressions A, B, and C, where $C \neq 0$,

$$\text{if} \qquad A = B$$
$$\text{then} \quad AC = BC$$

In Words: Multiplying both sides of an equation by the same nonzero quantity will not change the solution.

Note: Since subtraction is defined in terms of addition, and division is defined in terms of multiplication, we do not need to introduce separate properties for subtraction and division. The solution set for an equation will never be changed by subtracting the same amount from both sides or by dividing both sides by the same nonzero quantity.

Linear Equations

The following examples illustrate how we use the properties from Chapter 1 along with the addition property of equality and the multiplication property of equality to solve linear equations.

EXAMPLE 3   Solve: $\dfrac{3}{4}x + 5 = -4$.

SOLUTION   We begin by adding -5 to both sides of the equation. Once this has been done, we multiply both sides by the reciprocal of $\frac{3}{4}$, which is $\frac{4}{3}$.

$$\frac{3}{4}x + 5 = -4$$

$$\frac{3}{4}x + 5 + (-5) = -4 + (-5) \qquad \text{Add } -5 \text{ to both sides.}$$

$$\frac{3}{4}x = -9$$

$$\frac{4}{3}\left(\frac{3}{4}x\right) = \frac{4}{3}(-9) \qquad \text{Multiply both sides by } \tfrac{4}{3}.$$

$$x = -12 \qquad \tfrac{4}{3}(-9) = \tfrac{4}{3}(-\tfrac{9}{1}) = -\tfrac{36}{3} = -12 \qquad \blacktriangle$$

Our next example involves solving an equation that has variable terms on both sides of the equal sign.

EXAMPLE 4   Find the solution set for $3a - 5 = -6a + 1$.

SOLUTION   To solve for a we must isolate it on one side of the equation. Let's decide to isolate a on the left side by adding $6a$ to both sides of the equation.

$$3a - 5 = -6a + 1$$
$$3a + 6a - 5 = -6a + 6a + 1 \quad \text{Add } 6a \text{ to both sides.}$$
$$9a - 5 = 1$$
$$9a - 5 + 5 = 1 + 5 \qquad \text{Add } 5 \text{ to both sides.}$$
$$9a = 6$$
$$\frac{1}{9}(9a) = \frac{1}{9}(6) \qquad \text{Multiply both sides by } \tfrac{1}{9}.$$
$$a = \frac{2}{3} \qquad \tfrac{1}{9}(6) = \tfrac{6}{9} = \tfrac{2}{3} \qquad \blacktriangle$$

Note: From Chapter 1 we know that multiplication by a number and division by its reciprocal always produce the same result. Because of this fact, instead of multiplying each side of our equation by $\frac{1}{9}$, we could just as easily divide each side by 9. If we did so, the last two lines in our solution would look like this:

$$\frac{9a}{9} = \frac{6}{9}$$

$$a = \frac{2}{3}$$

There will be times when we solve equations and end up with a negative sign in front of the variable. The next example shows how to handle this situation.

EXAMPLE 5   Solve each equation.
(a) $-x = 4$ (b) $-y = -8$

SOLUTION   Neither equation can be considered solved because of the negative sign in front of the variable. To eliminate the negative signs we simply multiply both sides of each equation by -1.

(a) $-x = 4$ (b) $-y = -8$
$$-1(-x) = -1(4) \qquad\qquad -1(-y) = -1(-8) \quad \text{Multiply each side}$$
$$x = -4 \qquad\qquad\qquad\qquad y = 8 \qquad\quad \text{by } -1. \qquad \blacktriangle$$

The next example involves fractions. The least common denominator, which is the smallest expression that is divisible by each of the denominators, can be used with the multiplication property of equality to simplify equations containing fractions.

EXAMPLE 6   Solve: $\dfrac{2}{3}x + \dfrac{1}{2} = -\dfrac{3}{8}$.

SOLUTION   We can solve this equation by working with fractions, or we can begin by eliminating the fractions. Let's use both methods.

Method 1   Working with the fractions:

$$\frac{2}{3}x + \frac{1}{2} + \left(-\frac{1}{2}\right) = -\frac{3}{8} + \left(-\frac{1}{2}\right) \quad \text{Add } -\tfrac{1}{2} \text{ to each side.}$$

$$\frac{2}{3}x = -\frac{7}{8} \qquad\qquad -\tfrac{3}{8} + (-\tfrac{1}{2}) = -\tfrac{3}{8} + (-\tfrac{4}{8}) = -\tfrac{7}{8}$$

$$\frac{3}{2}\left(\frac{2}{3}x\right) = \frac{3}{2}\left(-\frac{7}{8}\right) \qquad \text{Multiply each side by } \tfrac{3}{2}.$$

$$x = -\frac{21}{16}$$

Method 2   Eliminating the fractions in the beginning: Our original equation has denominators of 3, 2, and 8. The least common denominator, abbreviated LCD, for these three denominators is 24, and it has the property that all three denominators will divide it evenly. Therefore, if we multiply both sides of our equation by 24, each denominator will divide into 24, and we will be left with an equation that does not contain any denominators other than 1:

$$24\left(\frac{2}{3}x + \frac{1}{2}\right) = 24\left(-\frac{3}{8}\right) \qquad \begin{array}{l}\text{Multiply each side by the} \\ \text{LCD } 24.\end{array}$$

$$24\left(\frac{2}{3}x\right) + 24\left(\frac{1}{2}\right) = 24\left(-\frac{3}{8}\right) \qquad \begin{array}{l}\text{Distributive property on} \\ \text{the left side}\end{array}$$

$$16x + 12 = -9 \qquad \text{Multiply.}$$

$$16x = -21 \qquad \text{Add } -12 \text{ to each side.}$$

$$x = -\frac{21}{16} \qquad \text{Multiply each side by } \tfrac{1}{16}.$$

As the third line above indicates, multiplying each side of the equation by the LCD eliminates all the fractions from the equation. Both methods yield the same solution.

Check   To check our solution, we substitute $x = -\frac{21}{16}$ back into our original equation to obtain

Note: We are placing question marks over the equal signs because we don't know yet if the expressions on the left will be equal to the expressions on the right.

$$\frac{2}{3}\left(-\frac{21}{16}\right) + \frac{1}{2} \stackrel{?}{=} -\frac{3}{8}$$

$$-\frac{7}{8} + \frac{1}{2} \stackrel{?}{=} -\frac{3}{8}$$

$$-\frac{7}{8} + \frac{4}{8} \stackrel{?}{=} -\frac{3}{8}$$

$$-\frac{3}{8} = -\frac{3}{8}$$

As we can see, our solution checks.

EXAMPLE 7 Solve the equation $0.06x + 0.05(10{,}000 - x) = 560$.

SOLUTION We can solve the equation in its original form by working with the decimals, or we can eliminate the decimals first by using the multiplication property of equality and then solve the resulting equation. Here are both methods.

Method 1 Working with the decimals:

$0.06x + 0.05(10{,}000 - x) = 560$	Original equation
$0.06x + 0.05(10{,}000) - 0.05x = 560$	Distributive property
$0.01x + 500 = 560$	Simplify the left side.
$0.01x + 500 + (-500) = 560 + (-500)$	Add -500 to each side.
$0.01x = 60$	
$\dfrac{0.01x}{0.01} = \dfrac{60}{0.01}$	Divide each side by **0.01**.
$x = 6{,}000$	

Method 2 Eliminating the decimals in the beginning: To move the decimal point two places to the right in $0.06x$ and 0.05, we multiply each side of the equation by 100.

$0.06x + 0.05(10{,}000 - x) = 560$	Original equation
$0.06x + 500 - 0.05x = 560$	Distributive property
$\mathbf{100}(0.06x) + \mathbf{100}(500) - \mathbf{100}(0.05x) = \mathbf{100}(560)$	Multiply each side by **100**.
$6x + 50{,}000 - 5x = 56{,}000$	
$x + 50{,}000 = 56{,}000$	Simplify the left side.
$x = 6{,}000$	Add $-50{,}000$ to each side.

Using either method, the solution to our equation is 6,000.

Check We check our work (to be sure we have not made a mistake in applying the properties or in arithmetic) by substituting 6,000 into our original equation and simplifying each side of the result separately, as shown at the top of the next page.

$$0.06(\mathbf{6,000}) + 0.05(10,000 - \mathbf{6,000}) \stackrel{?}{=} 560$$
$$0.06(6,000) + 0.05(4,000) \stackrel{?}{=} 560$$
$$360 + 200 \stackrel{?}{=} 560$$
$$560 = 560 \quad \text{A true statement}$$

Here is a list of steps to use as a guideline for solving linear equations in one variable.

Strategy for Solving Linear Equations in One Variable

Step 1 (a) Use the distributive property to separate terms, if necessary.
(b) If fractions are present, consider multiplying both sides by the LCD to eliminate the fractions. If decimals are present, consider multiplying both sides by a power of 10 to clear the equation of decimals.
(c) Combine similar terms on each side of the equation.

Step 2 Use the addition property of equality to get all variable terms on one side of the equation and all constant terms on the other side. A **variable term** is a term that contains the variable (for example, $5x$). A **constant term** is a term that does not contain the variable (the number 3, for example).

Step 3 Use the multiplication property of equality to get x (that is, $1x$) by itself on one side of the equation.

Step 4 Check your solution in the original equation to be sure that you have not made a mistake in the solution process.

As you will see as you work through the problems in the problem set, it is not always necessary to use all four steps when solving equations. The number of steps used depends upon the equation. In Example 8 there are no fractions or decimals in the original equation, so Step 1b will not be used.

EXAMPLE 8 Solve: $3(2y - 1) + y = 5y + 3$.

SOLUTION Applying the steps outlined above, we have

Step 1 (a) $\quad 3(2y - 1) + y = 5y + 3 \quad$ Distributive property
$$6y - 3 + y = 5y + 3$$
(c) $\quad\quad 7y - 3 = 5y + 3 \quad$ Simplify.

Step 2 $\quad 7y + (-5y) - 3 = 5y + (-5y) + 3 \quad$ Add $-5y$ to both sides.
$$2y - 3 = 3$$
$$2y - 3 + 3 = 3 + 3 \quad \text{Add } +3 \text{ to both sides.}$$
$$2y = 6$$

Step 3
$$\frac{1}{2}(2y) = \frac{1}{2}(6) \quad \text{Multiply by } \tfrac{1}{2}.$$
$$y = 3$$

Step 4 *Check* When $y = 3$:

the equation $3(2y - 1) + y = 5y + 3$

becomes $3(2 \cdot 3 - 1) + 3 \overset{?}{=} 5 \cdot 3 + 3$

$$3(5) + 3 \overset{?}{=} 15 + 3$$

$$18 = 18 \qquad \text{A true statement}$$

Our solution checks in the original equation.

EXAMPLE 9   Solve the equation $8 - 3(4x - 2) + 5x = 35$.

SOLUTION   We must begin by distributing the -3 across the quantity $4x - 2$. After we have simplified the left side of our equation, we apply the addition property and the multiplication property. In this example, we will show only the result:

It would be a mistake to subtract 3 from 8 first, since the rule for order of operations indicates we are to do multiplication before subtraction.

Step 1 (a) $8 - 3(4x - 2) + 5x = 35$ Original equation

$$8 - 12x + 6 + 5x = 35 \quad \text{Distributive property}$$

(c) $-7x + 14 = 35$ Simplify.

Step 2 $-7x = 21$ Add -14 to each side.

Step 3 $x = -3$ Multiply by $-\tfrac{1}{7}$.

Step 4   When x is replaced by -3 in the original equation, a true statement results. Therefore, -3 is the solution to our equation.

Quadratic Equations

In the rest of this section we will use our knowledge of factoring to solve equations. Most of the equations we will solve in this section are quadratic equations. Here is the definition of a **quadratic equation.**

DEFINITION

Any equation that can be written in the form

$$ax^2 + bx + c = 0$$

where a, b, and c are constants and a is not 0 ($a \neq 0$), is called a **quadratic equation.** The form $ax^2 + bx + c = 0$ is called **standard form** for quadratic equations.

Each of the following is a quadratic equation:

$$2x^2 = 5x + 3 \qquad 5x^2 = 75 \qquad 4x^2 - 3x + 2 = 0$$

The third equation is clearly a quadratic equation since it is in standard form. (Notice that a is 4, b is -3, and c is 2.) The first two equations are also quadratic because they could be put in the form $ax^2 + bx + c = 0$ by using the addition property of equality.

Notation: For a quadratic equation written in standard form, the first term, ax^2, is called the **quadratic term;** the second term, bx, is the **linear term;** and the last term, c, is called the **constant term.**

In the past we have noticed that the number 0 is a special number. There is another property of 0 that is the key to solving quadratic equations. It is called the **zero-factor property.**

ZERO-FACTOR PROPERTY

For all real numbers r and s,

$$r \cdot s = 0 \quad \text{if and only if} \quad r = 0 \quad \text{or} \quad s = 0 \quad \text{(or both)}$$

EXAMPLE 10 Solve: $x^2 - 2x - 24 = 0$.

SOLUTION We begin by factoring the left side as $(x - 6)(x + 4)$ and get

$$(x - 6)(x + 4) = 0$$

Now both $(x - 6)$ and $(x + 4)$ represent real numbers. We notice that their product is 0. By the zero-factor property, one or both of them must be 0:

$$x - 6 = 0 \quad \text{or} \quad x + 4 = 0$$

We have used factoring and the zero-factor property to rewrite our original quadratic equation as two linear equations connected by the word *or*. Completing the solution, we solve the two linear equations:

$$x - 6 = 0 \quad \text{or} \quad x + 4 = 0$$
$$x = 6 \quad \text{or} \quad x = -4$$

We check our solutions in the original equation as follows:

Check $x = 6$:   *Check* $x = -4$:

$6^2 - 2(6) - 24 \stackrel{?}{=} 0$ $(-4)^2 - 2(-4) - 24 \stackrel{?}{=} 0$

$36 - 12 - 24 \stackrel{?}{=} 0$ $16 + 8 - 24 \stackrel{?}{=} 0$

$0 = 0$ $0 = 0$

In both cases the result is a true statement, which means that both 6 and -4 are solutions to the original equation.

Although the next equation is not quadratic, the method we use to solve it is similar.

EXAMPLE 11   Solve: $\frac{1}{3} x^3 = \frac{5}{6} x^2 + \frac{1}{2} x$.

SOLUTION   We can simplify our work if we clear the equation of fractions. Multiplying both sides by the LCD, 6, we have

$$6 \cdot \frac{1}{3} x^3 = 6 \cdot \frac{5}{6} x^2 + 6 \cdot \frac{1}{2} x$$

$$2x^3 = 5x^2 + 3x$$

Next we add $-5x^2$ and $-3x$ to each side so that the right side will become 0:

$$2x^3 - 5x^2 - 3x = 0 \quad \text{Standard form}$$

We factor the left side:

$$x(2x^2 - 5x - 3) = 0 \quad \text{Factor out the greatest common factor.}$$

$$x(2x + 1)(x - 3) = 0 \quad \text{Continue factoring.}$$

Using the zero-factor property to set each factor to 0, we have

$$x = 0 \qquad \text{or} \qquad 2x + 1 = 0 \qquad \text{or} \qquad x - 3 = 0$$

Note: The checks for Examples 11–14 are left to the student.

Solving each of the resulting equations, we have

$$x = 0 \qquad \text{or} \qquad x = -\frac{1}{2} \qquad \text{or} \qquad x = 3$$

To generalize the preceding examples, here are the steps used in solving a quadratic equation by factoring.

> ### To Solve a Quadratic Equation by Factoring
>
> **Step 1** Write the equation in standard form.
> **Step 2** Factor the left side.
> **Step 3** Use the zero-factor property to set each factor equal to 0.
> **Step 4** Solve the resulting linear equations.
> **Step 5** Check the solutions in the original equation.

EXAMPLE 12   Solve: $100x^2 = 300x$.

SOLUTION   We begin by writing the equation in standard form and factoring:

$$100x^2 = 300x$$

$$100x^2 - 300x = 0 \qquad \text{Standard form}$$

$$100x(x - 3) = 0 \qquad \text{Factor}$$

Using the zero-factor property to set each factor to 0, we have

$$100x = 0 \qquad \text{or} \qquad x - 3 = 0$$

$$x = 0 \qquad \text{or} \qquad x = 3$$

The two solutions are 0 and 3.

EXAMPLE 13 Solve: $(x - 2)(x + 1) = 4$.

SOLUTION We begin by multiplying the two factors on the left side. (Notice that it would be incorrect to set each of the factors on the left side equal to 4. The fact that the product is 4 does not imply that either of the factors must be 4.)

$$(x - 2)(x + 1) = 4$$
$$x^2 - x - 2 = 4 \qquad \text{Multiply the left side.}$$
$$x^2 - x - 6 = 0 \qquad \text{Standard form}$$
$$(x - 3)(x + 2) = 0 \qquad \text{Factor.}$$
$$x - 3 = 0 \quad \text{or} \quad x + 2 = 0 \qquad \text{Zero-factor property}$$
$$x = 3 \quad \text{or} \qquad x = -2$$

EXAMPLE 14 Solve for x: $x^3 + 2x^2 - 9x - 18 = 0$.

SOLUTION We factored the left side of this equation previously in Section 1.7. We start with factoring by grouping:

$$x^3 + 2x^2 - 9x - 18 = 0$$
$$\left. \begin{array}{l} x^2(x + 2) - 9(x + 2) = 0 \\ (x + 2)(x^2 - 9) = 0 \end{array} \right\} \quad \begin{array}{l} \text{Factoring by} \\ \text{grouping} \end{array}$$
$$(x + 2)(x - 3)(x + 3) = 0 \qquad \begin{array}{l} \text{The difference of two} \\ \text{squares} \end{array}$$
$$x + 2 = 0 \quad \text{or} \quad x - 3 = 0 \quad \text{or} \quad x + 3 = 0 \qquad \text{Set factors to 0.}$$
$$x = -2 \quad \text{or} \quad x = 3 \quad \text{or} \qquad x = -3$$

We have three solutions: -2, 3, and -3.

Problem Set

2.1

Solve each of the following linear equations.

1. $2x - 4 = 6$

2. $3x - 5 = 4$

3. $-3 - 4x = 15$

4. $-8 - 5x = -6$

5. $-300y + 100 = 500$

6. $-20y + 80 = 30$

7. $-\frac{3}{5}a + 2 = 8$

8. $-\frac{5}{3}a + 3 = 23$

9. $9 - \frac{3}{4}t = 12$

10. $3 - \frac{2}{3}t = 1$

11. $-x = 2$

12. $-x = \frac{1}{2}$

13. $-a = -\frac{3}{4}$

14. $-a = -5$

15. $7y - 4 = 2y + 11$

16. $8y - 2 = 6y - 10$

17. $5y - 2 + 4y = 2y + 12$

18. $7y - 3 + 2y = 7y - 9$

19. $5(y + 2) - 4(y + 1) = 3$

20. $6(y - 3) - 5(y + 2) = 8$

21. $6 - 7(m - 3) = -1$

22. $3 - 5(2m - 5) = -2$

23. $5 = 7 - 2(3x - 1) + 4x$

24. $20 = 8 - 5(2x - 3) + 4x$

25. $10 - 4(2x + 1) - (3x - 4) = -9x + 4 - 4x$

26. $7 - 2(3x + 5) - (2x - 3) = -5x + 3 - 2x$

27. $\frac{1}{2}x + \frac{1}{4} = \frac{1}{3}x + \frac{5}{4}$

28. $\frac{2}{3}x - \frac{3}{4} = \frac{1}{6}x + \frac{21}{4}$

29. $\frac{1}{2}y - \frac{2}{7} = \frac{1}{7}y + \frac{11}{14}$

30. $\frac{1}{2}y - \frac{1}{8} = \frac{3}{8}y - \frac{5}{8}$

31. $0.08x + 0.09(9{,}000 - x) = 750$
32. $0.08x + 0.09(9{,}000 - x) = 500$
33. $0.42 - 0.18x = 0.48x - 0.24$
34. $0.3 - 0.12x = 0.18x + 0.06$

Solve each equation.

NOTE: Remember, you should check your work by substituting your solutions into the equation in place of the variable.

35. $x^2 - 5x - 6 = 0$ **36.** $x^2 + 5x - 6 = 0$
37. $x^3 - 5x^2 + 6x = 0$
38. $x^3 + 5x^2 + 6x = 0$
39. $60x^2 - 130x + 60 = 0$
40. $90x^2 + 60x - 80 = 0$
41. $100x^4 = 400x^3 + 2{,}100x^2$
42. $100x^4 = -400x^3 + 2{,}100x^2$
43. $\frac{1}{5}y^2 - 2 = -\frac{3}{10}y$ **44.** $\frac{1}{2}y^2 + \frac{5}{3} = \frac{17}{6}y$
45. $9x^2 - 12x = 0$ **46.** $4x^2 + 4x = 0$
47. $0.02r + 0.01 = 0.15r^2$
48. $0.02r - 0.01 = -0.08r^2$
49. $-100x = 10x^2$ **50.** $800x = 100x^2$
51. $(x + 6)(x - 2) = -7$
52. $(x - 7)(x + 5) = -20$
53. $(x + 1)^2 = 3x + 7$
54. $(x + 2)^2 = 9x$
55. $x^3 + 3x^2 - 4x - 12 = 0$
56. $x^3 + 5x^2 - 4x - 20 = 0$
57. $2x^3 + 3x^2 - 8x - 12 = 0$
58. $3x^3 + 2x^2 - 27x - 18 = 0$

Solve each equation.

NOTE: Some of the equations are linear, and some are not. You must be able to identify the type of equation you are trying to solve before you can solve it. You may want to write the equation type (linear, quadratic, or other) on your homework, before you try to solve it.

59. $5 - 2x = 3x + 1$
60. $7 - 3x = 8x - 4$
61. $\frac{1}{10}t^2 - \frac{5}{2} = 0$
62. $\frac{2}{7}t^2 - \frac{7}{2} = 0$
63. $7 + 3(x + 2) = 4(x + 1)$
64. $5 + 2(4x - 4) = 3(2x - 1)$
65. $-\frac{2}{5}x + \frac{2}{15} = \frac{2}{3}$
66. $-\frac{1}{6}x + \frac{2}{3} = \frac{1}{4}$

67. $(y - 4)(y + 1) = -6$
68. $(y - 6)(y + 1) = -12$
69. $x^3 + 2x^2 - 25x - 50 = 0$
70. $x^3 + 4x^2 - 9x - 36 = 0$
71. $\frac{1}{2}x + \frac{1}{3}x + \frac{1}{4}x = \frac{13}{12}$
72. $\frac{1}{3}x + \frac{1}{4}x + \frac{1}{5}x = \frac{47}{60}$
73. $(2r + 3)(2r - 1) = -(3r + 1)$
74. $(3r + 2)(r - 1) = -(7r - 7)$
75. $0.12x + 0.10(15{,}000 - x) = 1{,}600$
76. $0.09x + 0.11(11{,}000 - x) = 1{,}150$
77. $9a^3 = 16a$
78. $16a^3 = 25a$
79. $0.35x - 0.2 = 0.15x + 0.1$
80. $0.25x - 0.05 = 0.2x + 0.15$
81. $4x^3 + 12x^2 - 9x - 27 = 0$
82. $9x^3 + 18x^2 - 4x - 8 = 0$

In Words

83. There is no solution to the equation

$$6x - 2(x - 5) = 4x + 3$$

That is, there is no real number to use in place of x that will turn the equation into a true statement. What happens when you try to solve the equation?

84. The equation $5(x - 2) - 2x = 3x + 7$ has no solution. What happens when you try to solve the equation?

85. The equation $4x - 8 = 2(2x - 4)$ is called an **identity** because every real number is a solution. That is, replacing x with any real number will result in a true statement. Try to solve the equation.

86. Every real number is a solution to the equation $7(x + 2) - 4x = 3x + 14$. What happens when you try to solve the equation?

Solve each equation, if possible.

87. $3x - 6 = 3(x + 4)$
88. $7x - 14 = 7(x - 2)$
89. $4y + 2 - 3y + 5 = 3 + y + 4$
90. $7y + 5 - 2y - 3 = 6 + 5y - 4$
91. $2(4t - 1) + 3 = 5t + 4 + 3t$
92. $5(2t - 1) + 1 = 2t - 4 + 8t$

Review Problems

 From now on, each problem set will end with a series of review problems. In mathematics, it is very important to review. The more you review, the better you will understand the topics we cover and the longer you will remember them. Often, material that seemed confusing earlier will be less confusing the second time around.

The problems that follow review material we covered in Section 1.3.

Simplify each expression as much as possible.

93. $-9 \div \frac{3}{2}$

94. $-\frac{4}{5} \div (-4)$

95. $3 - 7(-6 - 3)$

96. $(3 - 7)(-6 - 3)$

97. $-4(-2)^3 - 5(-3)^2$

98. $4(2 - 5)^3 - 3(4 - 5)^5$

99. $\dfrac{2(-3) - 5(-6)}{-1 - 2 - 3}$

100. $\dfrac{4 - 8(3 - 5)}{2 - 4(3 - 5)}$

One Step Further

Solve for x.

101. $\dfrac{x + 4}{5} - \dfrac{x + 3}{3} = -\dfrac{7}{15}$

102. $\dfrac{x + 1}{7} - \dfrac{x - 2}{2} = \dfrac{1}{14}$

103. $\dfrac{1}{x} - \dfrac{2}{3} = \dfrac{2}{x}$

104. $\dfrac{1}{x} - \dfrac{3}{5} = \dfrac{2}{x}$

SECTION 2.2

Formulas

A **formula** in mathematics is an equation that contains more than one variable. There are probably some formulas that are already familiar to you—for example, the formula for the area A of a rectangle with length l and width w is $A = lw$.

To begin our work with formulas, we will consider some examples in which we are given numerical replacements for all but one of the variables.

EXAMPLE 1 Find y when x is 4 in the formula $3x - 4y = 2$.

SOLUTION We substitute 4 for x in the formula and then solve for y:

When $x = 4$:

the formula $\qquad 3x - 4y = 2$

becomes $\qquad 3(4) - 4y = 2$

$\qquad\qquad 12 - 4y = 2 \qquad$ Multiply 3 and 4.

$\qquad\qquad\quad -4y = -10 \qquad$ Add -12 to each side.

$\qquad\qquad\qquad y = \dfrac{5}{2} \qquad$ Divide each side by -4.

Note that, in the last line of the example above, we divided each side of the equation by -4. Remember that this is equivalent to multiplying each side of the equation by $-\frac{1}{4}$. For the rest of the examples in this section, it will be more convenient to think in terms of division rather than multiplication.

EXAMPLE 2 A store selling art supplies finds that they can sell x sketch pads each week at a price of p dollars each, according to the formula $x = 900 - 300p$. What price should they charge for each sketch pad if they want to sell 525 pads each week?

SOLUTION   We are given a formula, $x = 900 - 300p$, and asked to find the value of p if x is 525. To do so, we simply substitute 525 for x and solve for p:

When $x = 525$:

the formula $x = 900 - 300p$

becomes $525 = 900 - 300p$

$\qquad -375 = -300p$ Add -900 to each side.

$\qquad 1.25 = p$ Divide each side by -300.

In order to sell 525 sketch pads, the store should charge \$1.25 for each pad.

Our next example involves a formula that contains the number π. In this book, you can use either 3.14 or $\frac{22}{7}$ as an approximation for π. Generally speaking, if the problem contains decimals, use 3.14 to approximate π.

EXAMPLE 3   A company manufacturing coffee cans uses 486.7 square centimeters of material to make each can. Because of the way the cans are to be placed in boxes for shipping, the radius of each can must be 5 centimeters. If the formula for the surface area of a can is $S = \pi r^2 + 2\pi rh$ (the cans are open at the top), find the height of each can. (Use 3.14 as an approximation for π.)

— 5 cm

Surface area $= 486.7 \text{ cm}^2$

SOLUTION   Substituting 486.7 for S, 5 for r, and 3.14 for π into the formula $S = \pi r^2 + 2\pi rh$, we have

$486.7 = (\mathbf{3.14})(5^2) + 2(\mathbf{3.14})(5)h$

$486.7 = 78.5 + 31.4h$

$408.2 = 31.4h$ Add -78.5 to each side.

$13 = h$ Divide each side by 31.4.

The height of each can is 13 centimeters.

EXAMPLE 4   If an object is projected into the air with an initial vertical velocity v (in feet/second), its height h (in feet) after t seconds will be given by

$$h = vt - 16t^2$$

Find t if $v = 64$ feet/second and $h = 48$ feet.

SOLUTION   Substituting $v = 64$ and $h = 48$ into the formula, we have

$$48 = 64t - 16t^2$$

which is a quadratic equation. We write it in standard form and solve by factoring:

$16t^2 - 64t + 48 = 0$

$t^2 - 4t + 3 = 0$ Divide each side by 16.

$(t - 1)(t - 3) = 0$

$t - 1 = 0$ or $t - 3 = 0$

$t = 1$ or $t = 3$

Here is how we interpret our results: If an object is projected upward with an initial vertical velocity of 64 feet/second, it will be 48 feet above the ground after 1 second and after 3 seconds. That is, it passes 48 feet going up and also coming down.

Revenue

As we mentioned in Chapter 1, the most basic business formula is the formula for profit:

$$\text{Profit} = \text{Revenue} - \text{Cost}$$

where revenue is the total amount of money a company brings in by selling their product and cost is the total cost to produce the product they sell. Revenue itself can be broken down further by another formula common in the business world. The revenue obtained from selling all x items is the product of the number of items sold and the price per item. That is,

$$\text{Revenue} = (\text{Number of items sold})(\text{Price of each item})$$

For example, if 100 items are sold for $9 each, the revenue is $100(9) = \$900$. Likewise, if 500 items are sold for $11 each, then the revenue is $500(11) = \$5,500$. In general, if x is the number of items sold and p is the selling price of each item, then we can write

$$R = xp$$

EXAMPLE 5 A manufacturer of small portable radios knows that the number of radios she can sell each week is related to the price of the radios by the equation $x = 1,300 - 100p$, where x is the number of radios and p is the price per radio. What price should she charge for each radio if she wants the weekly revenue to be $4,000?

SOLUTION The formula for total revenue is $R = xp$. Since we want R in terms of p, we substitute $1,300 - 100p$ for x in the equation $R = xp$:

$$\begin{aligned} \text{If} \quad & R = xp \\ \text{and} \quad & x = 1,300 - 100p \\ \text{then} \quad & R = (1,300 - 100p)p \end{aligned}$$

We want to find p when R is 4,000. Substituting for R in the formula gives us

$$\begin{aligned} \mathbf{4,000} &= (1,300 - 100p)p \\ 4,000 &= 1,300p - 100p^2 \end{aligned}$$

This is a quadratic equation. To write it in standard form, we add $100p^2$ and $-1,300p$ to each side, giving us

$$\begin{aligned} 100p^2 - 1,300p + 4,000 &= 0 \\ p^2 - 13p + 40 &= 0 \qquad \text{Divide each side by 100.} \\ (p - 5)(p - 8) &= 0 \end{aligned}$$

$$p - 5 = 0 \quad \text{or} \quad p - 8 = 0$$
$$p = 5 \qquad\qquad\quad p = 8$$

If she sells the radios for $5 each or for $8 each, she will have a weekly revenue of $4,000.

 Facts from Geometry: Formulas for Area and Perimeter

To review, here are the formulas for the area and perimeter of some common geometric objects.

Note: The broken line labeled h in the triangle is its height, or altitude. It extends from the top of the triangle down to the base, meeting the base at an angle of 90°. The altitude of a triangle is always perpendicular to the base. The small square shown where the altitude meets the base is used to indicate that the angle formed is 90°.

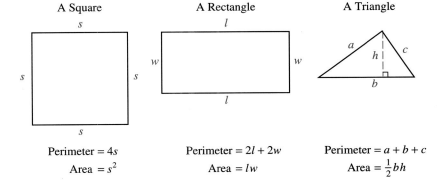

A Square
Perimeter = $4s$
Area = s^2

A Rectangle
Perimeter = $2l + 2w$
Area = lw

A Triangle
Perimeter = $a + b + c$
Area = $\frac{1}{2}bh$

The formula for perimeter gives us the distance around the outside of the object along its sides, while the formula for area gives us a measure of the amount of surface the object covers.

EXAMPLE 6 Given the formula $P = 2w + 2l$, solve for w.

SOLUTION To solve for w, we must isolate it on one side of the equation. We can accomplish this if we delete the $2l$ term and the coefficient 2 from the right side of the equation.

To begin, we add $-2l$ to both sides:

$$P + (-\mathbf{2l}) = 2w + 2l + (-\mathbf{2l})$$
$$P - 2l = 2w$$

Note: We know we are finished solving a formula for a specified variable when that variable appears alone on one side of the equal sign and not on the other.

To delete the 2 from the right side, we can multiply both sides by $\frac{1}{2}$:

$$\frac{1}{2}(P - 2l) = \frac{1}{2}(2w)$$
$$\frac{P - 2l}{2} = w$$

The two formulas

$$P = 2w + 2l \quad \text{and} \quad w = \frac{P - 2l}{2}$$

give the relationship between P, l, and w. They look different, but they both say the same thing about P, l, and w. The first formula gives P in terms of l and w, and the second formula gives w in terms of P and l.

EXAMPLE 7 Solve the formula $S = 2\pi rh + 2\pi r^2$ for h.

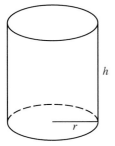

Surface area if open at one end:
$S = \pi r^2 + 2\pi rh$
Surface area if closed at both ends:
$S = 2\pi r^2 + 2\pi rh$
Volume: $V = \pi r^2 h$

SOLUTION This is the formula for the surface area of a right circular cylinder, with radius r and height h, that is closed at both ends. To isolate h, we first add $-2\pi r^2$ to both sides:

$$S = 2\pi rh + 2\pi r^2$$

$$S + (-2\pi r^2) = 2\pi rh + 2\pi r^2 + (-2\pi r^2) \quad \text{Add } -2\pi r^2 \text{ to both sides.}$$

$$S - 2\pi r^2 = 2\pi rh$$

$$\frac{S - 2\pi r^2}{2\pi r} = h \qquad\qquad\qquad \text{Divide each side by } 2\pi r.$$

or

$$h = \frac{S - 2\pi r^2}{2\pi r}$$

EXAMPLE 8 Solve for x: $ax - 3 = bx + 5$.

SOLUTION In this example, we begin by collecting all the variable terms on the left side of the equation and all the constant terms on the other side (just as we did when we were solving linear equations in Section 2.1):

$$ax - 3 = bx + 5$$

$$ax - bx - 3 = 5 \qquad \text{Add } -bx \text{ to each side.}$$

$$ax - bx = 8 \qquad\quad \text{Add 3 to each side.}$$

At this point we need to apply the distributive property to write the left side as $(a - b)x$. After that, we divide each side by $a - b$:

$$(a - b)x = 8 \qquad\qquad \text{Distributive property}$$

$$x = \frac{8}{a - b} \qquad \text{Divide each side by } a - b.$$

Note: We applied the distributive property in Example 8 in the same way we applied it when we first learned how to simplify $7x - 4x$. Recall that $7x - 4x = 3x$ because

$$7x - 4x = (7 - 4)x = 3x$$

We are using the same type of reasoning when we write

$$ax - bx = (a - b)x$$

Basic Percent Problems

The next examples in this section show how basic percent problems can be translated directly into equations. To understand these examples, we must recall that percent means "per hundred." That is, 75% is the same as 75/100, 0.75, and, in reduced fraction form, $\frac{3}{4}$. Likewise, the decimal 0.25 is equivalent to 25%. To change a decimal to a percent we move the decimal point two places to the right and write the % symbol. To change from a percent to a decimal, we drop the % symbol and move the decimal point two places to the left. The table at the right gives some of the most commonly used fractions and decimals and their equivalent percents.

Fraction	Decimal	Percent
$\frac{1}{2}$	0.5	50%
$\frac{1}{4}$	0.25	25%
$\frac{3}{4}$	0.75	75%
$\frac{1}{3}$	$0.\overline{3}$	$33\frac{1}{3}\%$
$\frac{2}{3}$	$0.\overline{6}$	$66\frac{2}{3}\%$
$\frac{1}{5}$	0.2	20%
$\frac{2}{5}$	0.4	40%
$\frac{3}{5}$	0.6	60%
$\frac{4}{5}$	0.8	80%

EXAMPLE 9 What number is 15% of 63?

SOLUTION To solve a problem like this, we let x = the number in question and then translate the sentence directly into an equation. Here is how it is done:

$$\underbrace{\text{What number}}_{x} \underset{\downarrow}{\text{is}} \underset{\downarrow}{15\%} \underset{\downarrow}{\text{of}} \underset{\downarrow}{63?}$$

$$x = 0.15 \cdot 63$$
$$= 9.45$$

The number 9.45 is 15% of 63.

EXAMPLE 10 What percent of 42 is 21?

SOLUTION We translate the sentence as follows:

$$\underbrace{\text{What percent}}_{x} \underset{\downarrow}{\text{of}} \underset{\downarrow}{42} \underset{\downarrow}{\text{is}} \underset{\downarrow}{21?}$$

$$x \cdot 42 = 21$$

Next, we divide each side by 42:

$$x = \frac{21}{42}$$
$$= 0.50 \quad \text{or } 50\%$$

EXAMPLE 11 25 is 40% of what number?

SOLUTION Again, we translate the sentence directly:

$$\underset{\downarrow}{25} \underset{\downarrow}{\text{is}} \underset{\downarrow}{40\%} \underset{\downarrow}{\text{of}} \underbrace{\text{what number?}}$$

$$25 = 0.40 \cdot x$$

We solve the equation by dividing both sides by 0.40:

$$\frac{25}{0.40} = \frac{0.40 \cdot x}{0.40}$$

$$62.5 = x$$

25 is 40% of 62.5.

Using Technology: Graphing Calculators

There are a variety of computer programs and graphing calculators currently available to help us perform many tasks in algebra. Throughout the text, discussions and examples will appear under the heading "Graphing Calculators." The instructions and keystroke sequences we give will be generic in form. You will have to use the manual that comes with your calculator to find the instructions that are appropriate to your calculator.

Entering, Recalling, and Editing Formulas

Right circular cylinder with radius r and height h

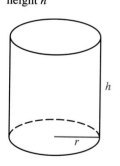

Graphing calculators can be useful in evaluating formulas. For example, the formula for the volume of a right circular cylinder is

$$V = \pi r^2 h$$

Suppose we want to compare the effect of the radius and the height on the volume. We can do so by finding the volumes of cylinders with a variety of values for the radius and height. Here are a few:

$$r = 3, \qquad h = 4 \qquad\qquad r = 10, \qquad h = 100$$
$$r = 4, \qquad h = 3 \qquad\qquad r = 100, \qquad h = 10$$

On your calculator begin by entering the formula for the volume produced by the first radius and height. Then, do the calculation and record the first volume. Here are the keys to press to find the first volume:

$$\boxed{\pi}\ \boxed{\times}\ \boxed{3}\ \boxed{\char`^}\ \boxed{2}\ \boxed{\times}\ \boxed{4}$$

Pressing the ENTER key will give the first volume as 113, to the nearest whole number. To find the volume for the second set of numbers, recall the original formula and simply change the numbers. That is, use the recall feature of your calculator to bring back the formula for the first volume ($\boxed{\text{2nd}}$ $\boxed{\text{ENTRY}}$ on a TI-82 calculator). Once you have it, move the cursor to the radius and replace it with a new radius. Do the same for the height. Pressing the ENTER key gives the new volume. Here are the volumes for all four sets of numbers:

$r = 3,$	$h = 4$	Formula: $\pi * 3 \char`^ 2 * 4$	$V = 113$
$r = 4,$	$h = 3$	Formula: $\pi * 4 \char`^ 2 * 3$	$V = 151$
$r = 10,$	$h = 100$	Formula: $\pi * 10 \char`^ 2 * 100$	$V = 31{,}416$
$r = 100,$	$h = 10$	Formula: $\pi * 100 \char`^ 2 * 10$	$V = 314{,}159$

Problem Set
2.2

The problems in this problem set are more realistic than some of the problems you have worked previously. You will find similar problems in some of the other classes you take.

Use the formula $3x - 4y = 12$ to find y if:

1. x is 0
2. x is -2
3. x is 4
4. x is -4

Use the formula $y = 2x - 3$ to find x when:

5. y is 0
6. y is -3
7. y is 5
8. y is -5

Pricing *A company that manufactures typewriter ribbons finds that they can sell x ribbons each week at a price of p dollars each, according to the formula $x = 1,300 - 100p$. Determine the price they should charge for each ribbon if they want to sell:*

9.   800 ribbons each week
10.   400 ribbons each week
11.   300 ribbons each week
12.   900 ribbons each week

Volume and Height of a Cylinder *The volume of a cylinder is given by the formula $V = \pi r^2 h$, where r is the radius and h is the height. Find the height if:*

13.   The volume is 308 cubic centimeters and the radius is 7 centimeters. (Use $\frac{22}{7}$ for π.)

14.   The volume is 308 cubic centimeters and the radius is $\frac{7}{2}$ centimeters.

15.   The volume is 628 cubic inches and the radius is 10 inches. (Use 3.14 for π.)

16.   The volume is 12.56 cubic inches and the radius is 5 inches.

Surface Area and Height of a Cylinder
The surface area of a cylinder that is closed at the top and bottom is given by the formula $S = 2\pi r^2 + 2\pi rh$. Find the height if:

17.   The surface area is 942 square feet and the radius is 10 feet. (Use 3.14 for π.)

18.   The surface area is 471 square feet and the radius is 5 feet.

Try doing this entire assignment without taking a break. You are building confidence and momentum as you work through these problems, and your momentum may be lost if you take a break. Then you can see what works best for you: Working through all the problems without a break, or doing the assignment in two or three sessions.

Height of a Projectile *The formula $h = vt - 16t^2$ gives the height h (in feet) of an object projected into the air with an initial vertical velocity v (in feet/second) after t seconds.*

19.   If an object is projected upward with an initial velocity of 48 feet/second, at what times will it reach a height of 32 feet above the ground?

20.   If an object is projected upward into the air with an initial velocity of 80 feet/second, at what times will it reach a height of 64 feet above the ground?

21.   An object is projected into the air with a vertical velocity of 24 feet/second. At what times will the object be on the ground? (It is on the ground when h is 0.)

22.   An object is projected into the air with a vertical velocity of 20 feet/second. At what times will the object be on the ground?

Surface Area of a Cylinder *The surface area of a right circular cylinder that is closed at the bottom but not at the top is given by $S = \pi r^2 + 2\pi rh$.*

23. Find the radius of a right circular cylinder with height 3 inches, if the surface area is 16π square inches.

24. Find the radius of a right circular cylinder with height 5 feet, if the surface area is 39π square feet.

25. **Revenue** A company that manufactures typewriter ribbons knows that the number of ribbons they can sell each week, x, is related to the price per ribbon, p, by the equation $x = 1,200 - 100p$. At what price should they sell the ribbons if they want the weekly revenue to be $3,200? (Remember: The equation for revenue is $R = xp$.)

26. **Revenue** A company manufactures diskettes for home computers. They know from past experience that the number of diskettes they can sell each day, x, is related to the price per diskette, p, by the equation $x = 800 - 100p$. At what price should they sell their diskettes if they want the daily revenue to be $1,200?

27. **Revenue** The relationship between the number of calculators a company sells per day, x, and the price of each calculator, p, is given by the equation $x = 1,700 - 100p$. At what price should the calculators be sold if the daily revenue is to be $7,000?

28. **Revenue** The relationship between the number of pencil sharpeners a company can sell each week, x, and the price of each sharpener, p, is given by the equation $x = 1,800 - 100p$. At what price should the sharpeners be sold if the weekly revenue is to be $7,200?

Solve each of the following formulas for the indicated variable.

29. $A = lw$; for l
30. $A = \frac{1}{2}bh$; for b
31. $I = prt$; for t
32. $I = prt$; for r
33. $A = P + Prt$; for r

34. $A = P + Prt$; for t
35. $C = \frac{5}{9}(F - 32)$; for F
36. $F = \frac{9}{5}C + 32$; for C
37. $h = vt + 16t^2$; for v
38. $h = vt - 16t^2$; for v
39. $A = a + (n - 1)d$; for d
40. $A = a + (n - 1)d$; for n
41. $2x + 3y = 6$; for y
42. $2x - 3y = 6$; for y
43. $-3x + 5y = 15$; for y
44. $-2x - 7y = 14$; for y
45. $9x - 3y = 6$; for y
46. $9x + 3y = 15$; for y
47. $2x - 6y + 12 = 0$; for y
48. $7x - 2y - 6 = 0$; for y
49. $ax + 4 = bx + 9$; for x
50. $ax - 5 = cx - 2$; for x
51. $A = P + Prt$; for P
52. $S = 2\pi r + \pi r^2 h$; for h
53. $ax + b = cx + d$; for x
54. $4x + 2y = 3x + 5y$; for y

Percent *Translate each of the following into a linear equation, and then solve the equation.*

55. What number is 54% of 38?
56. What number is 11% of 67?
57. What number is 5% of 10,000?
58. What number is 6% of 6,000?
59. What percent of 36 is 9?
60. What percent of 50 is 5?
61. 37 is 4% of what number?
62. 8 is 2% of what number?

Review Problems

NOTE: The following problems review some of the material we covered in Sections 1.1 and 1.2. Reviewing these problems will help you with the next section.

Translate each of the following into symbols. [1.1]

63. Twice the sum of x and 3.
64. The sum of twice x and 3.
65. Twice the sum of x and 3 is 16.
66. The sum of twice x and 3 is 16.

67. Five times the difference of x and 3.

68. Five times the difference of x and 3 is 10.

69. The sum of $3x$ and 2 is equal to the difference of x and 4.

70. The sum of x and $x + 2$ is 12 more than their difference.

Identify the property (or properties) that justifies each of the following statements. [1.2]

71. $ax = xa$
72. $5(\frac{1}{5}) = 1$
73. $3 + (x + y) = (3 + x) + y$
74. $3 + (x + y) = (x + y) + 3$
75. $3 + (x + y) = (3 + y) + x$
76. $7(3x - 5) = 21x - 35$
77. $2 + 0 = 2$
78. $2 + (-2) = 0$

One Step Further

79. Solve for x: $\dfrac{x}{a} + \dfrac{y}{b} = 1$.

80. Solve for y: $\dfrac{x}{a} + \dfrac{y}{b} = 1$.

81. Solve for a: $\dfrac{1}{a} + \dfrac{1}{b} = \dfrac{1}{c}$.

82. Solve for b: $\dfrac{1}{a} + \dfrac{1}{b} = \dfrac{1}{c}$.

83. Solve for R: $\dfrac{1}{R} = \dfrac{1}{a} + \dfrac{1}{b} + \dfrac{1}{c}$.

84. Solve for a: $\dfrac{1}{R} = \dfrac{1}{a} + \dfrac{1}{b} + \dfrac{1}{c}$.

Using Technology

Compound Interest *The formula below gives the amount of money A in an account in which P dollars has been invested for t years at an interest rate of r compounded n times a year:*

$$A = P\left(1 + \frac{r}{n}\right)^{nt}$$

Use the recall formula function on your calculator to answer the following questions.

85. How much money is in an account in which $500 has been invested at 5%, compounded quarterly, for 6 years? (That is, find A when $P = 500$, $r = 0.05$, $n = 4$, and $t = 6$.)

86. How much money is in an account in which $500 has been invested at 5%, compounded monthly, for 6 years?

87. How much money is in an account in which $500 has been invested at 5%, compounded daily, for 6 years?

88. How much money is in an account in which $500 has been invested at 6%, compounded monthly, for 6 years?

89. How much money is in an account in which $500 has been invested at 7%, compounded monthly, for 6 years?

90. In general, is it more desirable to increase the interest rate or the number of compounding periods to increase the amount of money in the account?

SECTION
2.3

Applications

In this section we use the skills we have developed for solving equations to solve problems written in words. You may find that some of the examples and problems are more realistic than others. Since we are just beginning our work with application problems, even the ones that seem unrealistic are good practice. What is important in this section is the *method* we use to solve application problems, not the applications themselves. The method, or strategy, that we use to solve application problems is called the *Blueprint for Problem Solving*. It is an outline that will overlay the solution process we use on all application problems.

Step 1 **Read** the problem, and then mentally **list** the items that are known and the items that are unknown.

Step 2 **Assign a variable** to one of the unknown items. (In most cases this will amount to letting $x =$ the item that is asked for in the problem.) Then **translate** the other **information** in the problem to expressions involving the variable.

Step 3 **Reread** the problem, and then **write an equation,** using the items and variables listed in Steps 1 and 2, that describes the situation.

Step 4 **Solve the equation** found in Step 3.

Step 5 **Write** your **answer** using a complete sentence.

Step 6 **Reread** the problem and **check** your solution with the original words in the problem.

There are a number of substeps within each of the steps in our blueprint. For instance, with Steps 1 and 2 it is always a good idea to draw a diagram or picture if it helps visualize the relationship between the items in the problem.

EXAMPLE 1 The length of a rectangle is 3 inches less than twice the width. The perimeter is 45 inches. Find the length and width.

SOLUTION When working problems that involve geometric figures, a sketch of the figure helps organize and visualize the problem.

Step 1 Read and list

Known items: The figure is a rectangle. The length is 3 inches less than twice the width. The perimeter is 45 inches.

Unknown items: The length and the width

Step 2 Assign a variable and translate information Since the length is given in terms of the width (the length is 3 less than twice the width), we let $x =$ the width of the rectangle. The length is 3 less than twice the width, so it must be $2x - 3$. The diagram in Figure 1 is a visual description of the relationships we have listed so far.

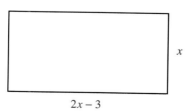

$2x - 3$

FIGURE 1

Step 3 Reread and write an equation The equation that describes the situation is

Twice the length	+ Twice the width	Is	The perimeter
$2(2x - 3)$	$+$ $\qquad 2x$	$=$	45

Step 4 Solve the equation

$$2(2x - 3) + 2x = 45$$
$$4x - 6 + 2x = 45$$
$$6x - 6 = 45$$
$$6x = 51$$
$$x = 8.5$$

Step 5 Write answer The width is 8.5 inches. The length is

$$2x - 3 = 2(\mathbf{8.5}) - 3 = 14 \text{ inches}$$

Step 6 Reread and check If the length is 14 inches and the width is 8.5 inches, then the perimeter must be $2(14) + 2(8.5) = 28 + 17 = 45$ inches. Also, the length, 14, is 3 less than twice the width.

Remember as you read through the steps in the solution to each example in this section that Step 1 is done mentally. Read the problem and then *mentally* list the items that you know and the items that you don't know. The purpose of Step 1 is to give you direction as you begin to work application problems. Finding the solution to an application problem is a process; it doesn't happen all at once. The first step is to read the problem with a purpose in mind. The purpose is to mentally note the items that are known and the items that are unknown.

EXAMPLE 2 In April, Pat bought a new Ford Mustang with a 5.0 liter engine. The total price, which includes the price of the car plus sales tax, was $17,481.75. If the sales tax rate is 7.25%, what was the price of the car?

SOLUTION

Step 1 Read and list
Known items: The total price is $17,481.75. The sales tax rate is 7.25%, which is 0.0725 in decimal form.
Unknown item: The price of the car

Step 2 Assign a variable and translate information If we let x = the price of the car, then to calculate the sales tax, we multiply the price of the car x by the sales tax rate:

$$\begin{aligned} \text{Sales tax} &= (\text{Price of the car})(\text{Sales tax rate}) \\ &= x \cdot 0.0725 \\ &= 0.0725x \end{aligned}$$

Step 3 Reread and write an equation

Car price	+ Sales tax	=	Total price
x	$+$ $\ 0.0725x$	$=$	$17,481.75$

Step 4 Solve the equation

$$x + 0.0725x = 17{,}481.75$$
$$1.0725x = 17{,}481.75$$
$$x = \frac{17{,}481.75}{1.0725}$$
$$= 16{,}300.00$$

Step 5 Write answer The price of the car is $16,300.00.
Step 6 Reread and check The price of the car is $16,300.00. The tax is $0.0725(16{,}300) = \$1{,}181.75$. Adding the retail price and the sales tax, we have a total bill of $17,481.75.

𝓕acts from 𝓖eometry: 𝒜ngles in 𝓖eneral

An angle is formed by two rays with the same endpoint. The common endpoint is called the **vertex** of the angle, and the rays are called the **sides** of the angle.

In Figure 2, angle θ (theta) is formed by the two rays OA and OB. The vertex of θ is O. Angle θ is also denoted as angle AOB, where the letter associated with the vertex is always the middle letter in the three letters used to denote the angle.

Degree Measure

The angle formed by rotating a ray through one complete revolution about its endpoint (Figure 3) has a measure of 360 degrees, which we write as 360°.

One degree of angle measure, written 1°, is $\frac{1}{360}$ of a complete rotation of a ray about its endpoint; there are 360° in one full rotation. (The number 360 was decided upon by early civilizations because it was believed that the earth was at the center of the universe and the sun would rotate once around the earth every 360 days.) Similarly, 180° is half of a complete rotation, and 90° is a quarter of a full rotation. Angles that measure 90° are called **right angles,** while angles that measure 180° are called **straight angles.** If an angle measures between 0° and 90° it is called an **acute angle,** while an angle that measures between 90° and 180° is an **obtuse angle.** Figure 4 illustrates.

FIGURE 2

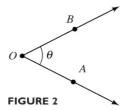

One complete revolution = 360°

FIGURE 3

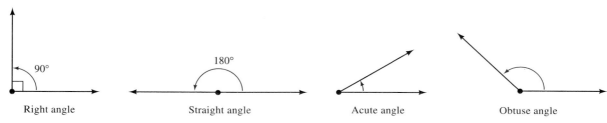

Right angle   Straight angle   Acute angle   Obtuse angle

FIGURE 4

Complementary Angles and Supplementary Angles

If two angles add up to 90°, then we call them **complementary angles,** and each is called the **complement** of the other. If two angles have a sum of 180°, then we call them **supplementary angles,** and each is called the **supplement** of the other. Figure 5 illustrates these relationships.

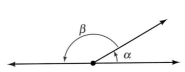

Complementary angles: $\alpha + \beta = 90°$ Supplementary angles: $\alpha + \beta = 180°$

FIGURE 5

EXAMPLE 3   Two complementary angles are such that one is twice as large as the other. Find the two angles.

SOLUTION   Applying the Blueprint for Problem Solving, we have:

Step 1 Read and list
    Known items: Two complementary angles. One is twice as large as the other.
    Unknown items: The sizes of the angles

Step 2 Assign a variable and translate information Let x = the smaller angle. The larger angle is twice the smaller so we represent the larger angle with $2x$.

Step 3 Reread and write an equation Since the two angles are complementary, their sum is 90°. Therefore,

$$x + 2x = 90$$

Step 4 Solve the equation

$$x + 2x = 90$$
$$3x = 90$$
$$x = 30$$

Step 5 Write answer The smaller angle is 30° and the larger angle is $2 \cdot 30 = 60°$.

Step 6 Reread and check The larger angle is twice the smaller angle and their sum is 90°.

Facts from Geometry: Special Triangles

The two special triangles we will discuss next are important triangles. If you go on to study trigonometry, you will see them often.

An **equilateral triangle** (Figure 6, p. 100) is a triangle with all three sides of equal length. If all three sides in a triangle have the same length, then the three interior angles in the triangle must also be equal. Since the sum of the interior angles in a triangle is always 180°, the three interior angles in any equilateral triangle must be 60°.

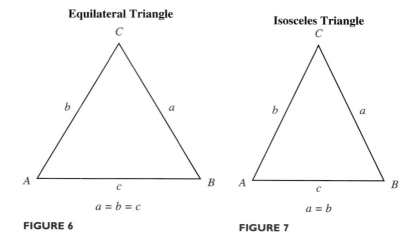

Equilateral Triangle

$a = b = c$

FIGURE 6

Isosceles Triangle

$a = b$

FIGURE 7

An **isosceles triangle** (Figure 7) is a triangle with two sides of equal length. Angles *A* and *B* in the isosceles triangle in Figure 7 are called the **base angles;** they are the angles opposite the two equal sides. In every isosceles triangle, the base angles are equal.

Note: As you can see from Figures 6 and 7, one way to label the important parts of a triangle is to label the vertices with capital letters and the sides with small letters: side *a* is opposite vertex *A*, side *b* is opposite vertex *B*, and side *c* is opposite vertex *C*.

Also, since each vertex is the vertex of one of the angles of the triangle, we refer to the three interior angles as *A, B,* and *C.*

Finally, the sum of the interior angles is 180° in any triangle, so for the triangles shown in Figures 6 and 7, this relationship is written

$$A + B + C = 180°$$

A **right triangle** (Figure 8) is a triangle with an interior angle of 90°. The side opposite the right angle is called the **hypotenuse.** The other two sides are called **legs.** The **Pythagorean Theorem** states a special relationship between the hypotenuse and legs of a right triangle.

PYTHAGOREAN THEOREM

In any right triangle, the square of the longest side (hypotenuse) is equal to the sum of the squares of the other two sides (legs).

$$c^2 = a^2 + b^2$$

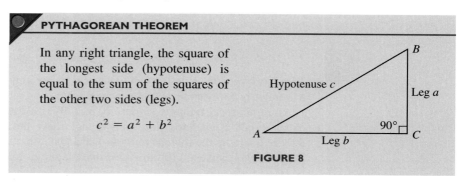

FIGURE 8

EXAMPLE 4   The base angles in an isosceles triangle are both twice as large as the third angle. Find the measures of all three angles.

> **SOLUTION**
> **Step 1 Read and list**
>   *Known items:* The three angles add to 180°. The triangle is isosceles, so the base angles are equal. The base angles are each twice as large as the third angle.
>   *Unknown items:* The measures of the three angles
> **Step 2 Assign a variable and translate information** Let x = the measure of the third angle. Each base angle is $2x$.
> **Step 3 Reread and write an equation** Since the sum of the interior angles of any triangle is 180°, we have
>
> $$x + 2x + 2x = 180$$
>
> **Step 4 Solve the equation**
>
> $$5x = 180$$
> $$x = 36$$
>
> **Step 5 Write answer** The third angle is 36°; the two base angles are each $2 \cdot 36 = 72°$.
> **Step 6 Reread and check** The base angles are each twice the third angle. The sum of the three angles is
>
> $$36° + 72° + 72° = 180°$$

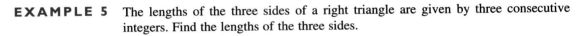

EXAMPLE 5   The lengths of the three sides of a right triangle are given by three consecutive integers. Find the lengths of the three sides.

> **SOLUTION**
> **Step 1 Read and list**
>   *Known items:* A right triangle. The three sides are three consecutive integers.
>   *Unknown items:* The three sides
> **Step 2 Assign a variable and translate information**
>   Let x = First integer (shortest side)
>   Then $x + 1$ = Next consecutive integer
> $x + 2$ = Last consecutive integer (longest side)

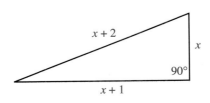

Step 3 Reread and write an equation By the Pythagorean Theorem, we have

$$(x + 2)^2 = (x + 1)^2 + x^2$$

Step 4 Solve the equation

$$x^2 + 4x + 4 = x^2 + 2x + 1 + x^2$$
$$x^2 - 2x - 3 = 0$$
$$(x - 3)(x + 1) = 0$$
$$x = 3 \quad \text{or} \quad x = -1$$

Step 5 Write answer Since x is the length of a side in a triangle, it must be a positive number. Therefore, $x = -1$ cannot be used.

The shortest side is 3. The other two sides are 4 and 5.

Step 6 Reread and check The three sides are given by consecutive integers. The square of the longest side is equal to the sum of the squares of the two shorter sides ($5^2 = 3^2 + 4^2$). ◢

Table Building

We can use our knowledge of formulas from Section 2.2 to build tables of paired data. As you will see, equations or formulas that contain exactly two variables produce pairs of numbers that can be used to construct tables.

EXAMPLE 6 A piece of string 12 inches long is to be formed into a rectangle. Build a table that gives the length of the rectangle if the width is 1, 2, 3, 4, or 5 inches. Then find the area of each of the rectangles formed.

12 inches

SOLUTION Since the formula for the perimeter of a rectangle is $P = 2l + 2w$, and our piece of string is 12 inches long, then the formula we will use to find the lengths for the given widths is $12 = 2l + 2w$. To solve this formula for l, we divide each side by 2 and then subtract w. The result is $l = 6 - w$. Table 1 organizes our work

TABLE I
Length, Width, and Area

Width (in.)	Length (in.)		Area (in.^{2})
w	$l = 6 - w$	l	$A = lw$
1	$l = 6 - 1$	5	5
2	$l = 6 - 2$	4	8
3	$l = 6 - 3$	3	9
4	$l = 6 - 4$	2	8
5	$l = 6 - 5$	1	5

so that the formula we use to find l for a given value of w is shown, and we have added a last column to give us the areas of the rectangles formed. The units for the first three columns are inches, while the units for the numbers in the last column are square inches.

 Using Technology: Graphing Calculators and Spreadsheet Programs

Graphing Calculators

A number of graphing calculators have table-building capabilities. We can let the calculator variable X represent the widths of the rectangles in Example 6. To find the lengths, we set variable Y_1 equal to $6 - X$. The area of each rectangle can be found by setting variable Y_2 equal to $X * Y_1$. To have the calculator produce the table automatically, we use a table minimum of 0 and a table increment of 1. Here is a summary of how the graphing calculator is set up:

TABLE SETUP

Table minimum = 0
Table increment = 1
Independent variable: Auto
Dependent variable: Auto

Y VARIABLES SETUP

$Y_1 = 6 - X$
$Y_2 = X * Y_1$

The table will look like this:

X	Y_1	Y_2
0	6	0
1	5	5
2	4	8
3	3	9
4	2	8
5	1	5
6	0	0

Spreadsheet Programs

A table similar to the one above can be created with almost any *spreadsheet program.* Using formulas to produce the numbers in the cells allows the spreadsheet to construct the table automatically. The advantage to using a spreadsheet program is that the program can create, label, and print a variety of graphics from the data in the table. Figure 9 (p. 104) shows the formulas in each cell of a spreadsheet program that will produce the numbers in Table 1 in Example 6.

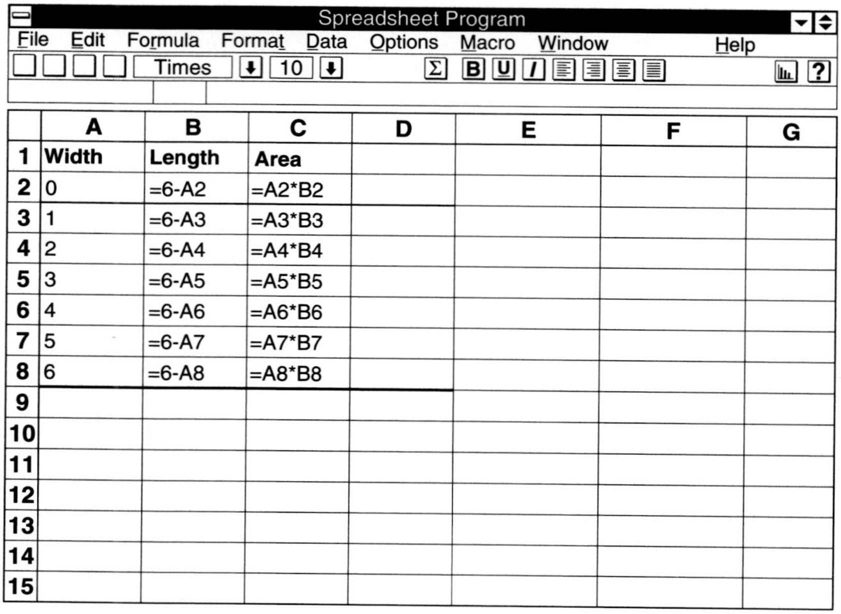

	A	B	C	D	E	F	G
1	**Width**	**Length**	**Area**				
2	0	=6-A2	=A2*B2				
3	1	=6-A3	=A3*B3				
4	2	=6-A4	=A4*B4				
5	3	=6-A5	=A5*B5				
6	4	=6-A6	=A6*B6				
7	5	=6-A7	=A7*B7				
8	6	=6-A8	=A8*B8				
9							
10							
11							
12							
13							
14							
15							

FIGURE 9
Computer screen showing a spreadsheet program

Problem Set 2.3

Solve each application problem. Be sure to follow the steps in the Blueprint for Problem Solving.

Geometry Problems

1. A rectangle is twice as long as it is wide. The perimeter is 60 feet. Find the dimensions.

2. The length of a rectangle is 5 times the width. The perimeter is 48 inches. Find the dimensions.

3. A square has a perimeter of 28 feet. Find the length of each side.

4. A square has a perimeter of 36 centimeters. Find the length of each side.

5. A triangle has a perimeter of 23 inches. The medium side is 3 inches more than the shortest side, and the longest side is twice the shortest side. Find the shortest side.

6. The longest side of a triangle is 2 times the shortest side, while the medium side is 3 meters more than the shortest side. The perimeter is 27 meters. Find the dimensions.

7. The length of a rectangle is 3 meters less than twice the width. The perimeter is 18 meters. Find the width.

8. The length of a rectangle is 1 foot more than twice the width. The perimeter is 20 feet. Find the dimensions.

9. A livestock pen is built in the shape of a rectangle that is twice as long as it is wide. The perimeter is 48 feet. If the material used to build the pen is $1.75 per foot for the longer sides, and $2.25 per foot for the shorter sides (the shorter sides have gates, which increase the cost per foot), find the cost to build the pen.

10. A garden is in the shape of a square with a perimeter of 42 feet. The garden is surrounded by two fences. One fence is around the perimeter of the garden, while the second fence is 3 feet from the first fence, as Figure 10 indicates. If the material used to build the two fences costs $1.28 per foot, what was the total cost of the fences?

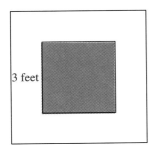

3 feet

FIGURE 10

Percent Problems

NOTE: Remember, one of your goals for this chapter is to increase your confidence with word problems. The word *increase* is the important part of this goal. When you are finished with this assignment, you should feel more confident about word problems.

11. Suppose the total price, including sales tax, of a used pickup truck is $10,039.43. If the sales tax rate is 7.5%, what was the price of the truck? Round your answer to the nearest cent.

12. Shane returned from a trip to Las Vegas with $300.00, which was 50% more money than she had at the beginning of the trip. How much money did Shane have at the beginning of her trip?

13. Every item in the Just a Dollar store is priced at $1.00. When Mary Jo opens the store, there is $125.50 in the cash register. When she counts the money in the cash register at the end of the day, the total is $1,058.60. If the sales tax rate is 8.5%, how many items were sold that day?

14. During one ratings period in 1998, it was estimated that 5.4 million people watched *The Tonight Show,* with Jay Leno. If the number of people watching Jay Leno was 1.8% more than the number of viewers watching *Nightline* with Ted Koppel, approximately how many viewers did Ted Koppel have? Round your answer to the nearest tenth of a million viewers.

15. On Monday, August 16, 1993, Continental Airlines announced it was laying off 2,500 workers, or 6% of its work force, by December 31 of that year. How many workers did Continental Airlines have on Friday, August 13, 1993?

16. Suppose a college bookstore buys a textbook from a publishing company, then marks up the price they paid for the book 33%, and sells it to a student at the marked-up price. If the student pays $45.00 for the textbook, what did the bookstore pay for it? Round your answer to the nearest cent.

17. An accountant earns $3,440 per month after receiving a 5.5% raise. What was the accountant's monthly income before the raise? Round your answer to the nearest cent.

18. A sheet-metal worker earns $26.80 per hour after receiving a 4.5% raise. What was the sheet-metal worker's hourly pay before the raise? Round your answer to the nearest cent.

More Geometry Problems

NOTE: Look for similarities among the problems. When you come across a problem that is similar to one you have already worked successfully, apply the same method of solution to the new problem. This will help you work toward your goal of increasing your effectiveness.

19. Two supplementary angles are such that one is eight times as large as the other. Find the two angles.

20. Two complementary angles are such that one is five times as large as the other. Find the two angles.

21. One angle is 12° less than four times another. Find the measure of each angle if:
(a) They are complements of each other.
(b) They are supplements of each other.

22. One angle is 4° more than three times another. Find the measure of each angle if:
(a) They are complements of each other.
(b) They are supplements of each other.

23. A triangle is such that the largest angle is three times the smallest angle. The third angle is 9° less than the largest angle. Find the measure of each angle.

24. The smallest angle in a triangle is half of the largest angle. The third angle is 15° less than the largest angle. Find the measure of all three angles.

25. The smallest angle in a triangle is one-third of the largest angle. The third angle is 10° more than the smallest angle. Find the measure of all three angles.

26. The third angle in an isosceles triangle is half as large as each of the two base angles. Find the measure of each angle.

27. The third angle in an isosceles triangle is 8° more than twice as large as each of the two base angles. Find the measure of each angle.

28. The third angle in an isosceles triangle is 4° more than one-fifth of each of the two base angles. Find the measure of each angle.

29. The lengths of the three sides of a right triangle are given by three consecutive even integers. Find the lengths of the three sides.

30. The longest side of a right triangle is 3 inches less than twice the shortest side. The third side measures 12 inches. Find the length of the shortest side.

31. The length of a rectangle is 2 feet more than three times the width. If the area is 16 square feet, find the width and the length.

32. The length of a rectangle is 4 yards more than twice the width. If the area is 70 square yards, find the width and the length.

33. The base of a triangle is 2 inches more than four times the height. If the area is 36 square inches, find the base and the height. (Refer to Figure 11.)

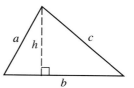

FIGURE 11
Triangle with base b and height h.
Perimeter is $P = a + b + c$.
Area is $A = \frac{1}{2}bh$.

34. The height of a triangle is 4 feet less than twice the base. If the area is 48 square feet, find the base and the height. (Refer to Figure 11.)

Miscellaneous Problems

35. Tickets for the father and son breakfast were $2.00 for fathers and $1.50 for sons. If a total of 75 tickets were sold for $127.50, how many fathers and how many sons attended the breakfast?

36. A Girl Scout troop sells 62 tickets to their mother and daughter dinner for a total of $216. If the tickets cost $4.00 for mothers and $3.00 for daughters, how many of each ticket did they sell?

37. A woman owns a small, cash-only business in a state that requires her to charge a 6% sales tax on each item she sells. At the beginning of the day she has $250 in the cash register. At the end of the day she has $1,204 in the register. How much money should she send to the state government for the sales tax she collected?

38. A store is located in a state that requires a 6% tax on all items sold. If the store brings in a total of $3,392 in one day, how much of that total was sales tax?

39. Patrick goes away to college. The first week he is away from home he calls his girlfriend, using his parents' telephone credit card, and talks for a long time. The telephone company charges 40 cents for the first minute and 30 cents for each additional minute, and then adds on a 50 cent service charge for using the credit card. If his parents receive a bill for $13.80 for Patrick's call, how long did he talk?

40. A person makes a long distance person-to-person call to Santa Barbara, California. The telephone company charges 41 cents for the first minute and 32 cents for each additional minute. Because the call is person-to-person, there is also a service charge of $3.00. If the cost of the call is $6.29, how many minutes did the person talk?

Table Building

Problems 41–44 may be solved using a graphing calculator.

41. A farmer buys 48 feet of fencing material to build a rectangular livestock pen. Build a table that gives the length of the pen if the width is 2, 4, 6, 8, 10, and 12 feet. Then find the area of each of the pens formed.

42. A store selling art supplies finds that they can sell x sketch pads each week at a price of p dollars each according to the formula $x = 900 - 300p$. Use this formula to build a table that gives the number of pads sold each week if the price per pad is $0.50, $1.00, $1.50, $2.00, and $2.50.

43. A ball is thrown straight up into the air with a velocity of 128 feet per second. The formula that gives the height h of the ball t seconds after it is tossed is

$$h = -16t^2 + 128t$$

Use this formula to find the height of the ball after 1, 2, 3, 4, 5, and 6 seconds.

If you were to buy a portable telephone in California and access it through GTE Mobilnet, you would have a choice of rates. Here is a problem that involves the cost of using a portable telephone.

44. One of the rates listed on the GTE Mobilnet rate card is a flat rate of $15 per month plus $1.50 for each minute you use the phone. Write an equation in two variables that will let you calculate the monthly charge for talking x minutes. Then build a table that shows the cost for talking 10, 15, 20, 25, and 30 minutes in one month.

Review Problems

> NOTE: The problems below review material we covered in Section 1.1.

For the set $\{-4, -\frac{2}{5}, 0, 2, \sqrt{5}, 3, \pi\}$, name all the:

45. Integers

46. Irrational numbers

47. Rational numbers

48. Whole numbers

If $A = \{3, 5, 7, 9\}$ and $B = \{4, 5, 6, 7\}$, find:

49. $A \cup B$

50. $A \cap B$

51. $\{x \mid x \in A \text{ and } x > 5\}$

52. $\{x \mid x \in B \text{ and } x \leq 5\}$

Using Technology

53. Use a graphing calculator or a spreadsheet program to reproduce Table 1 from this section (page 102). This time, start with a width of 0 and increase the width by 0.5 inch each time. What is the largest area produced by this table?

54. Use a graphing calculator or a spreadsheet program to reproduce Table 1 from this section (page 102). This time, start with a width of 0 and increase the width by 0.25 inch each time. What is the largest area produced by this table?

55. If you invest $100 in an account that earns 6% annual interest compounded monthly, the amount of money in that account t months later is given by the formula $A = 100(1.005)^t$. Use a graphing calculator or a spreadsheet program to produce a table that gives the amount of money in the account each month for the first 2 years it is on deposit. (On your graphing calculator, let X represent t, and Y_1 represent A. The minimum X is 0, and the table increment is 1.)

56. A spherical balloon is being inflated at a constant rate so that the radius increases by 1 centimeter every second. Since the balloon is a sphere, its volume when the radius is r is given by the formula $V = \frac{4}{3}\pi r^3$. Construct a table that gives the volume of the balloon every second, starting when the radius is 0 centimeters and ending when the radius is 10 centimeters. (Round to the nearest whole number.)

SECTION 2.4

Linear Inequalities in One Variable

A linear inequality in one variable is any inequality that can be put in the form

$$ax + b < c \qquad (a, b, \text{ and } c \text{ constants, } a \neq 0)$$

where the inequality symbol ($<$) can be replaced with any of the other three inequality symbols ($\leq$, $>$, or $\geq$).

Some examples of linear inequalities are

$$3x - 2 \geq 7 \qquad -5y < 25 \qquad 3(x - 4) > 2x$$

The first property for inequalities is similar to the addition property we used when solving equations.

◢ **ADDITION PROPERTY FOR INEQUALITIES**

For any algebraic expressions A, B, and C,

$$\text{if} \qquad A < B$$
$$\text{then} \quad A + C < B + C$$

In Words: Adding the same quantity to both sides of an inequality will not change the solution set.

Note: Since subtraction is defined as addition of the opposite, this new property holds for subtraction as well as addition. That is, we can subtract the same quantity from each side of an inequality and always be sure that we have not changed the solution.

EXAMPLE 1 Solve $3x + 3 < 2x - 1$ and graph the solution.

SOLUTION We use the addition property for inequalities to write all the variable terms on one side and all the constant terms on the other side:

$$3x + 3 < 2x - 1$$
$$3x + (-2x) + 3 < 2x + (-2x) - 1 \quad \text{Add } -2x \text{ to each side.}$$
$$x + 3 < -1$$
$$x + 3 + (-3) < -1 + (-3) \qquad \text{Add } -3 \text{ to each side.}$$
$$x < -4$$

The solution set is all real numbers that are less than -4. To show this we can use set notation and write

$$\{x \mid x < -4\}$$

Or we can graph the solution set on a number line using an open circle at -4 to show that -4 is not part of the solution set:

This graph gives rise to the following notation, called **interval notation,** which is an alternative way to write the solution set:

$$(-\infty, -4)$$

The English mathematician John Wallis (1616–1703) was the first person to use the ∞ symbol to represent infinity. When we encounter the interval $(3, \infty)$ we read it as "the interval from 3 to infinity," and we mean the set of real numbers that are greater than 3. Likewise, the interval $(-\infty, -4)$ is read "the interval from negative infinity to -4," which is all real numbers less than -4.

This expression indicates that the solution set is all real numbers from negative infinity up to, but not including, -4.

We have three equivalent representations for the solution set to our original inequality. Here are all three together.

Set Notation	Line Graph	Interval Notation
$\{x \mid x < -4\}$		$(-\infty, -4)$

Before we state the multiplication property for inequalities, we will take a look at what happens to an inequality statement when we multiply both sides by a positive number and what happens when we multiply by a negative number.

We begin by writing three true inequality statements:

$$3 < 5 \qquad\qquad -3 < 5 \qquad\qquad -5 < -3$$

We multiply both sides of each inequality by a positive number—say, 4:

$$4(3) < 4(5) \qquad 4(-3) < 4(5) \qquad 4(-5) < 4(-3)$$
$$12 < 20 \qquad\qquad -12 < 20 \qquad\qquad -20 < -12$$

Notice in each case that the resulting inequality symbol points in the same direction as the original inequality symbol. Multiplying both sides of an inequality by a positive number preserves the *sense* of the inequality.

Let's take the same three original inequalities and multiply both sides by -4:

$$3 < 5 \qquad\qquad -3 < 5 \qquad\qquad -5 < -3$$

$$-4(3) > -4(5) \qquad -4(-3) > -4(5) \qquad -4(-5) > -4(-3)$$
$$-12 > -20 \qquad\qquad 12 > -20 \qquad\qquad 20 > 12$$

Notice in this case that the resulting inequality symbol always points in the opposite direction from the original one. Multiplying both sides of an inequality by a negative number *reverses* the sense of the inequality. Keeping this in mind, we will now state the multiplication property for inequalities.

MULTIPLICATION PROPERTY FOR INEQUALITIES

Let A, B, and C represent algebraic expressions.

If $\qquad A < B$

then $\quad AC < BC \quad$ if $\quad C$ is positive ($C > 0$)

or $\quad AC > BC \quad$ if $\quad C$ is negative ($C < 0$)

In Words: Multiplying both sides of an inequality by a positive number always produces an equivalent inequality. Multiplying both sides of an inequality by a negative number reverses the sense of the inequality.

Note: Since division is defined as multiplication by the reciprocal, we can apply this new property to division as well as to multiplication. We can divide both sides of an inequality by any nonzero number as long as we reverse the direction of the inequality symbol when the number we are dividing by is a negative number.

The multiplication property for inequalities does not limit what we can do with inequalities. We are still free to multiply both sides of an inequality by any nonzero number we choose. If the number we multiply by happens to be *negative*, then we *must* also *reverse* the direction of the inequality symbol.

EXAMPLE 2 Solve: $3x - 5 \leq 7$.

SOLUTION We apply the addition property first, then the multiplication property:

$$3x - 5 \leq 7$$
$$3x - 5 + 5 \leq 7 + 5 \qquad \text{Add } 5 \text{ to both sides.}$$
$$3x \leq 12$$
$$\frac{1}{3}(3x) \leq \frac{1}{3}(12) \qquad \text{Multiply by } \tfrac{1}{3}.$$
$$x \leq 4$$

The solution set consists of all real numbers that are less than or equal to 4. Writing the solution set with set notation and interval notation, we have

$$\{x \mid x \le 4\} = (-\infty, 4]$$

Notice how a square bracket is used with interval notation to show that 4 is part of the solution set.

The graph of the solution set will contain a solid circle at 4, since 4 is included in the solution set. Here is the graph:

EXAMPLE 3 Find the solution set for $-2y - 3 < 7$.

SOLUTION We begin by adding 3 to each side of the inequality:

$$-2y - 3 < 7$$
$$-2y < 10 \qquad \text{Add } +3 \text{ to both sides.}$$

$$-\frac{1}{2}(-2y) > -\frac{1}{2}(10) \qquad \begin{array}{l}\text{Multiply by } -\frac{1}{2} \text{ and reverse the} \\ \text{direction of the inequality symbol.}\end{array}$$
$$y > -5$$

The solution set is all real numbers that are greater than -5. Below are three equivalent ways to represent this solution set.

Set Notation	Line Graph	Interval Notation
$\{y \mid y > -5\}$		$(-5, \infty)$

   When our inequalities become more complicated, we use the same basic steps we used in Section 2.1 when we were solving equations. That is, we simplify each side of the inequality before we apply the addition property or multiplication property. When we have solved the inequality, we graph the solution on a number line.

EXAMPLE 4 Solve: $3(2x - 4) - 7x \le -3x$.

SOLUTION We begin by using the distributive property to separate terms. Next, simplify both sides:

$$3(2x - 4) - 7x \le -3x \qquad \text{Original inequality}$$
$$6x - 12 - 7x \le -3x \qquad \text{Distributive property}$$

$$-x - 12 \le -3x \qquad 6x - 7x = (6 - 7)x = -x$$
$$-12 \le -2x \qquad \text{Add } x \text{ to both sides.}$$

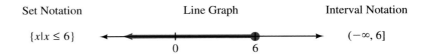

$$-\frac{1}{2}(-12) \ge -\frac{1}{2}(-2x) \qquad \text{Multiply both sides by } -\frac{1}{2} \text{ and}$$
$$\text{reverse the direction of the}$$
$$6 \ge x \qquad\qquad \text{inequality symbol.}$$

This last line is equivalent to $x \le 6$. The solution set can be represented in any of the three ways shown below.

Set Notation	Line Graph	Interval Notation
$\{x \mid x \le 6\}$		$(-\infty, 6]$

Note: In Examples 3 and 4, notice that each time we multiplied both sides of the inequality by a negative number we also reversed the direction of the inequality symbol. Failure to do so would cause our graph to lie on the wrong side of the endpoint.

Before we solve more inequalities, let's review some of the details involved with graphing more complicated inequalities.

Previously we defined the **union** of two sets A and B to be the set of all elements that are in either A or B. The word *or* is the key word in the definition. The **intersection** of two sets A and B is the set of all elements contained in both A and B, the key word here being *and*. We can use the words *and* and *or*, together with our methods of graphing inequalities, to graph some *compound inequalities*.

EXAMPLE 5   Graph: $\{x \mid x \le -2 \text{ or } x > 3\}$.

SOLUTION   The two inequalities connected by the word *or* are referred to as a **compound inequality.** We begin by graphing each inequality separately:

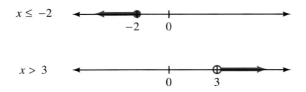

Since the two inequalities are connected by the word *or*, we graph their union. That is, we graph all points on either graph:

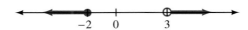

Notice that the square bracket indicates -2 is included in the solution set, and the parenthesis indicates 3 is not included.

To represent this set of numbers with interval notation we use two intervals connected with the symbol for the union of two sets. Here is the equivalent set of numbers described with interval notation:

$$(-\infty, -2] \cup (3, \infty)$$

EXAMPLE 6   Graph: $\{x | x > -1 \text{ and } x < 2\}$.

SOLUTION   We first graph each inequality separately:

Since the two inequalities are connected by the word *and,* we graph their intersection—the part they have in common:

This graph corresponds to the interval $(-1, 2)$, which is called an **open interval** since neither endpoint is included in the interval.

Notation: Sometimes compound inequalities that use the word *and* as the connecting word can be written in a shorter form. For example, the compound inequality $-3 \leq x$ and $x \leq 4$ can be written $-3 \leq x \leq 4$. The word *and* does not appear when an inequality is written in this form. It is implied. Inequalities of the form $-3 \leq x \leq 4$ are called **continued inequalities.** This new notation is useful because writing it takes fewer symbols. The graph of $-3 \leq x \leq 4$ is

The corresponding interval is $[-3, 4]$, which is called a **closed interval** since both endpoints are included in the interval.

EXAMPLE 7   Solve and graph: $-3 \leq 2x - 5 \leq 3$.

SOLUTION   We can extend the properties for addition and multiplication to cover this situation. If we add a number to the middle expression, we must add the same number to the outside expressions. If we multiply the center expression by a number, we must do the same to the outside expressions, remembering to reverse the direction of the inequality symbols if we multiply by a negative number. We begin by adding 5 to all three parts of the inequality:

$$-3 \le 2x - 5 \le 3$$

$$2 \le \quad 2x \quad \le 8 \quad \text{Add 5 to all three members.}$$

$$1 \le \quad x \quad \le 4 \quad \text{Multiply through by } \tfrac{1}{2}.$$

Here are three ways to represent this solution set:

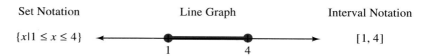

Set Notation	Line Graph	Interval Notation
$\{x \mid 1 \le x \le 4\}$		$[1, 4]$

EXAMPLE 8 Solve the compound inequality

$$3t + 7 \le -4 \quad \text{or} \quad 3t + 7 \ge 4$$

SOLUTION We solve each half of the compound inequality separately, then we graph the solution set:

$$3t + 7 \le -4 \qquad \text{or} \qquad 3t + 7 \ge 4$$

$$3t \le -11 \qquad \text{or} \qquad 3t \ge -3 \quad \text{Add } -7.$$

$$t \le -\frac{11}{3} \qquad \text{or} \qquad t \ge -1 \quad \text{Multiply by } \tfrac{1}{3}.$$

The solution set can be represented in any of the following ways:

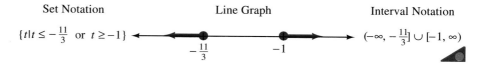

Set Notation	Line Graph	Interval Notation
$\{t \mid t \le -\frac{11}{3} \text{ or } t \ge -1\}$		$(-\infty, -\frac{11}{3}] \cup [-1, \infty)$

EXAMPLE 9 Suppose you have a part-time job that requires that you work at least 10 hours but no more than 20 hours each week. Use the letter t to write an inequality that shows the number of hours you work per week.

SOLUTION If t is at least 10 but no more than 20, then $10 \le t$ and $t \le 20$ or, equivalently, $10 \le t \le 20$. Note that the word *but,* as used here, has the same meaning as the word *and.*

EXAMPLE 10 If the highest temperature on Tuesday was 76°F and the lowest temperature was 55°F, write an inequality using the letter x that gives the range of temperatures on Tuesday.

SOLUTION   Since the smallest value of x is 55 and the largest value of x is 76, then $55 \leq x \leq 76$. We could say that the temperature on Tuesday was between 55°F and 76°F, inclusive.

EXAMPLE 11   A company that manufactures typewriter ribbons finds that they can sell x ribbons each week when they set the price at p dollars each, according to the formula $x = 1{,}300 - 100p$. What price should they charge for each ribbon if they want to sell at least 300 ribbons a week?

SOLUTION   Since x is the number of ribbons they sell each week, an inequality that corresponds to selling at least 300 ribbons a week is

$$x \geq 300$$

Substituting $1{,}300 - 100p$ for x gives us an inequality in the variable p:

$$
\begin{aligned}
1{,}300 - 100p &\geq 300 \\
-100p &\geq -1{,}000 \qquad \text{Add } -1{,}300 \text{ to each side.}
\end{aligned}
$$

$$p \leq 10 \qquad \begin{array}{l}\text{Divide each side by } -100 \text{ and} \\ \text{reverse the direction of the} \\ \text{inequality symbol.}\end{array}$$

In order to sell at least 300 ribbons each week, the price per ribbon should be no more than \$10. That is, selling the ribbons for \$10 or less will produce weekly sales of 300 or more ribbons.

EXAMPLE 12   The formula $F = \frac{9}{5}C + 32$ gives the relationship between the Celsius and Fahrenheit temperature scales. If the temperature range on a certain day is 86–104° Fahrenheit, what is the temperature range in degrees Celsius?

SOLUTION   From the given information we can write

$$86 \leq F \leq 104$$

But, since F is equal to $\frac{9}{5}C + 32$, we can also write

$$
\begin{aligned}
86 &\leq \tfrac{9}{5}C + 32 \leq 104 \\
54 &\leq \quad \tfrac{9}{5}C \quad \leq 72 \qquad \text{Add } -32 \text{ to each member.} \\
\tfrac{5}{9}(54) &\leq \tfrac{5}{9}\left(\tfrac{9}{5}C\right) \leq \tfrac{5}{9}(72) \qquad \text{Multiply by } \tfrac{5}{9}. \\
30 &\leq \quad C \quad \leq 40
\end{aligned}
$$

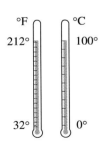

°F   °C

212°   100°

32°   0°

A temperature range of 86–104° Fahrenheit corresponds to a temperature range of 30–40° Celsius.

116

Problem Set 2.4

NOTE: When I work inequalities, I find myself making mistakes and then catching them moments later. You should expect to make some mistakes also. Write neatly, and use a pencil. That way, you can erase your mistakes, and start over.

Solve each of the following inequalities and graph each solution.

1. $2x \le 3$
2. $5x \ge -115$
3. $\frac{1}{2}x > 2$
4. $\frac{1}{3}x > 4$
5. $-5x \le 25$
6. $-7x \ge 35$
7. $-\frac{3}{2}x > -6$
8. $-\frac{2}{3}x < -8$
9. $-12 \le 2x$
10. $-20 \ge 4x$
11. $-1 \ge -\frac{1}{4}x$
12. $-1 \le -\frac{1}{5}x$
13. $-3x + 1 > 10$
14. $-2x - 5 \le 15$

15. $\frac{1}{2} - \frac{m}{12} \le \frac{7}{12}$
16. $\frac{1}{2} - \frac{m}{10} > -\frac{1}{5}$

17. $\frac{1}{2} \ge -\frac{1}{6} - \frac{2}{9}x$
18. $\frac{9}{5} > -\frac{1}{5} - \frac{1}{2}x$
19. $-40 \le 30 - 20y$
20. $-20 > 50 - 30y$
21. $\frac{2}{3}x - 3 < 1$
22. $\frac{3}{4}x - 2 > 7$
23. $10 - \frac{1}{2}y \le 36$
24. $8 - \frac{1}{3}y \ge 20$

NOTE: Remember, you want to list the problems that you have difficulty with, so you will know what to study before your next test. You may find some of the problems for your list right here. Work the difficult problems now, put them on your difficult problem list, then work them again before the next test.

Simplify each side first, then solve the following inequalities. Write your answers with interval notation.

25. $2(3y + 1) \le -10$
26. $3(2y - 4) > 0$
27. $-(a + 1) - 4a \le 2a - 8$
28. $-(a - 2) - 5a \le 3a + 7$
29. $\frac{1}{3}t - \frac{1}{2}(5 - t) < 0$
30. $\frac{1}{4}t - \frac{1}{3}(2t - 5) < 0$

31. $-2 \le 5 - 7(2a + 3)$
32. $1 < 3 - 4(3a - 1)$
33. $-\frac{1}{3}(x + 5) \le -\frac{2}{9}(x - 1)$
34. $-\frac{1}{2}(2x + 1) \le -\frac{3}{8}(x + 2)$
35. $5(y + 3) + 4 < 6y - 1 - 5y$
36. $4(y - 1) + 2 \ge 3y + 8 - 2y$

Solve the following continued inequalities. Use both a line graph and interval notation to write each solution set.

37. $-2 \le m - 5 \le 2$
38. $-3 \le m + 1 \le 3$
39. $-60 < 20a + 20 < 60$
40. $-60 < 50a - 40 < 60$
41. $0.5 \le 0.3a - 0.7 \le 1.1$
42. $0.1 \le 0.4a + 0.1 \le 0.3$
43. $3 < \frac{1}{2}x + 5 < 6$
44. $5 < \frac{1}{4}x + 1 < 9$
45. $4 < 6 + \frac{2}{3}x < 8$
46. $3 < 7 + \frac{4}{5}x < 15$

Graph the solution sets for the following compound inequalities. Then write each solution set using interval notation.

47. $x + 5 \le -2$ or $x + 5 \ge 2$
48. $3x + 2 < -3$ or $3x + 2 > 3$
49. $5y + 1 \le -4$ or $5y + 1 \ge 4$
50. $7y - 5 \le -2$ or $7y - 5 \ge 2$
51. $2x + 5 < 3x - 1$ or $x - 4 > 2x + 6$
52. $3x - 1 > 2x + 4$ or $5x - 2 < 3x + 4$

In Words

Translate each of the following phrases into an equivalent inequality statement.

53. x is at least 5.
54. x is at least -2.
55. x is no more than -3.
56. x is no more than 8.
57. x is at most 4.
58. x is at most -5.
59. x is between -4 and 4.
60. x is between -3 and 3.
61. x is between -4 and 4, inclusive.
62. x is between -3 and 3, inclusive.

Pricing *A store selling art supplies finds that they can sell x sketch pads each week at a price of p dollars each, according to the formula x = 900 − 300p. What price should they charge if they want to sell:*

63. At least 300 pads each week?

64. More than 600 pads each week?

65. Less than 525 pads each week?

66. At most 375 pads each week?

Solve each of the following formulas for y.

67. $3x + 2y < 6$ **68.** $2x + 3y > 6$

69. $4x - 5y \geq 20$ **70.** $5x - 3y \leq 15$

Each of the temperature ranges given below is in degrees Fahrenheit. Use the formula $F = \frac{9}{5}C + 32$ to find the corresponding temperature range in degrees Celsius.

71. 95° to 113°   **72.** 68° to 86°

73. − 13° to 14°   **74.** − 4° to 23°

75. Suppose that a number lies somewhere between 2 and 5 on the number line. (If we let x represent this number, then an inequality that describes this situation is $2 < x < 5$.) Between what two numbers does the opposite of the number lie? Write an inequality that shows where the opposite of the number lies.

76. A number lies between − 2 and 3 on the number line. Write an inequality that shows where the opposite of the number lies.

77. Is − 5 one of the numbers in the solution set for the inequality $|x| < 2$?

78. Is − 3 one of the numbers in the solution set for the inequality $|x| > 2$?

Review Problems

 The problems below review the different kinds of factoring we did in Sections 1.6 and 1.7.

Factor completely.

79. $6x^4y - 9xy^4 + 18x^3y^3$

80. $x^2y + 3xy + 4x + 12$

81. $x^2 - 5x + 6$ **82.** $2x^3 + 4x^2 - 30x$

83. $4x^2 - 20x + 25$ **84.** $2xb^3 - 8x^3b$

85. $x^4 - 16$ **86.** $8x^3 - 125$

87. $x^2 - 12x + 36 - y^2$

88. $49 - 25x^2$

89. $a^2 + 49$ **90.** $x^2 - 3x - 3$

One Step Further

Assume that a, b, and c are positive, and solve each formula for x.

91. $ax + b < c$ **92.** $\dfrac{x}{a} + \dfrac{y}{b} < 1$

93. $-c < ax + b < c$

94. $-1 < \dfrac{ax + b}{c} < 1$

Equations with Absolute Value

In Chapter 1 we defined the **absolute value** of x, $|x|$, to be the distance between x and 0 on the number line. The absolute value of a number measures its distance from 0.

EXAMPLE 1   Solve for x: $|x| = 5$.

SOLUTION   Using the definition of absolute value, we can read the equation as "The distance between x and 0 on the number line is 5." If x is 5 units from 0, then x can be 5 or − 5:

$$\text{If } |x| = 5 \quad \text{then} \quad x = 5 \quad \text{or} \quad x = -5$$

In general, then, we can see that any equation of the form $|a| = b$ is equivalent to the equations $a = b$ or $a = -b$, as long as $b > 0$.

EXAMPLE 2 Solve: $|2a - 1| = 7$.

SOLUTION We can read this equation as "$2a - 1$ is 7 units from 0 on the number line." The quantity $2a - 1$ must be equal to 7 or -7:

$$|2a - 1| = 7$$

$$2a - 1 = 7 \quad \text{or} \quad 2a - 1 = -7$$

We have transformed our absolute value equation into two equations that do not involve absolute value. We can solve each equation using the method in Section 2.1:

$$
\begin{array}{llll}
2a - 1 = 7 & \text{or} & 2a - 1 = -7 & \\
2a = 8 & \text{or} & 2a = -6 & \text{Add } +1 \text{ to both sides.} \\
a = 4 & \text{or} & a = -3 & \text{Multiply by } \frac{1}{2}.
\end{array}
$$

Our solution set is $\{4, -3\}$.

Check To check our solutions, we put each of them into the original absolute value equation:

When $a = 4$:

the equation	$|2a - 1| = 7$
becomes	$|2(\mathbf{4}) - 1| \stackrel{?}{=} 7$
	$|7| \stackrel{?}{=} 7$
	$7 = 7$

When $a = -3$:

the equation	$|2a - 1| = 7$
becomes	$|2(-\mathbf{3}) - 1| \stackrel{?}{=} 7$
	$|-7| \stackrel{?}{=} 7$
	$7 = 7$

EXAMPLE 3 Solve: $\left|\dfrac{2}{3}x - 3\right| + 5 = 12$.

SOLUTION In order to use the definition of absolute value to solve this equation, we must isolate the absolute value on the left side of the equal sign. To do so, we add -5 to both sides of the equation to obtain

$$\left|\frac{2}{3}x - 3\right| = 7$$

Now that the equation is in the correct form, we can write

$$
\begin{array}{llll}
\dfrac{2}{3}x - 3 = 7 & \text{or} & \dfrac{2}{3}x - 3 = -7 & \\
\dfrac{2}{3}x = 10 & \text{or} & \dfrac{2}{3}x = -4 & \text{Add } +3 \text{ to both sides.} \\
x = 15 & \text{or} & x = -6 & \text{Multiply by } \frac{3}{2}.
\end{array}
$$

The solution set is $\{15, -6\}$.

EXAMPLE 4   Solve: $|3a - 6| = -4$.

SOLUTION   The solution set is the null set, $\varnothing$, because the left side cannot be negative and the right side is negative. No matter what we try to substitute for the variable a, the quantity $|3a - 6|$ will always be positive or zero. It can never be -4.

Consider the statement $|a| = |b|$. What can we say about a and b? We know they are equal in absolute value. By the definition of absolute value, they are the same distance from 0 on the number line. They must be equal to each other or opposites of each other. In symbols, we write

$$|a| = |b| \quad \Leftrightarrow \quad a = b \ \text{ or } \ a = -b$$

$$\uparrow \qquad\qquad\qquad \uparrow \qquad\qquad \uparrow$$

Equal in   Equals   or   Opposites
absolute value

EXAMPLE 5   Solve: $|3a + 2| = |2a + 3|$.

SOLUTION   The quantities $3a + 2$ and $2a + 3$ have equal absolute values. They are therefore the same distance from 0 on the number line. They must be equals or opposites:

$$|3a + 2| = |2a + 3|$$

EQUALS		OPPOSITES
$3a + 2 = 2a + 3$	or	$3a + 2 = -(2a + 3)$
$a + 2 = 3$		$3a + 2 = -2a - 3$
$a = 1$		$5a + 2 = -3$
		$5a = -5$
		$a = -1$

The solution set is $\{1, -1\}$.

It makes no difference in the outcome of the problem if we take the opposite of the first or second expression. But it is very important, once we have decided which one to take the opposite of, that we take the opposite of both its terms and not just the first term. That is, the opposite of $2a + 3$ is $-(2a + 3)$, which we can think of as $-1(2a + 3)$. Distributing the -1 across *both* terms, we have

$$-1(2a + 3) = -2a - 3$$

EXAMPLE 6   Solve: $|x - 5| = |x - 7|$.

SOLUTION   As was the case in Example 5, the quantities $x - 5$ and $x - 7$ must be equals or opposites, because their absolute values are equal:

EQUALS		OPPOSITES
$x - 5 = x - 7$	or	$x - 5 = -(x - 7)$
$-5 = -7$		$x - 5 = -x + 7$
No solution here		$2x - 5 = 7$
		$2x = 12$
		$x = 6$

Since the first equation leads to a false statement, it will not give us a solution. In this case, our only solution is $x = 6$.

Note: The equation $x - 5 = x - 7$ that appears in the solution to Example 6 is a linear equation that does not have a unique solution. For all equations of this type, attempting to solve the equation eliminates the variable completely. The resulting statement is either a true statement, or a false statement (as in Example 6). A false statement indicates that there is no solution to the original equation. A true statement indicates that all real numbers are solutions to the equation. That is, substituting any real number for the variable in the equation leads to a true statement. Equations for which all real numbers are solutions are called **identities.**

Problem Set
2.5

Use the definition of absolute value to solve each of the following problems.

1. $|x| = 4$
2. $|x| = 7$
3. $2 = |a|$
4. $5 = |a|$
5. $|x| = -3$
6. $|x| = -4$
7. $|a| + 2 = 3$
8. $|a| - 5 = 2$
9. $|y| + 4 = 3$
10. $|y| + 3 = 1$
11. $4 = |x| - 2$
12. $3 = |x| - 5$
13. $|x - 2| = 5$
14. $|x + 1| = 2$
15. $|a - 4| = \frac{5}{3}$
16. $|a + 2| = \frac{7}{5}$
17. $1 = |3 - x|$
18. $2 = |4 - x|$
19. $\left|\frac{3}{5}a + \frac{1}{2}\right| = 1$
20. $\left|\frac{2}{7}a + \frac{3}{4}\right| = 1$
21. $60 = |20x - 40|$
22. $800 = |400x - 200|$
23. $|2x + 1| = -3$
24. $|2x - 5| = -7$
25. $\left|\frac{3}{4}x - 6\right| = 9$
26. $\left|\frac{4}{5}x - 5\right| = 15$
27. $\left|1 - \frac{1}{2}a\right| = 3$
28. $\left|2 - \frac{1}{3}a\right| = 10$

NOTE: As you approach the end of this chapter, keep to your schedule. Success in mathematics is related to how well you can discipline yourself to keep to the schedule and goals you have set for yourself.

Solve each equation.

29. $|3x + 4| + 1 = 7$
30. $|5x - 3| - 4 = 3$
31. $|3 - 2y| + 4 = 3$
32. $|8 - 7y| + 9 = 1$
33. $3 + |4t - 1| = 8$
34. $2 + |2t - 6| = 10$
35. $\left|9 - \frac{3}{5}x\right| + 6 = 12$
36. $\left|4 - \frac{2}{7}x\right| + 2 = 14$
37. $5 = \left|\frac{2x}{7} + \frac{4}{7}\right| - 3$
38. $7 = \left|\frac{3x}{5} + \frac{1}{5}\right| + 2$

39. $2 = -8 + |4 - \frac{1}{2}y|$

40. $1 = -3 + |2 - \frac{1}{4}y|$

41. $|3(x + 5) + 2| = 1$

42. $|2(x - 4) + 3| = 7$

43. $|1 + 2(x - 1)| = 7$

44. $|3 + 4(x + 2)| = 5$

45. $1 = |2(k + 4) - 3|$

46. $4 = |3(k - 2) + 1|$

Solve the following equations.

47. $|3a + 1| = |2a - 4|$

48. $|5a + 2| = |4a + 7|$

49. $|x - \frac{1}{3}| = |\frac{1}{2}x + \frac{1}{6}|$

50. $|\frac{1}{10}x - \frac{1}{2}| = |\frac{1}{5}x + \frac{1}{10}|$

51. $|y - 2| = |y + 3|$

52. $|y - 5| = |y - 4|$

53. $|3x - 1| = |3x + 1|$

54. $|5x - 8| = |5x + 8|$

55. $|3 - m| = |m + 4|$

56. $|5 - m| = |m + 8|$

57. $|0.03 - 0.01x| = |0.04 + 0.05x|$

58. $|0.07 - 0.01x| = |0.08 - 0.02x|$

59. $|x - 2| = |2 - x|$

60. $|x - 4| = |4 - x|$

61. $\left|\frac{x}{5} - 1\right| = \left|1 - \frac{x}{5}\right|$

62. $\left|\frac{x}{3} - 1\right| = \left|1 - \frac{x}{3}\right|$

63. Each of the equations in Problems 59–62 has the form $|a - b| = |b - a|$. The solution set for each of these equations is all real numbers. This means that the statement $|a - b| = |b - a|$ must be true no matter what numbers a and b are. The statement itself is a property of absolute value. Show that the statement is true when $a = 4$ and $b = -7$, as well as when $a = -5$ and $b = -8$.

64. Show that the statement $|ab| = |a||b|$ is true when $a = 3$ and $b = -6$, and when $a = -8$ and $b = -2$.

65. Name all the numbers in the set

$$\{-4, -3, -2, -1, 0, 1, 2, 3, 4\}$$

that are solutions to the equation $|x + 2| = x + 2$.

66. Name all the numbers in the set

$$\{-4, -3, -2, -1, 0, 1, 2, 3, 4\}$$

that are solutions to the equation $|x - 2| = x - 2$.

67. The equation $|a| = a$ is true only if $a \geq 0$. We can use this fact to actually solve equations like the one in Problem 65. The only way the equation $|x + 2| = x + 2$ can be true is if $x + 2 \geq 0$. Adding -2 to both sides of this last inequality we see that $x \geq -2$. Use the same kind of reasoning to solve the equation $|x - 2| = x - 2$.

68. Solve the equation $|x + 3| = x + 3$.

Review Problems

 The problems below review material we covered in Section 1.5.

Multiply.

69. $2x^2(5x^3 + 4x - 3)$

70. $3x^3(7x^2 - 4x - 8)$

71. $(3a - 1)(4a + 5)$ **72.** $(6a - 3)(2a + 1)$

73. $(x + 3)(x - 3)(x^2 + 9)$

74. $(2x - 3)(4x^2 + 6x + 9)$

75. $(4y - 5)^2$ **76.** $(2y - \frac{1}{2})^2$

77. $(3x + 7)(4y - 2)$ **78.** $(x + 2a)(2 - 3b)$

79. $(3 - t^2)^2$ **80.** $(2 - t^3)^2$

One Step Further

Solve each formula for x. (Assume a, b, and c are positive.)

81. $|x - a| = b$ **82.** $|x + a| - b = 0$

83. $|ax + b| = c$ **84.** $|ax - b| - c = 0$

85. $\left|\frac{x}{a} + \frac{y}{b}\right| = 1$ **86.** $\left|\frac{x}{a} + \frac{y}{b}\right| = c$

Inequalities Involving Absolute Value

In this section we will again apply the definition of absolute value to solve inequalities involving absolute value. Again, the **absolute value** of x, which is $|x|$, represents the distance that x is from 0 on the number line. We will begin by considering three absolute value expressions and their English translations:

EXPRESSION	IN WORDS
$|x| = 7$	x is exactly 7 units from 0 on the number line.
$|a| < 5$	a is less than 5 units from 0 on the number line.
$|y| \geq 4$	y is greater than or equal to 4 units from 0 on the number line.

Once we have translated the expression into words, we can use the translation to graph the original equation or inequality. The graph is then used to write a final equation or inequality that does not involve absolute value.

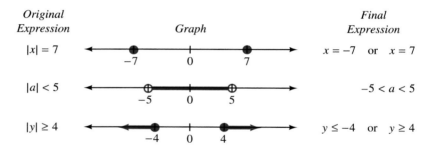

Original Expression	Graph	Final Expression
$|x| = 7$	$-7 \quad 0 \quad 7$	$x = -7$ or $x = 7$
$|a| < 5$	$-5 \quad 0 \quad 5$	$-5 < a < 5$
$|y| \geq 4$	$-4 \quad 0 \quad 4$	$y \leq -4$ or $y \geq 4$

Although we will not always write out the English translation of an absolute value inequality, it is important that we understand the translation. Our second expression, $|a| < 5$, means a is within 5 units of 0 on the number line. The graph of this relationship is

$$-5 \quad 0 \quad 5$$

which can be written with the following continued inequality:

$$-5 < a < 5$$

We can follow this same kind of reasoning to solve more complicated absolute value inequalities.

EXAMPLE 1 Graph the solution set: $|2x - 5| < 3$.

SOLUTION The absolute value of $2x - 5$ is the distance that $2x - 5$ is from 0 on the number line. We can translate the inequality as "$2x - 5$ is less than 3 units from

0 on the number line." That is, $2x - 5$ must appear between -3 and 3 on the number line. A picture of this relationship is

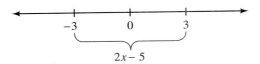

Using the picture, we can write an inequality without absolute value that describes the situation:

$$-3 < 2x - 5 < 3$$

Next, we solve the continued inequality by first adding $+5$ to all three members and then multiplying all three by $\frac{1}{2}$:

$$-3 < 2x - 5 < 3$$
$$2 < \quad 2x \quad < 8 \quad \text{Add } +5 \text{ to all three expressions.}$$
$$1 < \quad x \quad < 4 \quad \text{Multiply each expression by } \frac{1}{2}.$$

The graph of the solution set is

We can see from the solution that in order for the absolute value of $2x - 5$ to be within 3 units of 0 on the number line, x must be between 1 and 4. ◢

EXAMPLE 2   Solve and graph: $|3a + 7| \leq 4$.

SOLUTION   We can read the inequality as "The distance between $3a + 7$ and 0 is less than or equal to 4." Or, "$3a + 7$ is within 4 units of 0 on the number line." This relationship can be written without absolute value as

$$-4 \leq 3a + 7 \leq 4$$

Solving as usual, we have

$$-4 \leq 3a + 7 \leq 4$$
$$-11 \leq \quad 3a \quad \leq -3 \quad \text{Add } -7 \text{ to all three members.}$$
$$-\frac{11}{3} \leq \quad a \quad \leq -1 \quad \text{Multiply each expression by } \frac{1}{3}.$$

We can see from Examples 1 and 2 that in order to solve an inequality involving absolute value, we must be able to write an equivalent expression that does not involve absolute value.

EXAMPLE 3   Solve $|x - 3| > 5$ and graph the solution.

SOLUTION   We interpret the absolute value inequality to mean that $x - 3$ is more than 5 units from 0 on the number line. The quantity $x - 3$ must be either above $+5$ or below -5. Here is a picture of the relationship:

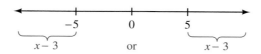

An inequality without absolute value that also describes this situation is

$$x - 3 < -5 \quad \text{or} \quad x - 3 > 5$$

Adding $+3$ to both sides of each inequality, we have

$$x < -2 \quad \text{or} \quad x > 8$$

The graph is

EXAMPLE 4   Graph the solution set: $|4t - 3| \geq 9$.

SOLUTION   The quantity $4t - 3$ is greater than or equal to 9 units from 0. It must be either above $+9$ or below -9.

$$4t - 3 \leq -9 \quad \text{or} \quad 4t - 3 \geq 9$$
$$4t \leq -6 \quad \text{or} \quad 4t \geq 12 \quad \text{Add } +3.$$
$$t \leq -\frac{6}{4} \quad \text{or} \quad t \geq \frac{12}{4} \quad \text{Multiply by } \tfrac{1}{4}.$$
$$t \leq -\frac{3}{2} \quad \text{or} \quad t \geq 3$$

We can use the results of our first few examples and the material in the previous section to summarize the information related to absolute value equations and inequalities.

Rewriting Absolute Value Equations and Inequalities

If c is a positive real number, then each statement on the left is equivalent to the corresponding statement on the right.

WITH ABSOLUTE VALUE	WITHOUT ABSOLUTE VALUE
$|x| = c$	$x = -c$ or $x = c$
$|x| < c$	$-c < x < c$
$|x| > c$	$x < -c$ or $x > c$
$|ax + b| = c$	$ax + b = -c$ or $ax + b = c$
$|ax + b| < c$	$-c < ax + b < c$
$|ax + b| > c$	$ax + b < -c$ or $ax + b > c$

EXAMPLE 5   Solve and graph: $|2x + 3| + 4 < 9$.

SOLUTION   Before we can apply the method of solution we used in the previous examples, we must isolate the absolute value on one side of the inequality. To do so, we add -4 to each side:

$$|2x + 3| + 4 < 9$$
$$|2x + 3| + 4 + (-4) < 9 + (-4)$$
$$|2x + 3| < 5$$

From this last line we know that $2x + 3$ must be between -5 and $+5$.

$$-5 < 2x + 3 < 5$$
$$-8 < \quad 2x \quad < 2 \quad \text{Add } -3 \text{ to each expression.}$$
$$-4 < \quad x \quad < 1 \quad \text{Multiply each expression by } \tfrac{1}{2}.$$

The graph is

EXAMPLE 6   Solve and graph: $|4 - 2t| > 2$.

SOLUTION   The inequality indicates that $4 - 2t$ is less than -2 or greater than $+2$. Writing this without absolute value symbols, we have

$$4 - 2t < -2 \qquad\qquad \text{or} \qquad\qquad 4 - 2t > 2$$

To solve these inequalities, we begin by adding -4 to each side:

$$4 + (-4) - 2t < -2 + (-4) \qquad \text{or} \qquad 4 + (-4) - 2t > 2 + (-4)$$
$$-2t < -6 \qquad\qquad\qquad \text{or} \qquad\qquad\qquad -2t > -2$$

Next we must multiply both sides of each inequality by $-\frac{1}{2}$. When we do so, we must also reverse the direction of each inequality symbol.

$$-2t < -6 \qquad \text{or} \qquad -2t > -2$$

$$-\frac{1}{2}(-2t) > -\frac{1}{2}(-6) \qquad \text{or} \qquad -\frac{1}{2}(-2t) < -\frac{1}{2}(-2)$$

$$t > 3 \qquad\qquad \text{or} \qquad\qquad t < 1$$

Although we are used to seeing the "less than" symbol written first in situations like this, the meaning of the solution is clear. We want to graph all real numbers that are either greater than 3 or less than 1. Here is the graph:

Since absolute value always results in a nonnegative quantity, we sometimes come across special solution sets when a negative number appears on the right side of an absolute value inequality.

EXAMPLE 7   Solve: $|7y - 1| < -2$.

SOLUTION   The *left* side is never negative, because it is an absolute value. The *right* side is negative. We have a nonnegative quantity less than a negative quantity, which is impossible. The solution set is the empty set, $\varnothing$. There is no real number to substitute for y to make the above inequality a true statement.

EXAMPLE 8   Solve: $|6x + 2| > -5$.

SOLUTION   This is the opposite case from that in Example 7. No matter what real number we use for x on the *left* side, the result will always be positive or zero. The *right* side is negative. We have a nonnegative quantity greater than a negative quantity. Every real number we choose for x gives us a true statement. The solution set is the set of all real numbers.

Problem Set

2.6

Solve each of the following inequalities using the definition of absolute value. Graph the solution set in each case.

1. $|x| < 3$

2. $|x| \leq 7$

3. $|x| \geq 2$

4. $|x| > 4$

5. $|x| + 2 < 5$

6. $|x| - 3 < -1$

7. $|t| - 3 > 4$

8. $|t| + 5 > 8$

9. $|y| < -5$

10. $|y| > -3$

11. $|x| \geq -2$

12. $|x| \leq -4$

13. $|x - 3| < 7$

14. $|x + 4| < 2$

15. $|a + 5| \geq 4$

16. $|a - 6| \geq 3$

NOTE: If you have trouble with the problems here, take advantage of the resources available to you. Watch the videotapes, try the tutorial software, join a study group, and visit your instructor during office hours.

Solve each inequality and graph the solution set.

17. $|a - 1| < -3$ **18.** $|a + 2| \geq -5$

19. $|2x - 4| < 6$ **20.** $|2x + 6| < 2$

21. $|3y + 9| \geq 6$ **22.** $|5y - 1| \geq 4$

23. $|2k + 3| \geq 7$ **24.** $|2k - 5| \geq 3$

25. $|x - 3| + 2 < 6$

26. $|x + 4| - 3 < -1$

27. $|2a + 1| + 4 \geq 7$ **28.** $|2a - 6| - 1 \geq 2$

29. $|3x + 5| - 8 < 5$ **30.** $|6x - 1| - 4 \leq 2$

NOTE: Keep in mind that if you multiply or divide both sides of an inequality by a negative number, you must reverse the sense of the inequality.

Solve each inequality and graph the solution set.

31. $|5 - x| > 3$ **32.** $|7 - x| > 2$

33. $|3 - \frac{2}{3}x| \geq 5$ **34.** $|3 - \frac{3}{4}x| \geq 9$

35. $|2 - \frac{1}{2}x| > 1$ **36.** $|3 - \frac{1}{3}x| > 1$

Solve each inequality.

37. $|x - 1| < 0.01$ **38.** $|x + 1| < 0.01$

39. $|2x + 1| \geq \frac{1}{5}$ **40.** $|2x - 1| \geq \frac{1}{8}$

41. $\left|\dfrac{3x - 2}{5}\right| \leq \dfrac{1}{2}$ **42.** $\left|\dfrac{4x - 3}{2}\right| \leq \dfrac{1}{3}$

43. $|2x - \frac{1}{5}| < 0.3$ **44.** $|3x - \frac{3}{5}| < 0.2$

45. Write the continued inequality $-4 \leq x \leq 4$ as a single inequality involving absolute value.

46. Write the continued inequality $-8 \leq x \leq 8$ as a single inequality involving absolute value.

47. Write $-1 \leq x - 5 \leq 1$ as a single inequality involving absolute value.

48. Write $-3 \leq x + 2 \leq 3$ as a single inequality involving absolute value.

Review Problems

NOTE: The problems below review material we covered in Section 1.4.

Simplify each expression. Assume all variables represent nonzero real numbers, and write your answer with positive exponents only.

49. 3^{-2} **50.** $\dfrac{x^6}{x^{-4}}$

51. $\dfrac{15x^3y^8}{5xy^{10}}$ **52.** $(2a^{-3}b^4)^2$

53. $\dfrac{(3x^{-3}y^5)^{-2}}{(9xy^{-2})^{-1}}$ **54.** $(3x^4y)^2(5x^3y^4)^3$

Write each number in scientific notation.

55. $54{,}000$ **56.** 0.0359

Write each number in expanded form.

57. 6.44×10^3 **58.** 2.5×10^{-2}

Simplify each expression as much as possible. Write all answers in scientific notation.

59. $(3 \times 10^8)(4 \times 10^{-5})$

60. $\dfrac{8 \times 10^5}{2 \times 10^{-8}}$

One Step Further

Assume a, b, and c are positive, and solve each formula for x.

61. $|x - a| < b$ **62.** $|x - a| > b$

63. $|ax - b| > c$ **64.** $|ax - b| < c$

REVIEW FOR CHAPTER 2

Examples

1. We can solve

$$x + 3 = 5$$

by adding -3 to both sides:

$$x + 3 + (-3) = 5 + (-3)$$
$$x = 2$$

2. We can solve $3x = 12$ by multiplying both sides by $\frac{1}{3}$:

$$3x = 12$$
$$\tfrac{1}{3}(3x) = \tfrac{1}{3}(12)$$
$$x = 4$$

3. Solve: $3(2x - 1) = 9$.

$$3(2x - 1) = 9$$
$$6x - 3 = 9$$
$$6x - 3 + 3 = 9 + 3$$
$$6x = 12$$
$$\tfrac{1}{6}(6x) = \tfrac{1}{6}(12)$$
$$x = 2$$

4. Solve: $x^2 - 5x = -6$.

$$x^2 - 5x + 6 = 0$$
$$(x - 3)(x - 2) = 0$$
$$x - 3 = 0 \text{ or } x - 2 = 0$$
$$x = 3 \text{ or } \quad x = 2$$

Chapter 2 Summary

Addition Property of Equality [2.1]

For algebraic expressions A, B, and C,

$$\text{if} \quad A = B$$
$$\text{then} \quad A + C = B + C$$

This property states that we can add the same quantity to both sides of an equation without changing the solution set.

Multiplication Property of Equality [2.1]

For algebraic expressions A, B, and C,

$$\text{if} \quad A = B$$
$$\text{then} \quad AC = BC \qquad C \neq 0$$

Multiplying both sides of an equation by the same nonzero quantity never changes the solution set.

Strategy for Solving Linear Equations in One Variable [2.1]

Step 1 (a) Use the distributive property to separate terms, if necessary.

(b) If fractions are present, consider multiplying both sides by the LCD to eliminate the fractions. If decimals are present, consider multiplying both sides by a power of 10 to clear the equation of decimals.

(c) Combine similar terms on each side of the equation.

Step 2 Use the addition property of equality to get all variable terms on one side of the equation and all constant terms on the other side. A variable term is a term that contains the variable (for example, $5x$). A constant term is a term that does not contain the variable (the number 3, for example).

Step 3 Use the multiplication property of equality to get x (that is, $1x$) by itself on one side of the equation.

Step 4 Check your solution in the original equation to be sure that you have not made a mistake in the solution process.

To Solve a Quadratic Equation by Factoring [2.1]

Step 1 Write the equation in standard form.
Step 2 Factor the left side.
Step 3 Use the zero-factor property to set each factor equal to 0.
Step 4 Solve the resulting linear equations.
Step 5 Check the solutions in the original equation.

5. Solve for w:

$$P = 2l + 2w$$
$$P - 2l = 2w$$
$$\frac{P - 2l}{2} = w$$

Formulas [2.2]

A **formula** in algebra is an equation involving more than one variable. To solve a formula for one of its variables, simply isolate that variable on one side of the equation.

6. The perimeter of a rectangle is 32 inches. If the length is 3 times the width, find the dimensions.

Step 1 This step is done mentally.

Step 2 Let $x = $ the width. Then the length is $3x$.

Step 3 The perimeter is 32; therefore

$$2x + 2(3x) = 32$$

Step 4 $8x = 32$

$$x = 4$$

Step 5 The width is 4 inches. The length is $3(4) = 12$ inches.

Step 6 The perimeter is $2(4) + 2(12)$, which is 32. The length is 3 times the width.

Blueprint for Problem Solving [2.3]

Step 1 **Read** the problem, and then mentally **list** the items that are known and the items that are unknown.

Step 2 **Assign a variable** to one of the unknown items. (In most cases this will amount to letting $x = $ the item that is asked for in the problem.) Then **translate** the other **information** in the problem to expressions involving the variable.

Step 3 **Reread** the problem, and then **write an equation,** using the items and variables listed in Steps 1 and 2, that describes the situation.

Step 4 **Solve the equation** found in Step 3.

Step 5 **Write** your **answer** using a complete sentence.

Step 6 **Reread** the problem and **check** your solution with the original words in the problem.

7. Adding 5 to both sides of the inequality $x - 5 < -2$ gives

$$x - 5 + 5 < -2 + 5$$
$$x < 3$$

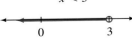

Addition Property for Inequalities [2.4]

For expressions A, B, and C,

$$\text{if} \qquad A < B$$
$$\text{then} \quad A + C < B + C$$

Adding the same quantity to both sides of an inequality never changes the solution set.

8. Multiplying both sides of $-2x \geq 6$ by $-\frac{1}{2}$ gives

$$-2x \geq 6$$

$$\downarrow$$

$$-\tfrac{1}{2}(-2x) \leq -\tfrac{1}{2}(6)$$

$$x \leq -3$$

Multiplication Property for Inequalities [2.4]

For expressions A, B, and C,

$$\text{if} \quad A < B$$
$$\text{then} \quad AC < BC \quad \text{if} \quad C > 0 \ (C \text{ is positive})$$
$$\text{or} \quad AC > BC \quad \text{if} \quad C < 0 \ (C \text{ is negative})$$

We can multiply both sides of an inequality by the same nonzero number without changing the solution set as long as each time we multiply by a negative number we also reverse the direction of the inequality symbol.

9. To solve $|2x - 1| + 2 = 7$, we first isolate the absolute value on the left side by adding -2 to each side to obtain

$$|2x - 1| = 5$$
$$2x - 1 = 5 \ \text{or} \ 2x - 1 = -5$$
$$2x = 6 \ \text{or} \quad 2x = -4$$
$$x = 3 \ \text{or} \quad x = -2$$

Absolute Value Equations [2.5]

To solve an equation that involves absolute value, we isolate the absolute value on one side of the equation and then rewrite the absolute value equation as two separate equations that do not involve absolute value. In general, if b is a positive real number, then

$$|a| = b \quad \text{is equivalent to} \quad a = b \quad \text{or} \quad a = -b$$

10. To solve $|x - 3| + 2 < 6$, we first add -2 to both sides to obtain

$$|x - 3| < 4$$

which is equivalent to

$$-4 < x - 3 < 4$$
$$-1 < \quad x \quad < 7$$

Absolute Value Inequalities [2.6]

To solve an inequality that involves absolute value, we first isolate the absolute value on the left side of the inequality symbol. Then we rewrite the absolute value inequality as an equivalent continued or compound inequality that does not contain absolute value symbols. In general, if b is a positive real number, then

$$|a| < b \quad \text{is equivalent to} \quad -b < a < b$$

and

$$|a| > b \quad \text{is equivalent to} \quad a < -b \quad \text{or} \quad a > b$$

Common Mistakes

A very common mistake in solving inequalities is to forget to reverse the direction of the inequality symbol when multiplying both sides by a negative number. When this mistake occurs, the graph of the solution set is always drawn on the wrong side of the endpoint.

Chapter 2 Review Problems

Solve each equation. [2.1]

1. $x - 3 = 7$
2. $x + 5 = 4$
3. $5x - 2 = 8$
4. $3x - 4 = 5$
5. $400 - 100a = 200$
6. $300 - 100a = -500$
7. $5 - \frac{2}{3}a = 7$
8. $3 - \frac{4}{5}a = -5$
9. $4x - 2 = 7x + 7$
10. $9x - 1 = -7x - 1$
11. $7y - 5 - 2y = 2y - 3$
12. $8y - 4 - 6y = 5y - 1$
13. $\dfrac{3y}{4} - \dfrac{1}{2} + \dfrac{3y}{2} = 2 - y$
14. $\dfrac{y}{4} - 1 + \dfrac{3y}{8} = \dfrac{3}{4} - y$
15. $8 - 3(2t + 1) = 5(t + 2)$
16. $7 - 2(8t - 3) = 4(t - 2)$
17. $6 + 4(1 - 3t) = -3(t - 4) + 2$
18. $8 + 5(4 - 2t) = -2(t - 1) - 4$
19. $a^2 - a - 6 = 0$
20. $a^2 - 4a - 5 = 0$
21. $2x^2 - 5x = 12$
22. $2x^2 - 11x = -12$
23. $10y^2 = 3y + 4$
24. $8y^2 = 14y - 5$
25. $9x^2 - 25 = 0$
26. $16x^2 - 9 = 0$
27. $81a^2 = 1$
28. $5x^2 = -10x$
29. $0.08x + 0.07(900 - x) = 67$
30. $0.07x + 0.06(1,400 - x) = 90$
31. $(x - 2)(x - 3) = 2$
32. $(x - 3)(x - 5) = 3$
33. $x^3 + 4x^2 - 9x - 36 = 0$
34. $9x^3 + 18x^2 - 4x - 8 = 0$

Solve each formula for the variable that does not have a numerical replacement. [2.2]

35. $A = \frac{1}{2}bh$: $A = 3$, $b = 6$
36. $A = \frac{1}{2}bh$: $A = 6$, $h = 3$
37. $P = 2b + 2h$: $P = 40$, $b = 3$
38. $P = 2b + 2h$: $P = 100$, $h = 10$
39. $A = P + Prt$: $A = 2,000$, $P = 1,000$, $r = 0.05$
40. $A = P + Prt$: $A = 1,000$, $P = 500$, $r = 0.1$
41. $A = a + (n - 1)d$: $A = 40$, $a = 4$, $d = 9$
42. $A = a + (n - 1)d$: $A = 32$, $a = 2$, $d = 10$

Solve each formula for the indicated variable. [2.2]

43. $I = prt$; for p
44. $I = prt$; for t
45. $y = mx + b$; for x
46. $y = mx + b$; for m
47. $4x - 3y = 12$; for y
48. $4x - 3y = 12$; for x
49. $d = vt + 16t^2$; for v
50. $d = vt - 16t^2$; for v
51. $C = \frac{5}{9}(F - 32)$; for F
52. $F = \frac{9}{5}C + 32$; for C

53. An object is thrown downward with an initial velocity of 4 feet per second. The relationship between the distance s (in feet) it travels and time t (in seconds) is given by $s = 4t + 16t^2$. How long does it take for the object to fall 72 feet? [2.2]

54. An object is tossed upward with an initial velocity of 40 feet per second. The height h (in feet) at time t (in seconds) is given by $h = 40t - 16t^2$. At what time will the object be 16 feet off the ground? [2.2]

Solve each application. In each case, be sure to show the equation that describes the situation. [2.3]

55. The length of a rectangle is 3 times the width. The perimeter is 32 feet. Find the length and width.

56. The length of a rectangle is 4 less than 3 times the width. The perimeter is 32 inches. Find the dimensions.

57. The three sides of a triangle are given by three consecutive integers. If the perimeter is 12 meters, find the length of each side.

58. The three sides of a triangle are given by three consecutive even integers. If the perimeter is 24 yards, find the length of each side.

59. A triangle is such that the largest angle is 5 times the smallest angle. The third angle is 7° less than the largest angle. Find the measure of each angle.

REVIEW FOR CHAPTER 2

60. The third angle in an isosceles triangle is one-fourth as large as each of the two base angles. Find the measure of each angle.

61. A teacher has a salary of $25,920 for her second year on the job. If this is 4.2% more than her first-year salary, how much did she earn her first year?

62. Two angles are complementary. If the larger angle is 15° more than twice the smaller angle, find the measure of each angle.

63. The lengths of the three sides of a right triangle are given by three consecutive integers. Find the three sides.

64. The lengths of the three sides of a right triangle are given by three consecutive even integers. Find the three sides.

Solve each inequality. Write your answer using interval notation. [2.4]

65. $-8a > -4$

66. $-9a > -3$

67. $6 - a \geq -2$

68. $7 - a \geq -6$

69. $\frac{3}{4}x + 1 \leq 10$

70. $\frac{2}{5}x - 1 \leq 9$

71. $800 - 200x < 1,000$

72. $600 - 300x < 900$

73. $\frac{1}{3} \leq \frac{1}{6}x \leq 1$

74. $-\frac{1}{2} \leq \frac{1}{6}x \leq \frac{1}{3}$

75. $5t + 1 \leq 3t - 2$ or $-7t \leq -21$

76. $6t - 3 \leq t + 1$ or $-8t \leq -16$

Solve each equation. [2.5]

77. $|a| - 3 = 1$

78. $|a| - 2 = 3$

79. $|x - 3| = 1$

80. $|x - 2| = 3$

81. $|2y - 3| = 5$

82. $|3y - 2| = 7$

83. $|4x - 3| + 2 = 11$

84. $|6x - 2| + 4 = 16$

85. $\left| \frac{7}{3} - \frac{x}{3} \right| + \frac{4}{3} = 2$

86. $|4 - \frac{1}{2}x| - 1 = \frac{3}{2}$

87. $|5t - 3| = |3t - 5|$

88. $|6t - 2| = |4t + 8|$

89. $|\frac{1}{2} - x| = |x + \frac{1}{2}|$

90. $|1 - \frac{2}{3}x| = |\frac{2}{3}x + 1|$

Solve each inequality and graph the solution set. [2.6]

91. $|x| < 5$

92. $|a| > 2$

93. $|0.01a| \geq 5$

94. $|0.01a| \geq 2$

95. $|x| \leq 0$

96. $|y - 2| < 3$

97. $|y + 5| \geq 0.02$

98. $|5x - 1| > 3$

99. $|2t + 1| - 3 < 2$

100. $|2t + 1| - 1 < 5$

Solve each equation or inequality, if possible. [2.1, 2.4, 2.5, 2.6]

101. $2x - 3 = 2(x - 3)$

102. $4(x + 1) = 4x + 1$

103. $3(5x - \frac{1}{2}) = 15x + 2$

104. $4(\frac{1}{2}x - 1) = 2x - 4$

105. $|4y + 8| = -1$

106. $|5y - 2| = -3$

107. $|x| > 0$

108. $|x| \geq 0$

109. $|5 - 8t| + 4 \leq 1$

110. $|7 - 9t| + 5 \leq 3$

111. $|2x + 1| \geq -4$

112. $|5x - 8| \geq -2$

Chapter 2 Test

Solve the following equations. [2.1]

1. $5 - \frac{4}{7}a = -11$
2. $\frac{1}{5}x - \frac{1}{2} - \frac{1}{10}x + \frac{2}{5} = \frac{3}{10}x + \frac{1}{2}$
3. $3x^2 = 5x + 2$
4. $100x^3 = 500x^2$
5. $5(x - 1) - 2(2x + 3) = 5x - 4$
6. $0.07 - 0.02(3x + 1) = -0.04x + 0.01$
7. $(x + 1)(x + 2) = 12$
8. $x^3 + 2x^2 - 16x - 32 = 0$

Solve for the indicated variable. [2.2]

9. $P = 2l + 2w$; for w
10. $A = \frac{1}{2}h(b + B)$; for B

Solve each of the following. [2.2, 2.3]

11. A rectangle is twice as long as it is wide. The perimeter is 36 inches. Find the dimensions.

12. The third angle in an isosceles triangle is 8° less than twice as large as each of the two base angles. Find the measure of each angle.

13. At the beginning of the day, the cash register at a coffee shop contains $75. At the end of the day, it contains $881.25. If the sales tax rate is 7.5%, how much of the total is sales tax?

14. Two angles are supplementary. If the larger angle is 15° more than twice the smaller angle, find the measure of each angle.

15. The longest side of a right triangle is 4 inches more than the shortest side. The third side is 2 inches more than the shortest side. Find the length of each side.

16. If an object is thrown straight up into the air with an initial velocity of 32 feet per second, then its height h (in feet) above the ground at any time t (in seconds) is given by the formula $h = 32t - 16t^2$. Find the times at which the object is on the ground by letting $h = 0$ in the equation and solving for t.

Solve the following inequalities. Write the solution set using interval notation, then graph the solution set. [2.4]

17. $5 - \frac{3}{2}x > -1$
18. $3(2y + 4) \geq 5(y - 8)$

Solve the following equations. [2.5]

19. $\left|\frac{1}{4}x - 1\right| = \frac{1}{2}$
20. $\left|\frac{2}{3}a + 4\right| = 6$
21. $|3 - 2x| + 5 = 2$
22. $5 = |3y + 6| - 4$

Solve the following inequalities and graph the solutions. [2.6]

23. $|6x - 1| > 7$
24. $|3x - 5| - 4 \leq 3$
25. $|5 - 4x| \geq -7$
26. $|4t - 1| < -3$

3

Equations and Inequalities in Two Variables

Contents

Introduction

The solution to each problem below is connected to an equation in two variables. The figure below each problem gives a visual representation of the information contained in each equation.

A 100 watt light bulb is d feet above the surface of a table. The intensity, I, of light that falls on the table is given by the equation

$$I = \frac{120}{d^2}$$

A 12 inch piece of string is formed into a rectangle. The area, A, of the rectangle formed is a function of its width, w.

$$A = w(6 - w)$$

A painting is purchased for $150. Its value increases continuously and doubles every 3 years. Its value t years after it is purchased is

$$V = 150 \cdot 2^{t/3}$$

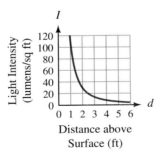

FIGURE 1

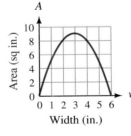

FIGURE 2

FIGURE 3

Each of the curves shown in Figures 1, 2, and 3 is drawn on a *rectangular coordinate system.* A rectangular coordinate system allows us to connect algebra and geometry by associating geometric shapes (the curves shown in the diagrams) with algebraic equations. The French philosopher and mathematician René Descartes (1596–1650) is the person usually credited with the invention of the rectangular coordinate system, which is often referred to as the *Cartesian coordinate system* in his honor. As a philosopher, Descartes is responsible for the statement "I think, therefore I am." Until Descartes invented his coordinate system in 1637, algebra and geometry were treated as separate subjects.

Overview

Most of this chapter is concerned with the graphs of linear equations in two variables. Equations of this type have graphs that are straight lines. We begin by finding the graph of a line from its equation. Then we define some important characteristics of lines: their slope and intercepts. We conclude our work with straight lines by finding the equation of a line in a variety of situations. Later in the chapter we look at graphs of inequalities in two variables, as well as functions, function notation, and inverse and direct variation. The background material needed for this chapter is in Chapter 2. You need to know how to solve a linear equation in one variable.

Study Skills

The study skills for this chapter are about attitude. They are points of view that point toward success.

1. **Be focused, not distracted** I have students who begin their assignments by asking themselves "Why am I taking this class?" If you are asking yourself similar questions, you are distracting yourself away from doing the things that will produce the results you want in this course. Don't dwell on questions and evaluations of the class that can be used as excuses for not doing well. If you want to succeed in this course, focus your energy and efforts toward success, rather than distracting yourself away from your goals.

2. **Be resilient** Don't let setbacks keep you from your goals. You want to put yourself on the road to becoming a person who can succeed in this class, or any class in college. Failing a test or quiz, or having a difficult time on some topics, is normal. No one goes through college without some setbacks. Don't let a temporary disappointment keep you from succeeding in this course. A low grade on a test or quiz is simply a signal that some reevaluation of your study habits needs to take place.

3. **Intend to succeed** I have a few students who simply go through the motions of studying without intending to master the material. It is more important to them to look like they are studying than to actually study. You need to study with the intention of being successful in the course. Intend to master the material, no matter what it takes.

SECTION
3.1

Paired Data and the Rectangular Coordinate System

In this section we begin our work with the visual component of algebra. We are setting the stage for two important topics: *functions* and *graphs*. Both topics have a wide variety of applications and are found in all the math classes that follow this one.

Table 1 gives the net price of a popular intermediate algebra text at the beginning of each year in which a new edition was published. (The net price is the price the bookstore pays for the book, not the price you pay for it.)

The information in Table 1 is shown visually in Figures 1 and 2. The diagram in Figure 1 is called a **bar chart.** The diagram in Figure 2 is called a **line graph.**

TABLE I Price of Textbook		
Edition	Year Published	Net Price ($)
First	1982	16.95
Second	1985	23.50
Third	1989	30.50
Fourth	1993	39.25
Fifth	1997	47.50

In both figures, the horizontal line that shows years is called the **horizontal axis,** while the vertical line that shows prices is called the **vertical axis.**

The data in Table 1 are called **paired data** because each number in the years column is paired with a specific number in the price column. Each of these pairs of numbers from Table 1 is associated with one of the solid bars in the bar chart (Figure 1) and one of the dots in the line graph (Figure 2).

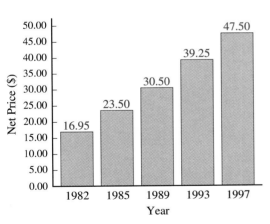

FIGURE 1

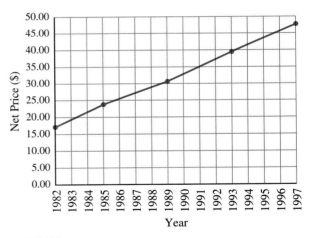

FIGURE 2

 Using Technology: Spreadsheet Programs

When I put together the manuscript for this book, I used a spreadsheet program to draw the bar chart and line graph shown in Figures 1 and 2.

Figure 3 (p. 138) shows what the screen of my computer looked like when I was preparing Figure 1. The bar chart was drawn by the computer from the data in the table. This was just one of many ways I could choose to display the data.

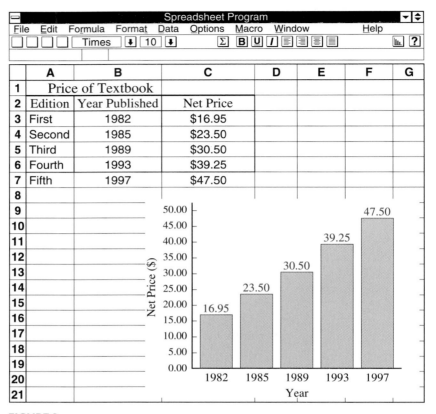

FIGURE 3
Computer screen showing spreadsheet program

Ordered Pairs

Paired data play a very important role in equations that contain two variables. Working with these equations is easier if we standardize the terminology and notation associated with paired data. So here is a definition that will do just that:

DEFINITION

A pair of numbers enclosed in parentheses and separated by a comma, such as $(-2, 1)$, is called an **ordered pair** of numbers. The first number in the pair is called the **x-coordinate** of the ordered pair, while the second number is called the **y-coordinate.** For the ordered pair $(-2, 1)$, the x-coordinate is -2 and the y-coordinate is 1.

In order to standardize the way in which we display paired data visually, we use a **rectangular coordinate system.** A rectangular coordinate system is made by drawing two real number lines at right angles to each other. The two number lines, called **axes,** cross each other at 0. This point is called the **origin.** Positive directions are to

the right and up. Negative directions are down and to the left. The
rectangular coordinate system is
shown in Figure 4.

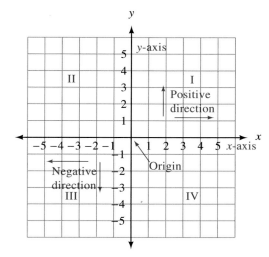

FIGURE 4

The horizontal number line is
called the **x-axis** and the vertical
number line is called the **y-axis.**
The two number lines divide the
coordinate system into four **quadrants,** which we number I through
IV in a counterclockwise direction
and refer to as QI, QII, QIII, and
QIV. Points on the axes are not
considered as being in any quadrant.

To graph the ordered pair (a, b)
on a rectangular coordinate system we start at the origin and
move a units right or left (right if
a is positive, left if a is negative). Then we move b units up or down (up if b is
positive and down if b is negative). The point where we end up is the graph of the
ordered pair (a, b).

EXAMPLE I Plot (graph) the ordered pairs $(2, 5)$, $(-2, 5)$, $(-2, -5)$, and $(2, -5)$.

SOLUTION To graph the ordered pair $(2, 5)$, we start at the origin and move
2 units to the right, then 5 units up. We are now at the point whose coordinates are $(2, 5)$. We graph the other three ordered pairs in a similar manner (see
Figure 5).

Note that any point in
quadrant I has both its x-
and y-coordinates positive
$(+, +)$. Points in quadrant
II have negative x-
coordinates and positive y-
coordinates $(-, +)$. In
quadrant III both
coordinates are negative
$(-, -)$. In quadrant IV the
form is $(+, -)$.

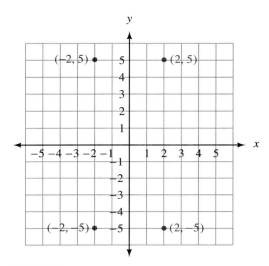

FIGURE 5

EXAMPLE 2 Graph the ordered pairs $(1, -3)$, $(\frac{1}{2}, 2)$, $(3, 0)$, $(0, -2)$, $(-1, 0)$, and $(0, 5)$.

SOLUTION

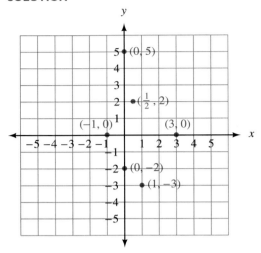

FIGURE 6

From Example 2 we see that any point on the x-axis has a y-coordinate of 0 (it has no vertical displacement), and any point on the y-axis has an x-coordinate of 0 (no horizontal displacement).

Linear Equations

We can plot a single point from an ordered pair, but to draw a line, we need two points or an equation in two variables.

DEFINITION

Any equation that can be put in the form $ax + by = c$, where a, b, and c are real numbers and a and b are not both 0, is called a **linear equation in two variables.** The graph of any equation of this form is a straight line (that is why these equations are called "linear"). The form $ax + by = c$ is called **standard form.**

To graph a linear equation in two variables, we simply graph its solution set. That is, we draw a line through all the points whose coordinates satisfy the equation.

EXAMPLE 3 Graph: $y = 2x - 3$.

SOLUTION Since $y = 2x - 3$ can be put in the form $ax + by = c$, it is a linear equation in two variables. Hence, the graph of its solution set is a straight line. We can find some specific solutions by substituting numbers for x and then solving for the corresponding values of y. We are free to choose any convenient numbers for x, so let's use the numbers -1, 0, and 2:

When $x = -1$:

the equation $y = 2x - 3$

becomes $\quad y = 2(\mathbf{-1}) - 3$

$$y = -5$$

The ordered pair $(-1, -5)$ is a solution.

When $x = 0$:

we have $\quad y = 2(\mathbf{0}) - 3$

$$y = -3$$

The ordered pair $(0, -3)$ is also a solution.

Using $x = 2$:

we have $\quad y = 2(\mathbf{2}) - 3$

$$y = 1$$

The ordered pair $(2, 1)$ is another solution.

Graphing these three ordered pairs and drawing a line through them, we have the graph of $y = 2x - 3$ shown in Figure 7.

In table form:

x	y
-1	-5
0	-3
2	1

Note: It actually takes only two points to determine a straight line. We have included a third point for "insurance." If all three points do not line up in a straight line, we have made a mistake.

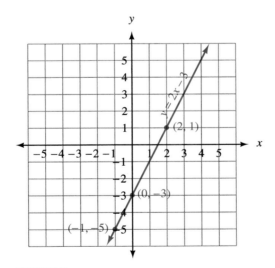

FIGURE 7

Example 3 illustrates again the connection between algebra and geometry that we mentioned in the introduction to this chapter. Descartes' rectangular coordinate system allows us to associate the equation $y = 2x - 3$ (an algebraic concept) with a specific straight line (a geometric concept). The study of the relationship between equations in algebra and their associated geometric figures is called *analytic geometry*.

EXAMPLE 4   Graph the equation: $y = -\dfrac{1}{3}x + 2$.

SOLUTION   We want to find three ordered pairs that satisfy the equation. To do so, we can let x equal any numbers we choose and find corresponding values of y. Since every value of x we substitute into the equation is going to be multiplied by $-\frac{1}{3}$, let's use numbers for x that are divisible by 3, such as -3, 0, and 3. That way, when we multiply them by $-\frac{1}{3}$, the result will be an integer.

$$\text{Let } x = -3: \quad y = -\frac{1}{3}(-3) + 2$$
$$y = 1 + 2$$
$$y = 3$$

The ordered pair $(-3, 3)$ is one solution.

$$\text{Let } x = 0: \quad y = -\frac{1}{3}(0) + 2$$
$$y = 0 + 2$$
$$y = 2$$

In table form:

x	y
-3	3
0	2
3	1

The ordered pair $(0, 2)$ is a second solution.

$$\text{Let } x = 3: \quad y = -\frac{1}{3}(3) + 2$$
$$y = -1 + 2$$
$$y = 1$$

The ordered pair $(3, 1)$ is a third solution.

Plotting the ordered pairs $(-3, 3)$, $(0, 2)$, and $(3, 1)$ and drawing a straight line through their graphs, we have the graph of the equation $y = -\frac{1}{3}x + 2$, as shown in Figure 8.

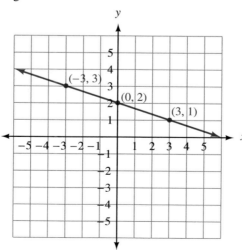

FIGURE 8

Note: In Example 4 the values of x we used, -3, 0, and 3, are referred to as convenient values of x because they are easier to work with than some other numbers. For instance, if we let $x = 2$ in our original equation, we would have to add $-\frac{2}{3}$ and 2 to find the corresponding value of y. Not only would the arithmetic be more difficult, but the ordered pair we would obtain would have a fraction for its y-coordinate, making it more difficult to graph accurately.

Intercepts

Two important points on the graph of a straight line, if they exist, are the points where the graph crosses the axes.

> **DEFINITION**
>
> The **x-intercept** of the graph of an equation is the x-coordinate of the point where the graph crosses the x-axis. The **y-intercept** is defined similarly.

Since any point on the x-axis has a y-coordinate of 0, we can find the x-intercept by letting $y = 0$ and solving the equation for x. We find the y-intercept by letting $x = 0$ and solving for y.

EXAMPLE 5   Find the x- and y-intercepts for $2x + 3y = 6$; then graph the solution set.

SOLUTION
To find the y-intercept, we let $x = 0$:

$$\text{When} \quad x = 0:$$
$$\text{we have} \quad 2(\mathbf{0}) + 3y = 6$$
$$3y = 6$$
$$y = 2$$

The y-intercept is 2, and the graph crosses the y-axis at the point $(0, 2)$.

To find the x-intercept, we let $y = 0$:

$$\text{When} \quad y = 0:$$
$$\text{we have} \quad 2x + 3(\mathbf{0}) = 6$$
$$2x = 6$$
$$x = 3$$

The x-intercept is 3, so the graph crosses the x-axis at the point $(3, 0)$.

We use these results to graph the solution set for $2x + 3y = 6$. The graph is shown in Figure 9.

Graphing straight lines by finding the intercepts works best when the coefficients of x and y are factors of the constant term.

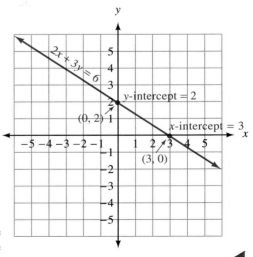

FIGURE 9

EXAMPLE 6 Graph the line $x = 3$ and the line $y = -2$.

SOLUTION The line $x = 3$ is the set of all points whose x-coordinate is 3. The variable y does not appear in the equation, so the y-coordinate can be any number.

The line $y = -2$ is the set of all points whose y-coordinate is -2. The x-coordinate can be any number. Here are the graphs:

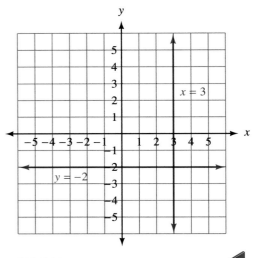

FIGURE 10

 Using Technology: Graphing Calculators and Computer Graphing Programs

A variety of computer programs and graphing calculators are currently available to help us graph equations and then obtain information from those graphs much faster than we could with paper and pencil. As we mentioned earlier, we will not give the instructions for all the calculators available. Most of the instructions we give are generic in form. You will have to use the manual that came with your calculator to find the instructions that are specific to your calculator.

Graphing with Trace and Zoom

All graphing calculators have the ability to graph a function and then trace over the points on the graph, giving their coordinates as it goes. Further, all graphing calculators can zoom in and out on a graph that has been drawn. To graph a linear equation on a graphing calculator, we first set the graph window. Most calculators call the smallest value of x Xmin and the largest value of x Xmax. The counterpart values of y are Ymin and Ymax. We will use the notation

$$\text{Window:}\quad \text{X from } -5 \text{ to } 4, \text{ Y from } -3 \text{ to } 2$$

to stand for a window in which

$$\text{Xmin} = -5 \qquad \text{Ymin} = -3$$
$$\text{Xmax} = 4 \qquad \text{Ymax} = 2$$

Set your calculator with the following window:

Window: X from -10 to 10, Y from -10 to 10

Graph the equation Y $= -$X $+$ 8. On the TI-82/83, you use the $\boxed{Y=}$ key to enter the equation; you enter a negative sign with the $\boxed{(-)}$ key, and a subtraction sign with the $\boxed{-}$ key. The graph will be similar to the one shown in Figure 11.

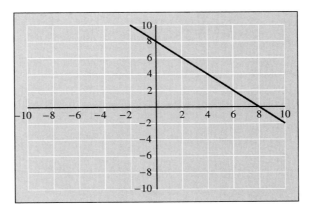

FIGURE 11

Use the Trace feature of your calculator to name three points on the graph. Next, use the Zoom feature of your calculator to zoom out so your window is twice as large.

Solving for y First

To graph the equation from Example 5, $2x + 3y = 6$, on a graphing calculator, you must first solve it for y. When you do so, you will get $y = -\frac{2}{3}x + 2$, which you enter into your calculator as Y $= -(2/3)$X $+$ 2. Graph this equation in the window described below and compare your results with the graph in Figure 9.

Window: X from -6 to 6, Y from -6 to 6

Hint on Tracing

If you are going to use the Trace feature and you want the x-coordinates to be exact numbers, set your window so that the range of X inputs is a multiple of the number of horizontal pixels on your calculator screen. On the TI-82/83, the screen has 94 pixels across. Here are a few convenient trace windows:

X from -4.7 to 4.7	To trace to the nearest tenth
X from -47 to 47	To trace to the nearest integer
X from 0 to 9.4	To trace to the nearest tenth
X from 0 to 94	To trace to the nearest integer
X from -94 to 94	To trace to the nearest even integer

Problem Set

3.1

NOTE: If you are drawing graphs by hand, use a ruler or some other straight edge to draw straight lines. Also, take time to include the proper labels and scales on the x-axis and y-axis. The better your work looks when you finish it, the greater the sense of accomplishment you will feel.

1. **Hourly Wages** Suppose you have a job that pays $7.50 per hour, and you work anywhere from 0 to 40 hours per week. Table 2 gives the amount of money you will earn in 1 week for working various hours. Construct a line graph from the information in Table 2. You can copy the grid in Figure 12 to use as a guide.

TABLE 2
Weekly Wages

Hours Worked	Pay ($)
0	0
10	75
20	150
30	225
40	300

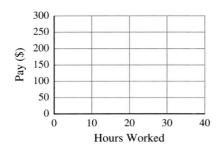

FIGURE 12
Template for constructing a line graph

2. **Softball Toss** Chaudra is tossing a softball into the air with an underhand motion. It takes exactly 2 seconds for the ball to come back to her. Table 3 shows the distance the ball is above her hand at quarter-second intervals. Construct a line graph from the information in the table. Use the grid in Figure 13 as a guide.

TABLE 3
Tossing a Softball into the Air

Time (sec)	Distance (ft)
0	0
0.25	7
0.5	12
0.75	15
1	16
1.25	15
1.5	12
1.75	7
2	0

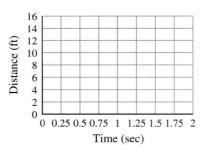

FIGURE 13
Template for constructing a line graph

3. **Intensity of Light** Table 4 gives the intensity of light that falls on a surface at various distances from a 100 watt light bulb. Use the template shown in Figure 14 to construct a bar chart from the information in Table 4.

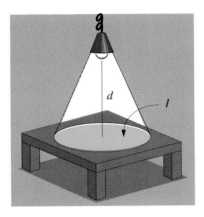

4. **Value of a Painting** A piece of abstract art was purchased in 1990 for \$125. Table 5 shows the value of the painting at various times, assuming that it doubles in value every 5 years. Construct a bar chart from the information in the table, using the template in Figure 15.

TABLE 5 Value of a Painting	
Year	Value (\$)
1990	125
1995	250
2000	500
2005	1,000
2010	2,000

TABLE 4 Light Intensity from a 100 Watt Light Bulb	
Distance above Surface (ft)	Intensity (lumens/sq ft)
1	120.0
2	30.0
3	13.3
4	7.5
5	4.8
6	3.3

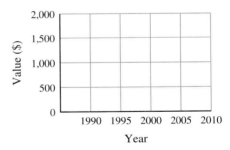

FIGURE 15
Template for constructing a bar chart

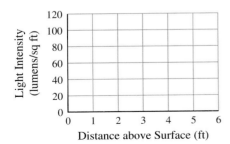

FIGURE 14
Template for constructing a bar chart

Graph each of the following ordered pairs on a rectangular coordinate system.

5. $(1, 2)$ 6. $(-1, 2)$

7. $(-1, -2)$ 8. $(1, -2)$

9. $(5, 0)$ 10. $(0, -3)$

11. $(0, 2)$ 12. $(4, 0)$

13. $(-5, -5)$ 14. $(-4, -1)$

15. $(\frac{1}{2}, 2)$ 16. $(3, \frac{1}{4})$

148

17–28. Give the coordinates of each of the points shown in the following rectangular coordinate system:

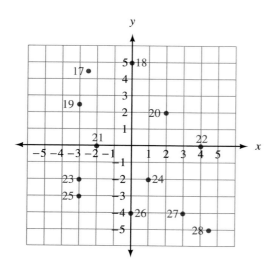

Graph each of the following linear equations by first finding the intercepts.

29. $2x - 3y = 6$
30. $y - 2x = 4$
31. $4x - 5y = 20$
32. $5x - 3y - 15 = 0$
33. $y = 2x + 3$
34. $y = 3x - 2$

35. Which of the following tables could be produced from the equation $y = 2x - 6$?

(a)

x	y
0	6
1	4
2	2
3	0

(b)

x	y
0	-6
1	-4
2	-2
3	0

(c)

x	y
0	-6
1	-5
2	-4
3	-3

36. Which of the following tables could be produced from the equation $3x - 5y = 15$?

(a)

x	y
0	5
-3	0
10	3

(b)

x	y
0	-3
5	0
10	3

(c)

x	y
0	-3
-5	0
10	-3

Graph each of the following straight lines.

37. $y = \frac{1}{3}x$ **38.** $y = \frac{1}{2}x$
39. $-2x + y = -3$ **40.** $-3x + y = -2$
41. $y = -\frac{2}{3}x + 1$ **42.** $y = -\frac{2}{3}x - 1$

43. $\dfrac{x}{3} + \dfrac{y}{4} = 1$ **44.** $\dfrac{x}{-2} + \dfrac{y}{3} = 1$

45. The graph shown below is the graph of which of the following equations?
(a) $3x - 2y = 6$
(b) $2x - 3y = 6$
(c) $2x + 3y = 6$

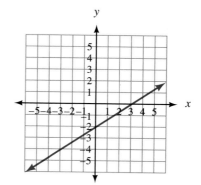

46. The graph shown below is the graph of which of the following equations?
  (a) $3x - 2y = 8$
  (b) $2x - 3y = 8$
  (c) $2x + 3y = 8$

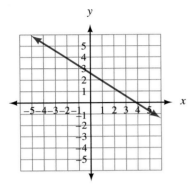

47. Graph the straight line: $0.02x + 0.03y = 0.06$.

48. Graph the straight line: $0.05x - 0.03y = 0.15$.

49. Graph the lines $x = -3$ and $y = 5$ on the same coordinate system.

50. Graph the lines $x = 4$ and $y = -3$ on the same coordinate system.

51. The ordered pairs that satisfy the equation $y = 3x$ all have the form $(x, 3x)$, because y is always 3 times x. Graph all ordered pairs of the form $(x, 3x)$.

52. Graph all ordered pairs of the form $(x, -3x)$.

Applying the Concepts

NOTE: For each of the following applied problems, first build a table of values. Make sure each input is accompanied by an output.

53. **Cost of a Phone Call** If the cost of a long-distance phone call is 50¢ for the first minute and 25¢ for each additional minute, then the total cost y (in cents) of a call that goes x minutes past the first minute is $y = 25x + 50$. Let 1 unit on the x-axis equal 1 minute, and 1 unit on the y-axis equal 25¢, and graph this equation.

54. **Cost of a Taxi Ride** If the cost of a taxi ride in Las Vegas is $1.50 for the first mile and $0.50 for each tenth of a mile after the first mile, then the total cost y (in cents) of a ride that goes x tenths of a mile after the first mile is $y = 50x + 150$. Let each unit on the x-axis equal one-tenth of a mile and each unit on the y-axis equal 50¢, and graph this equation.

55. **Demand Equation** A company that manufactures typewriter ribbons knows that the number of ribbons they can sell each week, x, is related to the price of each ribbon, p, by the equation $x + 100p = 1,200$. Note any restrictions on the variables; then graph the relationship between x and p, with the values of x along the horizontal axis and the values of p on the vertical axis.

56. **Demand Equation** A company that manufactures diskettes for home computers finds that they can sell x diskettes each day at p dollars per diskette according to the equation $x + 100p = 800$. Note any restrictions on the variables; then graph the relationship between x and p.

57. **A Snail's Pace** A snail is climbing straight up a brick wall at a constant rate. It takes the snail 4 hours to climb up 6 feet. After climbing for 4 hours, the snail rests for 4 hours, during which time it slides 2 feet down the wall. If the snail starts at the bottom of the wall and repeats this climbing up/sliding down process continuously, construct a table that gives the snail's height above the ground every 4 hours, starting at 0 and ending at 24 hours. Then construct a line graph from the information in the table.

58. **Air Temperature** A pilot checks the weather conditions before flying and finds that the air temperature drops 3.5°F every 1,000 feet. If the air temperature is 41°F when the plane reaches 10,000 feet, construct a table that gives the air temperature every 2,000 feet, starting at 10,000 feet and ending at 20,000 feet. Then construct a line graph from the information in the table.

Using Technology

59. Price of a Textbook The table on textbook prices shown at the beginning of this section is repeated below, along with an equation that approximates the pairs of numbers in the table.

Year of New Edition	Net Price ($)
x	y
1982	16.95
1985	23.50
1989	30.50
1993	39.25
1997	47.50

Approximating Equation: $y = 2.022x - 3991$

(a) Use your calculator to graph the equation using the following window:

Window: X from 1980 to 2003.5,
Y from 0 to 50

(b) Trace along the graph to find the value of y that corresponds to each of the following values of x:

From Approximating Equation	
x	y
1982	
1985	
1989	
1993	
1997	

60. Price of a Textbook If we let the year 1980 correspond to $x = 0$, then the table on textbook prices shown at the beginning of this section can be rewritten as shown at the top of the next column. Below the table is an equation that approximates the pairs of numbers in the table.

Year of New Edition	If x = 0 at 1980	Net Price ($)
	x	y
1982	2	16.95
1985	5	23.50
1989	9	30.50
1993	13	39.25
1997	17	47.50

Approximating Equation: $y = 2.022x + 12.94$

(a) Use your calculator to graph the equation using the following window:

Window: X from 0 to 23.5,
Y from 0 to 50

(b) Trace along the graph to find the value of y that corresponds to each of the following values of x:

Year of New Edition	From Approximating Equation	
	x	y
1982	2	
1985	5	
1989	9	
1993	13	
1997	17	

(c) Of the two approximating equations shown in this problem and the previous problem, which gives the best approximations to the actual textbook prices?

Drag Racing *Jim Rizzoli lives in San Luis Obispo, California. He owns and operates an alcohol-fueled dragster. The information in Table 6 was recorded by a computer in the dragster during one of his races at the 1993 Winternationals. It shows the time and speed of the dragster, along with the distance traveled past* *the starting line and the front axle RPM (revolutions per minute). Use a spreadsheet program to work these problems.*

TABLE 6
Speed, Distance, and Front Axle RPM for a Race Car

Time (sec)	Speed (mph)	Distance Traveled (ft)	Front Axle RPM
0	0.0	0	0
1	72.7	69	1,107
2	129.9	231	1,978
3	162.8	439	2,486
4	192.2	728	2,919
5	212.4	1,000	3,233
6	228.1	1,373	3,473

61. Construct a bar chart that shows the relationship between time and distance.

62. Construct a line graph that shows the relationship between time and front axle RPM.

63. Construct a line graph that shows the relationship between speed and distance traveled.

64. Construct a line graph that shows the relationship between speed and front axle RPM. Does your line graph suggest that the relationship between speed and front axle RPM is a linear one?

Review Problems

 The problems that follow review material we covered in Section 2.1.

Solve each equation.

65. $5x - 4 = -3x + 12$

66. $\dfrac{1}{2} - \dfrac{y}{5} = -\dfrac{9}{10} + \dfrac{y}{2}$

67. $2x^2 - 5x = 12$

68. $8x^2 = 14x - 5$

69. $\frac{1}{2} - \frac{1}{8}(3t - 4) = -\frac{7}{8}t$

70. $3(5t - 1) - (3 - 2t) = 5t - 8$

71. $4t^2 + 12t = 0$

72. $(x - 3)(x - 5) = 3$

One Step Further

Find the x- and y-intercepts for each equation. Your answers will contain the constants a, b, and c.

73. $ax + by = c$

74. $ax - by = c$

75. $\dfrac{x}{a} + \dfrac{y}{b} = 1$

76. $y = ax + b$

The Slope of a Line

A highway sign tells us we are approaching a 6% downgrade. As we drive down this hill, each 100 feet we travel horizontally is accompanied by a 6 foot drop in elevation.

Highway Sign   Mathematical Model

In mathematics we say the *slope* of the highway is $-0.06 = -\frac{6}{100} = -\frac{3}{50}$. The **slope** is the ratio of the vertical change to the accompanying horizontal change.

In defining the slope of a straight line, we are looking for a number to associate with a straight line that does two things. First of all, we want the slope of a line to measure the "steepness" of the line. That is, in comparing two lines, the slope of the steeper line should have the larger numerical value. Secondly, we want a line that *rises* going from left to right to have a **positive slope.** We want a line that *falls* going from left to right to have a **negative slope.** (A line that neither rises nor falls going from left to right must therefore have 0 slope.)

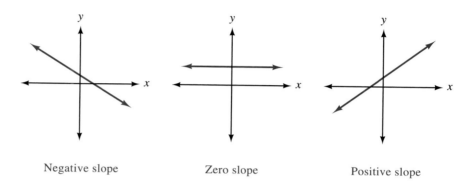

Negative slope   Zero slope   Positive slope

Geometrically, we can define the **slope** of a line as the ratio of the vertical change to the horizontal change encountered when moving from one point to another on the line. The vertical change is sometimes called the **rise.** The horizontal change is called the **run.**

EXAMPLE 1 Find the slope of the line $y = 2x - 3$.

SOLUTION In order to make use of our geometric definition, we first graph $y = 2x - 3$. We then pick any two convenient points and find the ratio of rise to run. By "convenient points" we mean points with integer coordinates. If we let $x = 2$ in the equation, then $y = 1$. Likewise, if we let $x = 4$, then y is 5.

Figure 1 shows that the line has a slope of $\frac{4}{2} = 2$.

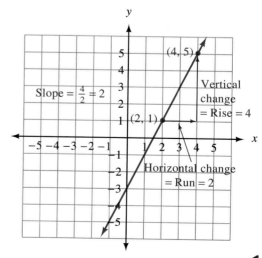

FIGURE 1

Notice that we can measure the vertical change (rise) by subtracting the y-coordinates of the two points shown in Figure 1: $5 - 1 = 4$. The horizontal change (run) is the difference of the x-coordinates: $4 - 2 = 2$. This gives us a second way of defining the slope of a line.

DEFINITION

The **slope** of the line between two points (x_1, y_1) and (x_2, y_2) is given by

$$\text{Slope} = m = \frac{\text{Rise}}{\text{Run}} = \frac{y_2 - y_1}{x_2 - x_1}$$

Geometric form   Algebraic form

EXAMPLE 2   Find the slope of the line through $(-2, -3)$ and $(-5, 1)$.

SOLUTION $m = \dfrac{y_2 - y_1}{x_2 - x_1} = \dfrac{1 - (-3)}{-5 - (-2)} = \dfrac{4}{-3} = -\dfrac{4}{3}$

Looking at the graph of the line between the two points (Figure 2), we can see that our geometric approach does not conflict with our algebraic approach.

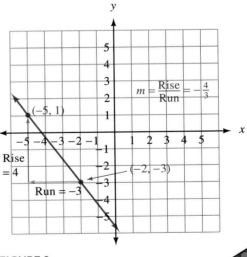

FIGURE 2

We should note here that it does not matter which ordered pair we call (x_1, y_1) and which we call (x_2, y_2). If we were to reverse the order of subtraction of both the x- and y-coordinates in Example 2, we would have

$$m = \frac{-3 - 1}{-2 - (-5)} = \frac{-4}{3} = -\frac{4}{3}$$

which is the same as our previous result.

Note: The two most common mistakes students make when first working with the formula for the slope of a line are:

1. Putting the difference of the *x*-coordinates over the difference of the *y*-coordinates.
2. Subtracting in one order in the numerator and then subtracting in the opposite order in the denominator. You would make this mistake in Example 2 if you wrote $1 - (-3)$ in the numerator and then $-2 - (-5)$ in the denominator.

EXAMPLE 3   Find the slope of the line containing $(3, -1)$ and $(3, 4)$.

SOLUTION   Using the definition for slope, we have

$$m = \frac{-1 - 4}{3 - 3} = \frac{-5}{0}$$

But the expression $\frac{-5}{0}$ is undefined. That is, there is no real number to associate with it. In this case, we say the line has *no slope*.

The graph of the line is shown in Figure 3.

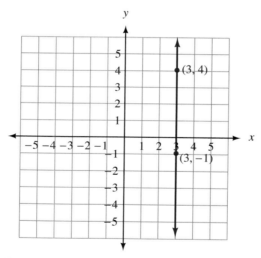

FIGURE 3

Our line with no slope is a vertical line. All vertical lines have no slope. (And all horizontal lines, as we mentioned earlier, have 0 slope.)

Slopes of Parallel and Perpendicular Lines

In geometry we call lines in the same plane that never intersect *parallel.* In order for two lines to be nonintersecting, they must rise or fall at the same rate. In other words, two lines are **parallel** if and only if they have the *same slope.*

Although it is not as obvious, it is also true that two nonvertical lines are **perpendicular** if and only if the *product of their slopes is* -1. This is the same as saying their slopes are negative reciprocals.

We can state these facts with symbols as follows: If line l_1 has slope m_1 and line l_2 has slope m_2, then

$$l_1 \text{ is parallel to } l_2 \Leftrightarrow m_1 = m_2$$

and

$$l_1 \text{ is perpendicular to } l_2 \Leftrightarrow m_1 \cdot m_2 = -1$$
$$\left(\text{or } m_1 = \frac{-1}{m_2} \right)$$

For example, if a line has a slope of $\frac{2}{3}$, then any line parallel to it has a slope of $\frac{2}{3}$. Any line perpendicular to it has a slope of $-\frac{3}{2}$ (the negative reciprocal of $\frac{2}{3}$).

Although we cannot give a formal proof of the relationship between the slopes of perpendicular lines at this level of mathematics, we can offer some justification for the relationship. Figure 4 shows the graphs of two lines. One of the lines has a slope of $\frac{2}{3}$, while the other has a slope of $-\frac{3}{2}$. As you can see, the lines are perpendicular.

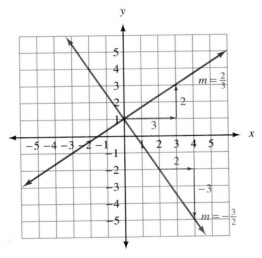

FIGURE 4

EXAMPLE 4 Find a if the line through $(3, a)$ and $(-2, -8)$ is perpendicular to a line with slope $-\frac{4}{5}$.

SOLUTION The slope of the line through the two points is

$$m = \frac{a - (-8)}{3 - (-2)} = \frac{a + 8}{5}$$

Since the line through the two points is perpendicular to a line with slope $-\frac{4}{5}$, we can also write its slope as $\frac{5}{4}$:

$$\frac{a + 8}{5} = \frac{5}{4}$$

156

Multiplying both sides by 20, we have

$$4(a + 8) = 5 \cdot 5$$
$$4a + 32 = 25$$
$$4a = -7$$
$$a = -\frac{7}{4}$$

 Using Technology: Graphing Calculators

We can use a graphing calculator to investigate the effects of the numbers a and b on the graph of $y = ax + b$. To see how the number b affects the graph, we can hold a constant and let b vary. Doing so will give us a *family* of curves. Suppose we set $a = 1$ and then let b take on integer values from -3 to 3. The equations we obtain are

$$y = x - 3$$
$$y = x - 2$$
$$y = x - 1$$
$$y = x$$
$$y = x + 1$$
$$y = x + 2$$
$$y = x + 3$$

There are two ways to graph this set of equations on a graphing calculator. The first is to use the list of Y variables. The second is to write a program that substitutes the given values of b into $y = x + b$, and then graph the result. To use the Y variables list, enter each equation at one of the Y variables, set the graph window, then graph. The calculator will graph the equations in order, starting with Y_1 and ending with Y_7. Following is the Y variables list, an appropriate window, and a sample of the type of graph obtained.

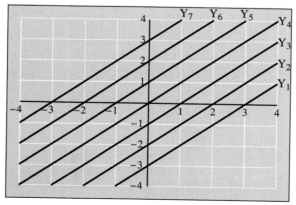

$$Y_1 = X - 3$$
$$Y_2 = X - 2$$
$$Y_3 = X - 1$$
$$Y_4 = X$$
$$Y_5 = X + 1$$
$$Y_6 = X + 2$$
$$Y_7 = X + 3$$

Window: X from -4 to 4,
Y from -4 to 4

FIGURE 5

The same result can be obtained by programming your calculator to graph $y = x + b$ for $b = -3, -2, -1, 0, 1, 2,$ and 3. Here is an outline of a program that will do this. Check the manual that came with your calculator to find the commands for your calculator.

Step 1 Clear screen
Step 2 Set window for X from -4 to 4 and Y from -4 to 4
Step 3 $-3 \rightarrow B$
Step 4 Label 1
Step 5 Graph $Y = X + B$
Step 6 $B + 1 \rightarrow B$
Step 7 If $B < 4$, Goto 1
Step 8 End

Running this program will produce graphs similar to those in Figure 5.

Problem Set
3.2

Find the slope of each of the following lines from the given graph.

1.

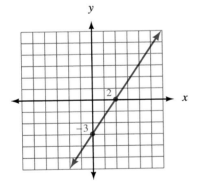

2.

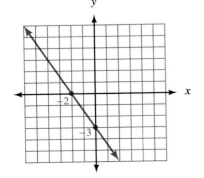

3.

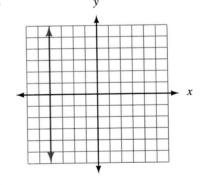

4.

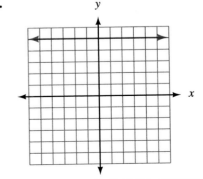

5.

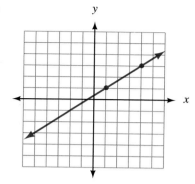

6.

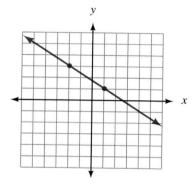

Find the slope of the line through each of the following pairs of points. Then plot each pair of points, draw a line through them, and indicate the rise and run in the graph in the manner shown in Example 2.

7. (2, 1), (4, 4)   **8.** (3, 1), (5, 4)
9. (1, 4), (5, 2)   **10.** (1, 3), (5, 2)
11. (1, −3), (4, 2)   **12.** (2, −3), (5, 2)
13. (−3, −2), (1, 3)   **14.** (−3, −1), (1, 4)
15. (−3, 2), (3, −2)   **16.** (−3, 3), (3, −1)

For each equation below, complete the table and then use the results to find the slope of the graph of the equation.

17. $2x + 3y = 6$ **18.** $3x - 2y = 6$

x	y
0	
	0

x	y
0	
	0

19. $y = \frac{2}{3}x - 5$

x	y
0	
3	

20. $y = -\frac{3}{4}x + 2$

x	y
0	
4	

NOTE: Remember, mistakes are part of the learning process. Your mistakes will be a benefit to you when you find them and then correct them.

Solve for the indicated variable if the line through the two given points has the given slope.

21. (5, a), (4, 2); $m = 3$
22. (3, a), (1, 5); $m = -4$
23. (2, 6), (3, y); $m = -7$
24. (−4, 9), (−5, y); $m = 3$

25. Graph the line with x-intercept 4 and y-intercept 2. What is the slope of this line?

26. Graph the line with x-intercept −4 and y-intercept −2. What is the slope of this line?

27. Find the slope of any line parallel to the line through (2, 3) and (−8, 1).

28. Find the slope of any line parallel to the line through (2, 5) and (5, −3).

29. Line *l* contains the points (5, −6) and (5, 2). Give the slope of any line perpendicular to *l*.

30. Line *l* contains points (3, 4) and (−3, 1). Give the slope of any line perpendicular to *l*.

31. Line *l* has a slope of $\frac{2}{3}$. A horizontal change of 12 will always be accompanied by how much of a vertical change?

32. For any line with slope $\frac{4}{5}$, a vertical change of 8 is always accompanied by how much of a horizontal change?

33. **Slope of a Sand Pile** A pile of sand at a construction site is in the shape of a cone. If the slope of the side of the pile is $\frac{2}{3}$ and the pile is 8 feet high, how wide is the diameter of the base of the pile?

34. **Slope of a Pyramid** The slope of the sides of one of the ancient pyramids in Egypt is $\frac{13}{10}$. If the base of the pyramid is 750 feet, how tall is the pyramid?

Heating a Block of Ice *A block of ice with an initial temperature of* $-20°C$ *is heated at a steady rate. The graph below shows how the temperature changes as the ice melts to become water and the water boils to become steam and water.*

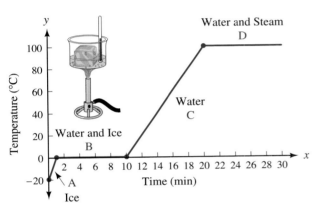

35. How long does it take all the ice to melt?

36. From the time the heat is applied to the block of ice, how long is it before the water boils?

37. Find the slope of the line segment labeled A. What units would you attach to this number?

38. Find the slope of the line segment labeled C. Be sure to attach units to your answer.

39. Is the temperature changing faster during the 1st minute or the 16th minute?

40. Line segments B and D both have 0 slope. Explain what this means in terms of the melting ice.

Value of a Used Car *The 1998 edition of a popular consumer car price book gives the values shown in the table at the top of the next column for Volkswagen Jettas in good condition. The line graph below the table was drawn from the information in the table.*

Year	Age in 1998	Value ($)
	x	y
1994	4	8,525
1995	3	9,575
1996	2	11,950
1997	1	13,200
1998	0	15,250

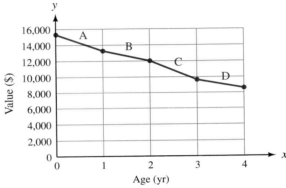

41. Find the slope of the line segment labeled B. What units should you attach to this number?

42. Find the slope of line segment C. Be sure to include units with your answer.

43. From the graph, does the value of the car decrease more from 1 to 2 years, or from 2 to 3 years?

44. From the graph, does the value of the car decrease more from 2 to 3 years, or from 3 to 4 years?

Using Technology

45. Use your Y variables list, or write a program, to graph the family of curves $Y = 2X + B$ for $B = -3, -2, -1, 0, 1, 2$, and 3.

46. Use your Y variables list, or write a program, to graph the family of curves $Y = -2X + B$ for $B = -3, -2, -1, 0, 1, 2$, and 3.

47. Use your Y variables list, or write a program, to graph the family of curves $Y = AX$ for $A = -3, -2, -1, 0, 1, 2$, and 3.

48. Use your Y variables list, or write a program, to graph the family of curves Y = AX + 2 for A = −3, −2, −1, 0, 1, 2, and 3.

49. Use your Y variables list, or write a program, to graph the family of curves Y = AX for A = $\frac{1}{4}$, $\frac{1}{3}$, $\frac{1}{2}$, 1, 2, and 3.

50. Use your Y variables list, or write a program, to graph the family of curves Y = AX − 2 for A = $\frac{1}{4}$, $\frac{1}{3}$, $\frac{1}{2}$, 1, 2, and 3.

 Are you reading each section in the book before going to class? It is important that you do so, especially if you are covering material that is difficult for you. Even if you understand very little of what you read, it is still important to read

each section before class. In the long run, you will have much less trouble with the class when you read ahead consistently.

Review Problems

 The problems below review material we covered in Section 2.2.

51. If 3x + 2y = 12, find y when x is 4.

52. If y = 3x − 1, find x when y is 0.

53. Solve the formula 3x + 2y = 12 for y.

54. Solve the formula y = 3x − 1 for x.

55. Solve the formula A = P + Prt for t.

56. Solve the formula S = $\pi r^2 + 2\pi rh$ for h.

SECTION 3.3

The Equation of a Line

The table and illustrations below show some corresponding temperatures on the Fahrenheit and Celsius temperature scales. For example, water freezes at 32°F and 0°C, and boils at 212°F and 100°C.

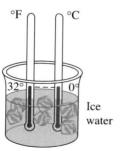

Degrees Celsius	Degrees Fahrenheit
0	32
25	77
50	122
75	167
100	212

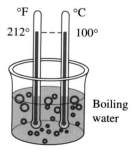

If we plot all the points in the table using the x-axis for temperatures on the Celsius scale and the y-axis for temperatures on the Fahrenheit scale, we see that they line up in a straight line (Figure 1). This means that there is a linear equation in two variables that will give a perfect description of the relationship between the two scales. That equation is

$$F = \frac{9}{5}C + 32$$

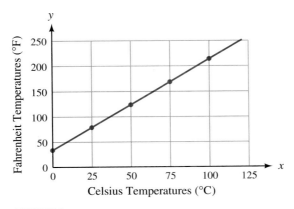

FIGURE 1

The techniques we use to find the equation of a line from a set of points is what this section is all about.

Suppose line l has slope m and y-intercept b. What is the equation of l? Since the y-intercept is b, we know the point $(0, b)$ is on the line. If (x, y) is any other point on l, then using the definition for slope, we have

$$\frac{y - b}{x - 0} = m \qquad \text{Definition of slope}$$

$$y - b = mx \qquad \text{Multiply both sides by } x.$$

$$y = mx + b \qquad \text{Add } b \text{ to both sides.}$$

This last equation is known as the **slope-intercept form** of the equation of a straight line.

Slope-Intercept Form of the Equation of a Line

The equation of any line with slope m and y-intercept b is given by

$$y = mx + b$$

Slope y-intercept

When the equation is in this form, the *slope* of the line is always the *coefficient of x*, and the *y-intercept* is always the *constant term*.

EXAMPLE 1   Find the equation of the line with slope $-\frac{4}{3}$ and y-intercept 5. Then graph the line.

SOLUTION   Substituting $m = -\frac{4}{3}$ and $b = 5$ into $y = mx + b$, we have

$$y = -\frac{4}{3}x + 5$$

Finding the equation from the slope and y-intercept is just that easy. If the slope is m and the y-intercept is b, then the equation is always $y = mx + b$. Now, let's graph the line.

Since the y-intercept is 5, the graph goes through the point $(0, 5)$. To find a second point on the graph, we start at $(0, 5)$ and move 4 units down (that's a rise of -4) and 3 units to the right (a run of 3), as indicated in Figure 2. The point we end up at is $(3, 1)$. Drawing a line that passes through $(0, 5)$ and $(3, 1)$, we have the graph of the equation.

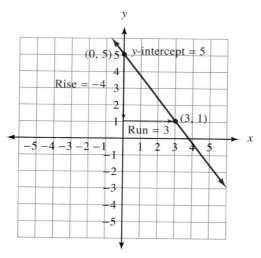

FIGURE 2

EXAMPLE 2   Find the slope and y-intercept for the line $2x - 3y = 5$.

SOLUTION   To use the slope-intercept form we must solve the equation for y in terms of x:

$$2x - 3y = 5$$
$$-3y = -2x + 5 \quad \text{Add } -2x \text{ to both sides.}$$
$$y = \frac{2}{3}x - \frac{5}{3} \quad \text{Divide by } -3.$$

The last equation has the form $y = mx + b$. The slope must be $m = \frac{2}{3}$, and the y-intercept is $b = -\frac{5}{3}$.

EXAMPLE 3   Graph the equation $2x + 3y = 6$ using the slope and y-intercept.

SOLUTION   Although we could graph this equation using the methods developed in Section 3.1 (by finding ordered pairs that are solutions to the equation and drawing a line through their graphs), it is sometimes easier to graph a line using the slope-intercept form of the equation.

Solving the equation for y, we have

$$2x + 3y = 6$$

$$3y = -2x + 6 \qquad \text{Add } -2x \text{ to both sides.}$$

$$y = -\frac{2}{3}x + 2 \qquad \begin{array}{l}\text{Divide by 3. This is the slope-intercept} \\ \text{form of the equation.}\end{array}$$

The slope is $m = -\frac{2}{3}$ and the y-intercept is $b = 2$. Therefore, the point $(0, 2)$ is on the graph, and the ratio rise/run going from $(0, 2)$ to any other point on the line is $-\frac{2}{3}$. If we start at $(0, 2)$ and move 2 units up (that's a rise of 2) and 3 units to the left (a run of -3), we arrive at another point on the graph. Then we draw the line that connects this new point, $(-3, 4)$, with our other point, $(0, 2)$, as shown in Figure 3. (We could also go down 2 units and right 3 units and still be assured of ending up at another point on the line, since $\frac{2}{-3}$ is the same as $\frac{-2}{3}$.)

As we mentioned in the introduction to this chapter, the rectangular coordinate system is the tool we use to connect algebra and geometry. Example 3 illustrates this connection, as do the many other examples in this chapter. In Example 3, Descartes' rectangular coordinate system allows us to associate the equation $2x + 3y = 6$ (an algebraic concept) with the straight line (a geometric concept) shown in Figure 3.

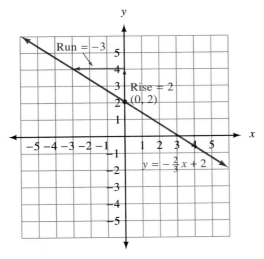

FIGURE 3

A second useful form of the equation of a straight line is the *point-slope form.*

Suppose line l contains the point (x_1, y_1) and has slope m. If (x, y) is any other point on line l, then by the definition of slope we have

$$\frac{y - y_1}{x - x_1} = m$$

Multiplying both sides by $(x - x_1)$ gives us

$$(x - x_1) \cdot \frac{y - y_1}{x - x_1} = m(x - x_1)$$

$$y - y_1 = m(x - x_1)$$

This last equation is known as the **point-slope form** of the equation of a straight line.

> ### Point-Slope Form of the Equation of a Line
>
> The equation of the line through (x_1, y_1) with slope m is given by
>
> $$y - y_1 = m(x - x_1)$$

This form of the equation of a straight line is used to find the equation of a line, either given one point on the line and the slope, or given two points on the line.

EXAMPLE 4 Find the equation of the line with slope -2 that contains the point $(-4, 3)$. Write the answer in slope-intercept form, and graph the equation.

SOLUTION

$$\text{Using} \quad (x_1, y_1) = (-4, 3) \text{ and } m = -2:$$

$y - y_1 = m(x - x_1)$	Point-slope form
gives us $\quad y - 3 = -2(x + 4)$	$x - (-4) = x + 4$
$y - 3 = -2x - 8$	Multiply out right side.
$y = -2x - 5$	Add 3 to each side. This is the slope-intercept form of the equation.

Figure 4 is the graph of the line that contains $(-4, 3)$ and has a slope of -2. Notice that the y-intercept on the graph can be read from the equation we found, since it is in slope-intercept form.

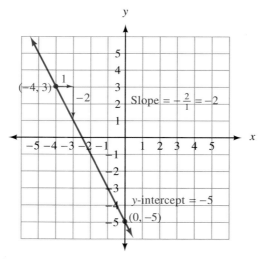

FIGURE 4

EXAMPLE 5 Find the equation of the line that passes through the points $(-3, 3)$ and $(3, -1)$, and graph it.

SOLUTION We begin by finding the slope of the line:

$$m = \frac{3 - (-1)}{-3 - 3} = \frac{4}{-6} = -\frac{2}{3}$$

Using $(x_1, y_1) = (3, -1)$ and $m = -\frac{2}{3}$ in $y - y_1 = m(x - x_1)$ yields

$$y + 1 = -\frac{2}{3}(x - 3)$$

$$y + 1 = -\frac{2}{3}x + 2 \qquad \text{Multiply out right side.}$$

$$y = -\frac{2}{3}x + 1 \qquad \text{Add } -1 \text{ to each side.}$$

Figure 5 shows the graph of the line that passes through the points $(-3, 3)$ and $(3, -1)$. As you can see, the slope and y-intercept are $-\frac{2}{3}$ and 1, respectively.

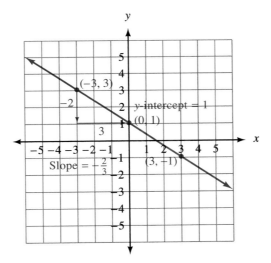

FIGURE 5

Note: In Example 5 we could have used the point $(-3, 3)$ instead of $(3, -1)$ and obtained the same equation. That is, using $(x_1, y_1) = (-3, 3)$ and $m = -\frac{2}{3}$ in $y - y_1 = m(x - x_1)$ gives us

$$y - 3 = -\frac{2}{3}(x + 3)$$

$$y - 3 = -\frac{2}{3}x - 2$$

$$y = -\frac{2}{3}x + 1$$

which is the same result we obtained using $(3, -1)$.

The last form of the equation of a line that we will consider in this section is called **standard form.** It is used mainly to write equations in a form that is free of fractions and is easy to compare with other equations.

> ### Standard Form for the Equation of a Line
>
> If a, b, and c are integers, then the equation of a line is in standard form when it is written as
>
> $$ax + by = c$$

If we want to rewrite the equation

$$y = -\frac{2}{3}x + 1$$

in standard form, we first multiply both sides by 3 to obtain

$$3y = -2x + 3$$

Then we add $2x$ to each side, yielding

$$2x + 3y = 3 \quad \text{Standard form}$$

EXAMPLE 6   Give the equation of the line through $(-1, 4)$ whose graph is perpendicular to the graph of $2x - y = -3$. Write the answer in standard form.

SOLUTION   First, to find the slope of $2x - y = -3$, we solve for y:

$$2x - y = -3$$
$$y = 2x + 3$$

The slope of this line is 2. The line we are interested in is perpendicular to the line with slope 2, so it must have a slope of $-\frac{1}{2}$.

Now, using $(x_1, y_1) = (-1, 4)$ and $m = -\frac{1}{2}$, we have

$$y - y_1 = m(x - x_1)$$
$$y - 4 = -\frac{1}{2}(x + 1)$$

Since we want our answer in standard form, we multiply each side by 2:

$$2y - 8 = -1(x + 1)$$
$$2y - 8 = -x - 1$$
$$x + 2y - 8 = -1$$
$$x + 2y = 7 \qquad \text{Standard form}$$

As a final note, we should mention again that all horizontal lines have equations of the form $y = b$ and slopes of 0. Vertical lines have no slope and have equations of the form $x = a$. These two special cases do not lend themselves to either the slope-intercept form or the point-slope form of the equation of a line.

Using Technology: Graphing Calculators

One advantage of using a graphing calculator to graph lines is that a calculator does not care whether the equation has been simplified or not. To illustrate, in Example 5 we found that the equation of the line with slope $-\frac{2}{3}$ that passes through the point $(3, -1)$ is

$$y + 1 = -\frac{2}{3}(x - 3)$$

Normally, to graph this equation we would simplify it first. With a graphing calculator we add -1 to each side and enter the equation this way:

$$Y_1 = -(2/3)(X - 3) - 1$$

No simplification is necessary. We can graph the equation in this form and the graph will be the same as the simplified form of the equation, which is $y = -\frac{2}{3}x + 1$. To convince yourself that this is true, graph both the simplified form for the equation and the unsimplified form in the same window. As you will see, the two graphs coincide.

Problem Set

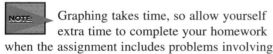
3.3

NOTE: Graphing takes time, so allow yourself extra time to complete your homework when the assignment includes problems involving graphing.

Give the equation of the line with the following slope and y-intercept.

1. $m = 2, b = 3$
2. $m = -4, b = 2$
3. $m = 1, b = -5$
4. $m = -5, b = -3$
5. $m = \frac{1}{2}, b = \frac{3}{2}$
6. $m = \frac{2}{3}, b = \frac{5}{6}$
7. $m = 0, b = 4$
8. $m = 0, b = -2$

Give the slope and y-intercept for each of the following equations. Sketch the graph using the slope and y-intercept. Give the slope of any line perpendicular to the given line.

9. $y = 3x - 2$
10. $y = 2x + 3$
11. $2x - 3y = 12$
12. $3x - 2y = 12$

13. $4x + 5y = 20$
14. $5x - 4y = 20$

For each of the following lines, name the slope and y-intercept. Then write the equation of the line in slope-intercept form.

15.

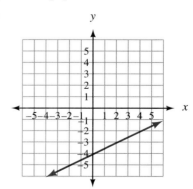

16.

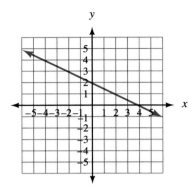

17.

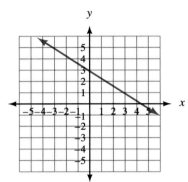

18.

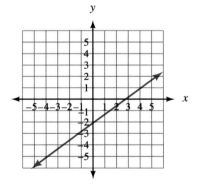

Write each equation in slope-intercept form. Then graph each line using the slope and y-intercept.

19. $-2x + y = 4$
20. $-2x + y = 2$
21. $3x + y = 3$
22. $3x + y = 6$
23. $-2x - 5y = 10$
24. $-4x + 5y = 20$

For each problem below, the slope and one point on a line are given. In each case, find the equation of that line. (Write the equation for each line in slope-intercept form.)

25. $(-2, -5); m = 2$
26. $(-1, -5); m = 2$
27. $(-4, 1); m = -\frac{1}{2}$
28. $(-2, 1); m = -\frac{1}{2}$
29. $(-\frac{1}{3}, 2); m = -3$
30. $(-\frac{2}{3}, 5); m = -3$

Find the equation of the line that passes through each pair of points. Write your answers in standard form.

31. $(-2, -4), (1, -1)$
32. $(2, 4), (-3, -1)$
33. $(-1, -5), (2, 1)$
34. $(-1, 6), (1, 2)$
35. $(\frac{1}{3}, -\frac{1}{5}), (-\frac{1}{3}, -1)$
36. $(-\frac{1}{2}, -\frac{1}{2}), (\frac{1}{2}, \frac{1}{10})$

For each line below, name the coordinates of any two points on the line. Then use those two points to find the equation of the line.

37.

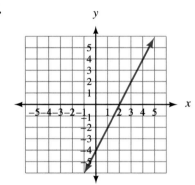

38.

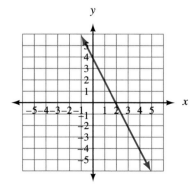

39.

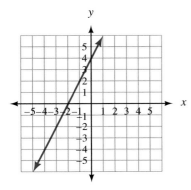

40.

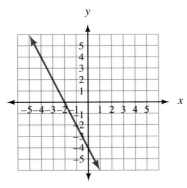

41. Give the slope and y-intercept of $y = -2$. Sketch the graph.

42. For the line $x = -3$, sketch the graph, give the slope, and name any intercepts.

43. Find the equation of the line parallel to the graph of $3x - y = 5$ that contains the point $(-1, 4)$.

44. Find the equation of the line parallel to the graph of $2x - 4y = 5$ that contains the point $(0, 3)$.

45. Line l is perpendicular to the graph of the equation $2x - 5y = 10$ and contains the point $(-4, -3)$. Find the equation for l.

46. Line l is perpendicular to the graph of the equation $-3x - 5y = 2$ and contains the point $(2, -6)$. Find the equation for l.

47. Give the equation of the line perpendicular to the graph of $y = -4x + 2$ that has an x-intercept of -1.

48. Write the equation of the line parallel to the graph of $7x - 2y = 14$ that has an x-intercept of 5.

49. Find the equation of the line with x-intercept 3 and y-intercept 2.

50. Find the equation of the line with x-intercept 2 and y-intercept 3.

51. **Deriving the Temperature Equation** The table from the introduction to this section is repeated below. The rows of the table give us ordered pairs (C, F).

Degrees Celsius	Degrees Fahrenheit
C	F
0	32
25	77
50	122
75	167
100	212

(a) Use any two of the ordered pairs from the table to derive the equation $F = \frac{9}{5}C + 32$.
(b) Use the equation from part (a) to find the Fahrenheit temperature that corresponds to a Celsius temperature of 30°.

52. **Maximum Heart Rate** The table below gives the maximum heart rate for adults 30, 40, 50, and 60 years old. Each row of the table gives us an ordered pair (A, M).

Age (years)	Maximum Heart Rate (beats per minute)
A	M
30	190
40	180
50	170
60	160

(a) Use any two of the ordered pairs from the

table to derive the equation $M = 220 - A$ that gives the maximum heart rate M for an adult whose age is A.

(b) Use the equation from part (a) to find the maximum heart rate for a 25-year-old adult.

Review Problems

 The problems that follow review material we covered in Section 2.3.

53. The length of a rectangle is 3 inches more than 4 times the width. The perimeter is 56 inches. Find the length and width.

54. One angle is 10° less than four times another. Find the measure of each angle if:
   (a) The two angles are complementary.
   (b) The two angles are supplementary.

55. The cash register in a candy shop contains $66 at the beginning of the day. At the end of the day, it contains $732.50. If the sales tax rate is 7.5%, how much of the total is sales tax?

56. The third angle in an isosceles triangle is 20° less than twice as large as each of the two base angles. Find the measure of each angle.

One Step Further

57. Label the units on a sheet of graph paper in multiples of 10, and graph the line $2x + 5y = 100$.

58. Label the units on a sheet of graph paper in multiples of 20, and graph the line $-4x + 10y = 100$.

59. Label the y-axis in multiples of 10 and the x-axis in multiples of 1, and graph the equation $y = 20x - 50$.

60. Label the y-axis in multiples of 10 and the x-axis in multiples of 1, and graph the equation $y = -20x + 30$.

Write each equation in slope-intercept form. Then name the slope, the y-intercept, and the x-intercept.

61. $\dfrac{x}{2} + \dfrac{y}{3} = 1$

62. $\dfrac{x}{5} + \dfrac{y}{4} = 1$

63. $\dfrac{x}{-2} + \dfrac{y}{3} = 1$

64. $\dfrac{x}{2} + \dfrac{y}{-3} = 1$

65. When a linear equation is written in the form

$$\frac{x}{a} + \frac{y}{b} = 1$$

it is said to be in *two-intercept form*. Find the x-intercept, the y-intercept, and the slope of this line.

SECTION 3.4

Linear Inequalities in Two Variables

A small movie theater holds 100 people. The owner charges more for adults than for children, so it is important to know the different combinations of adults and children that can be seated at one time. The shaded region in Figure 1 contains all the seating combinations. The line $x + y = 100$ shows the combinations for a full theater: The y-intercept corresponds to a theater full of adults, and the x-intercept corresponds to a theater full of children. In the shaded region below the line $x + y = 100$ are the combinations that occur if the theater is not full.

Shaded regions like the one shown in Figure 1 are produced by *linear inequalities in two variables,* which is the topic of this section.

A **linear inequality in two variables** is any expression that can be put in the form

$$ax + by < c$$

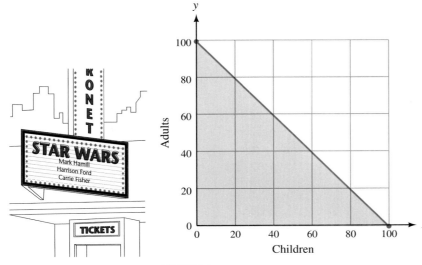

FIGURE I

where *a, b,* and *c* are real numbers (*a* and *b* not both 0). The inequality symbol can be any one of the following four: $<, \leq, >, \geq$.

Some examples of linear inequalities are

$$2x + 3y < 6 \qquad y \geq 2x + 1 \qquad x - y \leq 0$$

Although not all of these examples have the form $ax + by < c$, each one can be put in that form.

The solution set for a linear inequality is a *section of the coordinate plane.* The *boundary* for the section is found by replacing the inequality symbol with an equal sign and graphing the resulting equation. The boundary is included in the solution set (and represented with a *solid line*) if the inequality symbol used originally is $\leq$ or $\geq$. The boundary is not included (and is represented with a *broken line*) if the original symbol is $<$ or $>$.

EXAMPLE I Graph the solution set for $x + y \leq 4$.

SOLUTION The boundary for the graph is the graph of $x + y = 4$. The boundary is included in the solution set because the inequality symbol is $\leq$. The graph of the boundary is shown in Figure 2 (p. 172).

The boundary separates the coordinate plane into two sections, or regions—the region above the boundary and the region below the boundary. The solution set for $x + y \leq 4$ is one of these two regions, including the boundary. To find the correct region, we simply choose any convenient point that is *not* on the boundary. We then substitute the coordinates of the point into the original inequality $x + y \leq 4$. If the point we choose satisfies the inequality, then it is a member of the solution set, and we can assume that all points on the same side of the boundary as the chosen point are also in the solution set. If the coordinates of our point do not satisfy the original inequality, then the solution set lies on the other side of the boundary.

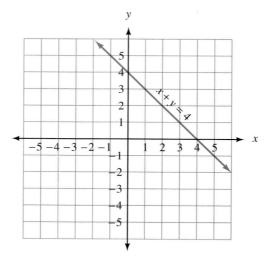

FIGURE 2

In this example, a convenient point that is not on the boundary is the origin.

Substituting   (0, 0):

into $x + y \leq 4$

gives us $0 + 0 \leq 4$

$0 \leq 4$ A true statement

Since the origin is a solution to the inequality $x + y \leq 4$, and the origin is below the boundary, all other points below the boundary are also solutions.

The graph of $x + y \leq 4$ is shown in Figure 3.

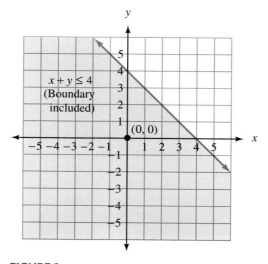

FIGURE 3

The region above the boundary is described by the inequality $x + y > 4$.

Here is a list of steps to follow when graphing the solution set for linear inequalities in two variables.

Strategy for Graphing a Linear Inequality in Two Variables

Step 1 Replace the inequality symbol with an equal sign. The resulting equation represents the boundary for the solution set.

Step 2 Graph the boundary found in Step 1 using a *solid line* if the boundary is included in the solution set (that is, if the original inequality symbol was either ≤ or ≥). Use a *broken line* to graph the boundary if it is *not* included in the solution set. (It is not included if the original inequality was either < or >.)

Step 3 Choose any convenient point not on the boundary and substitute the coordinates into the *original* inequality. If the resulting statement is *true,* the graph lies on the *same* side of the boundary as the chosen point. If the resulting statement is *false,* the solution set lies on the *opposite* side of the boundary.

EXAMPLE 2   Graph the solution set for $y < 2x - 3$.

SOLUTION   The boundary is the graph of $y = 2x - 3$, which is a line with slope 2 and y-intercept -3. The boundary is not included since the original inequality symbol is $<$. Therefore, we use a broken line to represent the boundary in Figure 4.

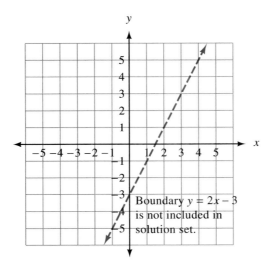

Boundary $y = 2x - 3$ is not included in solution set.

FIGURE 4

A convenient test point is again the origin:

$$\text{Using } (0, 0) \text{ in } \quad y < 2x - 3$$
$$\text{we have} \qquad 0 < 2(0) - 3$$
$$0 < -3 \qquad \text{A false statement}$$

Since our test point gives us a false statement and it lies above the boundary, the solution set must lie on the other side of the boundary, as shown in Figure 5.

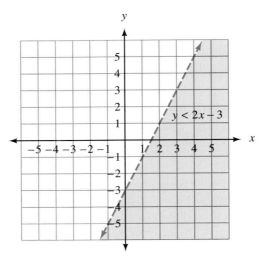

FIGURE 5

EXAMPLE 3   Graph the solution set for $x \leq 5$.

SOLUTION   The boundary is $x = 5$, which is a vertical line. All points to the left have x-coordinates less than 5 and all points to the right have x-coordinates greater than 5.

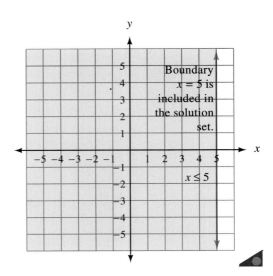

Using Technology: Graphing Calculators

Most graphing calculators have a Shade command that allows a portion of a graphing screen to be shaded. With this command we can visualize the solution sets to linear inequalities in two variables. However, since most graphing calculators cannot draw a dotted line, we are not actually "graphing" the solution set, only visualizing it.

Strategy for Visualizing a Linear Inequality in Two Variables on a Graphing Calculator

Step 1 Solve the inequality for y.
Step 2 Replace the inequality symbol with an equal sign. The resulting equation represents the boundary for the solution set.
Step 3 Graph the equation in an appropriate viewing window.
Step 4 Use the Shade command to indicate the solution set:
For inequalities having the $<$ or $\leq$ sign, use Shade(Xmin, Y_1).
For inequalities having the $>$ or $\geq$ sign, use Shade(Y_1, Xmax).

Note: On the TI-83/86, Step 4 can be done by manipulating the icons in the left column in the list of Y variables.

Figures 6 and 7 show the graphing calculator screens that help us visualize the solution set to the inequality $y < 2x - 3$ that we graphed in Example 2.

Windows: X from -5 to 5, Y from -5 to 5

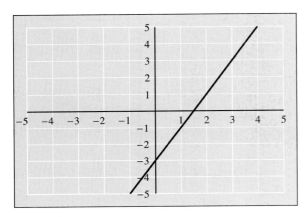

FIGURE 6
$Y_1 = 2X - 3$

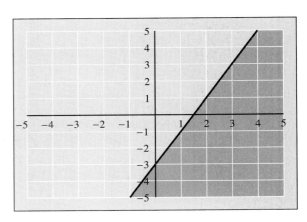

FIGURE 7
Shade (Xmin, Y_1)

Problem Set

3.4

NOTE: Remember to keep your goals for this class in front of you. Questioning the usefulness of a topic wastes time and moves you away from your goals. Take the time you might spend questioning the usefulness of a topic and use it to work more problems.

Graph the solution set for each of the following.

1. $x + y < 5$
2. $x + y \leq 5$
3. $x - y \geq -3$
4. $x - y > -3$
5. $2x + 3y < 6$
6. $2x - 3y > -6$
7. $-x + 2y > -4$
8. $-x - 2y < 4$
9. $2x + y < 5$
10. $2x + y < -5$
11. $y < 2x - 1$
12. $y \geq 2x - 1$

For each graph below, name the linear inequality in two variables that is represented by the shaded region.

13.

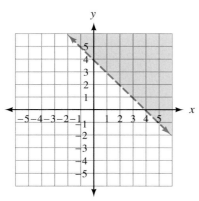

14.

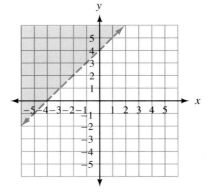

15.

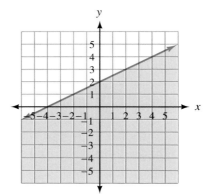

16.

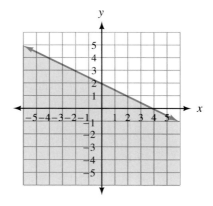

Graph each inequality.

17. $x \geq 3$
18. $x < -2$
19. $y \leq 4$
20. $y > -5$
21. $y < 2x$
22. $y > -3x$
23. $y \geq \frac{1}{2}x$
24. $y \leq \frac{1}{3}x$
25. $y \geq \frac{3}{4}x - 2$
26. $y > -\frac{2}{3}x + 3$
27. $\frac{x}{3} + \frac{y}{2} > 1$
28. $\frac{x}{5} + \frac{y}{4} < 1$

NOTE: Look back at Figure 1 in the introduction to this section. You will notice that the shading does not extend below the x-axis or to the left of the y-axis. The shading is restricted to the first quadrant only, because the problem represented by the graph is an applied problem in which neither x nor y can be negative. The same is true of the problems that follow.

29. Number of People in a Dance Club A dance club holds a maximum of 200 people. The club charges one price for students and a higher price for nonstudents. If the number of students in the club at any time is x and the number of nonstudents is y, shade the region in the first quadrant that contains all combinations of students and nonstudents that are in the club at any time.

30. Many Perimeters Suppose you have 500 feet of fencing that you will use to build a rectangular livestock pen. Let x represent the length of the pen, while y represents the width. Shade the region in the first quadrant that contains all possible values of x and y that will give you a rectangle from 500 feet of fencing. (You don't have to use all of the fencing, so the perimeter of the pen could be less than 500 feet.)

31. Gas Mileage You have two cars. The first car travels an average of 12 miles on a gallon of gasoline, while the second averages 22 miles per gallon. Suppose you can afford to buy up to 30 gallons of gasoline this month. If the first car is driven x miles this month, and the second car is driven y miles this month, shade the region in the first quadrant that gives all the possible values of x and y that will keep you from buying more than 30 gallons of gasoline this month. (The number of gallons the first car uses is $\frac{x}{12}$.)

32. Number Problem The sum of two positive numbers is at most 20. If the two numbers are represented by x and y, shade the region in the first quadrant that shows all the possibilities for the two numbers.

Review Problems

NOTE: The problems that follow review material we covered in Section 2.4.

Solve each of the following inequalities.

33. $\frac{1}{3} + \frac{y}{5} \leq \frac{26}{15}$

34. $-\frac{1}{3} \geq \frac{1}{6} - \frac{y}{2}$

35. $5t - 4 > 3t - 8$
36. $-3(t - 2) < 6 - 5(t + 1)$
37. $-9 < -4 + 5t < 6$
38. $-3 < 2t + 1 < 3$

One Step Further

Graph each inequality.

39. $y < |x + 2|$
40. $y > |x - 2|$
41. $y > |x - 3|$
42. $y < |x + 3|$

Introduction to Functions

The ad shown in the margin appeared in the help wanted section of the local newspaper the day I was writing this section of the book. We can use the information in the ad to start an informal discussion of our next topic: functions.

An Informal Look at Functions

To begin, suppose you have a job that pays $7.50 per hour and that you work anywhere from 0 to 40 hours per week. The amount of money you make in one week depends on the number of hours you work that week. In mathematics we say that your weekly earnings are a **function** of the number of hours you work. If we let the variable x represent hours and the variable y represent the money you make, then the relationship between x and y can be written as

$$y = 7.5x \qquad \text{for} \quad 0 \le x \le 40$$

Table 1 gives some of the paired data that satisfy the equation $y = 7.5x$. Figure 1 is the graph of the equation with the restriction $0 \le x \le 40$.

Function Rule: $y = 7.5x$ $\qquad$ for $0 \le x \le 40$

TABLE 1
Weekly Wages

Hours Worked	Rule	Pay
x	$y = 7.5x$	y
0	$y = 7.5(0)$	0
10	$y = 7.5(10)$	75
20	$y = 7.5(20)$	150
30	$y = 7.5(30)$	225
40	$y = 7.5(40)$	300

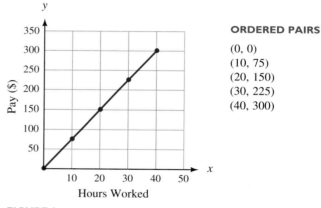

ORDERED PAIRS

(0, 0)
(10, 75)
(20, 150)
(30, 225)
(40, 300)

FIGURE 1
Weekly wages at $7.50 per hour

The equation $y = 7.5x$ with the restriction $0 \le x \le 40$, Table 1, and Figure 1 are three ways to describe the same relationship between the number of hours you work in one week and your gross pay for that week. In all three, we *input* values of x, and then use the function rule to *output* values of y.

Domain and Range of a Function

We began this discussion by saying that the number of hours worked during the week was from 0 to 40, so these are the values that x can assume. From the line graph in Figure 1, we see that the values of y range from 0 to 300. We call the complete set of values that x can assume the **domain** of the function. The values that are assigned to y are called the **range** of the function.

Function Maps

Another way to visualize the relationship between x and y is with the following diagram, which we call a **function map:**

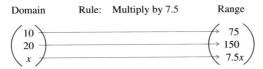

FIGURE 2
A function map

Although the diagram in Figure 2 does not show all the values that x and y can assume, it does give us a visual description of how x and y are related. It shows that values of y in the range come from values of x in the domain according to a specific rule (multiply by 7.5 each time).

A Formal Look at Functions

What is apparent from the discussion above is that we are working with paired data. The solutions to the equation $y = 7.5x$ are pairs of numbers; the points on the line graph in Figure 1 come from paired data; and the diagram in Figure 2 pairs numbers in the domain with numbers in the range. We are now ready for the formal definition of a function.

> **DEFINITION**
>
> A **function** is a rule that pairs each element in one set, called the **domain,** with exactly one element from a second set, called the **range.**

In other words, a function is a rule for which each input is paired with exactly one output.

Functions as Ordered Pairs

The function rule $y = 7.5x$ from the introduction to this section produces ordered pairs of numbers (x, y), as shown next to Figure 1. The same thing happens with all functions: The function rule produces ordered pairs of numbers. We use this result to write the alternative definition for a function given at the top of the next page.

A **function** is a set of ordered pairs in which no two different ordered pairs have the same first coordinate. The set of all first coordinates is called the **domain** of the function. The set of all second coordinates is called the **range** of the function.

The restriction on first coordinates in the alternative definition keeps us from assigning a number in the domain to more than one number in the range.

A Relationship That Is Not a Function

You may be wondering if any sets of paired data fail to qualify as functions. The answer is yes.

Table 2 shows the prices of used Ford Mustangs that were listed in the local newspaper. The diagram in Figure 3 is called a *scatter diagram*. It gives a visual representation of the data in Table 2.

TABLE 2
Used Mustang Prices

Year	Price ($)
x	y
1997	13,925
1997	11,850
1997	9,995
1996	10,200
1996	9,600
1995	9,525
1994	8,675
1994	7,900
1993	6,975

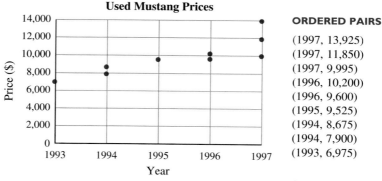

ORDERED PAIRS

(1997, 13,925)
(1997, 11,850)
(1997, 9,995)
(1996, 10,200)
(1996, 9,600)
(1995, 9,525)
(1994, 8,675)
(1994, 7,900)
(1993, 6,975)

FIGURE 3
Scatter diagram of data in Table 2

In Table 2, the year 1997 is paired with three different prices, $13,925, $11,850, and $9,995. That is enough to disqualify the data from belonging to a function. For a set of paired data to be considered a function, each number in the domain must be paired with exactly one number in the range.

Still, there is a relationship between the first coordinates and second coordinates in the used-car data. It is not a function relationship, but it is a relationship. In order to classify all relationships specified by ordered pairs, whether they are functions or not, we include the following two definitions.

A **relation** is a rule that pairs each element in one set, called the **domain,** with **one or more elements** from a second set, called the **range.**

Here are some facts that will help clarify the distinction between relations and functions.

1. Any rule that assigns numbers from one set to numbers in another set is a relation. If that rule does the assignment so that no input has more than one output, then it is also a function.
2. Any set of ordered pairs is a relation. If none of the first coordinates of those ordered pairs is repeated, the set of ordered pairs is also a function.
3. Every function is a relation.
4. Not every relation is a function.

Graphing Relations and Functions

To give ourselves a wider perspective on functions and relations, we consider some equations whose graphs are not straight lines.

EXAMPLE 1 Chaudra is tossing a softball into the air with an underhand motion. The distance of the ball above her hand at any time is given by the function

$$h = 32t - 16t^2 \qquad \text{for} \quad 0 \le t \le 2$$

where h is the height of the ball in feet, and t is the time in seconds. Construct a table that gives the height of the ball at quarter-second intervals, starting with $t = 0$ and ending with $t = 2$. Construct a line graph from the table.

SOLUTION We construct Table 3 (p. 182) using the following values of t: $0, \frac{1}{4}$, $\frac{1}{2}, \frac{3}{4}, 1, \frac{5}{4}, \frac{3}{2}, \frac{7}{4}, 2$. The values of h come from substituting these values of t into the equation $h = 32t - 16t^2$. (This equation comes from physics. If you take a physics class, you will learn how to derive this equation.) Then we construct the graph in Figure 4 (p. 182) from the table. The graph appears only in the first quadrant because neither t nor h can be negative.

Here is a summary of what we know about functions as it applies to this example: We input values of t and output values of h according to the function rule

$$h = 32t - 16t^2 \qquad \text{for} \quad 0 \le t \le 2$$

The domain is given by the inequality that follows the equation; it is

$$\text{Domain} = \{t \mid 0 \le t \le 2\}$$

The range is the set of all outputs that are possible by substituting the values of t from the domain into the equation. From our table and graph, it seems that the range is

$$\text{Range} = \{h \mid 0 \le h \le 16\}$$

TABLE 3
Tossing a Softball into the Air

Time (sec)	Function Rule	Distance (ft)
t	$h = 32t - 16t^2$	h
0	$h = 32(0) - 16(0)^2 = 0 - 0 = 0$	0
$\frac{1}{4}$	$h = 32(\frac{1}{4}) - 16(\frac{1}{4})^2 = 8 - 1 = 7$	7
$\frac{1}{2}$	$h = 32(\frac{1}{2}) - 16(\frac{1}{2})^2 = 16 - 4 = 12$	12
$\frac{3}{4}$	$h = 32(\frac{3}{4}) - 16(\frac{3}{4})^2 = 24 - 9 = 15$	15
1	$h = 32(1) - 16(1)^2 = 32 - 16 = 16$	16
$\frac{5}{4}$	$h = 32(\frac{5}{4}) - 16(\frac{5}{4})^2 = 40 - 25 = 15$	15
$\frac{3}{2}$	$h = 32(\frac{3}{2}) - 16(\frac{3}{2})^2 = 48 - 36 = 12$	12
$\frac{7}{4}$	$h = 32(\frac{7}{4}) - 16(\frac{7}{4})^2 = 56 - 49 = 7$	7
2	$h = 32(2) - 16(2)^2 = 64 - 64 = 0$	0

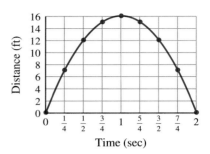

FIGURE 4

Using Technology: More about Example 1

Most graphing calculators can easily produce the information in Table 3. Simply set Y_1 equal to $32X - 16X^2$. Then set up the table so it starts at 0 and increases by an increment of 0.25 each time. (On a TI-82/83, use the TBLSET key to set up the table.)

TABLE SETUP

Table minimum = 0
Table increment = .25
Dependent variable: Auto
Independent variable: Auto

Y VARIABLES SETUP

$Y_1 = 32X - 16X^2$

The table will look like this:

X	Y_1
0	0
.25	7
.5	12
.75	15
1	16
1.25	15
1.5	12

EXAMPLE 2 Sketch the graph of $x = y^2$.

SOLUTION Without going into much detail, we graph the equation $x = y^2$ by finding a number of ordered pairs that satisfy the equation, plotting these points, then drawing a smooth curve that connects them. A table of values for x and y that satisfy the equation follows, along with the graph of $x = y^2$.

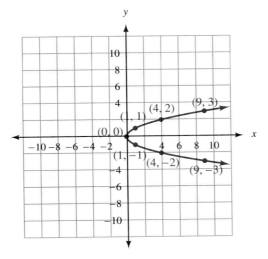

x	y
0	0
1	1
1	−1
4	2
4	−2
9	3
9	−3

FIGURE 5

As you can see from looking at the table and the graph in Figure 5, several ordered pairs whose graphs lie on the curve have repeated first coordinates. For instance, (1, 1) and (1, − 1), (4, 2) and (4, − 2), as well as (9, 3) and (9, − 3). Therefore, the graph is not the graph of a function.

Vertical Line Test

Look back at the scatter diagram for used Mustang prices shown in Figure 3. Notice that some of the points on the diagram lie above and below each other along vertical lines. This is an indication that the data do not constitute a function. Two data points that lie on the same vertical line must have come from two ordered pairs with the same first coordinates.

Now, look at the graph shown in Figure 5. The reason this graph is the graph of a relation, but not of a function, is that there are points on the graph that have the same first coordinates; for example, the points (4, 2) and (4, − 2). Furthermore, any time two points on a graph have the same first coordinates, those points must lie on a vertical line. [To convince yourself, connect the points (4, 2) and (4, − 2) with a straight line. You will see that it must be a vertical line.] This allows us to write the following test that uses the graph to determine whether a relation is also a function.

Vertical Line Test

If a vertical line crosses the graph of a relation in more than one place, the relation cannot be a function. If no vertical line can be found that crosses a graph in more than one place, then the graph is the graph of a function.

If we look back to the graph of $h = 32t - 16t^2$ as shown in Figure 4, we see that no vertical line can be found that crosses this graph in more than one place. Therefore, the graph shown in Figure 4 is the graph of a function.

EXAMPLE 3 Graph $y = |x|$. Use the graph to determine whether we have the graph of a function. State the domain and range.

SOLUTION We let x take on values of $-4, -3, -2, -1, 0, 1, 2, 3$, and 4. The corresponding values of y are shown in the table. The graph is shown in Figure 6.

x	y
-4	4
-3	3
-2	2
-1	1
0	0
1	1
2	2
3	3
4	4

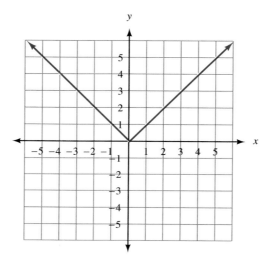

FIGURE 6

Since no vertical line can be found that crosses the graph in more than one place, $y = |x|$ is a function. The domain is all real numbers. The range is $\{y \mid y \geq 0\}$.

Problem Set

3.5

1. Suppose you have a job that pays $8.50 per hour and you work anywhere from 10 to 40 hours per week.
   (a) Write an equation, with a restriction on the variable x, that gives the amount of money, y, you will earn for working x hours in one week.
   (b) Use the function rule you have written in part (a) to complete Table 4.
   (c) Use the template in Figure 7 to construct a line graph from the information in Table 4.

TABLE 4
Weekly Wages

Hours Worked	Function Rule	Gross Pay ($)
x	$y =$	y
10		
20		
30		
40		

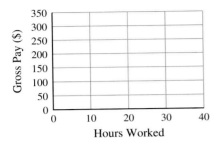

FIGURE 7
Template for line graph

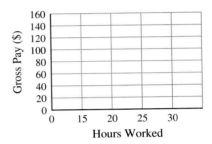

TABLE 5
Weekly Wages

Hours Worked	Function Rule	Gross Pay ($)
x	$y =$	y
15		
20		
25		
30		

(d) State the domain and range of this function.
(e) What is the minimum amount you can earn in a week with this job? What is the maximum amount?

2. The ad shown below was in the local newspaper. Suppose you are hired for the job described in the ad.

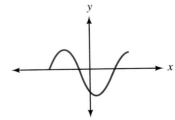

312 Help Wanted

ESPRESSO BAR
OPERATOR
Must be dependable, honest, service-oriented. Coffee exp desired. 15 – 30 hrs per wk. $5.25/hr. Start 5/31. Apply in person: Espresso Yourself, Central Coast Mall. Deadline 5/23.

(a) If x is the number of hours you work per week and y is your weekly gross pay, write the equation for y. (Be sure to include any restrictions on the variable x that are given in the ad.)
(b) Use the function rule you have written in part (a) to complete Table 5.
(c) Use the template in Figure 8 to construct a line graph from the information in Table 5.
(d) State the domain and range of this function.
(e) What is the minimum amount you can earn in a week with this job? What is the maximum amount?

FIGURE 8
Template for line graph

For each of the following relations, give the domain and range, and indicate which are also functions.

3. $\{(1, 3), (2, 5), (4, 1)\}$
4. $\{(3, 1), (5, 7), (2, 3)\}$
5. $\{(-1, 3), (1, 3), (2, -5)\}$
6. $\{(3, -4), (-1, 5), (3, 2)\}$
7. $\{(7, -1), (3, -1), (7, 4)\}$
8. $\{(5, -2), (3, -2), (5, -1)\}$

State whether each of the following graphs represents a function.

9.

10.

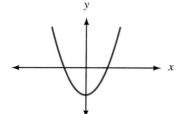

11.

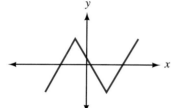

12.

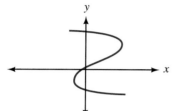

13.

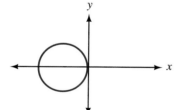

14.

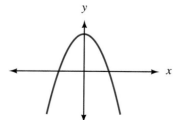

15.

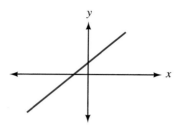

16.

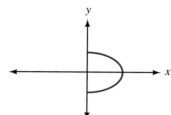

17.

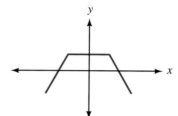

18.

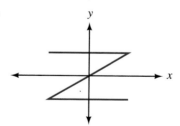

19. Tossing a Coin Hali is tossing a quarter into the air with an underhand motion. The distance the quarter is above her hand at any time is given by the function

$$h = 16t - 16t^2 \qquad \text{for} \quad 0 \le t \le 1$$

where h is the height of the quarter in feet, and t is the time in seconds.

(a) Construct a table that gives the height of the quarter every tenth of a second, starting at $t = 0$ and ending at $t = 1$.

(b) State the domain and range of this function.

(c) Use the data from the table to graph the function.

20. **Intensity of Light** The formula below gives the intensity of light that falls on a surface at various distances from a 100 watt light bulb:

$$I = \frac{120}{d^2} \qquad \text{for} \quad d > 0$$

where I is the intensity of light (in lumens per square foot), and d is the distance (in feet) from the light bulb to the surface.

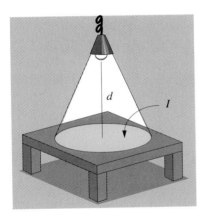

(a) Construct a table that gives the intensity of light when the bulb is 1, 2, 3, 4, 5, and 6 feet above the surface.
(b) Use the data from the table in part (a) to construct a graph of the function.

Graph each of the following relations. In each case, use the graph to find the domain and range, and indicate whether the graph is the graph of a function.

21. $y = x^2 - 1$
22. $y = x^2 + 1$
23. $y = x^2 + 4$
24. $y = x^2 - 9$
25. $x = y^2 - 1$
26. $x = y^2 + 1$
27. $x = y^2 + 4$
28. $x = y^2 - 9$
29. $y = |x - 2|$
30. $y = |x + 2|$
31. $y = |x| - 2$
32. $y = |x| + 2$
33. $x = |y|$
34. $x = |y| + 2$

Area of a Circle *The formula for the area A of a circle with radius r is given by $A = \pi r^2$. The formula shows that A is a function of r.*

35. Graph the function $A = \pi r^2$ for $0 \leq r \leq 3$. (On the graph, let the horizontal axis be the r-axis, and let the vertical axis be the A-axis.)

36. State the domain and range of the function $A = \pi r^2$, $0 \leq r \leq 3$. (Use $\pi \approx 3.14$.)

Area and Perimeter of a Rectangle *A rectangle is 2 inches longer than it is wide. Let x = the width, P = the perimeter, and A = the area of the rectangle.*

37. Write an equation that will give the perimeter P in terms of the width, x, of the rectangle. Are there any restrictions on the values that x can assume?

38. Graph the relationship between P and x.

39. Write an equation that will give the area A in terms of the width, x, of the rectangle. Are there any restrictions on the values that x can assume?

40. Graph the relationship between A and x.

41. **Tossing a Ball** A ball is thrown straight up into the air from ground level. The relationship between the height h of the ball at any time t is illustrated by the following graph:

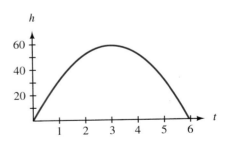

The horizontal axis represents time t, and the vertical axis represents height h.
(a) Is this graph the graph of a function?
(b) State the domain and range.
(c) At what time does the ball reach its maximum height?
(d) What is the maximum height of the ball?
(e) At what time does the ball hit the ground?

42. The following graph shows the relationship between a company's profits P and the number of items it sells, x. (P is in dollars.)

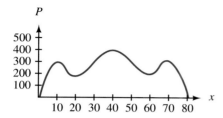

(a) Is this graph the graph of a function?
(b) State the domain and range.
(c) How many items must the company sell to make their maximum profit?
(d) What is their maximum profit?

Review Problems

 The problems that follow review material we covered in Section 2.2. Reviewing these problems will help you with the next section.

For the equation $y = 3x - 2$:

43. Find y if x is 4.   **44.** Find y if x is 0.
45. Find y if x is -4.   **46.** Find y if x is -2.

For the equation $y = x^2 - 3$:

47. Find y if x is 2.   **48.** Find y if x is -2.
49. Find y if x is 0.   **50.** Find y if x is -4.

Pricing *A company manufactures and sells prerecorded videotapes. They find that they can sell x videotapes each day at p dollars per tape, according to the equation $x = 230 - 20p$.*

51. How many videotapes will they sell if $p = \$5$?

52. Use the revenue equation $R = xp$ to find the revenue they will obtain from selling the tapes for $p = \$5$ each.

Function Notation

Let's return to the discussion that introduced us to functions. If a job pays $7.50 per hour for working from 0 to 40 hours a week, then the amount of money earned in one week, y, is a function of the number of hours worked, x. The exact relationship between x and y is written

$$y = 7.5x \qquad \text{for} \quad 0 \le x \le 40$$

Since the amount of money earned, y, depends on the number of hours worked, x, we call y the **dependent variable** and x the **independent variable.** Furthermore, if we let f represent all the ordered pairs produced by the equation, then we can write

$$f = \{(x, y) \mid y = 7.5x \text{ and } 0 \le x \le 40\}$$

Once we have named a function with a letter, we can use an alternative notation to represent the dependent variable y. The alternative notation for y is $f(x)$. It is read "f of x" and can be used instead of the variable y when working with functions. The notation y and the notation $f(x)$ are equivalent. That is,

$$y = 7.5x \quad \Leftrightarrow \quad f(x) = 7.5x$$

When we use the notation $f(x)$ we are using **function notation.** The benefit of using function notation is that we can write more information with fewer symbols than we can by using just the variable y. For example, asking how much money a person will make for working 20 hours is simply a matter of asking for $f(20)$. Without function notation, we would have to say "find the value of y that corresponds to a value of $x = 20$." To illustrate further, using the variable y, we can say "y is 150 when x is 20." Using the notation $f(x)$, we simply say "$f(20) = 150$." Each expression indicates that you will earn \$150 for working 20 hours.

EXAMPLE I If $f(x) = 7.5x$, find $f(0)$, $f(10)$, and $f(20)$.

SOLUTION To find $f(0)$ we substitute 0 for x in the expression $7.5x$ and simplify. We find $f(10)$ and $f(20)$ in a similar manner—by substitution.

$$\text{If} \qquad f(x) = 7.5x$$
$$\text{then} \qquad f(\mathbf{0}) = 7.5(\mathbf{0}) = 0$$
$$f(\mathbf{10}) = 7.5(\mathbf{10}) = 75$$
$$f(\mathbf{20}) = 7.5(\mathbf{20}) = 150$$

If we changed the example in the discussion that opened this section so that the hourly wage was \$6.50 per hour, we would have a new equation to work with, namely,

$$y = 6.5x \qquad \text{for} \quad 0 \le x \le 40$$

Suppose we name this new function with the letter g. Then

$$g = \{(x, y) \mid y = 6.5x \text{ and } 0 \le x \le 40\}$$

and

$$g(x) = 6.5x$$

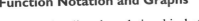

Input x

Function machine

Output
$f(x)$

Some students like to think of functions as machines. Values of x are put into the machine, which transforms them into values of $f(x)$, which are then output by the machine.

If we want to talk about both functions in the same discussion, having two different letters, f and g, makes it easy to distinguish between them. For example, since $f(x) = 7.5x$ and $g(x) = 6.5x$, asking how much money a person makes for working 20 hours is simply a matter of asking for $f(20)$ or $g(20)$, avoiding any confusion over which hourly wage we are talking about.

The diagrams shown in Figure 1 further illustrate the similarities and differences between the two functions we have been discussing.

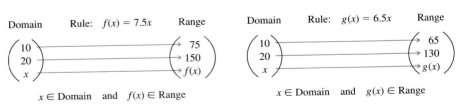

Domain   Rule: $f(x) = 7.5x$ Range   Domain   Rule: $g(x) = 6.5x$ Range

$$\begin{matrix} 10 \\ 20 \\ x \end{matrix} \qquad \begin{matrix} 75 \\ 150 \\ f(x) \end{matrix} \qquad\qquad \begin{matrix} 10 \\ 20 \\ x \end{matrix} \qquad \begin{matrix} 65 \\ 130 \\ g(x) \end{matrix}$$

$x \in$ Domain   and $f(x) \in$ Range $x \in$ Domain   and $g(x) \in$ Range

FIGURE I
Function maps

Function Notation and Graphs

We can visualize the relationship between x and $f(x)$ or $g(x)$ on the graphs of the two functions. Figure 2 shows the graph of $f(x) = 7.5x$ along with two additional line segments. The horizontal line segment corresponds to $x = 20$, while the vertical line segment corresponds to $f(20)$. Figure 3 shows the graph of $g(x) = 6.5x$ along with the horizontal line segment that corresponds to $x = 20$, and the vertical line segment that corresponds to $g(20)$. (Note that the domain in each case is restricted to $0 \leq x \leq 40$.)

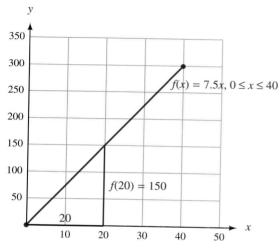

FIGURE 2

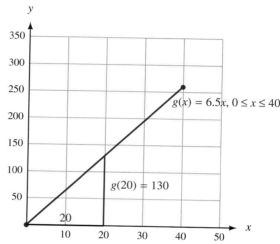

FIGURE 3

Using Function Notation

The remaining examples in this section show a variety of ways to use and interpret function notation.

EXAMPLE 2 If it takes Lorena t minutes to run a mile, then her average speed, $s(t)$, in miles per hour, is given by the formula

$$s(t) = \frac{60}{t} \qquad \text{for} \quad t > 0$$

Find $s(10)$ and $s(8)$, and then explain what they mean.

SOLUTION   To find $s(10)$, we substitute 10 for t in the equation and simplify:

$$s(\mathbf{10}) = \frac{60}{\mathbf{10}} = 6$$

In words: When Lorena runs a mile in 10 minutes, her average speed is 6 miles per hour.

We calculate $s(8)$ by substituting 8 for t in the equation. Doing so gives us

$$s(\mathbf{8}) = \frac{60}{\mathbf{8}} = 7.5$$

In words: Running a mile in 8 minutes is running at a rate of 7.5 miles per hour.

EXAMPLE 3   A painting is purchased as an investment for $125. If its value increases continuously so that it doubles every 5 years, then its value is given by the function

$$V(t) = 125 \cdot 2^{t/5} \qquad \text{for} \quad t \geq 0$$

where t is the number of years since the painting was purchased, and $V(t)$ is its value (in dollars) at time t. Find $V(5)$ and $V(10)$, and explain what they mean.

SOLUTION   The expression $V(5)$ is the value of the painting when $t = 5$ (5 years after it is purchased). We calculate $V(5)$ by substituting 5 for t in the equation $V(t) = 125 \cdot 2^{t/5}$. Here is our work:

$$V(5) = 125 \cdot 2^{5/5} = 125 \cdot 2^1 = 125 \cdot 2 = 250$$

In words: After 5 years, the painting is worth $250.

The expression $V(10)$ is the value of the painting after 10 years. To find this number, we substitute 10 for t in the equation:

$$V(10) = 125 \cdot 2^{10/5} = 125 \cdot 2^2 = 125 \cdot 4 = 500$$

In words: The value of the painting 10 years after it is purchased is $500.

EXAMPLE 4   A balloon has the shape of a sphere with a radius of 3 inches. Use the formulas below to find the volume and surface area of the balloon.

$$V(r) = \frac{4}{3}\pi r^3 \qquad S(r) = 4\pi r^2$$

SOLUTION   As you can see, we have used function notation to write the two formulas for volume and surface area, because each quantity is a function of the radius. To find these quantities when the radius is 3 inches, we evaluate $V(3)$ and $S(3)$:

$$V(3) = \frac{4}{3}\pi 3^3 = \frac{4}{3}\pi 27 = 36\pi \text{ cubic inches} \quad \text{or } 113 \text{ cubic inches} \qquad \text{To the nearest whole number}$$

$$S(3) = 4\pi 3^2 = 36\pi \text{ square inches} \quad \text{or } 113 \text{ square inches} \qquad \text{To the nearest whole number}$$

The fact that $V(3) = 36\pi$ means that the ordered pair $(3, 36\pi)$ belongs to the function V. Likewise, the fact that $S(3) = 36\pi$ tells us that the ordered pair $(3, 36\pi)$ is a member of function S.

We can generalize the discussion at the end of Example 4 this way:

$$(a, b) \in f \qquad \text{if and only if} \qquad f(a) = b$$

192

 Using Technology: More about Example 4

If we look back at Example 4, we see that when the radius of a sphere is 3, the numerical values of the volume and surface area are equal. How unusual is this? Are there other values of r for which $V(r)$ and $S(r)$ are equal? We can answer this question by looking at the graphs of both V and S.

To graph the function $V(r) = \frac{4}{3}\pi r^3$, set $Y_1 = 4\pi X^3/3$. To graph $S(r) = 4\pi r^2$, set $Y_2 = 4\pi X^2$. Graph the two functions in each of the following windows:

Window 1: X from -4 to 4, Y from -2 to 10

Window 2: X from 0 to 4, Y from 0 to 50

Window 3: X from 0 to 4, Y from 0 to 150

Then use the Trace and Zoom features of your calculator to locate the point in the first quadrant where the two graphs intersect. How do the coordinates of this point compare with the results in Example 4?

EXAMPLE 5   If $f(x) = 3x^2 + 2x - 1$, find $f(0), f(3),$ and $f(-2)$.

SOLUTION   Since $f(x) = 3x^2 + 2x - 1$, we have

$$f(0) = 3(0)^2 + 2(0) - 1 = 0 + 0 - 1 = -1$$
$$f(3) = 3(3)^2 + 2(3) - 1 = 27 + 6 - 1 = 32$$
$$f(-2) = 3(-2)^2 + 2(-2) - 1 = 12 - 4 - 1 = 7$$

In Example 5, the function f is defined by the equation $f(x) = 3x^2 + 2x - 1$. We could just as easily have said $y = 3x^2 + 2x - 1$. That is, $y = f(x)$. Saying $f(-2) = 7$ is exactly the same as saying y is 7 when x is -2.

EXAMPLE 6   If $f(x) = 4x - 1$ and $g(x) = x^2 + 2$, then

$f(5) = 4(5) - 1 = 19$	and	$g(5) = 5^2 + 2 = 27$
$f(-2) = 4(-2) - 1 = -9$	and	$g(-2) = (-2)^2 + 2 = 6$
$f(0) = 4(0) - 1 = -1$	and	$g(0) = 0^2 + 2 = 2$
$f(z) = 4z - 1$	and	$g(z) = z^2 + 2$
$f(a) = 4a - 1$	and	$g(a) = a^2 + 2$

 Using Technology: More about Example 6

Most graphing calculators can use tables to evaluate functions. To work Example 6 using a graphing calculator table, set Y_1 equal to $4X - 1$ and Y_2 equal to $X^2 + 2$. Then set the independent variable in the table to Ask instead of Auto. Go to your table and input 5, -2, and 0. Under Y_1 in the table, you will find $f(5), f(-2),$ and $f(0)$. Under Y_2, you will find $g(5), g(-2),$ and $g(0)$.

TABLE SETUP	Y VARIABLES SETUP
Table minimum $= 0$	$Y_1 = 4X - 1$
Table increment $= 1$	$Y_2 = X^2 + 2$
Independent variable: Auto	
Dependent variable: Ask	

The table will look like this:

X	Y_1	Y_2
5	19	27
-2	-9	6
0	-1	2

Although the calculator asks us for a table increment, the increment doesn't matter since we are inputting the X values ourselves.

EXAMPLE 7 If the function f is given by

$$f = \{(-2, 0), (3, -1), (2, 4), (7, 5)\}$$

then $f(-2) = 0, f(3) = -1, f(2) = 4,$ and $f(7) = 5.$

EXAMPLE 8 If $f(x) = 2x^2$ and $g(x) = 3x - 1$, find:
(a) $f[g(2)]$ (b) $g[f(2)]$

SOLUTION   The expression $f[g(2)]$ is read "f of g of 2."
(a) Since $g(2) = 3(2) - 1 = 5,$

$$f[g(2)] = f(5) = 2(5)^2 = 50$$

(b) Since $f(2) = 2(2)^2 = 8,$

$$g[f(2)] = g(8) = 3(8) - 1 = 23$$

Many of the equations and formulas we have worked with previously can be written in terms of function notation. For example, if a company sells x items at a price of p dollars per item, then in function notation:

■ $R(x)$ is the revenue function that gives the revenue R in terms of the number of items, x.
■ $R(p)$ is the revenue function that gives the revenue R in terms of the price per item, p.

With function notation we can see exactly which variables we want our formulas written in terms of.

In the next two examples, we will use function notation to combine a number of problems we have worked previously.

EXAMPLE 9 A company manufactures and sells prerecorded videotapes. They find that they can sell x videotapes each day at p dollars per tape, according to the equation $x = 230 - 20p$. Find $R(x)$ and $R(p)$.

SOLUTION The notation $R(p)$ tells us we are to write the revenue equation in terms of the variable p. To do so, we use the formula $R(p) = xp$ and substitute $230 - 20p$ for x to obtain

$$R(p) = xp = (230 - 20p)p = 230p - 20p^2$$

The notation $R(x)$ indicates we are to write the revenue equation in terms of the variable x. We need to solve the equation $x = 230 - 20p$ for p. Let's begin by interchanging the two sides of the equation:

$$230 - 20p = x$$
$$-20p = -230 + x \qquad \text{Add } -230 \text{ to each side.}$$
$$p = \frac{-230 + x}{-20} \qquad \text{Divide each side by } -20.$$
$$p = 11.5 - 0.05x \qquad \tfrac{-230}{-20} = 11.5 \text{ and } \tfrac{1}{-20} = -0.05$$

Now we can find $R(x)$ by substituting $11.5 - 0.05x$ for p in the formula $R(x) = xp$:

$$R(x) = xp = x(11.5 - 0.05x) = 11.5x - 0.05x^2$$

Our two revenue functions are actually equivalent. To offer some justification for this, suppose that the company decides to sell each tape for $5. The equation $x = 230 - 20p$ indicates that, at $5 per tape, they will sell $x = 230 - 20(5) = 230 - 100 = 130$ tapes per day. To find the revenue from selling the tape for $5 each, we use $R(p)$ with $p = 5$:

$$\text{If } \quad p = 5:$$
$$\text{then} \quad R(p) = R(5)$$
$$= 230(\mathbf{5}) - 20(\mathbf{5})^2$$
$$= 1{,}150 - 500$$
$$= \$650$$

On the other hand, to find the revenue from selling 130 tapes, we use $R(x)$ with $x = 130$:

$$\text{If } \quad x = 130:$$
$$\text{then} \quad R(x) = R(130)$$
$$= 11.5(\mathbf{130}) - 0.05(\mathbf{130})^2$$
$$= 1{,}495 - 845$$
$$= \$650$$

EXAMPLE 10 Suppose the daily cost function for the videotapes in Example 9 is $C(x) = 200 + 2x$. Find the profit function $P(x)$ and then find $P(130)$.

SOLUTION Since profit is equal to the difference of the revenue and the cost, we have

$$P(x) = R(x) - C(x)$$
$$= 11.5x - 0.05x^2 - (200 + 2x)$$
$$= -0.05x^2 + 9.5x - 200$$

Notice that we used the formula for $R(x)$ from Example 9 instead of the formula for $R(p)$. We did so because we were asked to find $P(x)$, meaning we want the profit P only in terms of the variable x.

Next, we use the formula we just obtained to find $P(130)$:

$$P(130) = -0.05(\mathbf{130})^2 + 9.5(\mathbf{130}) - 200$$
$$= -0.05(16,900) + 9.5(130) - 200$$
$$= -845 + 1,235 - 200$$
$$= \$190$$

Since $P(130) = \$190$, the company will make a profit of $190 per day by selling 130 tapes per day.

 Using Technology: More about Example 10

We can visualize the three functions $P(x)$, $R(x)$, and $C(x)$ from Example 10 if we set up the functions list and graphing window on a calculator this way:

$$Y_1 = 11.5X - 0.05X^2 \quad \text{This gives the graph of } R(x).$$
$$Y_2 = 200 + 2X \qquad\qquad \text{This gives the graph of } C(x).$$
$$Y_3 = Y_1 - Y_2 \qquad\qquad \text{This gives the graph of } P(x).$$
$$\text{Window:} \quad \text{X from 0 to 250, Y from 0 to 750}$$

The graphs in Figure 4 show the results.

Now, find the value of $P(x)$ when $R(x)$ and $C(x)$ intersect.

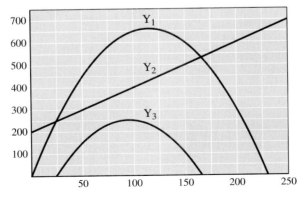

FIGURE 4

Problem Set

3.6

1. **Investing in Art** A painting is purchased as an investment for $150. If its value increases continuously so that it doubles every 3 years, then its value is given by the function

$$V(t) = 150 \cdot 2^{t/3} \qquad \text{for} \quad t \geq 0$$

where t is the number of years since the painting was purchased, and $V(t)$ is its value (in dollars) at time t. Find $V(3)$ and $V(6)$, and then explain what they mean.

2. **Average Speed** If it takes Minke t minutes to run a mile, then her average speed, $s(t)$, in miles per hour, is given by the formula

$$s(t) = \frac{60}{t} \qquad \text{for} \quad t > 0$$

Find $s(4)$ and $s(5)$, and then explain what they mean.

3. **Dimensions of a Rectangle** The length of a rectangle is 3 inches more than twice the width. Let x represent the width of the rectangle and $P(x)$ represent the perimeter of the rectangle. Use function notation to write the relationship between x and $P(x)$, noting any restrictions on the variable x.

4. **Dimensions of a Rectangle** The length of a rectangle is 3 inches more than twice the width. Let x represent the width of the rectangle and $A(x)$ represent the area of the rectangle. Use function notation to write the relationship between x and $A(x)$, noting any restrictions on the variable x.

Area of a Circle *The formula for the area A of a circle with radius r can be written with function notation as* $A(r) = \pi r^2$.

5. Find $A(2)$, $A(5)$, and $A(10)$. (Use $\pi \approx 3.14$.)

6. Why doesn't it make sense to ask for $A(-10)$?

Let $f(x) = 2x - 5$ *and* $g(x) = x^2 + 3x + 4$.
Evaluate the following.

7. $f(2)$ 8. $f(3)$ 9. $f(-3)$
10. $g(-2)$ 11. $g(-1)$ 12. $f(-4)$
13. $g(-3)$ 14. $g(2)$
15. $g(4) + f(4)$ 16. $f(2) - g(3)$
17. $f(3) - g(2)$ 18. $g(-1) + f(-1)$

Let $f(x) = 3x^2 - 4x + 1$ *and* $g(x) = 2x - 1$.
Evaluate the following.

19. $f(0)$ 20. $g(0)$ 21. $g(-4)$
22. $f(1)$ 23. $f(-1)$ 24. $g(-1)$
25. $g(10)$ 26. $f(10)$ 27. $f(3)$
28. $g(3)$ 29. $g(\frac{1}{2})$ 30. $g(\frac{1}{4})$
31. $f(a)$ 32. $g(b)$

33. **Cost of a Phone Call** Suppose a phone company charges 33¢ for the first minute and 24¢ for each additional minute to place a long-distance call between 5 P.M. and 11 P.M. If x is the number of additional minutes and $f(x)$ is the cost of the call, then $f(x) = 24x + 33$.
    (a) How much does it cost to talk for 10 minutes?
    (b) What does $f(5)$ represent in this problem?
    (c) If a call costs $1.29, how long was it?

34. **Cost of a Phone Call** The same phone company mentioned in Problem 33 charges 52¢ for the first minute and 36¢ for each additional minute to place a long-distance call between 8 A.M. and 5 P.M.
    (a) Let $g(x)$ be the total cost of a long-distance call between 8 A.M. and 5 P.M., and write an equation for $g(x)$.
    (b) Find $g(5)$.
    (c) Find the difference in price between a 10 minute call made between 8 A.M. and 5 P.M. and the same call made between 5 P.M. and 11 P.M.

If f = {(1, 4), (− 2, 0), (3, ½), (π, 0)} and g = {(1, 1), (− 2, 2), (½, 0)}, find each of the following values of f and g:

35. $f(1)$　　**36.** $g(1)$　　**37.** $g(\frac{1}{2})$

38. $f(3)$　　**39.** $g(-2)$　　**40.** $f(\pi)$

Let $f(x) = 2x^2 − 8$ and $g(x) = \frac{1}{2}x + 1$. Evaluate each of the following:

41. $f(0)$　　**42.** $g(0)$　　**43.** $g(-4)$

44. $f(1)$　　**45.** $f(a)$　　**46.** $g(z)$

47. $f(b)$　　**48.** $g(t)$　　**49.** $f[g(2)]$

50. $g[f(2)]$　　**51.** $g[f(-1)]$　　**52.** $f[g(-2)]$

53. $g[f(0)]$　　**54.** $f[g(0)]$

55. Graph the function $f(x) = \frac{1}{2}x + 2$. Then draw and label the line segments that represent $x = 4$ and $f(4)$.

56. Graph the function $f(x) = -\frac{1}{2}x + 6$. Then draw and label the line segments that represent $x = 4$ and $f(4)$.

57. For the function $f(x) = \frac{1}{2}x + 2$, find the value of x for which $f(x) = x$.

58. For the function $f(x) = -\frac{1}{2}x + 6$, find the value of x for which $f(x) = x$.

59. Graph the function $f(x) = x^2$. Then draw and label the line segments that represent $x = 1$ and $f(1)$, $x = 2$ and $f(2)$, and, finally, $x = 3$ and $f(3)$.

60. Graph the function $f(x) = x^2 − 2$. Then draw and label the line segments that represent $x = 2$ and $f(2)$, and the line segments corresponding to $x = 3$ and $f(3)$.

61. **Revenue** A company selling diskettes for home computers finds that they can sell x diskettes per day at p dollars per diskette, according to the formula $x = 800 − 100p$. Find $R(p)$ and $R(x)$.

62. **Revenue** A company sells an inexpensive accounting program for home computers. If they can sell x programs per week at p dollars per program, according to the formula $x = 350 − 10p$, find formulas for $R(p)$ and $R(x)$.

63. **Profit** If the cost to produce the x diskettes in Problem 61 is $C(x) = 2x + 200$, find $P(x)$ and $P(40)$.

64. **Profit** If the cost to produce the x programs in Problem 62 is $C(x) = 5x + 500$, find $P(x)$ and $P(60)$.

Review Problems

 The problems that follow review material we covered in Section 2.5.

Solve each equation.

65. $|3x − 5| = 7$

66. $|0.04 − 0.03x| = 0.02$

67. $|4y + 2| − 8 = −2$

68. $4 = |3 − 2y| − 5$

69. $5 + |6t + 2| = 3$

70. $7 + |3 − \frac{3}{4}t| = 10$

71. $|\frac{1}{10}x − \frac{1}{5}| = |\frac{1}{2} − \frac{1}{10}x|$

72. $|3x + 4| = |2 − 5x|$

One Step Further

*A function is called an **even function** if $f(−x) = f(x)$ for every value of x. Show that each of the following functions is an even function.*

73. $f(x) = x^2 − 4$

74. $f(x) = x^4 + x^2$

75. $f(x) = x^{-2}$

76. $f(x) = x^2 − x^{-2}$

*A function is called an **odd function** if $f(−x) = −f(x)$ for every value of x. Show that each of the following functions is an odd function.*

77. $f(x) = 3x$

78. $f(x) = x^3$

79. $f(x) = x^3 − x$

80. $f(x) = \dfrac{5}{x}$

Variation

If you are a runner and you average t minutes for every mile you run during one of your workouts, then your speed s in miles per hour is given by the equation and graph below (the graph is shown in the first quadrant only because both t and s are positive):

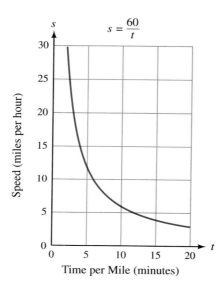

You know intuitively that as your average time per mile (t) increases, your speed (s) decreases. Likewise, lowering your time per mile will increase your speed. The equation and the graph also show this to be true: increasing t, decreases s; and decreasing t, increases s. Quantities that are connected in this way are said to *vary inversely* with each other. Inverse variation is one of the topics we will study in this section.

There are two main types of variation—*direct variation* and *inverse variation*. Variation problems are most common in the sciences, particularly in chemistry and physics.

Direct Variation

When we say the variable y **varies directly** with the variable x, we mean that the relationship can be written in symbols as $y = Kx$, where K is a nonzero constant called the **constant of variation** (or **proportionality constant**).

Another way of saying y varies directly with x is to say y is **directly proportional** to x.

Study the following list. It gives the mathematical equivalent of some direct variation statements.

English Phrase	Algebraic Equation
y varies directly with x	$y = Kx$
s varies directly with the square of t	$s = Kt^2$
y is directly proportional to the cube of z	$y = Kz^3$
u is directly proportional to the square root of v	$u = K\sqrt{v}$

EXAMPLE 1 Suppose y varies directly with x. If y is 15 when x is 5, find y when x is 7.

SOLUTION The first sentence gives us the general relationship between x and y. The equation that represents the statement "y varies directly with x" is

$$y = Kx$$

The first part of the second sentence in our example gives us the information necessary to evaluate the constant K:

When $y = 15$ and $x = 5$:

the equation $y = Kx$

becomes $15 = K \cdot 5$

or $K = 3$

The equation can now be written specifically as

$$y = 3x$$

Letting $x = 7$, we have

$$y = 3 \cdot 7$$
$$y = 21$$

EXAMPLE 2 The distance a body falls from rest toward the earth is directly proportional to the square of the time it has been falling. If a body falls 64 feet in 2 seconds, how far will it fall in 3.5 seconds?

SOLUTION We will let d = distance and t = time. Since distance is directly proportional to the square of time, we have

$$d = Kt^2$$

Next, we evaluate the constant K:

When $t = 2$ and $d = 64$:

the equation $d = Kt^2$

becomes $64 = K(2)^2$

or $64 = 4K$

and $K = 16$

Specifically, then, the relationship between d and t is

$$d = 16t^2$$

Finally, we find d when $t = 3.5$:

$$d = 16(3.5)^2$$
$$d = 16(12.25)$$
$$d = 196 \text{ feet}$$

Note: The equation $d = 16t^2$ is a specific example of the equation used in physics to find the distance a body falls from rest during a time t. In physics, the equation is $d = \frac{1}{2}gt^2$, where g is the acceleration of gravity. On earth the acceleration of gravity is $g = 32$ feet/second^2. The equation, then, is $d = \frac{1}{2} \cdot 32t^2$ or $d = 16t^2$.

Inverse Variation

If two variables are related so that an *increase* in one produces a proportional *decrease* in the other, then the variables are said to **vary inversely.** If y varies inversely with x, then

$$y = K\frac{1}{x} \qquad \text{or} \qquad y = \frac{K}{x}$$

We can also say y is **inversely proportional** to x. The constant K is again called the *constant of variation* or *proportionality constant.*

English Phrase	Algebraic Equation
y is inversely proportional to x	$y = \dfrac{K}{x}$
s varies inversely with the square of t	$s = \dfrac{K}{t^2}$
y is inversely proportional to x^4	$y = \dfrac{K}{x^4}$
z varies inversely with the cube root of t	$z = \dfrac{K}{\sqrt[3]{t}}$

EXAMPLE 3 Suppose y varies inversely with the square of x. If y is 4 when x is 5, find y when x is 10.

SOLUTION Since y is inversely proportional to the square of x, we can write

$$y = \frac{K}{x^2}$$

Evaluating K using the information given, we have

When $x = 5$ and $y = 4$:

the equation $y = \dfrac{K}{x^2}$

becomes $4 = \dfrac{K}{5^2}$

or $4 = \dfrac{K}{25}$

and $K = 100$

Now we write the equation again as

$$y = \dfrac{100}{x^2}$$

We finish by substituting $x = 10$ into the last equation:

$$y = \dfrac{100}{10^2} = \dfrac{100}{100} = 1$$

EXAMPLE 4 The volume of a gas is inversely proportional to the pressure of the gas on its container. If a pressure of 48 pounds per square inch corresponds to a volume of 50 cubic feet, what pressure is needed to produce a volume of 100 cubic feet?

SOLUTION We can represent volume with V and pressure with P: $V = \dfrac{K}{P}$

Using $P = 48$ and $V = 50$, we have

$$50 = \dfrac{K}{48}$$
$$K = 50(48)$$
$$K = 2{,}400$$

The equation that describes the relationship between P and V is

$$V = \dfrac{2{,}400}{P}$$

Substituting $V = 100$ into this last equation, we get

$$100 = \dfrac{2{,}400}{P}$$
$$100P = 2{,}400$$
$$P = \dfrac{2{,}400}{100}$$
$$P = 24$$

A volume of 100 cubic feet is produced by a pressure of 24 pounds per square inch.

Note: The relationship between pressure and volume as given in this example is known as *Boyle's Law* and applies to situations such as those encountered in a piston–cylinder arrangement. It was Robert Boyle who, in 1662, published the results of some of his experiments that showed, among other things, that the volume of a gas decreases as the pressure increases. This is an example of inverse variation.

Joint Variation and Other Variation Combinations

Many times relationships among different quantities are described in terms of more than two variables. If the variable y varies directly with *two* other variables, say x and z, then we say y **varies jointly** with x and z. In addition to joint variation, there are many other combinations of direct and inverse variation involving more than two variables. The following table is a list of some variation statements and their equivalent mathematical forms:

English Phrase	Algebraic Equation
y varies jointly with x and z	$y = Kxz$
z varies jointly with r and the square of s	$z = Krs^2$
V is directly proportional to T and inversely proportional to P	$V = \dfrac{KT}{P}$
F varies jointly with m_1 and m_2 and inversely with the square of r	$F = \dfrac{Km_1 \cdot m_2}{r^2}$

EXAMPLE 5 Suppose y varies jointly with x and the square of z. When x is 5 and z is 3, y is 180. Find y when x is 2 and z is 4.

SOLUTION   The general equation is given by

$$y = Kxz^2$$

Substituting $x = 5$, $z = 3$, and $y = 180$, we have

$$180 = K(5)(3)^2$$
$$180 = 45K$$
$$K = 4$$

The specific equation is

$$y = 4xz^2$$

When $x = 2$ and $z = 4$, the last equation becomes

$$y = 4(2)(4)^2$$
$$y = 128$$

EXAMPLE 6 In electricity, the resistance of a cable is directly proportional to its length and inversely proportional to the square of the diameter. If a 100 foot cable 0.5 inch in diameter has a resistance of 0.2 ohm, what will be the resistance of a cable made from the same material if it is 200 feet long with a diameter of 0.25 inch?

SOLUTION   Let R = resistance, l = length, and d = diameter. The equation is

$$R = \frac{Kl}{d^2}$$

When $R = 0.2$, $l = 100$, and $d = 0.5$, the equation becomes

$$0.2 = \frac{K(100)}{(0.5)^2}$$

or $$K = 0.0005$$

Using this value of K in our original equation, the result is

$$R = \frac{0.0005l}{d^2}$$

When $l = 200$ and $d = 0.25$, the equation becomes

$$R = \frac{0.0005(200)}{(0.25)^2}$$

$$R = 1.6 \text{ ohms}$$

Graphing Direct and Inverse Variation Statements

When y varies directly with x, the relationship is always written $y = Kx$. For any value of K, the graph of $y = Kx$ is a straight line with slope K and y-intercept 0 (in slope-intercept form the equation is $y = Kx + 0$). Since the y-intercept is 0, the graph of every direct variation statement of the form $y = Kx$ will pass through the origin.

EXAMPLE 7   Graph the direct variation statements $y = 2x$ and $y = -2x$ on the same coordinate system.

SOLUTION   The graph of each equation is a straight line that passes through the origin. The graph of $y = 2x$ has a slope of 2, while the graph of $y = -2x$ has a slope of -2. Both graphs are shown in Figure 1.

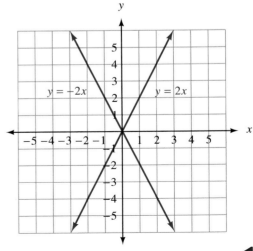

FIGURE 1

Next, we will graph two simple inverse variation statements.

EXAMPLE 8   Graph the inverse variation statement $y = \dfrac{1}{x}$.

SOLUTION   Since this is the first time we have graphed an equation of this form, we will make a table of values for x and y that satisfy the equation. Before we do, let's make some generalizations about the graph.

First, notice that, since y is equal to 1 divided by x, y will be positive when x is positive. (The quotient of two positive numbers is a positive number.) Likewise, when x is negative, y will be negative. In other words, x and y will always have the same sign. Thus, our graph will appear in quadrants 1 and 3 only, because in those quadrants x and y have the same sign.

Next, notice that the expression $\frac{1}{x}$ will be undefined when x is 0, meaning that there is no value of y corresponding to $x = 0$. Because of this, the graph will not cross the y-axis. Further, the graph will not cross the x-axis either. If we try to find the x-intercept by letting $y = 0$, we have

$$0 = \frac{1}{x}$$

But there is no value of x to divide into 1 to obtain 0. Therefore, since there is no solution to this equation, our graph will not cross the x-axis.

To summarize, we can expect to find the graph in quadrants 1 and 3 only, and the graph will cross neither axis. The graph is shown in Figure 2.

x	y
-3	$-\frac{1}{3}$
-2	$-\frac{1}{2}$
-1	-1
$-\frac{1}{2}$	-2
$-\frac{1}{3}$	-3
0	Undefined
$\frac{1}{3}$	3
$\frac{1}{2}$	2
1	1
2	$\frac{1}{2}$
3	$\frac{1}{3}$

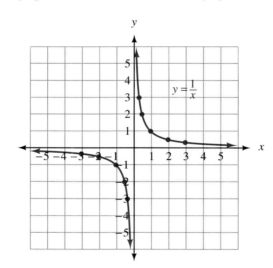

FIGURE 2

EXAMPLE 9   Graph the inverse variation statement $y = \dfrac{-6}{x}$.

SOLUTION   Since y is -6 divided by x, when x is positive, y will be negative (a negative divided by a positive is negative), and when x is negative, y will be positive (a negative divided by a negative). Thus, the graph will appear in quadrants 2 and 4 only. As was the case in Example 8, the graph will not cross either axis. The graph is shown in Figure 3.

x	y
−6	1
−3	2
−2	3
−1	6
0	Undefined
1	−6
2	−3
3	−2
6	−1

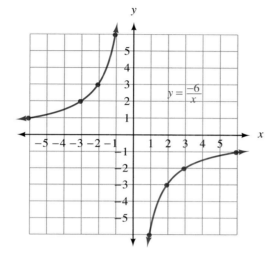

FIGURE 3

We'll have more to say about the functions graphed in Examples 8 and 9 later.

Problem Set

3.7

For the following problems, y varies directly with x.

1. If y is 10 when x is 2, find y when x is 6.

2. If y is 20 when x is 5, find y when x is 3.

3. If y is $−32$ when x is 4, find x when y is $−40$.

4. If y is $−50$ when x is 5, find x when y is $−70$.

For the following problems, r is inversely proportional to s.

5. If r is $−3$ when s is 4, find r when s is 2.

6. If r is $−10$ when s is 6, find r when s is $−5$.

7. If r is 8 when s is 3, find s when r is 48.

8. If r is 12 when s is 5, find s when r is 30.

For the following problems, d varies directly with the square of r.

9. If $d = 10$ when $r = 5$, find d when $r = 10$.

10. If $d = 12$ when $r = 6$, find d when $r = 9$.

11. If $d = 100$ when $r = 2$, find d when $r = 3$.

12. If $d = 50$ when $r = 5$, find d when $r = 7$.

For the following problems, y varies inversely with the square of x.

13. If $y = 45$ when $x = 3$, find y when x is 5.

14. If $y = 12$ when $x = 2$, find y when x is 6.

15. If $y = 18$ when $x = 3$, find y when x is 2.

16. If $y = 45$ when $x = 4$, find y when x is 5.

For the following problems, z varies jointly with x and the square of y.

17. If z is 54 when x and y are 3, find z when $x = 2$ and $y = 4$.

18. If z is 80 when x is 5 and y is 2, find z when $x = 2$ and $y = 5$.

19. If z is 64 when $x = 1$ and $y = 4$, find x when $z = 32$ and $y = 1$.

20. If z is 27 when $x = 6$ and $y = 3$, find x when $z = 50$ and $y = 4$.

21. **Length of a Spring** The length a spring stretches is directly proportional to the force applied. If a force of 5 pounds stretches a spring 3 inches, how much force is necessary to stretch the same spring 10 inches?

22. **Weight and Surface Area** The weight of a certain material varies directly with the surface area of that material. If 8 square feet weighs half a pound, how much will 10 square feet weigh?

23. **Volume and Pressure** The volume of a gas is inversely proportional to the pressure. If a pressure of 36 pounds per square inch corresponds to a volume of 25 cubic feet, what pressure is needed to produce a volume of 75 cubic feet?

24. **Wave Frequency** The frequency of an electromagnetic wave varies inversely with the wavelength. If a wavelength of 200 meters has a frequency of 800 kilocycles per second, what frequency will be associated with a wavelength of 500 meters?

25. **Surface Area of a Cylinder** The surface area of a hollow cylinder varies jointly with the height and radius of the cylinder. If a cylinder with radius 3 inches and height 5 inches has a surface area of 94 square inches, what is the surface area of a cylinder with radius 2 inches and height 8 inches?

26. **Capacity of a Cylinder** The capacity of a cylinder varies jointly with its height and the square of its radius. If a cylinder with a radius of 3 centimeters and a height of 6 centimeters has a capacity of 169.56 cubic centimeters, what will be the capacity of a cylinder with radius 4 centimeters and height 9 centimeters?

27. **Electrical Resistance** The resistance of a wire varies directly with its length and inversely with the square of its diameter. If 100 feet of wire with diameter 0.01 inch has a resistance of 10 ohms, what is the resistance of 60 feet of the same type of wire if its diameter is 0.02 inch?

28. **Volume and Temperature** The volume of a gas varies directly with its temperature and inversely with the pressure. If the volume of a certain gas is 30 cubic feet at a temperature of 300 K and a pressure of 20 pounds per square inch, what is the volume of the same gas at 340 K when the pressure is 30 pounds per square inch?

Graph each pair of direct variation statements on the same coordinate system.

29. $y = 4x$ and $y = -4x$

30. $y = \frac{1}{4}x$ and $y = -\frac{1}{4}x$

31. $y = 3x$ and $y = -\frac{1}{3}x$

32. $y = 2x$ and $y = -\frac{1}{2}x$

33. $y = \frac{2}{3}x$ and $y = -\frac{2}{3}x$

34. $y = \frac{3}{5}x$ and $y = -\frac{5}{3}x$

Graph each of the following inverse variation statements.

35. $y = \dfrac{2}{x}$

36. $y = \dfrac{-2}{x}$

37. $y = \dfrac{-4}{x}$

38. $y = \dfrac{4}{x}$

39. $y = \dfrac{8}{x}$

40. $y = \dfrac{-8}{x}$

41. Graph

$$y = \frac{3}{x} \quad \text{and} \quad x + y = 4$$

on the same coordinate system. At what points do the two graphs intersect?

42. Graph

$$y = \frac{4}{x} \quad \text{and} \quad x - y = 3$$

on the same coordinate system. At what points do the two graphs intersect?

Using Technology

Graph each pair of equations in the same window. Then find the coordinates of the points where the two graphs intersect.

43. $Y_1 = 2X, Y_2 = 2/X$
44. $Y_1 = -2X, Y_2 = -2/X$
45. $Y_1 = 0.5X, Y_2 = 0.5/X$
46. $Y_1 = -0.5X, Y_2 = -0.5/X$

Review Problems

 The following problems review material we covered in Section 2.6.

Solve each inequality and graph the solution set.

47. $\left| \frac{x}{5} + 1 \right| \geq \frac{4}{5}$

48. $|x - 6| \geq 0.01$

49. $|3 - 4t| > -5$

50. $|2 - 6t| < -5$

51. $-8 + |3y + 5| < 5$

52. $|6y - 1| - 4 \leq 2$

REVIEW FOR CHAPTER 3

Examples

Chapter 3 Summary

1. The equation $3x + 2y = 6$ is an example of a linear equation in two variables.

Linear Equations in Two Variables [3.1, 3.3]

A **linear equation in two variables** is any equation that can be put in **standard form** $ax + by = c$. The graph of every linear equation is a straight line.

2. To find the x-intercept for $3x + 2y = 6$, we let $y = 0$ and get

$$3x = 6$$
$$x = 2$$

In this case the x-intercept is 2, and the graph crosses the x-axis at $(2, 0)$.

Intercepts [3.1]

The **x-intercept** of an equation is the **x-coordinate** of the point where the graph crosses the x-axis. The **y-intercept** is the y-coordinate of the point where the graph crosses the y-axis. We find the y-intercept by substituting $x = 0$ into the equation and solving for y. The x-intercept is found by letting $y = 0$ and solving for x.

3. The slope of the line through $(6, 9)$ and $(1, -1)$ is

$$m = \frac{9 - (-1)}{6 - 1} = \frac{10}{5} = 2$$

The Slope of a Line [3.2]

The **slope** of the line containing points (x_1, y_1) and (x_2, y_2) is given by

$$\text{Slope} = m = \frac{\text{Rise}}{\text{Run}} = \frac{y_2 - y_1}{x_2 - x_1}$$

Horizontal lines have 0 slope, and vertical lines have no slope.
Parallel lines have equal slopes, and perpendicular lines have slopes that are negative reciprocals.

4. The equation of the line with slope 5 and y-intercept 3 is

$$y = 5x + 3$$

The Slope-Intercept Form of a Line [3.3]

The equation of a line with slope m and y-intercept b is given by

$$y = mx + b$$

5. The equation of the line through $(3, 2)$ with slope -4 is

$$y - 2 = -4(x - 3)$$

which can be simplified to

$$y = -4x + 14$$

The Point-Slope Form of a Line [3.3]

The equation of the line through (x_1, y_1) that has slope m can be written as

$$y - y_1 = m(x - x_1)$$

6. The graph of

$$x - y \leq 3$$

is

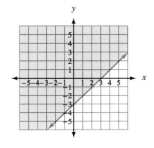

Linear Inequalities in Two Variables [3.4]

An inequality of the form $ax + by < c$ is a **linear inequality in two variables.** The equation for the boundary of the solution set is given by $ax + by = c$. (This equation is found by simply replacing the inequality symbol with an equal sign.)

To graph a linear inequality, first graph the boundary. Next, choose any point not on the boundary and substitute its coordinates into the original inequality. If the resulting statement is true, the graph lies on the same side of the boundary as the test point. A false statement indicates that the solution set lies on the other side of the boundary.

7. The relation

$$\{(8, 1), (6, 1), (-3, 0)\}$$

is also a function since no ordered pairs have the same first coordinates. The domain is $\{8, 6, -3\}$ and the range is $\{1, 0\}$.

Relations and Functions [3.5]

A **function** is a rule that pairs each element in one set, called the **domain,** with exactly one element from a second set, called the **range.**

A **relation** is any set of ordered pairs. The set of all first coordinates is called the **domain** of the relation, and the set of all second coordinates is the **range** of the relation. A function is a relation in which no two different ordered pairs have the same first coordinates.

If the domain for a relation or a function is not specified, it is assumed to be all real numbers for which the relation (or function) is defined. Since we are concerned only with real number functions, a function is not defined for those values of x that give 0 in the denominator or the square root of a negative number.

8. The graph of $x = y^2$ shown in Figure 5 on page 183 fails the vertical line test. It is not the graph of a function.

Vertical Line Test [3.5]

If a vertical line crosses the graph of a relation in more than one place, the relation cannot be a function. If no vertical line can be found that crosses the graph in more than one place, the relation must be a function.

9. If $f(x) = 5x - 3$ then

$$f(0) = 5(0) - 3$$
$$= -3$$
$$f(1) = 5(1) - 3$$
$$= 2$$
$$f(-2) = 5(-2) - 3$$
$$= -13$$
$$f(a) = 5a - 3$$

Function Notation [3.6]

The alternative notation for y is $f(x)$. It is read "f of x," and can be used instead of the variable y when working with functions. The notation y and the notation $f(x)$ are equivalent. That is, $y = f(x)$.

10. If y varies directly with x, then

$$y = Kx$$

Then if y is 18 when x is 6,

$$18 = K \cdot 6$$

or $K = 3$

So the equation can be written more specifically as

$$y = 3x$$

If we want to know what y is when x is 4, we simply substitute:

$$y = 3 \cdot 4$$
$$y = 12$$

Variation [3.7]

If y **varies directly** with x (y is directly proportional to x), then we say

$$y = Kx$$

If y **varies inversely** with x (y is inversely proportional to x), then we say

$$y = \frac{K}{x}$$

If z **varies jointly** with x and y (z is directly proportional to both x and y), then we say

$$z = Kxy$$

In each case, K is called the **constant of variation.**

Common Mistakes

1. When graphing ordered pairs, the most common mistake is to associate the first coordinate with the y-axis and the second with the x-axis. If you make this mistake you would graph (3, 1) by going up 3 and to the right 1, which is just the reverse of what you should do. Remember, the first coordinate is always associated with the horizontal axis, and the second coordinate is always associated with the vertical axis.

2. The two most common mistakes students make when first working with the formula for the slope of a line are:
   (a) Putting the difference of the x-coordinates over the difference of the y-coordinates.
   (b) Subtracting in one order in the numerator and then subtracting in the opposite order in the denominator.

3. When graphing linear inequalities in two variables, remember to graph the boundary with a broken line when the inequality symbol is $<$ or $>$. The only time you use a solid line for the boundary is when the inequality symbol is $\leq$ or $\geq$.

Chapter 3 Review Problems

Graph each line. [3.1]

1. $3x + 2y = 6$ **2.** $-3x + 2y = 6$
3. $5x - 2y = 10$ **4.** $y = 2x - 3$
5. $y = -\frac{3}{2}x + 1$ **6.** $x = 3$

Find the slope of the line through the following pairs of points. [3.2]

7. $(5, 2), (3, 6)$ **8.** $(3, -2), (-1, 2)$
9. $(-4, 2), (3, 2)$ **10.** $(-6, 4), (-6, 8)$

Find x if the line through the two given points has the given slope. [3.2]

11. $(4, x), (1, -3); m = 2$
12. $(-3, x), (-1, -2); m = -2$
13. $(-4, 7), (2, x); m = -\frac{1}{3}$
14. $(-3, 5), (1, x); m = -\frac{1}{2}$

Solve the following problems. [3.2]

15. Find the slope of any line parallel to the line through $(3, 8)$ and $(5, -2)$.
16. Find the slope of any line perpendicular to the line through $(-3, 8)$ and $(5, 2)$.
17. The line through $(5, 3y)$ and $(2, y)$ is parallel to a line with slope 4. What is the value of y?
18. The line through $(2, y)$ and $(-2, 4y)$ is perpendicular to a line with slope $-\frac{1}{3}$. What is the value of y?

Give the equation of the line with the following slope and y-intercept. [3.3]

19. $m = 3, b = 5$ **20.** $m = 5, b = 3$
21. $m = -2, b = 0$ **22.** $m = \frac{1}{3}, b = -\frac{2}{3}$

Give the slope and y-intercept of each equation. [3.3]

23. $3x - y = 6$ **24.** $2x - 3y = 6$
25. $2x - 3y = 9$ **26.** $3x - 2y = 4$

Find the equation of the line that contains the given point and has the given slope. [3.3]

27. $(2, 4), m = 2$ **28.** $(3, 2), m = 3$
29. $(-3, 1), m = -\frac{1}{3}$ **30.** $(\frac{1}{3}, \frac{2}{3}), m = -\frac{1}{2}$

Find the equation of the line that contains the given pair of points. [3.3]

31. $(2, 5), (-3, -5)$ **32.** $(1, 4), (-1, -2)$
33. $(-3, 7), (4, 7)$ **34.** $(-2, 6), (-2, 3)$
35. $(-5, -1), (-3, -4)$
36. $(-8, -2), (1, -3)$

37. Find the equation of the line that is parallel to $2x - y = 4$ and contains the point $(2, -3)$. [3.3]
38. Find the equation of the line that is parallel to the graph of $3x - 2y = 4$ and contains the point $(0, -1)$. [3.3]
39. Find the equation of the line perpendicular to $y = -3x + 1$ that has an x-intercept of 2. [3.3]
40. Find the equation of the line perpendicular to $2x - 4y = 5$ that has an x-intercept of -2. [3.3]

Graph each linear inequality. [3.4]

41. $y \le 2x - 3$ **42.** $2x - 3y > 6$
43. $x \ge -1$ **44.** $y > 4$

State the domain and range of each relation, and then indicate which relations are also functions. [3.5]

45. $\{(2, 4), (3, 3), (4, 2)\}$
46. $\{(-5, 2), (3, 4), (-5, -2)\}$
47. $\{(6, 3), (-4, 3), (-2, 0)\}$
48. $\{(1, -1), (3, 0), (-2, 0)\}$

Let $f(x) = 3x - 2$ and $g(x) = x^2 - 2x + 4$, and evaluate the following. [3.6]

49. $f(0)$ **50.** $g(-3)$
51. $f(3) - g(1)$ **52.** $g(-1) - f(-1)$

If $f = \{(2, -1), (-3, 0), (4, \frac{1}{2}), (\pi, 2)\}$ and $g = \{(2, 2), (-1, 4), (0, 0)\}$, find the following. [3.6]

53. $f(-3)$ **54.** $g(2)$
55. $f(2) + g(2)$
56. $f(-3) + g(-1) + g(0)$

REVIEW FOR CHAPTER 3

Let $f(x) = 2x^2 - 4x + 1$ and $g(x) = 3x + 2$, and evaluate each of the following. [3.6]

57. $f(0)$ **58.** $f(-1)$

59. $g(a)$ **60.** $g(a - 2)$

61. $f[g(0)]$ **62.** $g[f(-2)]$

63. $f[g(1)]$ **64.** $g[f(1)]$

65. The length of a rectangle is 3 inches more than 2 times the width. If x is the width of the rectangle, use function notation to write the perimeter $P(x)$ in terms of the width x. [3.5, 3.6]

66. Graph the function you found in Problem 65. Then draw a vertical line, starting at 2 on the x-axis, that corresponds to $P(2)$. [3.5, 3.6]

The formula for the volume of a right circular cylinder is $V = \pi r^2 h$, where r is the radius of the base and h is the height of the cylinder. Use this information to work Problems 67 and 68. [3.5, 3.6]

67. If the height is twice the radius, use function notation to write the volume $V(r)$ as a function of just r.

68. If the radius is twice the height, use function notation to write the volume $V(h)$ as a function of just h.

For the following problems, y varies directly with x. [3.7]

69. If y is 6 when x is 2, find y when x is 8.

70. If y is -3 when x is 5, find y when x is -10.

For the following problems, y varies inversely with the square of x. [3.7]

71. If y is 9 when x is 2, find y when x is 3.

72. If y is 4 when x is 5, find y when x is 2.

For the following problems, z varies jointly with x and the cube of y. [3.7]

73. If z is 6 when x is 2 and y is -1, find z when x is 3 and y is 2.

74. If z is -48 when x is 3 and y is 2, find z when x is 2 and y is 3.

75. If z is -20 when x is -5 and y is 1, find x when z is 64 and y is 2.

76. If z is -108 when x is 4 and y is 3, find x when z is 8 and y is -2.

Solve each application problem. [3.7]

77. The tension t in a spring varies directly with the distance d the spring is stretched. If the tension is 42 pounds when the spring is stretched 2 inches, find the tension when the spring is stretched twice as far.

78. The power P (in watts) in an electric circuit varies directly with the square of the current I (in amperes). If P is 30 when I is 2, find P when I is 6.

79. The intensity of a light source varies inversely with the square of the distance from the source. Four feet from the source the intensity is 9 foot-candles. What is the intensity 3 feet from the source?

80. The weight of a body varies inversely with the square of its distance from the center of the earth. If a man weighs 150 pounds 4,000 miles from the center of the earth, how much will he weigh 5,000 miles from the center of the earth?

Graph each equation. [3.7]

81. $y = 3x$ **82.** $y = -2x$

83. $y = \dfrac{8}{x}$ **84.** $y = \dfrac{-4}{x}$

Chapter 3 Test

For each of the following straight lines, identify the x-intercept, y-intercept, and slope, and sketch the graph. [3.1–3.3]

1. $2x + y = 6$

2. $y = -2x - 3$

3. $y = \frac{3}{2}x + 4$

4. $x = -2$

Find the equation for each line. [3.3]

5. Give the equation of the line through $(-1, 3)$ that has slope $m = 2$.

6. Give the equation of the line through $(-3, 2)$ and $(4, -1)$.

7. Line l contains the point $(5, -3)$ and has a graph parallel to the graph of $2x - 5y = 10$. Find the equation for l.

8. Line l contains the point $(-1, -2)$ and has a graph perpendicular to the graph of $y = 3x - 1$. Find the equation for l.

9. Give the equation of the vertical line through $(4, -7)$.

Graph the following linear inequalities. [3.4]

10. $3x - 4y < 12$

11. $y \le -x + 2$

State the domain and range for the following relations, and indicate which relations are also functions. [3.5]

12. $\{(-2, 0), (-3, 0), (-2, 1)\}$

13. $y = x^2 - 9$

Let $f(x) = x - 2$, $g(x) = 3x + 4$, and $h(x) = 3x^2 - 2x - 8$, and find the following. [3.6]

14. $f(3) + g(2)$

15. $h(0) + g(0)$

16. $f[g(2)]$

17. $g[f(2)]$

A piece of cardboard is in the shape of a square with each side 8 inches long. You are going to make a box out of the piece of cardboard by cutting four equal squares from each corner and then folding up the sides, as shown in Figure 1. Use this information in Problems 18–20. [3.5, 3.6]

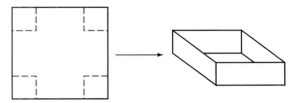

FIGURE 1

18. If the length of each side of the squares you are cutting from the corners is represented by the variable x, state the restrictions on x.

19. Use function notation to write the volume of the resulting box, $V(x)$, as a function of x.

20. Find $V(2)$ and explain what it represents.

Solve the following variation problems. [3.7]

21. Quantity y varies directly with the square of x. If y is 50 when x is 5, find y when x is 3.

22. Quantity z varies jointly with x and the cube of y. If z is 15 when x is 5 and y is 2, find z when x is 2 and y is 3.

23. The maximum load (L) a horizontal beam can safely hold varies jointly with the width (w) and the square of the depth (d) and inversely with the length (l). If a 10 foot beam with width 3 feet and depth 4 feet will safely hold up to 800 pounds, how many pounds will a 12 foot beam with width 3 feet and depth 4 feet hold?

Graph each of the following. [3.7]

24. $y = \dfrac{4}{x}$

25. $y = \dfrac{-4}{x}$

Rational Expressions

Contents

Introduction

If you have ever put yourself on a weight-loss diet, you know that you lose more weight at the beginning of the diet than you do later on. If we let $W(x)$ represent a person's weight (in pounds) after x weeks on the diet, then the function

$$W(x) = \frac{80(2x + 15)}{x + 6}$$

is a mathematical model showing the weekly progress of a person on a diet intended to take them from 200 pounds to about 160 pounds. This function is called a *rational function* because the expression on the right side of the equal sign is a *rational expression*—that is, it is the ratio of two polynomials. Rational functions are good models for quantities that fall off rapidly at the beginning and then level out over time. The figure shows the graph of this function from $x = 0$ weeks to $x = 24$ weeks.

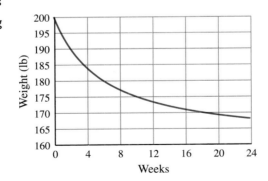

As you can see, the graph falls off rapidly during the first few weeks and then levels off. This means that the person on the diet loses more weight at the beginning of the diet than they do later on.

Overview

This chapter is concerned mostly with simplifying rational expressions. Rational expressions are to algebra what rational numbers are to arithmetic. Most of the work we will do with rational expressions parallels the work you have done in previous math classes with numerical fractions. Once you have learned to add, subtract, multiply, and divide rational expressions, we will turn our attention to equations involving rational expressions. The single most important tool needed for success in this chapter is factoring. Almost every problem encountered in this chapter involves factoring at one point or another. You may be able to understand all the theory and steps involved in solving the problems, but unless you can factor the polynomials in the problems, you will be unable to work any of them.

Study Skills

The study skills for this chapter are concerned with getting ready to take an exam.

1. **Getting ready to take an exam** Try to arrange your daily study habits so that you have very little studying to do the night before your next exam. The next two goals will help you achieve goal number 1.
2. **Review with the exam in mind** Review material that will be covered on the next exam every day. Your review should consist of working problems. Preferably, the problems you work should be problems from the list of difficult problems you have been recording.
3. **Continue to list difficult problems** This study skill was started in Chapter 2. You should continue to list and rework the problems that give you the most difficulty. It is this list that you will use to study for the next exam. Your goal is to go into the next exam knowing that you can successfully work any problem from this list.
4. **Pay attention to instructions** Taking a test is different from doing homework. When you take a test, the problems will be mixed up. When you do your homework, you usually work a number of similar problems together. Sometimes students who do very well on their homework become confused when they see the same problems on a test, because they have not paid attention to the instructions. For example, look at the two problems below:

$$\textbf{Problem 1:} \quad \frac{5}{x^2 - 3x + 2} - \frac{1}{x - 2}$$

$$\textbf{Problem 2:} \quad \frac{5}{x^2 - 3x + 2} - \frac{1}{x - 2} = \frac{1}{3x - 3}$$

Without instructions to accompany them, we don't know what to do with these problems. Here are two valid instructions that could accompany either problem:

List any restrictions on the variable.

Find the least common denominator.

On the other hand, "Subtract" is an instruction that could accompany Problem 1, but not Problem 2. Likewise, "Solve for x" could accompany Problem 2, but not Problem 1. Train yourself to pay attention to the instructions that accompany the problems you work on your assignments, so that you will know what to do with those same problems when you see them on your next test.

Basic Properties and Reducing to Lowest Terms

We will begin this section with the definition of a rational expression. We will then state the two basic properties associated with rational expressions, and go on to apply one of the properties to reduce rational expressions to lowest terms.

Recall from Chapter 1 that a **rational number** is any number that can be expressed as the ratio of two integers:

$$\text{Rational numbers} = \left\{ \frac{a}{b} \,\middle|\, a \text{ and } b \text{ are integers, } b \neq 0 \right\}$$

A **rational expression** is defined similarly as any expression that can be written as the ratio of two polynomials:

$$\text{Rational expressions} = \left\{ \frac{P}{Q} \,\middle|\, P \text{ and } Q \text{ are polynomials, } Q \neq 0 \right\}$$

Some examples of rational expressions are

$$\frac{2x - 3}{x + 5} \qquad \frac{x^2 - 5x - 6}{x^2 - 1} \qquad \frac{a - b}{b - a}$$

Basic Properties

For rational expressions, multiplying the numerator and denominator by the same nonzero expression may change the form of the rational expression, but it will always produce an expression equivalent to the original one. The same is true when dividing the numerator and denominator by the same nonzero quantity.

PROPERTIES OF RATIONAL EXPRESSIONS

If P, Q, and K are polynomials with $Q \neq 0$ and $K \neq 0$, then

$$\frac{P}{Q} = \frac{PK}{QK} \qquad \text{and} \qquad \frac{P}{Q} = \frac{P/K}{Q/K}$$

Reducing to Lowest Terms

The fraction $\frac{6}{8}$ can be written in lowest terms as $\frac{3}{4}$. The process is shown here:

$$\frac{6}{8} = \frac{3 \cdot \overset{1}{\cancel{2}}}{4 \cdot \underset{1}{\cancel{2}}} = \frac{3}{4}$$

Reducing $\frac{6}{8}$ to $\frac{3}{4}$ involves dividing the numerator and denominator by 2, the factor they have in common. Before dividing out the common factor 2, we must notice that the common factor *is* 2! (This may not be obvious since we are very familiar with the numbers 6 and 8 and therefore do not have to put much thought into finding what number divides both of them.)

Similarly, we reduce rational expressions to lowest terms by first factoring the numerator and denominator and then dividing both the numerator and denominator by any factors they have in common.

EXAMPLE 1 Reduce $\dfrac{x^2 - 9}{x - 3}$ to lowest terms.

SOLUTION Factoring, we have

$$\frac{x^2 - 9}{x - 3} = \frac{(x + 3)(x - 3)}{x - 3}$$

The numerator and denominator have the factor $x - 3$ in common. Dividing the numerator and denominator by $x - 3$, we have

Note that the lines drawn through the $x - 3$ in the numerator and denominator indicate that we have divided through by $x - 3$. As the problems become more involved, these lines will help keep track of which factors have been divided out and which have not.

$$\frac{(x + 3)\overset{1}{\cancel{(x - 3)}}}{\underset{1}{\cancel{x - 3}}} = \frac{x + 3}{1} = x + 3$$

Note: For the problem in Example 1, there is an implied restriction on the variable x: It cannot be 3. If x were 3, the expression $(x^2 - 9)/(x - 3)$ would become 0/0, an expression that we cannot associate with a real number. For all problems involving rational expressions, we restrict the variable to only those values that result in a nonzero denominator. For example, when we state the relationship

$$\frac{x^2 - 9}{x - 3} = x + 3$$

we are assuming that it is true for all values of x except $x = 3$.

Here are some other examples of reducing rational expressions to lowest terms.

EXAMPLES Reduce to lowest terms.

2. $\dfrac{y^2 - 5y - 6}{y^2 - 1} = \dfrac{(y - 6)\cancel{(y + 1)}}{(y - 1)\cancel{(y + 1)}}$ Factor numerator and denominator.

$\qquad\qquad = \dfrac{y - 6}{y - 1}$ Divide out common factor $y + 1$.

3. $\dfrac{2a^3 - 16}{4a^2 - 12a + 8} = \dfrac{2(a^3 - 8)}{4(a^2 - 3a + 2)}$

$$= \dfrac{\overset{1}{2}(a - 2)(a^2 + 2a + 4)}{\underset{2}{4}(a - 2)(a - 1)}$$

Factor numerator and denominator.

$$= \dfrac{a^2 + 2a + 4}{2(a - 1)}$$

Divide out common factor $2(a - 2)$.

4. $\dfrac{x^2 - 3x + ax - 3a}{x^2 - ax - 3x + 3a} = \dfrac{x(x - 3) + a(x - 3)}{x(x - a) - 3(x - a)}$

$$= \dfrac{(x - 3)(x + a)}{(x - a)(x - 3)}$$

Factor numerator and denominator.

$$= \dfrac{x + a}{x - a}$$

Divide out common factor $x - 3$.

Note: The answer to Example 4 is $(x + a)/(x - a)$. The problem cannot be reduced further. It is a fairly common mistake to attempt to divide out an x or an a in this last expression. Remember, we can divide out only the factors common to the numerator and denominator of a rational expression. For the last expression in Example 4, neither the numerator nor the denominator can be factored further; x is not a factor of the numerator or the denominator and neither is a. The expression is in lowest terms.

The next two examples involve a special trick. The trick is to reverse the order of the terms in a difference by factoring -1 from each term. Pay close attention to see how this is done — and why.

EXAMPLE 5 Reduce to lowest terms: $\dfrac{a - b}{b - a}$.

SOLUTION The relationship between $a - b$ and $b - a$ is that they are opposites. We can show this by factoring -1 from each term in the numerator:

$$\dfrac{a - b}{b - a} = \dfrac{-1(-a + b)}{b - a}$$

Factor -1 from each term in the numerator.

$$= \dfrac{-1(b - a)}{b - a}$$

Reverse the order of the terms in the numerator.

$$= -1$$

Divide out common factor $b - a$.

EXAMPLE 6 Reduce to lowest terms: $\dfrac{x^2 - 25}{5 - x}$.

SOLUTION We begin by factoring the numerator:

$$\dfrac{x^2 - 25}{5 - x} = \dfrac{(x - 5)(x + 5)}{5 - x}$$

The factors $x - 5$ and $5 - x$ are similar but are not exactly the same. We can reverse the order of either by factoring -1 from it. That is, $5 - x = -1(-5 + x) = -1(x - 5)$.

$$\frac{(x - 5)(x + 5)}{5 - x} = \frac{(x - 5)(x + 5)}{-1(x - 5)}$$

$$= \frac{x + 5}{-1}$$

$$= -(x + 5)$$

Ratios   You may recall from previous math classes that the ratio of a to b is the same as the fraction $\frac{a}{b}$. To illustrate, if the ratio of men to women in a math class is 3 to 2, then $\frac{3}{5}$ of the class are men and $\frac{2}{5}$ of the class are women.

Here are two ratios that are used frequently in mathematics:

1. The number π is defined as the ratio of the circumference of a circle to the diameter of a circle. That is,

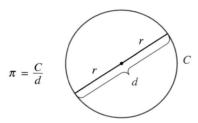

$$\pi = \frac{C}{d}$$

Multiplying both sides of this formula by d, we have the more common form $C = \pi d$. Using function notation, we could write $C(d) = \pi d$.

Note: Since the diameter of a circle is twice the radius, the formula for circumference is sometimes written $C(r) = 2\pi r$. Of course, neither formula should be confused with the formula for the area of a circle, which is $A(r) = \pi r^2$.

2. The **average speed** of a moving object is defined to be the ratio of distance to time. If you drive your car for 5 hours and travel a distance of 200 miles, then your average rate of speed is

$$\text{Average speed} = \frac{200 \text{ miles}}{5 \text{ hours}} = 40 \text{ miles/hour}$$

The formula we use for the relationship between average rate of speed r, distance d, and time t is

$$r = \frac{d}{t}$$

The formula is sometimes called the **rate equation.** Multiplying both sides by t, we have the equivalent form of the rate equation: $d = rt$.

Our next example involves both the formula for the circumference of a circle and the rate equation.

EXAMPLE 7 The first Ferris wheel was designed and built by George Ferris in 1893. The diameter of the wheel was 250 feet. It had 36 carriages, equally spaced around the wheel, each of which held a maximum of 40 people. One trip around the wheel took 20 minutes. Find the average speed of a rider on the first Ferris wheel. (Use 3.14 as an approximation for π.)

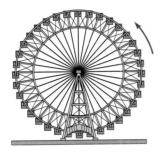

SOLUTION The distance traveled is the circumference of the wheel, which is

$$C = 250\pi = 250(3.14) = 785 \text{ feet}$$

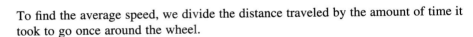

To find the average speed, we divide the distance traveled by the amount of time it took to go once around the wheel.

$$r = \frac{d}{t} = \frac{785 \text{ feet}}{20 \text{ minutes}} = 39.3 \text{ feet/minute} \quad \text{To the nearest tenth}$$

Later in this chapter we will convert this ratio into an equivalent ratio that gives the speed of the rider in miles per hour.

The Ferris wheel in Example 7 has a circumference of 785 feet. In the next example we graph the relationship between the average speed of a person riding this wheel and the amount of time it takes the wheel to complete one revolution.

EXAMPLE 8 A Ferris wheel has a circumference of 785 feet. If one complete revolution of the wheel takes from 10 to 30 minutes, then the relationship between the average speed of a rider on the wheel and the amount of time it takes the wheel to complete one revolution is given by the function

$$r(t) = \frac{785}{t} \qquad 10 \le t \le 30$$

where $r(t)$ is the average speed (in feet per minute) and t is the amount of time (in minutes) it takes the wheel to complete one revolution. Graph the function.

SOLUTION Since the variables r and t represent speed and time, both must be positive quantities. Therefore, the graph of this function will lie in the first quadrant only. Below is a table of values of t and $r(t)$ found from the function, along with the graph of the function (Figure 1). (Some of the numbers in the table have been rounded to the nearest tenth.)

Time to Complete 1 Revolution	Speed (ft/min)
t	$r(t)$
10	78.5
15	52.3
20	39.3
25	31.4
30	26.2

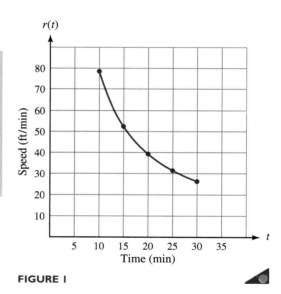

FIGURE I

Using Technology: More about Example 8

If we use a graphing calculator to graph the equation in Example 8, it is not necessary to construct the table first. In fact, if we graph

$$Y_1 = 785/X \qquad \text{Window:} \quad X \text{ from 0 to 40, Y from 0 to 90}$$

we can use the Trace and Zoom features together to produce the numbers in the table next to Figure 1. Graph the equation above and zoom in on the point with x-coordinate 20 until you are convinced that the table values for x and y are correct.

Rational Functions

The function shown in Example 8 is called a **rational function** because the right side, $785/t$, is a rational expression (the numerator, 785, is a polynomial of degree 0). We can extend our knowledge of rational expressions to functions with the following definition:

> **DEFINITION**
>
> A **rational function** is any function that can be written in the form
>
> $$f(x) = \frac{P(x)}{Q(x)}$$
>
> where $P(x)$ and $Q(x)$ are polynomials and $Q(x) \neq 0$.

EXAMPLE 9 For the rational function $f(x) = \dfrac{x - 4}{x - 2}$, find $f(0), f(-4), f(4), f(-2),$ and $f(2)$.

SOLUTION   To find these function values, we substitute the given value of x into the rational expression, and then simplify if possible.

$$f(0) = \frac{0 - 4}{0 - 2} = \frac{-4}{-2} = 2 \qquad\qquad f(-2) = \frac{-2 - 4}{-2 - 2} = \frac{-6}{-4} = \frac{3}{2}$$

$$f(-4) = \frac{-4 - 4}{-4 - 2} = \frac{-8}{-6} = \frac{4}{3} \qquad\qquad f(2) = \frac{2 - 4}{2 - 2} = \frac{-2}{0} \qquad \text{Undefined}$$

$$f(4) = \frac{4 - 4}{4 - 2} = \frac{0}{2} = 0$$

Because the rational function in Example 9 is not defined when x is 2, the domain of that function does not include 2. We have more to say about the domain of a rational function next.

The Domain of a Rational Function

In Example 8, the domain of the rational function is specified as $10 \leq t \leq 30$, and the function is defined for all values of t in that domain. If the domain of a rational function is not specified, it is assumed to be all real numbers for which the function is defined. That is, the domain of the rational function $f(x) = \dfrac{P(x)}{Q(x)}$ is all x for which $Q(x)$ is nonzero. For example:

The domain for $r(t) = \dfrac{785}{t}$, $10 \leq t \leq 30$, is $\{t \mid 10 \leq t \leq 30\}$.

The domain for $f(x) = \dfrac{x - 4}{x - 2}$ is $\{x \mid x \neq 2\}$.

The domain for $g(x) = \dfrac{x^2 + 5}{x + 1}$ is $\{x \mid x \neq -1\}$.

The domain for $h(x) = \dfrac{x}{x^2 - 9}$ is $\{x \mid x \neq -3, x \neq 3\}$.

Notice that, for the functions above, $f(2), g(-1), h(-3),$ and $h(3)$ are all undefined, and that is why the domains are written as shown.

EXAMPLE 10 Graph the equation $y = \dfrac{x^2 - 9}{x - 3}$. How is this graph different from the graph of $y = x + 3$?

SOLUTION We know from the discussion in Example 1 that

$$y = \frac{x^2 - 9}{x - 3} = \frac{(x + 3)(x - 3)}{x - 3} = x + 3$$

This relationship is true for all x except $x = 3$, because the rational expressions with $x - 3$ in the denominator are undefined when x is 3. However, for all other values of x, the expressions

$$\frac{x^2 - 9}{x - 3} \qquad \text{and} \qquad x + 3$$

are equal. Therefore, the graphs of

$$y = \frac{x^2 - 9}{x - 3} \qquad \text{and} \qquad y = x + 3$$

will be the same except when x is 3. In the first equation, there is no value of y to correspond to $x = 3$. In the second equation, $y = x + 3$, so y is 6 when x is 3.

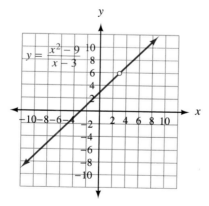
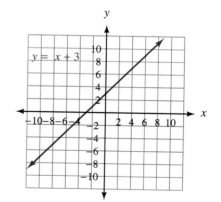

FIGURE 2
Graphs of $y = \dfrac{x^2 - 9}{x - 3}$ (left) and $y = x + 3$ (right)

Now you can see the difference in the graphs of the two equations. To show that there is no y value for $x = 3$ in the graph on the left in Figure 2, we draw an open circle at that point on the line.

Notice that the two graphs shown in Figure 2 are both graphs of functions. Suppose we use function notation to designate them as follows:

$$f(x) = \frac{x^2 - 9}{x - 3} \quad \text{and} \quad g(x) = x + 3$$

The two functions, f and g, are equivalent except when $x = 3$, because $f(3)$ is undefined, while $g(3) = 6$. The domain of the function f is all real numbers except $x = 3$, while the domain for g is all real numbers, with no restrictions.

Problem Set 4.1

NOTE: If you have had any difficulty with factoring in the past, now is the time to straighten it out. The problems that follow include all types of factoring that you have learned. If you can work the problems here successfully, then your factoring skills are right where they should be.

Reduce each fraction to lowest terms.

1. $-\dfrac{12}{36}$

2. $-\dfrac{45}{60}$

3. $\dfrac{2a^2b^3}{4a^2}$

4. $\dfrac{3a^3b^2}{6b^2}$

5. $-\dfrac{24x^3y^5}{16x^4y^2}$

6. $-\dfrac{36x^6y^8}{24x^3y^9}$

7. $\dfrac{144a^2b^3c^4}{56a^4b^3c^2}$

8. $\dfrac{108a^5b^2c^5}{27a^2b^5c^2}$

Reduce each rational expression to lowest terms.

9. $\dfrac{x^2 - 16}{6x + 24}$

10. $\dfrac{5x + 25}{x^2 - 25}$

11. $\dfrac{12x - 9y}{3x^2 + 3xy}$

12. $\dfrac{x^3 - xy^2}{4x + 4y}$

13. $\dfrac{a^4 - 81}{a - 3}$

14. $\dfrac{a + 4}{a^2 - 16}$

15. $\dfrac{a^2 - 4a - 12}{a^2 + 8a + 12}$

16. $\dfrac{a^2 - 7a + 12}{a^2 - 9a + 20}$

17. $\dfrac{20y^2 - 45}{10y^2 - 5y - 15}$

18. $\dfrac{54y^2 - 6}{18y^2 - 60y + 18}$

19. $\dfrac{20x^2 - 93x + 34}{4x^2 - 9x - 34}$

20. $\dfrac{15x^2 - 59x + 52}{5x^2 - 33x + 52}$

21. $\dfrac{12y - 2xy - 2x^2y}{6y - 4xy - 2x^2y}$

22. $\dfrac{250a + 100ax + 10ax^2}{50a - 2ax^2}$

23. $\dfrac{(x - 3)^2(x + 2)}{(x + 2)^2(x - 3)}$

24. $\dfrac{(x - 4)^3(x + 3)}{(x + 3)^2(x - 4)}$

25. $\dfrac{a^3 + b^3}{a^2 - b^2}$

26. $\dfrac{a^2 - b^2}{a^3 - b^3}$

27. $\dfrac{8x^4 - 8x}{4x^4 + 4x^3 + 4x^2}$

28. $\dfrac{6x^5 - 48x^2}{12x^3 + 24x^2 + 48x}$

29. $\dfrac{6x^2 + 7xy - 3y^2}{6x^2 + xy - y^2}$

30. $\dfrac{4x^2 - y^2}{4x^2 - 8xy - 5y^2}$

31. $\dfrac{ax + 2x + 3a + 6}{ay + 2y - 4a - 8}$

32. $\dfrac{ax - x - 5a + 5}{ax + x - 5a - 5}$

33. $\dfrac{x^2 + bx - 3x - 3b}{x^2 - 2bx - 3x + 6b}$

34. $\dfrac{x^2 - 3ax - 2x + 6a}{x^2 - 3ax + 2x - 6a}$

35. $\dfrac{x^3 + 3x^2 - 4x - 12}{x^2 + x - 6}$

36. $\dfrac{x^3 + 5x^2 - 4x - 20}{x^2 + 7x + 10}$

37. $\dfrac{3x^3 + 21x^2 + 36x}{3x^4 + 12x^3 - 27x^2 - 108x}$

38. $\dfrac{2x^4 + 14x^3 + 20x^2}{2x^5 + 4x^4 - 50x^3 - 100x^2}$

39. $\dfrac{4x^4 - 25}{6x^3 - 4x^2 + 15x - 10}$

40. $\dfrac{16x^4 - 49}{8x^3 - 12x^2 + 14x - 21}$

Refer to Examples 5 and 6 in this section and reduce the following to lowest terms.

41. $\dfrac{x - 4}{4 - x}$

42. $\dfrac{6 - x}{x - 6}$

43. $\dfrac{y^2 - 36}{6 - y}$

44. $\dfrac{1 - y}{y^2 - 1}$

45. $\dfrac{1 - 9a^2}{9a^2 - 6a + 1}$

46. $\dfrac{1 - a^2}{a^2 - 2a + 1}$

Reduce each rational expression to lowest terms; then subtract.

47. $\dfrac{28x^2 - 41x + 15}{7x - 5} - \dfrac{12x^2 - 41x + 24}{3x - 8}$

48. $\dfrac{42x^2 + 47x - 55}{7x - 5} - \dfrac{18x^2 - 15x - 88}{3x - 8}$

49. $\dfrac{x^3 - 8}{x - 2} - \dfrac{x^3 + 8}{x + 2}$

50. $\dfrac{x^4 - 16}{x + 2} - \dfrac{x^4 - 16}{x - 2}$

51. **Diet** The rational function below is the one we mentioned in the introduction to this chapter. The quantity $W(x)$ is the weight (in pounds) of the person after x weeks of dieting. Use the function to fill in the table. Then compare your results with the graph in the chapter introduction.

$$W(x) = \dfrac{80(2x + 15)}{x + 6}$$

Weeks	Weight
x	$W(x)$
0	
1	
4	
12	
24	

52. **Drag Racing** The rational function below gives the speed $V(x)$, in miles per hour, of a dragster at each second, x, during a quarter-mile race. Use the function to fill in the table.

$$V(x) = \dfrac{340x}{x + 3}$$

Time (sec)	Speed (mi/hr)
x	$V(x)$
0	
1	
2	
3	
4	
5	
6	

53. If $g(x) = \dfrac{x + 3}{x - 1}$, find $g(0)$, $g(-3)$, $g(3)$, $g(-1)$, and $g(1)$, if possible.

54. If $g(x) = \dfrac{x-2}{x-1}$, find $g(0)$, $g(-2)$, $g(2)$, $g(-1)$, and $g(1)$, if possible.

55. If $h(t) = \dfrac{t-3}{t+1}$, find $h(0)$, $h(-3)$, $h(3)$, $h(-1)$, and $h(1)$, if possible.

56. If $h(t) = \dfrac{t-2}{t+1}$, find $h(0)$, $h(-2)$, $h(2)$, $h(-1)$, and $h(1)$, if possible.

State the domain for each rational function.

57. $f(x) = \dfrac{x-3}{x-1}$

58. $f(x) = \dfrac{x+4}{x-2}$

59. $g(x) = \dfrac{x^2-4}{x-2}$

60. $g(x) = \dfrac{x^2-9}{x-3}$

61. $h(t) = \dfrac{t-4}{t^2-16}$

62. $h(t) = \dfrac{t-5}{t^2-25}$

Applying the Concepts

For Problems 63–66, round all answers to the nearest tenth.

63. Average Speed A jogger covers 3.5 miles in 29.75 minutes. Find the average speed of the jogger in miles per minute.

64. Average Speed A bullet fired from a gun travels a distance of 4,750 feet in 3.2 seconds. Find the average speed of the bullet in feet per second.

65. Fuel Consumption A pickup truck travels 175.8 miles on 16.3 gallons of gas. Give the average rate of fuel consumption of the truck in miles per gallon.

66. Fuel Consumption A luxury car travels 200 miles on 16.5 gallons of gas. Give the average fuel consumption of the car in miles per gallon.

Average Speed *For Problems 67 and 68, use 3.14 as an approximation for π. Round answers to the nearest tenth.*

67. A person riding a Ferris wheel with a diameter of 65 feet travels once around the wheel in 30

seconds. What is the average speed of the rider in feet per second?

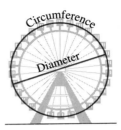

68. A person riding a Ferris wheel with a diameter of 102 feet travels once around the wheel in 3.5 minutes. What is the average speed of the rider in feet per minute?

Average Speed *The abbreviation "rpm" stands for revolutions per minute. If a point on a circle rotates at 300 rpm, then it rotates through one complete revolution 300 times every minute. The length of time it takes to rotate once around the circle is $\frac{1}{300}$ minute. Use 3.14 as an approximation for π.*

69. A $3\frac{1}{2}$ inch diskette, when placed in the disk drive of a computer, rotates at 300 rpm (1 revolution takes $\frac{1}{300}$ minute). Find the average speed of a point 2 inches from the center of the diskette. Then find the average speed of a point 1.5 inches from the center of the diskette.

70. A 5 inch fixed disk in a computer rotates at 3,600 rpm. Find the average speed of a point 2 inches from the center of the disk. Then find the average speed of a point 1.5 inches from the center.

Let $f(x) = \dfrac{x^2 - 4}{x - 2}$ and $g(x) = x + 2$, *and evaluate the following expressions, if possible.*

71. $f(0)$ and $g(0)$ **72.** $f(1)$ and $g(1)$
73. $f(2)$ and $g(2)$ **74.** $f(3)$ and $g(3)$

Let $f(x) = \dfrac{x^2 - 1}{x - 1}$ and $g(x) = x + 1$, *and evaluate the following expressions, if possible.*

75. $f(0)$ and $g(0)$ **76.** $f(1)$ and $g(1)$
77. $f(2)$ and $g(2)$ **78.** $f(3)$ and $g(3)$

79. Graph the equation $y = \dfrac{x^2 - 4}{x - 2}$. Then explain how this graph is different from the graph of $y = x + 2$.

80. Graph the equation $y = \dfrac{x^2 - 1}{x - 1}$. Then explain how this graph is different from the graph of $y = x + 1$.

81. Average Speed The Ferris wheel in Problem 67 has a circumference of 204 feet (to the nearest foot). If a ride on the wheel takes from 20 to 50 seconds, then the relationship between the average speed of a rider and the amount of time it takes to complete one revolution is given by the function

$$r(t) = \frac{204}{t} \qquad 20 \le t \le 50$$

where $r(t)$ is in feet per second and t is in seconds.
(a) State the domain for this function.
(b) Graph the function.

82. Average Speed The Ferris wheel in Problem 68 has a circumference of 320 feet (to the nearest foot). If a ride on the wheel takes from 3 to 5 minutes, then the relationship between the average speed of a rider and the amount of time it takes to complete one revolution is given by the function

$$r(t) = \frac{320}{t} \qquad 3 \le t \le 5$$

where $r(t)$ is in feet per minute and t is in minutes.
(a) State the domain for this function.
(b) Graph the function.

83. Intensity of Light The relationship between the intensity of light that falls on a surface from a 100 watt light bulb and the distance from that surface is given by the rational function

$$I(d) = \frac{120}{d^2} \qquad \text{for } 1 \le d \le 6$$

where $I(d)$ is the intensity of light (in lumens per square foot) and d is the distance (in feet) from the light bulb to the surface.

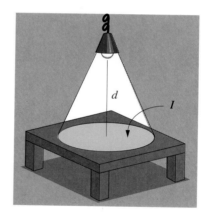

(a) State the domain for this function.
(b) Use the template in Figure 3 to graph this function.

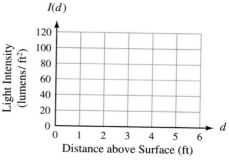

FIGURE 3
Template for graphing a rational function

84. Average Speed If it takes Maria t minutes to run a mile, then her average speed, $s(t)$, is given by the rational function

$$s(t) = \frac{60}{t} \qquad \text{for } 6 \leq t \leq 12$$

where $s(t)$ is in miles per hour and t is in minutes.

(a) State the domain for this function.
(b) Use the template in Figure 4 to graph this function.

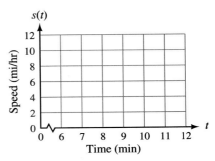

FIGURE 4
Template for graphing a rational function

Review Problems

 NOTE: The problems below review material we covered in Section 1.5. Reviewing these problems will help you with the next section.

Subtract as indicated.

85. Subtract $x^2 + 2x + 1$ from $4x^2 - 5x + 5$.
86. Subtract $3x^2 - 5x + 2$ from $7x^2 + 6x + 4$.
87. Subtract $10x - 20$ from $10x - 11$.
88. Subtract $-6x - 18$ from $-6x + 5$.
89. Subtract $4x^3 - 8x^2$ from $4x^3$.
90. Subtract $2x^2 + 6x$ from $2x^2$.

SECTION 4.2

Division of Polynomials

We begin this section by considering division of a polynomial by a monomial. This is the simplest kind of polynomial division. The rest of the section is devoted to division of a polynomial by a polynomial. This kind of division is similar to long division with whole numbers.

Dividing a Polynomial by a Monomial

To divide a polynomial by a monomial, we use the definition of division and apply the distributive property. The following example illustrates the procedure.

EXAMPLE 1   Divide: $\dfrac{10x^5 - 15x^4 + 20x^3}{5x^2}$.

    SOLUTION   Writing division as multiplication by the reciprocal, we have

$$\frac{10x^5 - 15x^4 + 20x^3}{5x^2}$$

$$= (10x^5 - 15x^4 + 20x^3) \cdot \frac{1}{5x^2} \qquad \text{Dividing by } 5x^2 \text{ is the same as multiplying by } \frac{1}{5x^2}.$$

$$= 10x^5 \cdot \frac{1}{5x^2} - 15x^4 \cdot \frac{1}{5x^2} + 20x^3 \cdot \frac{1}{5x^2} \qquad \text{Distributive property}$$

$$= \frac{10x^5}{5x^2} - \frac{15x^4}{5x^2} + \frac{20x^3}{5x^2} \qquad \text{Multiplying by } \frac{1}{5x^2} \text{ is the same as dividing by } 5x^2.$$

$$= 2x^3 - 3x^2 + 4x \qquad \text{Divide coefficients and subtract exponents.}$$

    Notice that division of a polynomial by a monomial is accomplished by dividing each term of the polynomial by the monomial. The first two steps are usually not shown in a problem like this. They are included in Example 1 to justify distributing $5x^2$ under all three terms of the polynomial $10x^5 - 15x^4 + 20x^3$.

    Here are some more examples of this kind of division.

EXAMPLES   Divide. Write all results with positive exponents.

2. $\dfrac{8x^3y^5 - 16x^2y^2 + 4x^4y^3}{-2x^2y} = \dfrac{8x^3y^5}{-2x^2y} + \dfrac{-16x^2y^2}{-2x^2y} + \dfrac{4x^4y^3}{-2x^2y}$

$$= -4xy^4 + 8y - 2x^2y^2$$

3. $\dfrac{10a^4b^2 + 8ab^3 - 12a^3b + 6ab}{4a^2b^2} = \dfrac{10a^4b^2}{4a^2b^2} + \dfrac{8ab^3}{4a^2b^2} - \dfrac{12a^3b}{4a^2b^2} + \dfrac{6ab}{4a^2b^2}$

$$= \frac{5a^2}{2} + \frac{2b}{a} - \frac{3a}{b} + \frac{3}{2ab}$$

    Notice in Example 3 that the result is not a polynomial because of the last three terms. If we were to write each as a product, some of the variables would have negative exponents. For example, the second term would be

$$\frac{2b}{a} = 2a^{-1}b \qquad \text{Not a polynomial}$$

The divisor in each of the examples above was a monomial. We now want to turn our attention to division of polynomials in which the divisor has two or more terms.

Dividing a Polynomial by a Polynomial

EXAMPLE 4   Divide: $\dfrac{x^2 - 6xy - 7y^2}{x + y}$.

SOLUTION   In this case, we can factor the numerator and perform division by simply dividing out common factors, just like we did in the previous section:

$$\frac{x^2 - 6xy - 7y^2}{x + y} = \frac{(x + y)(x - 7y)}{x + y}$$
$$= x - 7y$$

The diagram in Figure 1 is an important diagram from calculus. Although it may look complicated, the point of it is simple: The slope of the line passing through the points P and Q is given by the formula

$$\text{Slope of line through } PQ = m = \frac{f(x) - f(a)}{x - a}$$

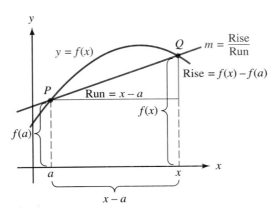

FIGURE 1

If $f(x)$ is a polynomial, then the expression $\dfrac{f(x) - f(a)}{x - a}$ will be a rational expression. In our next example, we explore this situation.

232

EXAMPLE 5 If $f(x) = x^2 - 4$, find $\dfrac{f(x) - f(a)}{x - a}$ and simplify.

SOLUTION Since $f(x) = x^2 - 4$ and $f(a) = a^2 - 4$, we have

$$\frac{f(x) - f(a)}{x - a} = \frac{(x^2 - 4) - (a^2 - 4)}{x - a}$$

$$= \frac{x^2 - 4 - a^2 + 4}{x - a}$$

$$= \frac{x^2 - a^2}{x - a}$$

$$= \frac{(x + a)(x - a)}{x - a} \quad \text{Factor and divide out common factor.}$$

$$= x + a$$

For the type of division shown in Examples 4 and 5, the denominator must be a factor of the numerator. When the denominator is not a factor of the numerator, or in the case where we can't factor the numerator, the method used in Examples 4 and 5 won't work. We need to develop a new method for these cases. Since this new method is very similar to long division with whole numbers, we will review the method of long division here.

EXAMPLE 6 Divide: $25\overline{)4{,}628}$.

SOLUTION

$$\begin{array}{r} 1 \\ 25\overline{)4{,}628} \\ 2\,5 \\ \hline 2\,1 \end{array}$$ ← Estimate 25 into 46.
← Multiply $1 \times 25 = 25$.
← Subtract $46 - 25 = 21$.

$$\begin{array}{r} 1 \\ 25\overline{)4{,}628} \\ 2\,5\downarrow \\ \hline 2\,12 \end{array}$$ ← Bring down the 2.

These are the four basic steps in long division: estimate, multiply, subtract, and bring down the next term. To complete the problem, we simply repeat the same four steps:

$$\begin{array}{r} 18 \\ 25\overline{)4{,}628} \\ 2\,5 \\ \hline 2\,12 \\ 2\,00\downarrow \\ \hline 128 \end{array}$$ ← 8 is the estimate.
← Multiply to get 200.
← Subtract to get 12; then bring down the 8.

Chapter 4 Rational Expressions

One more time:

$$
\begin{array}{r}
185 \leftarrow 5 \text{ is the estimate.} \\
25\overline{)4{,}628} \\
\underline{2\,5} \\
2\,12 \\
\underline{2\,00}\downarrow \\
128 \\
\underline{125} \leftarrow \text{Multiply to get 125.} \\
3 \leftarrow \text{Subtract to get 3.}
\end{array}
$$

Since 3 is less than 25 and we have no more terms to bring down, we have our answer:

$$
\frac{4{,}628}{25} = 185 + \frac{3}{25}
$$

Check   To check our answer, we multiply 185 by 25 and then add 3 to the result:

$$
25(185) + 3 = 4{,}625 + 3 = 4{,}628
$$

Long division with polynomials is very similar to long division with whole numbers. Both use the same four basic steps: estimate, multiply, subtract, and bring down the next term. We use long division with polynomials when the denominator has two or more terms and is not a factor of the numerator. Here is an example.

EXAMPLE 7   Divide: $\dfrac{2x^2 - 7x + 9}{x - 2}$.

SOLUTION

$$
\begin{array}{r}
2x \qquad\qquad \leftarrow \text{Estimate } 2x^2 \div x = 2x. \\
x - 2\overline{)\,2x^2 - 7x + 9} \\
-\quad\ + \\
\underline{\mp\,2x^2 \mp 4x} \qquad \leftarrow \text{Multiply } 2x(x - 2) = 2x^2 - 4x. \\
- 3x \qquad \leftarrow \text{Subtract } (2x^2 - 7x) - (2x^2 - 4x) = -3x.
\end{array}
$$

$$
\begin{array}{r}
2x \qquad\qquad\quad \\
x - 2\overline{)\,2x^2 - 7x + 9} \\
-\quad\ + \qquad \downarrow \\
\underline{\mp\,2x^2 \mp 4x} \qquad\quad \\
- 3x + 9 \leftarrow \text{Bring down the 9.}
\end{array}
$$

Notice that we change the signs on $2x^2 - 4x$ and add in the subtraction step, since subtracting a polynomial is equivalent to adding its opposite.

We repeat the four steps:

$$
\begin{array}{r}
2x \;-\; 3 \quad \leftarrow -3 \text{ is the estimate: } -3x \div x = -3. \\
x - 2 \overline{)\; 2x^2 - 7x + 9\;} \\
\end{array}
$$

$$
\begin{array}{r}
\; - \quad + \\
\cancel{+}\; 2x^2 \;\cancel{-}\; 4x \\
\hline
-3x + 9 \\
+ \quad - \\
\cancel{-}\; 3x \;\cancel{+}\; 6 \leftarrow \text{Multiply } -3(x - 2) = -3x + 6. \\
\hline
3 \leftarrow \text{Subtract } (-3x + 9) - (-3x + 6) = 3.
\end{array}
$$

Since we have no other term to bring down, we have our answer:

$$
\frac{2x^2 - 7x + 9}{x - 2} = 2x - 3 + \frac{3}{x - 2}
$$

Check   To check, we multiply $(2x - 3)(x - 2)$ to get $2x^2 - 7x + 6$; then, adding the remainder 3 to this result, we have $2x^2 - 7x + 9$.   ◄

In setting up a division problem involving two polynomials, we must remember two things: (1) both polynomials should be in decreasing powers of the variable, and (2) neither should skip any powers from the highest power down to the constant term. If there are any missing terms, they can be filled in using a coefficient of 0.

EXAMPLE 8   Divide: $2x - 4 \overline{)\, 4x^3 - 6x - 11\,}$.

SOLUTION   Since the trinomial is missing a term in x^2, we fill it in with $0x^2$:

$$
4x^3 - 6x - 11 = 4x^3 + 0x^2 - 6x - 11
$$

Adding $0x^2$ does not change the original problem.

$$
\begin{array}{r}
2x^2 + 4x \;+\; 5 \\
2x - 4 \overline{)\; 4x^3 + 0x^2 - \; 6x - 11\;} \\
\end{array}
$$

$$
\begin{array}{r}
- \qquad + \\
\cancel{+}\; 4x^3 \;\cancel{-}\; 8x^2 \\
\hline
+ 8x^2 - \;6x \\
- \qquad + \\
\cancel{+}\; 8x^2 \;\cancel{-}\; 16x \\
\hline
+ 10x - 11 \\
- \qquad + \\
\cancel{+}\; 10x \;\cancel{-}\; 20 \\
\hline
+ 9
\end{array}
$$

Notice: Adding the $0x^2$ term gives us a column in which to write $+8x^2$.

$$
\frac{4x^3 - 6x - 11}{2x - 4} = 2x^2 + 4x + 5 + \frac{9}{2x - 4}
$$

Check   To check this result, we multiply $2x - 4$ and $2x^2 + 4x + 5$:

$$\begin{array}{r} 2x^2 + 4x + 5 \\ 2x - 4 \\ \hline 4x^3 + 8x^2 + 10x \\ -8x^2 - 16x - 20 \\ \hline 4x^3 \qquad\quad - 6x - 20 \end{array}$$

Adding 9 (the remainder) to this result gives us the polynomial $4x^3 - 6x - 11$. Our answer checks.

For our next example in this section, let's do Example 4 again, but this time using long division.

EXAMPLE 9   Divide: $\dfrac{x^2 - 6xy - 7y^2}{x + y}$.

SOLUTION

$$\begin{array}{r} x - 7y \\ x + y\overline{)\ x^2 - 6xy - 7y^2} \\ \underline{\mp x^2 \mp xy} \\ -7xy - 7y^2 \\ \underline{\mp 7xy \mp 7y^2} \\ 0 \end{array}$$

In this case, the remainder is 0 and we have

$$\frac{x^2 - 6xy - 7y^2}{x + y} = x - 7y$$

Check   This is easy to check, since

$$(x + y)(x - 7y) = x^2 - 6xy - 7y^2$$

EXAMPLE 10   Factor $x^3 + 9x^2 + 26x + 24$ completely if $x + 2$ is one of its factors.

SOLUTION   Since $x + 2$ is one of the factors of the polynomial we are trying to factor, it must divide that polynomial evenly—that is, without a remainder. Therefore, we begin by dividing the polynomial by $x + 2$:

$$\begin{array}{r} x^2 + 7x + 12 \\ x + 2\overline{)\ x^3 + 9x^2 + 26x + 24} \\ \underline{\mp x^3 \mp 2x^2} \\ +7x^2 + 26x \\ \underline{\mp 7x^2 \mp 14x} \\ +12x + 24 \\ \underline{\mp 12x \mp 24} \\ 0 \end{array}$$

Now we know that the polynomial we are trying to factor is equal to the product of $x + 2$ and $x^2 + 7x + 12$. To factor completely, we simply factor $x^2 + 7x + 12$:

$$x^3 + 9x^2 + 26x + 24 = (x + 2)(x^2 + 7x + 12)$$
$$= (x + 2)(x + 3)(x + 4)$$

Problem Set
4.2

Find the following quotients.

1. $\dfrac{4x^3 - 8x^2 + 6x}{2x}$

2. $\dfrac{6x^3 + 12x^2 - 9x}{3x}$

3. $\dfrac{10x^4 + 15x^3 - 20x^2}{-5x^2}$

4. $\dfrac{12x^5 - 18x^4 - 6x^3}{6x^3}$

5. $\dfrac{8y^5 + 10y^3 - 6y}{4y^3}$

6. $\dfrac{6y^4 - 3y^3 + 18y^2}{9y^2}$

7. $\dfrac{28a^3b^5 + 42a^4b^3}{7a^2b^2}$

8. $\dfrac{a^2b + ab^2}{ab}$

Divide by factoring numerators and then dividing out common factors.

9. $\dfrac{x^2 - x - 6}{x - 3}$

10. $\dfrac{x^2 - x - 6}{x + 2}$

11. $\dfrac{2a^2 - 3a - 9}{2a + 3}$

12. $\dfrac{2a^2 + 3a - 9}{2a - 3}$

13. $\dfrac{5x^2 - 14xy - 24y^2}{x - 4y}$

14. $\dfrac{5x^2 - 26xy - 24y^2}{5x + 4y}$

15. $\dfrac{x^3 - y^3}{x - y}$

16. $\dfrac{x^3 + 8}{x + 2}$

17. $\dfrac{y^4 - 16}{y - 2}$

18. $\dfrac{y^4 - 81}{y - 3}$

19. $\dfrac{x^3 + 2x^2 - 25x - 50}{x - 5}$

20. $\dfrac{x^3 + 2x^2 - 25x - 50}{x + 5}$

For each function below, find $\dfrac{f(x) - f(a)}{x - a}$ and simplify the result.

21. $f(x) = 2x - 3$

22. $f(x) = 5x - 1$

23. $f(x) = x^2 + 4$

24. $f(x) = x^2 - 1$

25. $f(x) = x^3 + 2$

26. $f(x) = x^3 - 2$

Divide using the long division method.

NOTE: Long division is a mechanical process. It is more important that you are able to do it, rather than understand it. Just follow the steps in the examples until you can successfully complete the problems on your own.

27. $\dfrac{x^2 - 5x - 7}{x + 2}$

28. $\dfrac{x^2 + 4x - 8}{x - 3}$

29. $\dfrac{6x^2 + 7x - 18}{3x - 4}$

30. $\dfrac{8x^2 - 26x - 9}{2x - 7}$

31. $\dfrac{2x^3 - 3x^2 - 4x + 5}{x + 1}$

32. $\dfrac{3x^3 - 5x^2 + 2x - 1}{x - 2}$

33. $\dfrac{2y^3 - 9y^2 - 17y + 39}{2y - 3}$

34. $\dfrac{3y^3 - 19y^2 + 17y + 4}{3y - 4}$

35. $\dfrac{2x^3 - 9x^2 + 11x - 6}{2x^2 - 3x + 2}$

36. $\dfrac{6x^3 + 7x^2 - x + 3}{3x^2 - x + 1}$

37. $\dfrac{6y^3 - 8y + 5}{2y - 4}$ **38.** $\dfrac{9y^3 - 6y^2 + 8}{3y - 3}$

39. $\dfrac{a^4 - 2a + 5}{a - 2}$ **40.** $\dfrac{a^4 + a^3 - 1}{a + 2}$

41. $\dfrac{y^4 - 16}{y - 2}$ **42.** $\dfrac{y^4 - 81}{y - 3}$

43. $\dfrac{x^4 + x^3 - 3x^2 - x + 2}{x^2 + 3x + 2}$

44. $\dfrac{2x^4 + x^3 + 4x - 3}{2x^2 - x + 3}$

45. Factor $x^3 + 6x^2 + 11x + 6$ completely if one of its factors is $x + 3$.

46. Factor $x^3 + 10x^2 + 29x + 20$ completely if one of its factors is $x + 4$.

47. Factor $x^3 + 5x^2 - 2x - 24$ completely if one of its factors is $x + 3$.

48. Factor $x^3 + 3x^2 - 10x - 24$ completely if one of its factors is $x + 2$.

49. Problems 17 and 41 are the same problem. Are the two answers you obtained equivalent?

50. Problems 18 and 42 are the same problem. Are the two answers you obtained equivalent?

51. Find $P(-2)$ if $P(x) = x^2 - 5x - 7$. Compare it with the remainder in Problem 27.

52. Find $P(3)$ if $P(x) = x^2 + 4x - 8$. Compare it with the remainder in Problem 28.

Review Problems

NOTE: The problems that follow review material we covered in Sections 1.3 and 3.1. Reviewing the problems from Section 1.3 will help you with the next section.

Divide. [1.3]

53. $\frac{3}{5} \div \frac{2}{7}$ **54.** $\frac{2}{7} \div \frac{3}{5}$

55. $\frac{3}{4} \div \frac{6}{11}$ **56.** $\frac{6}{8} \div \frac{3}{5}$

57. $\frac{4}{9} \div 8$ **58.** $\frac{3}{7} \div 6$

59. $8 \div \frac{1}{4}$ **60.** $12 \div \frac{2}{3}$

Find the x- and y-intercepts for each line. [3.1]

61. $5x - 4y = 10$ **62.** $12x - 5y = 15$

63. $y = \frac{2}{3}x + 4$ **64.** $y = \frac{3}{2}x - 6$

One Step Further

Divide.

65. $\dfrac{4x^5 - x^4 - 20x^3 + 8x^2 - 15}{x^2 - 5}$

66. $\dfrac{4x^5 + 2x^4 + x^3 - 20x^2 - 10x - 5}{x^3 - 5}$

67. $\dfrac{0.5x^3 - 0.3x^2 + 0.22x + 0.06}{x + 0.2}$

68. $\dfrac{0.6x^3 - 1.1x^2 - 0.1x + 0.6}{0.3x + 0.2}$

69. $\dfrac{3x^2 + x - 9}{2x + 4}$ **70.** $\dfrac{2x^2 - x - 9}{3x + 6}$

71. $\dfrac{2x^2 + \frac{1}{3}x + \frac{5}{3}}{3x - 1}$ **72.** $\dfrac{x^2 + \frac{3}{5}x + \frac{8}{5}}{5x - 2}$

SECTION 4.3

Multiplication and Division with Rational Expressions

If you have ever taken a home videotape to be duplicated, you know the amount you pay for the duplication service depends on the number of copies you have made: The more copies you have made, the lower the charge per copy. The demand function below gives the price (in dollars) per tape, $p(x)$, a company charges for making x copies of a 30 minute videotape. As you can see, it is a rational function.

$$p(x) = \frac{2(x + 60)}{x + 5}$$

The graph in Figure 1 shows this function from $x = 0$ to $x = 100$. As you can see, the more copies that are made, the lower the price per copy.

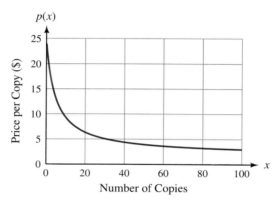

FIGURE I

If we were interested in finding the revenue function for this situation, we would multiply the number of copies made, x, by the price per copy, $p(x)$. This would involve multiplication with a rational expression, which is one of the topics we cover in this section.

In Section 4.1 we found the process of reducing rational expressions to lowest terms to be the same process used in reducing fractions to lowest terms. The similarity also holds for the process of multiplication or division of rational expressions.

Multiplication with fractions is the simplest of the four basic operations. To multiply two fractions we simply multiply numerators and multiply denominators. That is, if a, b, c, and d are real numbers, with $b \neq 0$ and $d \neq 0$, then

$$\frac{a}{b} \cdot \frac{c}{d} = \frac{ac}{bd}$$

EXAMPLE I   Multiply: $\dfrac{6}{7} \cdot \dfrac{14}{18}$.

SOLUTION $\dfrac{6}{7} \cdot \dfrac{14}{18} = \dfrac{6(14)}{7(18)}$ Multiply numerators and denominators.

$= \dfrac{2 \cdot 3(2 \cdot 7)}{7(2 \cdot 3 \cdot 3)}$ Factor.

$= \dfrac{2}{3}$ Divide out common factors.

Our next example is similar to some of the problems we worked in Chapter 1. We multiply fractions whose numerators and denominators are monomials by multiplying numerators and multiplying denominators and then reducing to lowest terms. Here is how it looks.

EXAMPLE 2 Multiply: $\dfrac{8x^3}{27y^8} \cdot \dfrac{9y^3}{12x^2}$.

SOLUTION We multiply numerators and denominators without actually carrying out the multiplication:

Once again, we should mention that the little slashes we have drawn through the factors are simply used to denote the factors we have divided out of the numerator and denominator.

$$\dfrac{8x^3}{27y^8} \cdot \dfrac{9y^3}{12x^2} = \dfrac{8 \cdot 9x^3y^3}{27 \cdot 12x^2y^8} \qquad \begin{array}{l} \text{Multiply numerators.} \\ \text{Multiply denominators.} \end{array}$$

$$= \dfrac{\cancel{4} \cdot 2 \cdot \cancel{9}x^3y^3}{\cancel{9} \cdot 3 \cdot \cancel{4} \cdot 3x^2y^8} \qquad \text{Factor coefficients.}$$

$$= \dfrac{2x}{9y^5} \qquad \text{Divide out common factors.}$$

The product of two rational expressions is the product of their numerators over the product of their denominators.

EXAMPLE 3 Multiply: $\dfrac{x - 3}{x^2 - 4} \cdot \dfrac{x + 2}{x^2 - 6x + 9}$.

SOLUTION We begin by multiplying numerators and denominators. We then factor all polynomials and divide out factors common to the numerator and denominator:

$$\dfrac{x - 3}{x^2 - 4} \cdot \dfrac{x + 2}{x^2 - 6x + 9} = \dfrac{(x - 3)(x + 2)}{(x^2 - 4)(x^2 - 6x + 9)} \qquad \text{Multiply.}$$

$$= \dfrac{\cancel{(x - 3)}\cancel{(x + 2)}}{\cancel{(x + 2)}(x - 2)\cancel{(x - 3)}(x - 3)} \qquad \text{Factor.}$$

$$= \dfrac{1}{(x - 2)(x - 3)} \qquad \text{Divide out common factors.}$$

The first two steps can be combined to save time. We can perform the multiplication and factoring steps together.

EXAMPLE 4 Multiply: $\dfrac{2y^2 - 4y}{2y^2 - 2} \cdot \dfrac{y^2 - 2y - 3}{y^2 - 5y + 6}$.

SOLUTION $\dfrac{2y^2 - 4y}{2y^2 - 2} \cdot \dfrac{y^2 - 2y - 3}{y^2 - 5y + 6} = \dfrac{2y\cancel{(y - 2)}\cancel{(y - 3)}(y + 1)}{2\cancel{(y + 1)}(y - 1)\cancel{(y - 3)}\cancel{(y - 2)}}$

$$= \dfrac{y}{y - 1}$$

Notice in both of the preceding examples that we did not actually multiply the polynomials as we did in Chapter 1. It would be senseless to do that since we would then have to factor each of the resulting products to reduce them to lowest terms.

The quotient of two rational expressions is the product of the first and the reciprocal of the second. That is, we find the quotient of two rational expressions the same way we find the quotient of two fractions. Here is an example that reviews division with fractions.

EXAMPLE 5   Divide: $\dfrac{6}{8} \div \dfrac{3}{5}$.

SOLUTION $\dfrac{6}{8} \div \dfrac{3}{5} = \dfrac{6}{8} \cdot \dfrac{5}{3}$ Write division in terms of multiplication.

$\qquad\qquad = \dfrac{6(5)}{8(3)}$ Multiply numerators and denominators.

$\qquad\qquad = \dfrac{\cancel{2} \cdot \cancel{3}(5)}{\cancel{2} \cdot 2 \cdot 2(\cancel{3})}$ Factor.

$\qquad\qquad = \dfrac{5}{4}$ Divide out common factors.

To divide one rational expression by another, we use the definition of division to multiply by the reciprocal of the expression that follows the division symbol.

EXAMPLE 6   Divide: $\dfrac{8x^3}{5y^2} \div \dfrac{4x^2}{10y^6}$.

SOLUTION   We rewrite the problem in terms of multiplication. Then we multiply.

$$\frac{8x^3}{5y^2} \div \frac{4x^2}{10y^6} = \frac{8x^3}{5y^2} \cdot \frac{10y^6}{4x^2}$$

$$= \frac{\overset{2}{\cancel{8}} \cdot \overset{2}{\cancel{10}}x^3y^6}{\cancel{4} \cdot \cancel{5}x^2y^2} = 4xy^4$$

EXAMPLE 7   Divide: $\dfrac{x^2 - y^2}{x^2 - 2xy + y^2} \div \dfrac{x^3 + y^3}{x^3 - x^2y}$.

SOLUTION   We begin by writing the problem as the product of the first and the reciprocal of the second, and then proceed as in the previous two examples:

$$\frac{x^2 - y^2}{x^2 - 2xy + y^2} \div \frac{x^3 + y^3}{x^3 - x^2y} = \frac{x^2 - y^2}{x^2 - 2xy + y^2} \cdot \frac{x^3 - x^2y}{x^3 + y^3}$$

Multiply by the reciprocal of the divisor.

$$= \frac{\cancel{(x - y)}\cancel{(x + y)}(x^2)\cancel{(x - y)}}{\cancel{(x - y)}\cancel{(x - y)}\cancel{(x + y)}(x^2 - xy + y^2)}$$

Factor and multiply.

$$= \frac{x^2}{x^2 - xy + y^2}$$

Divide out common factors.

Here are some more examples of multiplication and division with rational expressions. The first example involves factoring trinomials.

EXAMPLE 8   Perform the indicated operations:

$$\frac{a^2 - 8a + 15}{a + 4} \cdot \frac{a + 2}{a^2 - 5a + 6} \div \frac{a^2 - 3a - 10}{a^2 + 2a - 8}$$

SOLUTION   First we rewrite the division as multiplication by the reciprocal. Then we proceed as usual.

$$\frac{a^2 - 8a + 15}{a + 4} \cdot \frac{a + 2}{a^2 - 5a + 6} \div \frac{a^2 - 3a - 10}{a^2 + 2a - 8}$$

$$= \frac{(a^2 - 8a + 15)(a + 2)(a^2 + 2a - 8)}{(a + 4)(a^2 - 5a + 6)(a^2 - 3a - 10)}$$ Change division to multiplication by the reciprocal.

$$= \frac{(a - 5)(a - 3)(a + 2)(a + 4)(a - 2)}{(a + 4)(a - 3)(a - 2)(a - 5)(a + 2)}$$ Factor.

$$= 1$$ Divide out common factors.

Our next example involves factoring by grouping. As you may have noticed, working the problems in this chapter gives you a very detailed review of factoring. (If you need more practice, review Sections 1.6 and 1.7.)

EXAMPLE 9   Multiply: $\dfrac{xa + xb + ya + yb}{xa - xb - ya + yb} \cdot \dfrac{xa + xb - ya - yb}{xa - xb + ya - yb}.$

SOLUTION   We will factor each polynomial by grouping, which takes two steps.

$$\frac{xa + xb + ya + yb}{xa - xb - ya + yb} \cdot \frac{xa + xb - ya - yb}{xa - xb + ya - yb}$$

$$= \frac{x(a + b) + y(a + b)}{x(a - b) - y(a - b)} \cdot \frac{x(a + b) - y(a + b)}{x(a - b) + y(a - b)}$$ } Factor by grouping

$$= \frac{(a + b)(x + y)(a + b)(x - y)}{(a - b)(x - y)(a - b)(x + y)}$$

$$= \frac{(a + b)^2}{(a - b)^2}$$

EXAMPLE 10   Multiply: $(4x^2 - 36) \cdot \dfrac{12}{4x + 12}$.

SOLUTION   We can think of $4x^2 - 36$ as having a denominator of 1. Thinking of it in this way allows us to proceed as we did in the previous examples.

$$(4x^2 - 36) \cdot \frac{12}{4x + 12} = \frac{4x^2 - 36}{1} \cdot \frac{12}{4x + 12} \qquad \text{Write } 4x^2 - 36 \text{ with denominator 1.}$$

$$= \frac{4(x - 3)(x + 3)12}{4(x + 3)} \qquad \text{Factor.}$$

$$= 12(x - 3) \qquad \text{Divide out common factors.}$$

EXAMPLE 11   Multiply: $3(x - 2)(x - 1) \cdot \dfrac{5}{x^2 - 3x + 2}$.

SOLUTION   This problem is very similar to the problem in Example 10. Writing the first rational expression with a denominator of 1, we have

$$\frac{3(x - 2)(x - 1)}{1} \cdot \frac{5}{x^2 - 3x + 2} = \frac{3(x - 2)(x - 1)5}{(x - 2)(x - 1)}$$

$$= 3 \cdot 5$$

$$= 15$$

Problem Set

4.3

Perform the indicated operations involving fractions.

1. $\dfrac{2}{9} \cdot \dfrac{3}{4}$

2. $\dfrac{5}{6} \cdot \dfrac{7}{8}$

3. $\dfrac{3}{4} \div \dfrac{1}{3}$

4. $\dfrac{3}{8} \div \dfrac{5}{4}$

5. $\dfrac{3}{7} \cdot \dfrac{14}{24} \div \dfrac{1}{2}$

6. $\dfrac{6}{5} \cdot \dfrac{10}{36} \div \dfrac{3}{4}$

7. $\dfrac{10x^2}{5y^2} \cdot \dfrac{15y^3}{2x^4}$

8. $\dfrac{8x^3}{7y^4} \cdot \dfrac{14y^6}{16x^2}$

9. $\dfrac{11a^2b}{5ab^2} \div \dfrac{22a^3b^2}{10ab^4}$

10. $\dfrac{8ab^3}{9a^2b} \div \dfrac{16a^2b^2}{18ab^3}$

11. $\dfrac{6x^2}{5y^3} \cdot \dfrac{11z^2}{2x^2} \div \dfrac{33z^5}{10y^8}$

12. $\dfrac{4x^3}{7y^2} \cdot \dfrac{6z^5}{5x^6} \div \dfrac{24z^2}{35x^6}$

NOTE: This is your second chance, in this chapter, to bring your factoring skills up to speed. Every type of factoring is covered somewhere in these problems. If you work these problems successfully, you have learned how to factor.

Perform the indicated operations. Be sure to write all answers in lowest terms.

13. $\dfrac{x^2 - 9}{x^2 - 4} \cdot \dfrac{x - 2}{x - 3}$

14. $\dfrac{x^2 - 16}{x^2 - 25} \cdot \dfrac{x - 5}{x - 4}$

15. $\dfrac{y^2 - 1}{y + 2} \cdot \dfrac{y^2 + 5y + 6}{y^2 + 2y - 3}$

16. $\dfrac{y-1}{y^2-y-6} \cdot \dfrac{y^2+5y+6}{y^2-1}$

17. $\dfrac{3x-12}{x^2-4} \cdot \dfrac{x^2+6x+8}{x-4}$

18. $\dfrac{x^2+5x+1}{4x-4} \cdot \dfrac{x-1}{x^2+5x+1}$

19. $\dfrac{5x+2y}{25x^2-5xy-6y^2} \cdot \dfrac{20x^2-7xy-3y^2}{4x+y}$

20. $\dfrac{7x+3y}{42x^2-17xy-15y^2} \cdot \dfrac{12x^2-4xy-5y^2}{2x+y}$

21. $\dfrac{a^2-5a+6}{a^2-2a-3} \div \dfrac{a-5}{a^2+3a+2}$

22. $\dfrac{a^2+7a+12}{a-5} \div \dfrac{a^2+9a+18}{a^2-7a+10}$

23. $\dfrac{4t^2-1}{6t^2+t-2} \div \dfrac{8t^3+1}{27t^3+8}$

24. $\dfrac{9t^2-1}{6t^2+7t-3} \div \dfrac{27t^3+1}{8t^3+27}$

25. $\dfrac{2x^2-5x-12}{4x^2+8x+3} \div \dfrac{x^2-16}{2x^2+7x+3}$

26. $\dfrac{x^2-2x+1}{3x^2+7x-20} \div \dfrac{x^2+3x-4}{3x^2-2x-5}$

27. $\dfrac{6a^2b+2ab^2-20b^3}{4a^2b-16b^3} \cdot \dfrac{10a^2-22ab+4b^2}{27a^3-125b^3}$

28. $\dfrac{12a^2b-3ab^2-42b^3}{9a^2-36b^2} \cdot \dfrac{6a^2-15ab+6b^2}{8a^3b-b^4}$

29. $\dfrac{360x^3-490x}{36x^2+84x+49} \cdot \dfrac{30x^2+83x+56}{150x^3+65x^2-280x}$

30. $\dfrac{490x^2-640}{49x^2-112x+64} \cdot \dfrac{28x^2-95x+72}{56x^3-62x^2-144x}$

31. $\dfrac{x^5-x^2}{5x^5-5x} \cdot \dfrac{10x^4-10x^2}{2x^4+2x^3+2x^2}$

32. $\dfrac{2x^4-16x}{3x^6-48x^2} \cdot \dfrac{6x^5+24x^3}{4x^4+8x^3+16x^2}$

33. $\dfrac{a^2-16b^2}{a^2-8ab+16b^2} \cdot \dfrac{a^2-9ab+20b^2}{a^2-7ab+12b^2} \div \dfrac{a^2-25b^2}{a^2-6ab+9b^2}$

34. $\dfrac{a^2-6ab+9b^2}{a^2-4b^2} \cdot \dfrac{a^2-5ab+6b^2}{(a-3b)^2} \div \dfrac{a^2-9b^2}{a^2-ab-6b^2}$

35. $\dfrac{2y^2-7y-15}{42y^2-29y-5} \cdot \dfrac{12y^2-16y+5}{7y^2-36y+5} \div \dfrac{4y^2-9}{49y^2-1}$

36. $\dfrac{8y^2+18y-5}{21y^2-16y+3} \cdot \dfrac{35y^2-22y+3}{6y^2+17y+5} \div \dfrac{16y^2-1}{9y^2-1}$

37. $\dfrac{xy-2x+3y-6}{xy+2x-4y-8} \cdot \dfrac{xy+x-4y-4}{xy-x+3y-3}$

38. $\dfrac{ax+bx+2a+2b}{ax-3a+bx-3b} \cdot \dfrac{ax-bx-3a+3b}{ax-bx-2a+2b}$

39. $\dfrac{2x^3+10x^2-8x-40}{x^3+4x^2-9x-36} \cdot \dfrac{x^2+x-12}{2x^2+14x+20}$

40. $\dfrac{x^3+2x^2-9x-18}{x^4+3x^3-4x^2-12x} \cdot \dfrac{x^3+5x^2+6x}{x^2-x-6}$

Use the method shown in Examples 10 and 11 to find the following products.

NOTE: Remember, any polynomial can be thought of as a rational expression with a denominator of 1.

41. $(3x-6) \cdot \dfrac{x}{x-2}$

42. $(4x+8) \cdot \dfrac{x}{x+2}$

43. $(x^2-3x+2) \cdot \dfrac{3}{3x-3}$

44. $(x^2-3x+2) \cdot \dfrac{-1}{x-2}$

45. $(y-3)(y-4)(y+3) \cdot \dfrac{-1}{y^2-9}$

46. $(y+1)(y+4)(y-1) \cdot \dfrac{3}{y^2-1}$

47. $a(a+5)(a-5) \cdot \dfrac{a+1}{a^2+5a}$

48. $a(a+3)(a-3) \cdot \dfrac{a-1}{a^2-3a}$

Demand Equation *At the beginning of this section we introduced the demand equation given below. Use it to work Problems 49–52.*

$$p(x) = \frac{2(x + 60)}{x + 5}$$

49. Use the demand equation above to fill in the table. Then compare your results with the graph shown in Figure 1 of this section.

Number of Copies	Price per Copy
x	$p(x)$
1	
10	
20	
50	
100	

50. To find the revenue for selling 50 copies of a tape, we multiply the price per tape by 50. Find the revenue for selling 50 tapes.

51. Find the revenue for selling 100 tapes.

52. Find the revenue equation $R(x)$.

Review Problems

 The following problems review material we covered in Sections 1.2 and 3.2.

Find the slope of the line that contains the following pairs of points. [3.2]

53. $(-4, -1)$ and $(-2, 5)$
54. $(-2, -3)$ and $(-5, 1)$

55. Find y if the slope of the line through $(5, y)$ and $(4, 2)$ is 3. [3.2]

56. Find x if the slope of the line through $(4, 9)$ and $(x, -2)$ is $-\frac{7}{3}$. [3.2]

A line has a slope of $\frac{2}{3}$. Find the slope of any line:

57. Parallel to it [3.2]

58. Perpendicular to it [3.2]

59. Find the slope of the line with x-intercept 5 and y-intercept -2. [3.2]

60. Find the slope of the line with x-intercept -3 and y-intercept 6. [3.2]

Combine. [1.2]

61. $\dfrac{3}{14} + \dfrac{7}{30}$

62. $\dfrac{5}{12} + \dfrac{7}{18}$

63. $\dfrac{4}{15} - \dfrac{5}{21}$

64. $\dfrac{3}{14} - \dfrac{5}{22}$

One Step Further

Simplify each expression.

65. $(1 + 2^{-1})(1 + 3^{-1})(1 + 4^{-1})(1 + 5^{-1})$
66. $(1 - 2^{-1})(1 - 3^{-1})(1 - 4^{-1})(1 - 5^{-1})$

The dots in the problems below represent factors not written that are in the same pattern as the surrounding factors. Simplify.

67. $(1 + 2^{-1})(1 + 3^{-1})(1 + 4^{-1}) \cdots (1 + 100^{-1})$
68. $(1 - 2^{-1})(1 - 3^{-1})(1 - 4^{-1}) \cdots (1 - 100^{-1})$

SECTION 4.4

Addition and Subtraction with Rational Expressions

This section is concerned with addition and subtraction of rational expressions. In the first part of this section we will look at addition of expressions that have the same denominator. In the second part we will look at addition of expressions that have different denominators.

Addition and Subtraction with the Same Denominator

To add two expressions that have the same denominator, we simply add numerators and put the sum over the common denominator. Since the process we use to add and subtract rational expressions is the same process used to add and subtract fractions, we will begin with an example involving fractions.

EXAMPLE 1 Add: $\frac{4}{9} + \frac{2}{9}$.

SOLUTION Recall from Chapter 1 that we add fractions with the same denominator by using the distributive property. Here is a detailed look at the steps involved.

$$
\begin{aligned}
\frac{4}{9} + \frac{2}{9} &= 4(\tfrac{1}{9}) + 2(\tfrac{1}{9}) \\
&= (4 + 2)(\tfrac{1}{9}) \quad &\text{Distributive property} \\
&= 6(\tfrac{1}{9}) \\
&= \frac{6}{9} \\
&= \frac{2}{3} \quad &\text{Divide numerator and denominator} \\
& &\text{by the common factor 3.}
\end{aligned}
$$

Note that the important thing about the fractions in this example is that they each have a denominator of 9. If they did not have the same denominator, we could not have written them as two terms with a factor of $\frac{1}{9}$ in common. Without the $\frac{1}{9}$ common to each term, we couldn't apply the distributive property. And without the distributive property, we would not have been able to add the two fractions. ◣

In the examples that follow, we will not show all the steps we have shown in Example 1. The steps are shown in Example 1 so that you will see why both fractions must have the same denominator before we can add them. In actual practice we simply add numerators and place the result over the common denominator.

The similarities between operations on rational numbers and operations on rational expressions make this chapter more like an extension of rational numbers than a completely new set of topics.

We add and subtract rational expressions with the same denominator in the same way—by combining numerators and writing the result over the common denominator. Then we reduce the result to lowest terms if possible. Example 2 shows this process in detail.

EXAMPLE 2 Add: $\dfrac{x}{x^2 - 1} + \dfrac{1}{x^2 - 1}$.

SOLUTION Since the denominators are the same, we simply add numerators:

$$
\begin{aligned}
\frac{x}{x^2 - 1} + \frac{1}{x^2 - 1} &= \frac{x + 1}{x^2 - 1} \quad &\text{Add numerators.} \\
&= \frac{x + 1}{(x - 1)(x + 1)} \quad &\text{Factor the denominator.} \\
&= \frac{1}{x - 1} \quad &\text{Divide out common} \\
& &\text{factor } x + 1.
\end{aligned}
$$

◣

Our next example involves subtraction of rational expressions. Pay careful attention to what happens to the signs of the terms in the numerator of the second expression when we subtract it from the first expression.

EXAMPLE 3 Subtract: $\dfrac{2x-5}{x-2} - \dfrac{x-3}{x-2}$.

SOLUTION Since each expression has the same denominator we simply subtract the numerator in the second expression from the numerator in the first expression and write the difference over the common denominator $x-2$. We must be careful, however, that we subtract both terms in the second numerator. To ensure that we do, we will enclose that numerator in parentheses.

$$\frac{2x-5}{x-2} - \frac{x-3}{x-2} = \frac{2x-5-(x-3)}{x-2} \quad \text{Subtract numerators.}$$

$$= \frac{2x-5-x+3}{x-2} \quad \text{Remove parentheses.}$$

$$= \frac{x-2}{x-2} \quad \text{Combine similar terms in the numerator.}$$

$$= 1 \quad \text{Reduce (or divide).}$$

Note the $+3$ in the numerator of the second step. It is a very common mistake to write this as -3, by forgetting to subtract *both* terms in the numerator of the second expression. Whenever the expression we are subtracting has two or more terms in the numerator, we have to watch for this mistake.

Next we consider addition and subtraction of fractions and rational expressions that have different denominators.

Addition and Subtraction with Different Denominators

Before we look at an example of addition of fractions with different denominators, we need to review the definition for the least common denominator.

DEFINITION

The **least common denominator,** abbreviated **LCD,** for a set of denominators is the smallest expression that is divisible by each of the denominators.

The first step in combining two fractions is to find the LCD. Once we have the common denominator, we rewrite each fraction as an equivalent fraction with the common denominator. After that, we simply add or subtract as we did in our first three examples.

Example 4 is a review of the step-by-step procedure used to add two fractions with different denominators.

EXAMPLE 4 Add: $\dfrac{3}{14} + \dfrac{7}{30}$.

SOLUTION

Step 1 Find the LCD To do this we first factor both denominators into prime factors:

$$\text{Factor 14:} \quad 14 = 2 \cdot 7$$
$$\text{Factor 30:} \quad 30 = 2 \cdot 3 \cdot 5$$

Since the LCD must be divisible by 14, it must have factors of $2 \cdot 7$. It must also be divisible by 30, so it must have factors of $2 \cdot 3 \cdot 5$. We do not need to repeat the 2 that appears in both the factors of 14 and 30. Therefore,

$$\text{LCD} = 2 \cdot 3 \cdot 5 \cdot 7 = 210$$

Step 2 Change to equivalent fractions Since we want each fraction to have a denominator of 210 and at the same time keep its original value, we multiply each by 1 in the appropriate form.
Change $\frac{3}{14}$ to a fraction with denominator 210:

$$\frac{3}{14} \cdot \frac{\mathbf{15}}{\mathbf{15}} = \frac{45}{210}$$

Change $\frac{7}{30}$ to a fraction with denominator 210:

$$\frac{7}{30} \cdot \frac{\mathbf{7}}{\mathbf{7}} = \frac{49}{210}$$

Step 3 Add numerators of equivalent fractions found in Step 2

$$\frac{45}{210} + \frac{49}{210} = \frac{94}{210}$$

Step 4 Reduce to lowest terms if necessary

$$\frac{94}{210} = \frac{47}{105}$$

The main idea in adding fractions is to rewrite each fraction with the LCD for a denominator. In doing so, we must be sure not to change the value of either of the original fractions.

EXAMPLE 5 Add: $\dfrac{-2}{x^2 - 2x - 3} + \dfrac{3}{x^2 - 9}$.

SOLUTION

Step 1 Factor each denominator and build the LCD from the factors:

$$\left.\begin{array}{l} x^2 - 2x - 3 = (x - 3)(x + 1) \\ x^2 - 9 = (x - 3)(x + 3) \end{array}\right\} \quad \text{LCD} = (x - 3)(x + 3)(x + 1)$$

Step 2 Change each rational expression to an equivalent expression that has the LCD for a denominator:

$$\frac{-2}{x^2 - 2x - 3} = \frac{-2}{(x - 3)(x + 1)} \cdot \frac{(x + 3)}{(x + 3)} = \frac{-2x - 6}{(x - 3)(x + 3)(x + 1)}$$

$$\frac{3}{x^2 - 9} = \frac{3}{(x - 3)(x + 3)} \cdot \frac{(x + 1)}{(x + 1)} = \frac{3x + 3}{(x - 3)(x + 3)(x + 1)}$$

Step 3 Add numerators of the rational expressions found in Step 2:

$$\frac{-2x - 6}{(x - 3)(x + 3)(x + 1)} + \frac{3x + 3}{(x - 3)(x + 3)(x + 1)} = \frac{x - 3}{(x - 3)(x + 3)(x + 1)}$$

Step 4 Reduce to lowest terms by dividing out the common factor $x - 3$:

$$= \frac{1}{(x + 3)(x + 1)}$$

EXAMPLE 6 Subtract: $\dfrac{x + 4}{2x + 10} - \dfrac{5}{x^2 - 25}$.

SOLUTION We begin by factoring each denominator:

$$\frac{x + 4}{2x + 10} - \frac{5}{x^2 - 25} = \frac{x + 4}{2(x + 5)} - \frac{5}{(x + 5)(x - 5)}$$

The LCD is $2(x + 5)(x - 5)$. Completing the problem, we have

$$= \frac{x + 4}{2(x + 5)} \cdot \frac{(x - 5)}{(x - 5)} - \frac{5}{(x + 5)(x - 5)} \cdot \frac{2}{2}$$

$$= \frac{x^2 - x - 20}{2(x + 5)(x - 5)} - \frac{10}{2(x + 5)(x - 5)}$$

$$= \frac{x^2 - x - 30}{2(x + 5)(x - 5)}$$

To see if this expression can be reduced, we factor the numerator into $(x - 6)(x + 5)$:

$$= \frac{(x - 6)(x + 5)}{2(x + 5)(x - 5)}$$

$$= \frac{x - 6}{2(x - 5)}$$

EXAMPLE 7 Subtract: $\dfrac{2x - 2}{x^2 + 4x + 3} - \dfrac{x - 1}{x^2 + 5x + 6}$.

SOLUTION We factor each denominator and build the LCD from these factors:

$$\frac{2x - 2}{x^2 + 4x + 3} - \frac{x - 1}{x^2 + 5x + 6}$$

$$= \frac{2x - 2}{(x + 3)(x + 1)} - \frac{x - 1}{(x + 3)(x + 2)}$$

$$= \frac{2x - 2}{(x + 3)(x + 1)} \cdot \frac{(x + 2)}{(x + 2)} - \frac{x - 1}{(x + 3)(x + 2)} \cdot \frac{(x + 1)}{(x + 1)} \qquad \begin{array}{l}\text{The LCD is} \\ (x + 1)(x + 2)(x + 3).\end{array}$$

$$= \frac{2x^2 + 2x - 4}{(x + 1)(x + 2)(x + 3)} - \frac{x^2 - 1}{(x + 1)(x + 2)(x + 3)} \qquad \begin{array}{l}\text{Multiply out each} \\ \text{numerator.}\end{array}$$

$$\left.\begin{array}{l} = \dfrac{(2x^2 + 2x - 4) - (x^2 - 1)}{(x + 1)(x + 2)(x + 3)} \\[3mm] = \dfrac{x^2 + 2x - 3}{(x + 1)(x + 2)(x + 3)} \end{array}\right\} \qquad \text{Subtract numerators.}$$

$$= \frac{(x + 3)(x - 1)}{(x + 1)(x + 2)(x + 3)} \qquad \begin{array}{l}\text{Factor the numerator} \\ \text{to see if the fraction} \\ \text{can be reduced.}\end{array}$$

$$= \frac{x - 1}{(x + 1)(x + 2)} \qquad \text{Reduce.} \qquad \blacktriangle$$

EXAMPLE 8 Add: $\dfrac{x^2}{x - 7} + \dfrac{6x + 7}{7 - x}$.

SOLUTION In Section 4.1 we were able to reverse the terms in a factor such as $7 - x$ by factoring -1 from each term. In a problem like this, the same result can be obtained by multiplying the numerator and denominator by -1:

$$\frac{x^2}{x - 7} + \frac{6x + 7}{7 - x} \cdot \frac{-1}{-1} = \frac{x^2}{x - 7} + \frac{-6x - 7}{x - 7}$$

$$= \frac{x^2 - 6x - 7}{x - 7} \qquad \text{Add numerators.}$$

$$= \frac{(x - 7)(x + 1)}{(x - 7)} \qquad \text{Factor numerator.}$$

$$= x + 1 \qquad \text{Divide out } x - 7. \qquad \blacktriangle$$

For our next example we will look at a problem in which we combine a whole number and a rational expression.

EXAMPLE 9 Subtract: $2 - \dfrac{9}{3x + 1}$.

SOLUTION To subtract these two expressions, we think of 2 as a rational expression with a denominator of 1:

$$2 - \frac{9}{3x + 1} = \frac{2}{1} - \frac{9}{3x + 1}$$

The LCD is $3x + 1$. Multiplying the numerator and denominator of the first expression by $3x + 1$ gives us a rational expression equivalent to 2, but with a denominator of $3x + 1$:

$$\frac{2}{1} \cdot \frac{(3x + 1)}{(3x + 1)} - \frac{9}{3x + 1} = \frac{6x + 2 - 9}{3x + 1}$$

$$= \frac{6x - 7}{3x + 1}$$

The numerator and denominator of this last expression do not have any factors in common other than 1, so the expression is in lowest terms.

EXAMPLE 10 Write an expression for the sum of a number and twice its reciprocal. Then, simplify that expression.

SOLUTION If x is the number, then its reciprocal is $\frac{1}{x}$. Twice its reciprocal is $\frac{2}{x}$. The sum of the number and twice its reciprocal is

$$x + \frac{2}{x}$$

To combine these two expressions, we think of the first term, x, as a rational expression with a denominator of 1. The least common denominator is x:

$$x + \frac{2}{x} = \frac{x}{1} + \frac{2}{x}$$

$$= \frac{x}{1} \cdot \frac{x}{x} + \frac{2}{x}$$

$$= \frac{x^2 + 2}{x}$$

Problem Set

4.4

Combine the following fractions.

1. $\frac{3}{4} + \frac{1}{2}$

2. $\frac{5}{6} + \frac{1}{3}$

3. $\frac{2}{5} - \frac{1}{15}$

4. $\frac{5}{8} - \frac{1}{4}$

5. $\frac{5}{6} + \frac{7}{8}$

6. $\frac{3}{4} + \frac{2}{3}$

7. $\frac{9}{48} - \frac{3}{54}$

8. $\frac{6}{28} - \frac{5}{42}$

9. $\frac{3}{4} - \frac{1}{8} + \frac{2}{3}$

10. $\frac{1}{3} - \frac{5}{6} + \frac{5}{12}$

Combine the following rational expressions. Reduce all answers to lowest terms.

11. $\dfrac{x}{x+3} + \dfrac{3}{x+3}$

12. $\dfrac{5x}{5x+2} + \dfrac{2}{5x+2}$

13. $\dfrac{4}{y-4} - \dfrac{y}{y-4}$

14. $\dfrac{8}{y+8} + \dfrac{y}{y+8}$

15. $\dfrac{x}{x^2-y^2} - \dfrac{y}{x^2-y^2}$

16. $\dfrac{x}{x^2-y^2} + \dfrac{y}{x^2-y^2}$

17. $\dfrac{2x-3}{x-2} - \dfrac{x-1}{x-2}$

18. $\dfrac{2x-4}{x+2} - \dfrac{x-6}{x+2}$

19. $\dfrac{1}{a} + \dfrac{2}{a^2} - \dfrac{3}{a^3}$

20. $\dfrac{3}{a} + \dfrac{2}{a^2} - \dfrac{1}{a^3}$

21. $\dfrac{7x-2}{2x+1} - \dfrac{5x-3}{2x+1}$

22. $\dfrac{7x-1}{3x+2} - \dfrac{4x-3}{3x+2}$

23. $\dfrac{2}{t^2} - \dfrac{3}{2t}$

24. $\dfrac{5}{3t} - \dfrac{4}{t^2}$

25. $\dfrac{3x+1}{2x-6} - \dfrac{x+2}{x-3}$

26. $\dfrac{x+1}{x-2} - \dfrac{4x+7}{5x-10}$

27. $\dfrac{6x+5}{5x-25} - \dfrac{x+2}{x-5}$

28. $\dfrac{4x+2}{3x+12} - \dfrac{x-2}{x+4}$

29. $\dfrac{x+1}{2x-2} - \dfrac{2}{x^2-1}$

30. $\dfrac{x+7}{2x+12} + \dfrac{6}{x^2-36}$

31. $\dfrac{1}{a-b} - \dfrac{3ab}{a^3-b^3}$

32. $\dfrac{1}{a+b} + \dfrac{3ab}{a^3+b^3}$

33. $\dfrac{1}{2y-3} - \dfrac{18y}{8y^3-27}$

34. $\dfrac{1}{3y-2} - \dfrac{18y}{27y^3-8}$

35. $\dfrac{x}{x^2-5x+6} - \dfrac{3}{3-x}$

36. $\dfrac{x}{x^2+4x+4} - \dfrac{2}{2+x}$

37. $\dfrac{2}{4t-5} + \dfrac{9}{8t^2-38t+35}$

38. $\dfrac{3}{2t-5} + \dfrac{21}{8t^2-14t-15}$

39. $\dfrac{1}{a^2-5a+6} + \dfrac{3}{a^2-a-2}$

40. $\dfrac{-3}{a^2+a-2} + \dfrac{5}{a^2-a-6}$

41. $\dfrac{1}{8x^3-1} - \dfrac{1}{4x^2-1}$

42. $\dfrac{1}{27x^3-1} - \dfrac{1}{9x^2-1}$

43. $\dfrac{4}{4x^2-9} - \dfrac{6}{8x^2-6x-9}$

44. $\dfrac{9}{9x^2+6x-8} - \dfrac{6}{9x^2-4}$

45. $\dfrac{4a}{a^2+6a+5} - \dfrac{3a}{a^2+5a+4}$

46. $\dfrac{3a}{a^2+7a+10} - \dfrac{2a}{a^2+6a+8}$

47. $\dfrac{2x-1}{x^2+x-6} - \dfrac{x+2}{x^2+5x+6}$

48. $\dfrac{4x+1}{x^2+5x+4} - \dfrac{x+3}{x^2+4x+3}$

49. $\dfrac{2x-8}{3x^2+8x+4} + \dfrac{x+3}{3x^2+5x+2}$

50. $\dfrac{5x+3}{2x^2+5x+3} - \dfrac{3x+9}{2x^2+7x+6}$

51. $\dfrac{2}{x^2+5x+6} - \dfrac{4}{x^2+4x+3} + \dfrac{3}{x^2+3x+2}$

52. $\dfrac{-5}{x^2+3x-4} + \dfrac{5}{x^2+2x-3} + \dfrac{1}{x^2+7x+12}$

53. $\dfrac{2x + 8}{x^2 + 5x + 6} - \dfrac{x + 5}{x^2 + 4x + 3} - \dfrac{x - 1}{x^2 + 3x + 2}$

54. $\dfrac{2x + 11}{x^2 + 9x + 20} - \dfrac{x + 1}{x^2 + 7x + 12} - \dfrac{x + 6}{x^2 + 8x + 15}$

55. $2 + \dfrac{3}{2x + 1}$ **56.** $3 - \dfrac{2}{2x + 3}$

57. $5 + \dfrac{2}{4 - t}$ **58.** $7 + \dfrac{3}{5 - t}$

59. $x - \dfrac{4}{2x + 3}$ **60.** $x - \dfrac{5}{3x + 4} + 1$

61. $\dfrac{x}{x + 2} + \dfrac{1}{2x + 4} - \dfrac{3}{x^2 + 2x}$

62. $\dfrac{x}{x + 3} + \dfrac{7}{3x + 9} - \dfrac{2}{x^2 + 3x}$

63. $\dfrac{1}{x} + \dfrac{x}{2x + 4} - \dfrac{2}{x^2 + 2x}$

64. $\dfrac{1}{x} + \dfrac{x}{3x + 9} - \dfrac{3}{x^2 + 3x}$

65. **Optometry** The formula

$$P = \frac{1}{a} + \frac{1}{b}$$

is used by optometrists to help determine how strong to make the lenses for a pair of eyeglasses. If a is 10 and b is 0.2, find the corresponding value of P.

66. **Optometry** Show that the formula in Problem 65 can be written

$$P = \frac{a + b}{ab}$$

Then let $a = 10$ and $b = 0.2$ in this new form of the formula to find P.

67. Show that the expressions $(x + y)^{-1}$ and $x^{-1} + y^{-1}$ are not equal when $x = 3$ and $y = 4$.

68. Show that the expressions $(x + y)^{-1}$ and $x^{-1} + y^{-1}$ are not equal. (Begin by writing each with positive exponents only.)

Simplify each of the following expressions. (Change to positive exponents first.)

69. $(1 - 3^{-2}) \div (1 - 3^{-1})$
70. $(1 - 5^{-2}) \div (1 - 5^{-1})$
71. $(1 - x^{-2}) \div (1 - x^{-1})$
72. $(1 - x^{-3}) \div (1 - x^{-2})$

73. Write an expression for the sum of a number and four times its reciprocal. Then simplify that expression.

74. Write an expression for the sum of a number and three times its reciprocal. Then simplify that expression.

75. Write an expression for the sum of the reciprocals of two consecutive integers. Then simplify that expression.

76. Write an expression for the sum of the reciprocals of two consecutive even integers. Then simplify that expression.

Review Problems

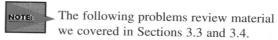 **NOTE:** The following problems review material we covered in Sections 3.3 and 3.4.

Solve the following problems. [3.3]

77. Give the slope and y-intercept of the line $2x - 3y = 6$.

78. Give the equation of the line with slope -3 and y-intercept 5.

79. Find the equation of the line with slope $\frac{2}{3}$ that contains the point $(-6, 2)$.

80. Find the equation of the line with slope 5 that contains the point $(3, -2)$.

81. Find the equation of the line through $(1, 3)$ and $(-1, -5)$.

82. Find the equation of the line through $(-1, 4)$ whose graph is perpendicular to the graph of $y = 2x + 3$.

Graph each inequality. [3.4]

83. $2x + 3y < 6$ **84.** $2x + y < -5$
85. $y \geq -3x - 4$ **86.** $y \geq 2x - 1$
87. $x \geq 3$ **88.** $y \geq -5$

One Step Further

Simplify. The dots in Problems 91–94 represent factors not written that are in the same pattern as the surrounding factors.

89. $\left(1 - \frac{1}{x}\right)\left(1 - \frac{1}{x+1}\right)\left(1 - \frac{1}{x+2}\right)\left(1 - \frac{1}{x+3}\right)$

90. $\left(1 + \frac{1}{x}\right)\left(1 + \frac{1}{x+1}\right)\left(1 + \frac{1}{x+2}\right)\left(1 + \frac{1}{x+3}\right)$

91. $\left(1 - \frac{1}{x}\right)\left(1 - \frac{1}{x+1}\right)\left(1 - \frac{1}{x+2}\right) \cdots \left(1 - \frac{1}{x+50}\right)$

92. $\left(1 + \frac{1}{x}\right)\left(1 + \frac{1}{x+1}\right)\left(1 + \frac{1}{x+2}\right) \cdots \left(1 + \frac{1}{x+99}\right)$

93. $\left(1 + \frac{1}{x}\right)\left(1 + \frac{1}{x-1}\right)\left(1 + \frac{1}{x-2}\right) \cdots \left(1 + \frac{1}{x-100}\right)$

94. $\left(1 - \frac{1}{x-1}\right)\left(1 - \frac{1}{x-2}\right)\left(1 - \frac{1}{x-3}\right) \cdots \left(1 - \frac{1}{x-49}\right)$

SECTION 4.5

Complex Fractions

The quotient of two fractions or two rational expressions is called a **complex fraction.** This section is concerned with the simplification of complex fractions.

EXAMPLE 1   Simplify: $\dfrac{\frac{3}{4}}{\frac{5}{8}}$.

SOLUTION   There are generally two methods that can be used to simplify complex fractions.

Method 1   We can multiply the numerator and denominator of the complex fraction by the LCD for both fractions, which in this case is 8:

$$\frac{\frac{3}{4}}{\frac{5}{8}} = \frac{\frac{3}{4} \cdot \mathbf{8}}{\frac{5}{8} \cdot \mathbf{8}}$$

$$= \frac{6}{5}$$

Method 2   Instead of dividing by $\frac{5}{8}$, we can multiply by $\frac{8}{5}$:

$$\frac{\frac{3}{4}}{\frac{5}{8}} = \frac{3}{4} \cdot \frac{8}{5}$$

$$= \frac{24}{20} = \frac{6}{5}$$

Here are some examples of complex fractions involving rational expressions. Most can be solved using either of the two methods shown in Example 1.

EXAMPLE 2 Simplify: $\dfrac{\dfrac{1}{x} + \dfrac{1}{y}}{\dfrac{1}{x} - \dfrac{1}{y}}$.

SOLUTION This problem is most easily solved using Method 1. We begin by multiplying both the numerator and denominator by the quantity xy, which is the LCD for all the fractions:

$$\frac{\dfrac{1}{x} + \dfrac{1}{y}}{\dfrac{1}{x} - \dfrac{1}{y}} = \frac{\left(\dfrac{1}{x} + \dfrac{1}{y}\right) \cdot xy}{\left(\dfrac{1}{x} - \dfrac{1}{y}\right) \cdot xy}$$

$$= \frac{\dfrac{1}{x}(xy) + \dfrac{1}{y}(xy)}{\dfrac{1}{x}(xy) - \dfrac{1}{y}(xy)} \qquad \begin{array}{l}\text{Apply the distributive property to} \\ \text{distribute } xy \text{ over both terms in} \\ \text{the numerator and denominator.}\end{array}$$

$$= \frac{y + x}{y - x}$$

EXAMPLE 3 Simplify: $\dfrac{\dfrac{x - 2}{x^2 - 9}}{\dfrac{x^2 - 4}{x + 3}}$.

SOLUTION Applying Method 2, we have

$$\frac{\dfrac{x - 2}{x^2 - 9}}{\dfrac{x^2 - 4}{x + 3}} = \frac{x - 2}{x^2 - 9} \cdot \frac{x + 3}{x^2 - 4}$$

$$= \frac{\cancel{(x - 2)}\cancel{(x + 3)}}{\cancel{(x + 3)}(x - 3)(x + 2)\cancel{(x - 2)}}$$

$$= \frac{1}{(x - 3)(x + 2)}$$

EXAMPLE 4 Simplify: $\dfrac{1 - \dfrac{4}{x^2}}{1 - \dfrac{1}{x} - \dfrac{6}{x^2}}$.

SOLUTION The easiest way to simplify this complex fraction is to multiply the numerator and denominator by the LCD, x^2:

$$\dfrac{1 - \dfrac{4}{x^2}}{1 - \dfrac{1}{x} - \dfrac{6}{x^2}} = \dfrac{x^2\left(1 - \dfrac{4}{x^2}\right)}{x^2\left(1 - \dfrac{1}{x} - \dfrac{6}{x^2}\right)}$$

Multiply the numerator and denominator by x^2.

$$= \dfrac{x^2 \cdot 1 - x^2 \cdot \dfrac{4}{x^2}}{x^2 \cdot 1 - x^2 \cdot \dfrac{1}{x} - x^2 \cdot \dfrac{6}{x^2}}$$

Distributive property

$$= \dfrac{x^2 - 4}{x^2 - x - 6}$$

Simplify.

$$= \dfrac{(x - 2)(x + 2)}{(x - 3)(x + 2)}$$

Factor.

$$= \dfrac{x - 2}{x - 3}$$

Reduce.

EXAMPLE 5 Simplify: $2 - \dfrac{3}{x + \frac{1}{3}}$.

SOLUTION First we simplify the expression that follows the subtraction sign:

$$2 - \dfrac{3}{x + \frac{1}{3}} = 2 - \dfrac{\mathbf{3 \cdot 3}}{\mathbf{3}(x + \frac{1}{3})} = 2 - \dfrac{9}{3x + 1}$$

Now we subtract by rewriting the first term, 2, with the LCD, $3x + 1$:

$$2 - \dfrac{9}{3x + 1} = \dfrac{2}{1} \cdot \dfrac{\mathbf{3x + 1}}{\mathbf{3x + 1}} - \dfrac{9}{3x + 1} = \dfrac{6x + 2 - 9}{3x + 1} = \dfrac{6x - 7}{3x + 1}$$

Problem Set

4.5

Simplify each of the following as much as possible.

1. $\dfrac{\frac{3}{4}}{\frac{2}{3}}$

2. $\dfrac{\frac{5}{9}}{\frac{7}{12}}$

7. $\dfrac{\dfrac{1}{x}}{1 + \dfrac{1}{x}}$

8. $\dfrac{1 - \dfrac{1}{x}}{\dfrac{1}{x}}$

3. $\dfrac{\frac{1}{3} - \frac{1}{4}}{\frac{1}{2} + \frac{1}{8}}$

4. $\dfrac{\frac{1}{6} - \frac{1}{3}}{\frac{1}{4} - \frac{1}{8}}$

5. $\dfrac{3 + \frac{2}{5}}{1 - \frac{3}{7}}$

6. $\dfrac{2 + \frac{5}{6}}{1 - \frac{7}{8}}$

9. $\dfrac{1 + \dfrac{1}{a}}{1 - \dfrac{1}{a}}$

10. $\dfrac{1 - \dfrac{2}{a}}{1 - \dfrac{3}{a}}$

11. $\dfrac{\dfrac{1}{x} - \dfrac{1}{y}}{\dfrac{1}{x} + \dfrac{1}{y}}$

12. $\dfrac{\dfrac{1}{x} + \dfrac{2}{y}}{\dfrac{2}{x} + \dfrac{1}{y}}$

13. $\dfrac{\dfrac{x-5}{x^2-4}}{\dfrac{x^2-25}{x+2}}$

14. $\dfrac{\dfrac{3x+1}{x^2-49}}{\dfrac{9x^2-1}{x-7}}$

15. $\dfrac{\dfrac{4a}{2a^3+2}}{\dfrac{8a}{4a+4}}$

16. $\dfrac{\dfrac{2a}{3a^3-3}}{\dfrac{4a}{6a-6}}$

17. $\dfrac{1 - \dfrac{9}{x^2}}{1 - \dfrac{1}{x} - \dfrac{6}{x^2}}$

18. $\dfrac{4 - \dfrac{1}{x^2}}{4 + \dfrac{4}{x} + \dfrac{1}{x^2}}$

19. $\dfrac{2 + \dfrac{5}{a} - \dfrac{3}{a^2}}{2 - \dfrac{5}{a} + \dfrac{2}{a^2}}$

20. $\dfrac{3 + \dfrac{5}{a} - \dfrac{2}{a^2}}{3 - \dfrac{10}{a} + \dfrac{3}{a^2}}$

21. $\dfrac{2 + \dfrac{3}{x} - \dfrac{18}{x^2} - \dfrac{27}{x^3}}{2 + \dfrac{9}{x} + \dfrac{9}{x^2}}$

22. $\dfrac{3 + \dfrac{5}{x} - \dfrac{12}{x^2} - \dfrac{20}{x^3}}{3 + \dfrac{11}{x} + \dfrac{10}{x^2}}$

23. $\dfrac{1 + \dfrac{1}{x+3}}{1 - \dfrac{1}{x+3}}$

24. $\dfrac{1 + \dfrac{1}{x-2}}{1 - \dfrac{1}{x-2}}$

25. $\dfrac{1 + \dfrac{1}{x+3}}{1 + \dfrac{7}{x-3}}$

26. $\dfrac{1 + \dfrac{1}{x-2}}{1 - \dfrac{3}{x+2}}$

27. $\dfrac{1 - \dfrac{1}{a+1}}{1 + \dfrac{1}{a-1}}$

28. $\dfrac{\dfrac{1}{a-1} + 1}{\dfrac{1}{a+1} - 1}$

29. $\dfrac{\dfrac{1}{x+3} + \dfrac{1}{x-3}}{\dfrac{1}{x+3} - \dfrac{1}{x-3}}$

30. $\dfrac{\dfrac{1}{x+a} + \dfrac{1}{x-a}}{\dfrac{1}{x+a} - \dfrac{1}{x-a}}$

31. $\dfrac{\dfrac{y+1}{y-1} + \dfrac{y-1}{y+1}}{\dfrac{y+1}{y-1} - \dfrac{y-1}{y+1}}$

32. $\dfrac{\dfrac{y-1}{y+1} - \dfrac{y+1}{y-1}}{\dfrac{y-1}{y+1} + \dfrac{y+1}{y-1}}$

33. $1 - \dfrac{x}{1 - \dfrac{1}{x}}$

34. $x - \dfrac{1}{x - \dfrac{1}{2}}$

35. $1 + \dfrac{1}{1 + \dfrac{1}{1+1}}$

36. $1 - \dfrac{1}{1 - \dfrac{1}{1 - \frac{1}{2}}}$

37. $\dfrac{1 - \dfrac{1}{x + \frac{1}{2}}}{1 + \dfrac{1}{x + \frac{1}{2}}}$

38. $\dfrac{2 + \dfrac{1}{x - \frac{1}{3}}}{2 - \dfrac{1}{x - \frac{1}{3}}}$

39. Optics The formula

$$f = \dfrac{ab}{a+b}$$

is used in optics to find the focal length of a lens. Show that the formula $f = (a^{-1} + b^{-1})^{-1}$ is equivalent to the above formula by rewriting it without the negative exponents and then simplifying the results.

40. Optics Show that the expression $(a^{-1} - b^{-1})^{-1}$ can be simplified to

$$\dfrac{ab}{b-a}$$

by first writing it without the negative exponents and then simplifying the result.

41. Show that the expression

$$\dfrac{1 - x^{-1}}{1 + x^{-1}}$$

can be written as $\dfrac{x-1}{x+1}$.

42. Show that the expression $(1 + x^{-1})^{-1}$ can be written as $\dfrac{x}{x + 1}$.

Applying the Concepts

43. Simplify the first three terms in the sequence below. Then use the results to predict the next three terms in the sequence.

$$1 + \frac{1}{1 + 1}, \; 1 + \frac{1}{1 + \dfrac{1}{1 + 1}}, \; 1 + \frac{1}{1 + \dfrac{1}{1 + \dfrac{1}{1 + 1}}}, \; \cdots$$

44. Use a calculator to find a decimal representation for each term in the sequence in Problem 43. Round your answers to the nearest thousandth. You will see these decimals again in Chapter 5 when we discuss the golden rectangle.

45. Simplify the first three terms in the sequence below.

$$1 + \frac{1}{2 + 1}, \; 1 + \frac{1}{2 + \dfrac{1}{2 + 1}}, \; 1 + \frac{1}{2 + \dfrac{1}{2 + \dfrac{1}{2 + 1}}}, \; \cdots$$

46. Use a calculator to find a decimal representation for each term in the sequence in Problem 45. Round your answers to the nearest thousandth. You will see these decimals again in Chapter 5 as approximations to $\sqrt{2}$.

Electric Circuitry *The diagrams at the top of the next column show part of two electric circuits. The jagged lines represent resistors, which are so-named because they resist the flow of electricity. The resistors that are connected one after the other are said to be connected "in series," while the other resistors are connected "in parallel." For each of the diagrams, R_T is the total resistance between points A and B. As you can see, R_T for the two resistors connected in series is the sum of the individual resistances, while R_T for the resistors connected in parallel is given by a complex fraction.*

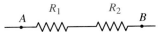

Resistors connected in series:
$R_T = R_1 + R_2$

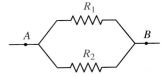

Resistors connected in parallel:

$$R_T = \frac{1}{\dfrac{1}{R_1} + \dfrac{1}{R_2}}$$

Suppose that the resistance R_1 is twice the resistance R_2.

47. Show that R_T for the resistors connected in series is $3R_2$.

48. Show that R_T for the resistors connected in parallel is $\frac{2}{3}R_2$.

Suppose that the resistance R_1 is three times the resistance R_2.

49. Show that R_T for the resistors connected in series is $4R_2$.

50. Show that R_T for the resistors connected in parallel is $\frac{3}{4}R_2$.

51. If $R_1 = kR_2$, where k is a positive integer, show that R_T for the resistors connected in series is $(k + 1)R_2$.

52. If $R_1 = kR_2$, where k is a positive integer, show that R_T for the resistors connected in parallel is

$$\frac{k}{k + 1}R_2$$

Review Problems

NOTE: The following problems review material we covered in Section 2.1. Reviewing these problems will help you with the next section.

Solve each equation.

53. $3x + 60 = 15$ **54.** $3x - 18 = 4$

55. $3(y - 3) = 2(y - 2)$

56. $5(y + 2) = 4(y + 1)$
57. $10 - 2(x + 3) = x + 1$
58. $15 - 3(x - 1) = x - 2$
59. $x^2 - x - 12 = 0$
60. $x^2 + x - 12 = 0$
61. $3x^2 + x - 10 = 0$
62. $10x^2 - x - 3 = 0$
63. $(x + 1)(x - 6) = -12$
64. $(x + 1)(x - 4) = -6$

One Step Further

Try both methods shown in this section on each of the following problems. (One of the two methods will be much easier than the other.)

65. $\dfrac{1 + \dfrac{1}{x - 3}}{x - \dfrac{10}{x + 3}}$

66. $\dfrac{1 + \dfrac{1}{x + 2}}{x - \dfrac{10}{x - 2}}$

67. $\dfrac{1 + \dfrac{1}{x + a}}{1 + \dfrac{2a + 1}{x - a}}$

68. $\dfrac{1 + \dfrac{1}{x - a}}{1 - \dfrac{2a - 1}{x + a}}$

Equations Involving Rational Expressions

The first step in solving an equation that contains one or more rational expressions is to find the LCD for all denominators in the equation. We then multiply both sides of the equation by the LCD to clear the equation of all fractions. That is, after we have multiplied through by the LCD, each term in the resulting equation will have a denominator of 1.

EXAMPLE 1 Solve: $\dfrac{x}{2} - 3 = \dfrac{2}{3}$.

SOLUTION The LCD for 2 and 3 is 6. Multiplying both sides by 6, we have

$$6\left(\frac{x}{2} - 3\right) = 6\left(\frac{2}{3}\right)$$

$$6\left(\frac{x}{2}\right) - 6(3) = 6\left(\frac{2}{3}\right)$$

$$3x - 18 = 4$$

$$3x = 22$$

$$x = \frac{22}{3}$$

Multiplying both sides of an equation by the LCD clears the equation of fractions because the LCD has the property that all the denominators divide it evenly.

EXAMPLE 2 Solve: $\dfrac{6}{a-4}=\dfrac{3}{8}$.

SOLUTION The LCD for $a-4$ and 8 is $8(a-4)$. Multiplying both sides by this quantity yields

$$8(a-4)\cdot\frac{6}{a-4}=8(a-4)\cdot\frac{3}{8}$$
$$48=(a-4)\cdot 3$$
$$48=3a-12$$
$$60=3a$$
$$20=a$$

The solution set is $\{20\}$, which checks in the original equation. (Try it.)

When we multiply both sides of an equation by an expression containing the variable, we must be sure to check our solutions. The multiplication property of equality does not allow multiplication by 0. If the expression we multiply by contains the variable, then it has the possibility of being 0. In the last example we multiplied both sides by $8(a-4)$. This gives a restriction $a\neq 4$ for any solution we find.

EXAMPLE 3 Solve: $\dfrac{x}{x-2}+\dfrac{2}{3}=\dfrac{2}{x-2}$.

SOLUTION The LCD is $3(x-2)$. We are assuming $x\neq 2$ when we multiply both sides of the equation by $3(x-2)$:

$$3(x-2)\cdot\left[\frac{x}{x-2}+\frac{2}{3}\right]=3(x-2)\cdot\frac{2}{x-2}$$
$$3x+(x-2)\cdot 2=3\cdot 2$$
$$3x+2x-4=6$$
$$5x-4=6$$
$$5x=10$$
$$x=2$$

The only possible solution is $x=2$. Checking this value back in the original equation gives

$$\frac{2}{2-2}+\frac{2}{3}\overset{?}{=}\frac{2}{2-2}$$
$$\frac{2}{0}+\frac{2}{3}\overset{?}{=}\frac{2}{0}$$

The first and last terms are undefined. The proposed solution, $x=2$, therefore does not check in the original equation. The solution set is the empty set. There is no solution to the original equation.

When the proposed solution to an equation is not actually a solution, it is called an **extraneous solution.** In the last example, $x=2$ is an extraneous solution.

EXAMPLE 4 Solve: $\dfrac{5}{x^2 - 3x + 2} - \dfrac{1}{x - 2} = \dfrac{1}{3x - 3}$.

SOLUTION Writing the equation again with the denominators in factored form, we have

$$\dfrac{5}{(x - 2)(x - 1)} - \dfrac{1}{x - 2} = \dfrac{1}{3(x - 1)}$$

The LCD is $3(x - 2)(x - 1)$. Multiplying through by the LCD, we have

$$3(x - 2)(x - 1) \cdot \dfrac{5}{(x - 2)(x - 1)} - 3(x - 2)(x - 1) \cdot \dfrac{1}{(x - 2)} = 3(x - 2)(x - 1) \cdot \dfrac{1}{3(x - 1)}$$

$$3 \cdot 5 - 3(x - 1) \cdot 1 = (x - 2) \cdot 1$$
$$15 - 3x + 3 = x - 2$$
$$-3x + 18 = x - 2$$
$$-4x + 18 = -2$$
$$-4x = -20$$
$$x = 5$$

Checking the proposed solution $x = 5$ in the original equation yields a true statement. Try it and see.

EXAMPLE 5 Solve: $3 + \dfrac{1}{x} = \dfrac{10}{x^2}$.

SOLUTION To clear the equation of denominators, we multiply both sides by x^2:

$$x^2\left(3 + \dfrac{1}{x}\right) = x^2\left(\dfrac{10}{x^2}\right)$$
$$3(x^2) + \left(\dfrac{1}{x}\right)(x^2) = \left(\dfrac{10}{x^2}\right)(x^2)$$
$$3x^2 + x = 10$$

Rewrite in standard form and solve:

Remember: We have to check *all solutions* any time we multiply both sides of the equation by an expression that contains the variable, just to be sure we haven't multiplied by 0.

$$3x^2 + x - 10 = 0$$
$$(3x - 5)(x + 2) = 0$$
$$3x - 5 = 0 \quad \text{or} \quad x + 2 = 0$$
$$x = \tfrac{5}{3} \quad \text{or} \quad x = -2$$

The solution set is $\{-2, \tfrac{5}{3}\}$. Both solutions check in the original equation.

EXAMPLE 6 Solve: $\dfrac{y - 4}{y^2 - 5y} = \dfrac{2}{y^2 - 25}$.

SOLUTION Factoring each denominator, we find the LCD is $y(y - 5)(y + 5)$. Multiplying each side of the equation by the LCD clears the equation of denominators and leads us to our possible solutions:

$$y(y - 5)(y + 5) \cdot \frac{y - 4}{y(y - 5)} = \frac{2}{(y - 5)(y + 5)} \cdot y(y - 5)(y + 5)$$

$$(y + 5)(y - 4) = 2y$$

$$y^2 + y - 20 = 2y \quad \text{Multiply out the left side.}$$

$$y^2 - y - 20 = 0 \quad \text{Add } -2y \text{ to each side.}$$

$$(y - 5)(y + 4) = 0$$

$$y - 5 = 0 \quad \text{or} \quad y + 4 = 0$$

$$y = 5 \quad \text{or} \qquad y = -4$$

The two possible solutions are 5 and -4. If we substitute -4 for y in the original equation, we find that it leads to a true statement. It is therefore a solution. On the other hand, if we substitute 5 for y in the original equation, we find that both sides of the equation are undefined. The only solution to our original equation is $y = -4$. The other possible solution, $y = 5$, is extraneous.

EXAMPLE 7 Solve for y: $x = \frac{y - 4}{y - 2}$.

SOLUTION To solve for y, we first multiply each side by $y - 2$ to obtain

$$x(y - 2) = y - 4$$

$$xy - 2x = y - 4 \qquad \text{Distributive property}$$

$$xy - y = 2x - 4 \qquad \text{Collect all terms containing } y \text{ on the left side.}$$

$$y(x - 1) = 2x - 4 \qquad \text{Factor } y \text{ from each term on the left side.}$$

$$y = \frac{2x - 4}{x - 1} \qquad \text{Divide each side by } x - 1.$$

EXAMPLE 8 Solve the formula $\frac{1}{x} = \frac{1}{b} + \frac{1}{a}$ for x.

SOLUTION We begin by multiplying both sides by the least common denominator, xab. As you can see from our previous examples, multiplying both sides of an equation by the LCD is equivalent to multiplying each term of both sides by the LCD:

$$xab \cdot \frac{1}{x} = \frac{1}{b} \cdot xab + \frac{1}{a} \cdot xab$$

$$ab = xa + xb$$

$$ab = (a + b)x \qquad \text{Factor } x \text{ from the right side.}$$

$$\frac{ab}{a + b} = x$$

We know we are finished because the variable we were solving for is alone on one side of the equation and does not appear on the other side.

Graphing Rational Functions

We graphed simple rational functions in Section 3.7 when we graphed the inverse variation statement $y = \frac{1}{x}$. In the first section of this chapter we looked at a graph that was a little more complicated when we graphed $r(t) = \frac{785}{t}$. Our next example continues our investigation of the graphs of rational functions.

EXAMPLE 9 Graph the rational function: $f(x) = \dfrac{6}{x - 2}$.

SOLUTION Unlike the graphs in Section 3.7, this graph will cross the y-axis. To find the y-intercept, we let x equal 0.

$$\text{When } x = 0: \quad y = \frac{6}{0 - 2} = \frac{6}{-2} = -3 \quad \text{y-intercept}$$

The graph will not cross the x-axis. If it did, we would have a solution to the equation

$$0 = \frac{6}{x - 2}$$

which has no solution because there is no number to divide 6 by to obtain 0.

The graph of our equation is shown in Figure 1 along with a table giving values of x and y that satisfy the equation. Notice that y is undefined when x is 2. This means that the graph will not cross the vertical line $x = 2$. (If it did, there would be a value of y for $x = 2$.) The line $x = 2$ is called a **vertical asymptote** of the graph. The graph will get very close to the vertical asymptote, but will never touch or cross it.

x	y
-4	-1
-1	-2
0	-3
1	-6
2	Undefined
3	6
4	3
5	2

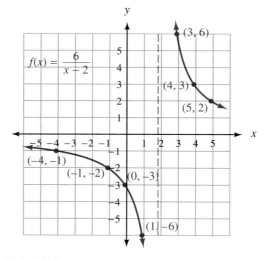

FIGURE 1

The graph of $f(x) = \dfrac{6}{x - 2}$

If you were to graph $y = \frac{6}{x}$ on the coordinate system in Figure 1, you would see that the graph of $y = \frac{6}{x-2}$ is the graph of $y = \frac{6}{x}$ with all points shifted 2 units to the right.

Using Technology: More about Example 9

We know the graph of $f(x) = \frac{6}{x-2}$ will not cross the vertical asymptote $x = 2$ because replacing x with 2 in the equation gives us an undefined expression, meaning there is no value of y to associate with $x = 2$. We can use a graphing calculator to explore the behavior of this function when x gets closer and closer to 2 by using the table function on the calculator. We want to put our own values for X into the table, so we set the independent variable to Ask. (On a TI-82/83, use the TBLSET key to set up the table.) To see how the function behaves as x gets close to 2, we let X take on values of 1.9, 1.99, and 1.999. Then we move to the other side of 2 and let X become 2.1, 2.01, and 2.001.

TABLE SETUP

Table minimum = 0
Table increment = 1
Independent variable: Ask
Dependent variable: Auto

Y VARIABLES

$Y_1 = 6/(x - 2)$

The table will look like this:

X	Y_1
1.9	−60
1.99	−600
1.999	−6000
2.1	60
2.01	600
2.001	6000

Again, the calculator asks us for a table increment. Because we are inputting the x values ourselves, the increment value does not matter.

As you can see, the values in the table support the shape of the curve in Figure 1 around the vertical asymptote $x = 2$.

EXAMPLE 10 Graph: $g(x) = \dfrac{6}{x + 2}$.

SOLUTION The only difference between this equation and the equation in Example 9 is in the denominator. This graph will have the same shape as the graph in Example 9, but the vertical asymptote will be $x = -2$ instead of $x = 2$. Figure 2 (at the top of the next page) shows the graph.

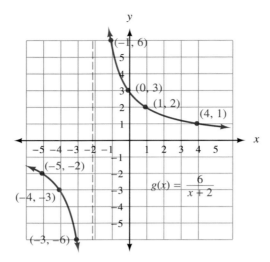

FIGURE 2

The graph of $g(x) = \dfrac{6}{x + 2}$

Notice that the graphs shown in Figures 1 and 2 are both graphs of functions because no vertical line will cross either graph in more than one place. Notice the similarities and differences in our two functions,

$$f(x) = \frac{6}{x - 2} \quad \text{and} \quad g(x) = \frac{6}{x + 2}$$

and their graphs. The vertical asymptotes shown in Figures 1 and 2 correspond to the fact that both $f(2)$ and $g(-2)$ are undefined. The domain for the function f is all real numbers except $x = 2$, while the domain for g is all real numbers except $x = -2$.

Problem Set

4.6

Solve each of the following equations:

1. $\dfrac{x}{5} + 4 = \dfrac{5}{3}$

2. $\dfrac{x}{5} = \dfrac{x}{2} - 9$

3. $\dfrac{a}{3} + 2 = \dfrac{4}{5}$

4. $\dfrac{a}{4} + \dfrac{1}{2} = \dfrac{2}{3}$

5. $\dfrac{y}{2} + \dfrac{y}{4} + \dfrac{y}{6} = 3$

6. $\dfrac{y}{3} - \dfrac{y}{6} + \dfrac{y}{2} = 1$

7. $\dfrac{5}{2x} = \dfrac{1}{x} + \dfrac{3}{4}$

8. $\dfrac{1}{2a} = \dfrac{2}{a} - \dfrac{3}{8}$

9. $\dfrac{1}{x} = \dfrac{1}{3} - \dfrac{2}{3x}$

10. $\dfrac{5}{2x} = \dfrac{2}{x} - \dfrac{1}{12}$

11. $\dfrac{2x}{x - 3} + 2 = \dfrac{2}{x - 3}$

12. $\dfrac{2}{x + 5} = \dfrac{2}{5} - \dfrac{x}{x + 5}$

13. $1 - \dfrac{1}{x} = \dfrac{12}{x^2}$

14. $2 + \dfrac{5}{x} = \dfrac{3}{x^2}$

15. $y - \dfrac{4}{3y} = -\dfrac{1}{3}$

16. $\dfrac{y}{2} - \dfrac{4}{y} = -\dfrac{7}{2}$

17. $\dfrac{x+2}{x+1} = \dfrac{1}{x+1} + 2$

18. $\dfrac{x+6}{x+3} = \dfrac{3}{x+3} + 2$

19. $\dfrac{3}{a-2} = \dfrac{2}{a-3}$

20. $\dfrac{5}{a+1} = \dfrac{4}{a+2}$

21. $6 - \dfrac{5}{x^2} = \dfrac{7}{x}$

22. $10 - \dfrac{3}{x^2} = -\dfrac{1}{x}$

NOTE: As you work your way through these problems, look back at the problems you worked in Problem Set 4.4. The answers to those problems were rational expressions, while the answers you are getting here are real numbers. When you are tested on this chapter, you will work both types of problems. Be sure you recognize the differences between them.

23. $\dfrac{1}{x-1} - \dfrac{1}{x+1} = \dfrac{3x}{x^2-1}$

24. $\dfrac{5}{x-1} + \dfrac{2}{x-1} = \dfrac{4}{x+1}$

25. $\dfrac{2}{x-3} + \dfrac{x}{x^2-9} = \dfrac{4}{x+3}$

26. $\dfrac{2}{x+5} + \dfrac{3}{x+4} = \dfrac{2x}{x^2+9x+20}$

27. $\dfrac{3}{2} - \dfrac{1}{x-4} = \dfrac{-2}{2x-8}$

28. $\dfrac{2}{x} - \dfrac{1}{x+1} = \dfrac{-2}{5x+5}$

29. $\dfrac{t-4}{t^2-3t} = \dfrac{-2}{t^2-9}$

30. $\dfrac{t+3}{t^2-2t} = \dfrac{10}{t^2-4}$

31. $\dfrac{3}{y-4} - \dfrac{2}{y+1} = \dfrac{5}{y^2-3y-4}$

32. $\dfrac{1}{y+2} - \dfrac{2}{y-3} = \dfrac{-2y}{y^2-y-6}$

33. $\dfrac{2}{1+a} = \dfrac{3}{1-a} + \dfrac{5}{a}$

34. $\dfrac{1}{a+3} - \dfrac{a}{a^2-9} = \dfrac{2}{3-a}$

35. $\dfrac{3}{2x-6} - \dfrac{x+1}{4x-12} = 4$

36. $\dfrac{2x-3}{5x+10} + \dfrac{3x-2}{4x+8} = 1$

37. $\dfrac{y+2}{y^2-y} - \dfrac{6}{y^2-1} = 0$

38. $\dfrac{y+3}{y^2-y} - \dfrac{8}{y^2-1} = 0$

39. $\dfrac{4}{2x-6} - \dfrac{12}{4x+12} = \dfrac{12}{x^2-9}$

40. $\dfrac{1}{x+2} + \dfrac{1}{x-2} = \dfrac{4}{x^2-4}$

41. $\dfrac{2}{y^2-7y+12} - \dfrac{1}{y^2-9} = \dfrac{4}{y^2-y-12}$

42. $\dfrac{1}{y^2+5y+4} + \dfrac{3}{y^2-1} = \dfrac{-1}{y^2+3y-4}$

43. Solve the equation $6x^{-1} + 4 = 7$ by multiplying both sides by x.
(*Remember:* $x^{-1} \cdot x = x^{-1} \cdot x^1 = x^0 = 1$.)

44. Solve the equation $3x^{-1} - 5 = 2x^{-1} - 3$ by multiplying both sides by x.

45. Solve the equation $1 + 5x^{-2} = 6x^{-1}$ by multiplying both sides by x^2.

46. Solve the equation $1 + 3x^{-2} = 4x^{-1}$ by multiplying both sides by x^2.

47. Solve the formula $\dfrac{1}{x} = \dfrac{1}{b} - \dfrac{1}{a}$ for x.

48. Solve $\dfrac{1}{x} = \dfrac{1}{a} - \dfrac{1}{b}$ for x.

49. Solve for R in the formula $\dfrac{1}{R} = \dfrac{1}{R_1} + \dfrac{1}{R_2}$.

50. Solve for R in the formula

$$\frac{1}{R} = \frac{1}{R_1} + \frac{1}{R_2} + \frac{1}{R_3}$$

Solve for y.

51. $x = \dfrac{y - 3}{y - 1}$

52. $x = \dfrac{y - 2}{y - 3}$

53. $x = \dfrac{2y + 1}{3y + 1}$

54. $x = \dfrac{3y + 2}{5y + 1}$

Graph each function. Show the vertical asymptote.

55. $f(x) = \dfrac{1}{x - 3}$

56. $f(x) = \dfrac{1}{x + 3}$

57. $f(x) = \dfrac{4}{x + 2}$

58. $f(x) = \dfrac{4}{x - 2}$

59. $g(x) = \dfrac{2}{x - 4}$

60. $g(x) = \dfrac{2}{x + 4}$

61. $g(x) = \dfrac{6}{x + 1}$

62. $g(x) = \dfrac{6}{x - 1}$

Let $f(x) = \dfrac{1}{x - 3}$ *and* $g(x) = \dfrac{1}{x + 3}$, *and evaluate the following:*

63. $f(0)$ and $f(6)$

64. $g(0)$ and $g(-6)$

65. $f(1)$ and $f(5)$

66. $g(1)$ and $g(-7)$

67. Give the domain for the function f as defined above.

68. Give the domain for the function g as defined above.

Let $f(x) = \dfrac{4}{x + 2}$ *and* $g(x) = \dfrac{4}{x - 2}$, *and evaluate the following:*

69. $f(0)$ and $f(-4)$

70. $g(0)$ and $g(4)$

71. $f(2)$ and $f(-6)$

72. $g(1)$ and $g(-5)$

73. Give the domain for the function f as defined above.

74. Give the domain for the function g as defined above.

Review Problems

 The problems that follow review material we covered in Section 3.7.

75. If y varies directly with the square of x, and y is 75 when x is 5, find y when x is 7.

76. Suppose y varies directly with the cube of x. If y is 16 when x is 2, find x when y is 128.

77. Suppose y varies inversely with the square of x. If y is 8 when x is 5, find x when y is 2.

78. If y varies inversely with the cube of x, and y is 2 when x is 2, find y when x is 4.

79. Suppose z varies jointly with x and the square of y. If z is 40 when x is 5 and y is 2, find z when x is 2 and y is 5.

80. Suppose z varies jointly with x and the cube of y. If z is 48 when x is 3 and y is 2, find z when x is 4 and y is $\frac{1}{2}$.

One Step Further

Solve each equation.

81. $\dfrac{12}{x} + \dfrac{8}{x^2} - \dfrac{75}{x^3} - \dfrac{50}{x^4} = 0$

82. $\dfrac{45}{x} + \dfrac{18}{x^2} - \dfrac{80}{x^3} - \dfrac{32}{x^4} = 0$

83. $\dfrac{1}{x^3} - \dfrac{1}{3x^2} - \dfrac{1}{4x} + \dfrac{1}{12} = 0$

84. $\dfrac{1}{x^3} - \dfrac{1}{2x^2} - \dfrac{1}{9x} + \dfrac{1}{18} = 0$

SECTION 4.7

Applications

We begin this section with some application problems, the solutions to which involve equations that contain rational expressions. As you will see, the solutions to the examples show only the essential steps from our Blueprint for Problem Solving (Section 2.3). Recall that Step 1 was done mentally; we read the problem and mentally list the items that are known and the items that are unknown. This is an essential part of problem solving. Now that you have had some experience with application problems, you should be doing Step 1 automatically.

Also in this section we will look at a method of solving conversion problems that is called *unit analysis*. With unit analysis, we can convert expressions with units of feet per minute to equivalent expressions in miles per hour. This method of converting between different units of measure is used often in chemistry, physics, and engineering.

EXAMPLE 1   One number is twice another. The sum of their reciprocals is 2. Find the numbers.

SOLUTION   Let $x =$ the smaller number. The larger number is $2x$. Their reciprocals are $\frac{1}{x}$ and $\frac{1}{2x}$. The equation that describes the situation is

$$\frac{1}{x} + \frac{1}{2x} = 2$$

Multiplying both sides by the LCD, $2x$, we have

$$2x \cdot \frac{1}{x} + 2x \cdot \frac{1}{2x} = 2x(2)$$
$$2 + 1 = 4x$$
$$3 = 4x$$
$$x = \frac{3}{4}$$

The smaller number is $\frac{3}{4}$. The larger is $2(\frac{3}{4}) = \frac{6}{4} = \frac{3}{2}$. Adding their reciprocals, we have

$$\frac{4}{3} + \frac{2}{3} = \frac{6}{3} = 2$$

The sum of the reciprocals of $\frac{3}{4}$ and $\frac{3}{2}$ is 2.

EXAMPLE 2   The speed of a boat in still water is 20 miles/hour. It takes the same amount of time for the boat to travel 3 miles downstream (with the current) as it does to travel 2 miles upstream (against the current). Find the speed of the current.

SOLUTION   The following table will be helpful in finding the equation necessary to solve this problem:

	d (distance)	r (rate)	t (time)
Upstream			
Downstream			

If we let x = the speed of the current, the speed (rate) of the boat upstream is $(20 - x)$ since it is traveling against the current. The rate downstream is $(20 + x)$ since the boat is then traveling with the current. The distance traveled upstream is 2 miles, while the distance traveled downstream is 3 miles. We put all this information into the table as shown.

	d	r	t
Upstream	2	20 − x	
Downstream	3	20 + x	

To fill in the last two spaces in the table, we must use the relationship $d = r \cdot t$. Since the spaces to be filled in are in the time column, we solve the equation $d = r \cdot t$ for t and get

$$t = \frac{d}{r}$$

The completed table then is

	d	r	$t = \dfrac{d}{r}$
Upstream	2	20 − x	$\dfrac{2}{20 - x}$
Downstream	3	20 + x	$\dfrac{3}{20 + x}$

Reading the problem again, we find that the time moving upstream is equal to the time moving downstream, or

$$\frac{2}{20 - x} = \frac{3}{20 + x}$$

Multiplying both sides by the LCD, $(20 - x)(20 + x)$, gives

$$(20 + x) \cdot 2 = 3(20 - x)$$
$$40 + 2x = 60 - 3x$$
$$5x = 20$$
$$x = 4$$

The speed of the current is 4 miles/hour.

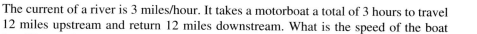

EXAMPLE 3 The current of a river is 3 miles/hour. It takes a motorboat a total of 3 hours to travel 12 miles upstream and return 12 miles downstream. What is the speed of the boat in still water?

SOLUTION This time we let x = the speed of the boat in still water. Then, we fill in as much of the table as possible using the information given in the problem. For instance, since we let x = the speed of the

	d	r	t
Upstream	12	x − 3	
Downstream	12	x + 3	

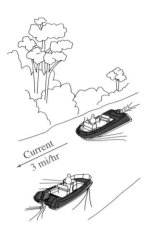

Current
3 mi/hr

boat in still water, the rate upstream (against the current) must be $x - 3$. The rate downstream (with the current) is $x + 3$.

The last two boxes can be filled in using the relationship

$$t = \frac{d}{r}$$

	d	r	$t = \dfrac{d}{r}$
Upstream	12	$x - 3$	$\dfrac{12}{x - 3}$
Downstream	12	$x + 3$	$\dfrac{12}{x + 3}$

The total time for the trip up and back is 3 hours:

Time upstream + Time downstream = Total time

$$\frac{12}{x - 3} \quad + \quad \frac{12}{x + 3} \quad = \quad 3$$

Multiplying both sides by $(x - 3)(x + 3)$, we have

$$12(x + 3) + 12(x - 3) = 3(x^2 - 9)$$
$$12x + 36 + 12x - 36 = 3x^2 - 27$$
$$3x^2 - 24x - 27 = 0$$
$$x^2 - 8x - 9 = 0 \qquad \text{Divide both sides by 3.}$$
$$(x - 9)(x + 1) = 0$$
$$x = 9 \quad \text{or} \quad x = -1$$

The speed of the motorboat in still water is 9 miles/hour. (We don't use $x = -1$ because the speed of the motorboat cannot be a negative number.)

EXAMPLE 4   An inlet pipe can fill a pool in 10 hours, while the drain can empty it in 12 hours. If the pool is empty and the inlet pipe and drain are both open, how long will it take to fill the pool?

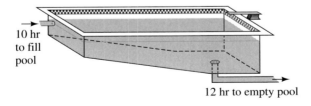

10 hr
to fill
pool

12 hr to empty pool

SOLUTION   It is helpful to think in terms of how much work is done by the inlet pipe and the drain in 1 hour.

Let $x =$ the time it takes to fill the pool with both the inlet pipe and the drain open.

If the inlet pipe can fill the pool in 10 hours, then in 1 hour it is $\frac{1}{10}$ full. If the drain empties the pool in 12 hours, then in 1 hour it is $\frac{1}{12}$ empty. If the pool can be filled in x hours with both the inlet pipe and the drain open, then in 1 hour it is $\frac{1}{x}$ full.

Here is the equation for 1 hour:

$$\left[\begin{array}{c}\text{Amount filled by}\\ \text{inlet pipe}\end{array}\right] - \left[\begin{array}{c}\text{Amount emptied by}\\ \text{the drain}\end{array}\right] = \left[\begin{array}{c}\text{Fraction of pool filled}\\ \text{with both open}\end{array}\right]$$

$$\frac{1}{10} \qquad - \qquad \frac{1}{12} \qquad = \qquad \frac{1}{x}$$

Multiplying through by $60x$, we have

$$60x \cdot \frac{1}{10} - 60x \cdot \frac{1}{12} = 60x \cdot \frac{1}{x}$$

$$6x - 5x = 60$$

$$x = 60$$

It takes 60 hours to fill the pool if both the inlet pipe and the drain are open.

Unit Analysis

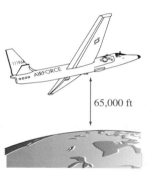

In the 1950s the United States had a spy plane, the U-2, that could fly at an altitude of 65,000 feet. Do you know how many miles are in 65,000 feet?

We can solve problems like this by using a method called **unit analysis.** With unit analysis, we analyze the units we are given and the units for which we are asked, and then multiply by the appropriate **conversion factor.** Since 1 mile is 5,280 feet, the conversion factor we use is

$$\frac{1 \text{ mile}}{5{,}280 \text{ feet}}$$

which is the number 1. Multiplying 65,000 feet by this conversion factor, we have the following:

$$65{,}000 \text{ feet} = \frac{65{,}000 \text{ feet}}{1} \cdot \frac{1 \text{ mile}}{5{,}280 \text{ feet}}$$

We treat the units common to the numerator and denominator in the same way we treat factors common to the numerator and denominator: We divide out common units, just as we divide out common factors. In the expression above, we have feet common to the numerator and denominator. Dividing them out leaves us with miles only. Here is the complete problem:

$$65{,}000 \text{ feet} = \frac{65{,}000 \, \cancel{\text{feet}}}{1} \cdot \frac{1 \text{ mile}}{5{,}280 \, \cancel{\text{feet}}}$$

$$= \frac{65{,}000}{5{,}280} \text{ mile}$$

$$= 12.3 \text{ miles} \quad \text{To the nearest tenth of a mile}$$

The key to solving a problem like this one lies in choosing the appropriate conversion factor. The fact that 1 mile = 5,280 feet yields two conversion factors, each of which is equal to the number 1. They are

$$\frac{1 \text{ mile}}{5,280 \text{ feet}} \quad \text{and} \quad \frac{5,280 \text{ feet}}{1 \text{ mile}}$$

The conversion factor we choose will depend on the units we are given and the units we want in our answer. Multiplying any expression by either of the two conversion factors will leave the value of the original expression unchanged because each of the conversion factors is simply the number 1.

EXAMPLE 5 In Section 4.1 we found that a rider on the first Ferris wheel was traveling at approximately 39.3 feet per minute. Convert 39.3 feet per minute to miles per hour.

SOLUTION We know that 5,280 feet = 1 mile and 60 minutes = 1 hour. Therefore, we have the following conversion factors, each of which is equal to 1:

$$\frac{5,280 \text{ feet}}{1 \text{ mile}} \qquad \frac{1 \text{ mile}}{5,280 \text{ feet}} \qquad \frac{60 \text{ minutes}}{1 \text{ hour}} \qquad \frac{1 \text{ hour}}{60 \text{ minutes}}$$

The conversion factors we choose to multiply by are the ones that will allow us to divide out the units we are converting from and leave us with the units we are converting to. Specifically, we want to get rid of feet and be left with miles. Likewise, we want to get rid of minutes and be left with hours. Here is the conversion process that will accomplish these goals:

$$39.3 \text{ feet/minute} = \frac{39.3 \text{ feet}}{1 \text{ minute}} \cdot \frac{1 \text{ mile}}{5,280 \text{ feet}} \cdot \frac{60 \text{ minutes}}{1 \text{ hour}}$$

$$= \frac{39.3 \cdot 60 \text{ miles}}{5,280 \text{ hours}}$$

$$= 0.45 \text{ mile/hour} \quad \text{To the nearest hundredth}$$

EXAMPLE 6 In 1993 a ski resort in Vermont advertised their new high-speed chair lift as "the world's fastest chair lift with a speed of 1,100 feet per second." Show why the speed cannot be correct.

SOLUTION To solve this problem, we can convert feet per second into miles per hour, a unit of measure we are more familiar with on an intuitive level.

$$1,100 \text{ feet/second} = \frac{1,100 \text{ feet}}{1 \text{ second}} \cdot \frac{1 \text{ mile}}{5,280 \text{ feet}} \cdot \frac{60 \text{ seconds}}{1 \text{ minute}} \cdot \frac{60 \text{ minutes}}{1 \text{ hour}}$$

$$= \frac{1,100 \cdot 60 \cdot 60 \text{ miles}}{5,280 \text{ hours}}$$

$$= 750 \text{ miles/hour}$$

Obviously, there is a mistake in the advertisement.

More about Graphing Rational Functions

We continue our investigation of the graphs of rational functions by considering the graph of a rational function with binomials in the numerator and denominator.

EXAMPLE 7 Graph the rational function: $y = \dfrac{x - 4}{x - 2}$.

SOLUTION In addition to making a table to find some points on the graph, we can analyze the graph as follows:

1. The graph will have a y-intercept of 2, because when $x = 0$, $y = \dfrac{-4}{-2} = 2$.

2. To find the x-intercept, we let $y = 0$ to get

$$0 = \frac{x - 4}{x - 2}$$

The only way this expression can be 0 is if the numerator is 0, which happens when $x = 4$. (If you want to solve the above equation, multiply both sides by $x - 2$. You will get the same solution, $x = 4$.)

3. The graph will have a vertical asymptote at $x = 2$, since $x = 2$ will make the denominator of the function 0, meaning y is undefined when x is 2.

4. The graph will have a **horizontal asymptote** at $y = 1$ because for very large values of x, $\dfrac{x - 4}{x - 2}$ is very close to 1. The larger x is, the closer $\dfrac{x - 4}{x - 2}$ is to 1. The same is true for very small values of x, such as $-1,000$ and $-10,000$.

Putting this information together with the ordered pairs in the table next to the figure, we have the graph shown in Figure 1.

x	y
-1	$\frac{5}{3}$
0	2
1	3
2	Undefined
3	-1
4	0
5	$\frac{1}{3}$

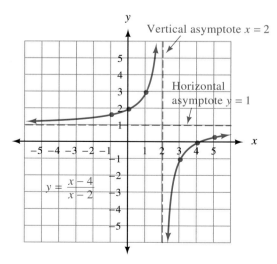

FIGURE I

Using Technology: More about Example 7

In the previous section we used technology to explore the graph of a rational function around a vertical asymptote. This time, we are going to explore the graph near the horizontal asymptote. In Figure 1, the horizontal asymptote is at $y = 1$. To show that the graph approaches this line as x becomes very large, we use the table function on our graphing calculator with X taking values of 100, 1,000, and 10,000. To show that the graph approaches the line $y = 1$ on the left side of the coordinate system, we let X become -100, $-1,000$, and $-10,000$.

TABLE SETUP

Table minimum = 0
Table increment = 1
Independent variable: Ask
Dependent variable: Auto

Y VARIABLES

$Y_1 = (x - 4)/(x - 2)$

The table will look like this:

X	Y_1
100	.97959
1000	.998
10000	.9998
-100	1.0196
-1000	1.002
-10000	1.0002

As you can see, as x becomes very large in the positive direction, the graph approaches the line $y = 1$ from below. As x becomes very small in the negative direction, the graph approaches the line $y = 1$ from above.

Problem Set

4.7

Solve each of the following applications. Be sure to show the equation in each case.

Number Problems

1. One number is 3 times another. The sum of their reciprocals is $\frac{20}{3}$. Find the numbers.

2. One number is 3 times another. The sum of their reciprocals is $\frac{4}{9}$. Find the numbers.

3. The sum of a number and its reciprocal is $\frac{10}{3}$. Find the number.

4. The sum of a number and twice its reciprocal is $\frac{27}{5}$. Find the number.

5. The sum of the reciprocals of two consecutive integers is $\frac{7}{12}$. Find the two integers.

6. Find two consecutive even integers, the sum of whose reciprocals is $\frac{3}{4}$.

7. If a certain number is added to the numerator and denominator of $\frac{7}{9}$, the result is $\frac{5}{6}$. Find the number.

8. Find the number you would add to both the numerator and denominator of $\frac{8}{11}$ so the result is $\frac{6}{7}$.

Rate Problems

9. The speed of a boat in still water is 5 miles per hour. If the boat travels 3 miles downstream in the same amount of time it takes to travel 1.5 miles upstream, what is the speed of the current?

10. A boat, which moves at 18 miles/hour in still water, travels 14 miles downstream in the same amount of time it takes to travel 10 miles upstream. Find the speed of the current.

11. The current of a river is 2 miles/hour. A boat travels to a point 8 miles upstream and back again in 3 hours. What is the speed of the boat in still water?

12. A motorboat travels at 4 miles/hour in still water. It goes 12 miles upstream and 12 miles back again in a total of 8 hours. Find the speed of the current of the river.

13. Train *A* has a speed 15 miles/hour greater than that of train *B*. If train *A* travels 150 miles in the same time train *B* travels 120 miles, what are the speeds of the two trains?

14. A train travels 30 miles/hour faster than a car. If the train covers 120 miles in the same time the car covers 80 miles, what is the speed of each of them?

15. A small airplane flies 810 miles from Los Angeles to Portland, Oregon, with an average speed of 270 miles/hour. An hour and a half after the plane leaves, a Boeing 747 leaves Los Angeles for Portland. Both planes arrive in Portland at the same time. What was the average speed of the 747?

16. Lou leaves for a cross-country excursion on a bicycle traveling at 20 miles/hour. His friends are driving the same route and will meet him at several rest stops along the way. The first stop is scheduled 30 miles from the original starting point. If the people in the car leave 15 minutes after Lou from the same place, how fast will they have to drive to reach the first rest stop at the same time as Lou?

17. A tour bus leaves Sacramento every Friday evening at 5:00 for a 270 mile trip to Las Vegas. This week, however, the bus leaves at 5:30 P.M. In order to arrive in Las Vegas on time, the driver drives 6 miles/hour faster than usual. What is the bus's usual speed?

18. A bakery delivery truck leaves the bakery at 5:00 A.M. each morning on its 140 mile route. One day the driver gets a late start and does not leave the bakery until 5:30 A.M. In order to finish her route on time, the driver drives 5 miles per hour faster than usual. At what speed does she usually drive?

Work Problems

19. A water tank can be filled by an inlet pipe in 8 hours. It takes twice that long for the outlet pipe to empty the tank. How long will it take to fill the tank if both pipes are open?

20. A sink can be filled from the faucet in 5 minutes. It takes only 3 minutes to empty the sink when the drain is open. If the sink is full and both the faucet and the drain are open, how long will it take to empty the sink?

21. It takes 10 hours to fill a pool with the inlet pipe. It can be emptied in 15 hours through the outlet pipe. If the pool is half full to begin with, how long will it take to fill the remaining half if both pipes are open?

22. A sink is $\frac{1}{4}$ full when both the faucet and the drain are opened. The faucet alone can fill the sink in 6 minutes, while it takes 8 minutes to empty it through the drain. How long will it take to fill the remaining $\frac{3}{4}$ of the sink?

23. A sink has two faucets, one for hot water and one for cold water. The sink can be filled by the cold-water faucet in 3.5 minutes. If both faucets are open, the sink is filled in 2.1 minutes. How long does it take to fill the sink with just the hot-water faucet open?

24. A water tank is being filled by two inlet pipes. Pipe *A* can fill the tank in $4\frac{1}{2}$ hours, while both pipes together can fill the tank in 2 hours. How long does it take to fill the tank using only pipe *B*?

Unit Analysis Problems

Give your answers to the following problems to the nearest tenth.

25. The South Coast Shopping Mall in Costa Mesa, California, covers an area of 2,224,750 square feet. If 1 acre = 43,560 square feet, how many acres does the South Coast Shopping Mall cover?

26. The relationship between liters and cubic inches, both of which are measures of volume, is 0.0164 liter ≈ 1 cubic inch. If a Ford Mustang has a motor with a displacement of 4.9 liters, what is the displacement in cubic inches?

27. The Forest chair lift at the Northstar ski resort in Lake Tahoe is 5,750 feet long. If a ride on this chair lift takes 11 minutes, what is the average speed of the lift in miles per hour?

28. The Bear Paw chair lift at the Northstar ski resort in Lake Tahoe is 790 feet long. If a ride on this chair lift takes 2.2 minutes, what is the average speed of the lift in miles per hour?

29. A sprinter runs 100 meters in 10.8 seconds. What is the sprinter's average speed in miles per hour? (1 meter ≈ 3.28 feet)

30. A runner covers 400 meters in 49.8 seconds. What is the average speed of the runner in miles per hour?

31. A person riding a Ferris wheel with a diameter of 65 feet travels once around the wheel in 30 seconds. What is the average speed of the rider in miles per hour?

32. A person riding a Ferris wheel with a diameter of 102 feet travels once around the wheel in 3.5 minutes. What is the average speed of the rider in miles per hour?

33. A $3\frac{1}{2}$ inch diskette, when placed in the disk drive of a computer, rotates at 300 rpm (1 revolution takes $\frac{1}{300}$ minute). Find the average speed of a point 2 inches from the center of the diskette in miles per hour.

34. A 5 inch fixed disk in a computer rotates at 3,600 rpm. Find the average speed of a point 2 inches from the center of the disk in miles per hour.

Graph each rational function. In each case, show the vertical asymptote and the horizontal asymptote, and any intercepts that exist.

35. $y = \dfrac{x - 3}{x - 1}$

36. $f(x) = \dfrac{x + 4}{x - 2}$

37. $f(x) = \dfrac{x + 3}{x - 1}$

38. $f(x) = \dfrac{x - 2}{x - 1}$

39. $g(x) = \dfrac{x - 3}{x + 1}$

40. $g(x) = \dfrac{x - 2}{x + 1}$

Review Problems

NOTE: The problems that follow review material we covered in Sections 4.3, 4.4, 4.5, and 4.6. Reviewing these problems will help clarify the different methods we have used in this chapter.

Perform the indicated operations. [4.3, 4.4]

41. $\dfrac{2a + 10}{a^3} \cdot \dfrac{a^2}{3a + 15}$

42. $\dfrac{4a + 8}{a^2 - a - 6} \div \dfrac{a^2 + 7a + 12}{a^2 - 9}$

43. $(x^2 - 9)\left(\dfrac{x + 2}{x + 3}\right)$

44. $\dfrac{1}{x + 4} + \dfrac{8}{x^2 - 16}$

45. $\dfrac{2x - 7}{x - 2} - \dfrac{x - 5}{x - 2}$

46. $2 + \dfrac{25}{5x - 1}$

Simplify each expression. [4.5]

47. $\dfrac{\dfrac{1}{x} - \dfrac{1}{3}}{\dfrac{1}{x} + \dfrac{1}{3}}$

48. $\dfrac{1 - \dfrac{9}{x^2}}{1 - \dfrac{1}{x} - \dfrac{6}{x^2}}$

Solve each equation. [4.6]

49. $\dfrac{x}{x - 3} + \dfrac{3}{2} = \dfrac{3}{x - 3}$

50. $1 - \dfrac{3}{x} = \dfrac{-2}{x^2}$

REVIEW FOR CHAPTER 4

Examples

1. $\frac{3}{4}$ is a rational number.

$\frac{x-3}{x^2-9}$ is a rational expression.

Chapter 4 Summary

Rational Numbers and Expressions [4.1]

A **rational number** is any number that can be expressed as the ratio of two integers.

$$\text{Rational numbers} = \left\{ \frac{a}{b} \,\middle|\, a \text{ and } b \text{ are integers, } b \neq 0 \right\}$$

A **rational expression** is any quantity that can be expressed as the ratio of two polynomials:

$$\text{Rational expressions} = \left\{ \frac{P}{Q} \,\middle|\, P \text{ and } Q \text{ are polynomials, } Q \neq 0 \right\}$$

Properties of Rational Expressions [4.1]

If P, Q, and K are polynomials with $Q \neq 0$ and $K \neq 0$, then

$$\frac{P}{Q} = \frac{PK}{QK} \qquad \text{and} \qquad \frac{P}{Q} = \frac{P/K}{Q/K}$$

which is to say that multiplying or dividing the numerator and denominator of a rational expression by the same nonzero quantity always produces an equivalent rational expression.

2. $\frac{x-3}{x^2-9} = \frac{x-3}{(x-3)(x+3)}$

$= \frac{1}{x+3}$

Reducing to Lowest Terms [4.1]

To reduce a rational expression to lowest terms, we first factor the numerator and denominator and then divide the numerator and denominator by any factors they have in common.

3. $\frac{15x^3 - 20x^2 + 10x}{5x}$

$= 3x^2 - 4x + 2$

Dividing a Polynomial by a Monomial [4.2]

To divide a polynomial by a monomial, divide each term of the polynomial by the monomial.

4.
```
          x  - 2
x - 3 ) x² - 5x + 8
        -    +
       ∓ x² ∓ 3x
        ─────────
            - 2x + 8
            +    -
           ∓ 2x ∓ 6
           ─────────
                 2
```

Long Division with Polynomials [4.2]

If division with polynomials cannot be accomplished by dividing out factors common to the numerator and denominator, then we use a process similar to long division with whole numbers. The steps in the process are: estimate, multiply, subtract, and bring down the next term.

5. $\dfrac{x+1}{x^2-4} \cdot \dfrac{x+2}{3x+3}$

$= \dfrac{(x+1)(x+2)}{(x-2)(x+2)(3)(x+1)}$

$= \dfrac{1}{3(x-2)}$

Multiplication [4.3]

To multiply two rational numbers or rational expressions, multiply numerators and multiply denominators. In symbols,

$$\frac{P}{Q} \cdot \frac{R}{S} = \frac{PR}{QS} \qquad Q \neq 0, S \neq 0$$

In actual practice, we don't really multiply, but rather, we factor and then divide out common factors.

6. $\dfrac{x^2-y^2}{x^3+y^3} \div \dfrac{x-y}{x^2-xy+y^2}$

$= \dfrac{x^2-y^2}{x^3+y^3} \cdot \dfrac{x^2-xy+y^2}{x-y}$

$= \dfrac{(x+y)(x-y)(x^2-xy+y^2)}{(x+y)(x^2-xy+y^2)(x-y)}$

$= 1$

Division [4.3]

To divide one rational expression by another, we use the definition of division to rewrite our division problem as an equivalent multiplication problem. Instead of dividing by a rational expression, we multiply by its reciprocal. In symbols,

$$\frac{P}{Q} \div \frac{R}{S} = \frac{P}{Q} \cdot \frac{S}{R} = \frac{PS}{QR} \qquad Q \neq 0, S \neq 0, R \neq 0$$

7. The LCD for $\dfrac{2}{x-3}$ and $\dfrac{3}{5}$ is $5(x-3)$.

Least Common Denominator [4.4]

The **least common denominator, LCD,** for a set of denominators is the smallest quantity divisible by each of the denominators.

8. $\dfrac{2}{x-3} + \dfrac{3}{5}$

$= \dfrac{2}{x-3} \cdot \dfrac{5}{5} + \dfrac{3}{5} \cdot \dfrac{x-3}{x-3}$

$= \dfrac{3x+1}{5(x-3)}$

Addition and Subtraction [4.4]

If P, Q, and R represent polynomials, $R \neq 0$, then

$$\frac{P}{R} + \frac{Q}{R} = \frac{P+Q}{R} \qquad \text{and} \qquad \frac{P}{R} - \frac{Q}{R} = \frac{P-Q}{R}$$

When adding or subtracting rational expressions with different denominators, we must find the LCD for all denominators and change each rational expression to an equivalent expression that has the LCD.

9. $\dfrac{\dfrac{1}{x} + \dfrac{1}{y}}{\dfrac{1}{x} - \dfrac{1}{y}} = \dfrac{xy\left(\dfrac{1}{x} + \dfrac{1}{y}\right)}{xy\left(\dfrac{1}{x} - \dfrac{1}{y}\right)}$

$= \dfrac{y+x}{y-x}$

Complex Fractions [4.5]

A rational expression that contains other rational expressions in its numerator or denominator is called a **complex fraction.** One method of simplifying a complex fraction is to multiply the numerator and denominator by the LCD for all denominators.

REVIEW FOR CHAPTER 4

10. Solve $\dfrac{x}{2} + 3 = \dfrac{1}{3}$.

$$6\left(\dfrac{x}{2}\right) + 6 \cdot 3 = 6 \cdot \dfrac{1}{3}$$

$$3x + 18 = 2$$

$$x = -\dfrac{16}{3}$$

Equations Involving Rational Expressions [4.6]

To solve an equation involving rational expressions, we first find the LCD for all denominators appearing on either side of the equation. We then multiply both sides by the LCD to clear the equation of all fractions and solve as usual.

Common Mistakes

1. Attempting to divide the numerator and denominator of a rational expression by a quantity that is not a factor of both is a very common mistake. For example,

$$\dfrac{x^2 \overset{3}{-} 9x + \overset{2}{20}}{x^2 \underset{1}{-} 3x \underset{1}{-} 10} \quad \text{Mistake}$$

This makes no sense at all. The numerator and denominator must be factored completely before any factors they have in common can be recognized:

$$\dfrac{x^2 - 9x + 20}{x^2 - 3x - 10} = \dfrac{(x - 5)(x - 4)}{(x - 5)(x + 2)}$$

$$= \dfrac{x - 4}{x + 2}$$

2. It is common to forget to check solutions to equations involving rational expressions. When we multiply both sides of an equation by a quantity containing the variable, we must be sure to check for extraneous solutions (see Section 4.6).

Chapter 4 Review Problems

Reduce to lowest terms. [4.1]

1. $\dfrac{125x^4yz^3}{35x^2y^4z^3}$

2. $\dfrac{28xy^3z^2}{14x^2y^3z}$

3. $\dfrac{a^3 - ab^2}{4a + 4b}$

4. $\dfrac{12a - 9b}{16a^2 - 9b^2}$

5. $\dfrac{x^2 - 25}{x^2 + 10x + 25}$

6. $\dfrac{x^2 - 9x + 20}{x^2 - 7x + 12}$

7. $\dfrac{ax + x - 5a - 5}{ax - x - 5a + 5}$

8. $\dfrac{x^3 + 2x^2 - 9x - 18}{x^2 - x - 6}$

9. $\dfrac{6 - x - x^2}{x^2 - 5x + 6}$

10. $\dfrac{x^2 + 5x - 6}{6 - 5x - x^2}$

Divide. If the denominator is a factor of the numerator, as in Problem 17, you may want to factor the numerator and divide out the common factor first. [4.2]

11. $\dfrac{12x^3 + 8x^2 + 16x}{4x^2}$

12. $\dfrac{10x^4 - 5x^3 + 15x^2}{5x^2}$

13. $\dfrac{27a^2b^3 - 15a^3b^2 + 21a^4b^4}{-3a^2b^2}$

14. $\dfrac{18a^4b^2 - 9a^2b^2 + 27a^2b^4}{-9a^2b^2}$

15. $\dfrac{x^{6n} - x^{5n}}{x^{3n}}$

16. $\dfrac{10x^{3n} - 15x^{4n}}{5x^n}$

17. $\dfrac{x^2 - x - 6}{x - 3}$

18. $\dfrac{x^2 - x - 6}{x + 2}$

19. $\dfrac{5x^2 - 14xy - 24y^2}{x - 4y}$

20. $\dfrac{5x^2 - 26xy - 24y^2}{5x + 4y}$

21. $\dfrac{y^4 - 16}{y - 2}$

22. $\dfrac{y^4 - 81}{y - 3}$

23. $\dfrac{8x^2 - 26x - 9}{2x - 7}$

24. $\dfrac{9x^2 + 9x - 18}{3x - 4}$

25. $\dfrac{2y^3 - 9y^2 - 17y + 39}{2y - 3}$

26. $\dfrac{3y^3 - 19y^2 + 17y + 4}{3y - 4}$

27. $\dfrac{a^4 - 2a + 5}{a - 2}$

28. $\dfrac{a^4 + a^3 - 1}{a + 2}$

Multiply and divide as indicated. [4.3]

29. $\frac{3}{4} \cdot \frac{12}{15} \div \frac{1}{3}$

30. $\frac{2}{3} \cdot \frac{9}{8} \div \frac{1}{4}$

31. $\dfrac{15x^2y}{8xy^2} \div \dfrac{10xy}{4x}$

32. $\dfrac{27x^3y^2}{13x^2y^4} \div \dfrac{9xy}{26y}$

33. $\dfrac{x^3 - 1}{x^4 - 1} \cdot \dfrac{x^2 - 1}{x^2 + x + 1}$

34. $\dfrac{x^4 - 16}{x^3 - 8} \cdot \dfrac{x^2 + 2x + 4}{x^2 + 4}$

35. $\dfrac{a^2 + 5a + 6}{a + 1} \cdot \dfrac{a + 5}{a^2 + 2a - 3} \div \dfrac{a^2 + 7a + 10}{a^2 - 1}$

36. $\dfrac{2a^2 - a - 1}{a^2 - 2a - 15} \cdot \dfrac{a^2 - 9}{a - 1} \div \dfrac{4a^2 - 1}{a - 5}$

37. $\dfrac{ax + bx + 2a + 2b}{ax - 3a + bx - 3b} \div \dfrac{ax - bx - 2a + 2b}{ax - bx - 3a + 3b}$

38. $\dfrac{xy + 2x + y + 2}{xy - 3y + 2x - 6} \cdot \dfrac{xy - 3y + 4x - 12}{xy - y + 4x - 4}$

39. $(4x^2 - 9) \cdot \dfrac{x + 3}{2x + 3}$

40. $(9x^2 - 25) \cdot \dfrac{x + 5}{3x - 5}$

41. $\dfrac{2x^3 + 3x^2 - 18x - 27}{10x^2 + 13x - 3} \cdot \dfrac{25x^2 - 1}{5x^2 - 14x - 3}$

42. $\dfrac{3x^3 + 2x^2 - 12x - 8}{12x^2 + 5x - 2} \cdot \dfrac{16x^2 - 1}{4x^2 + 9x + 2}$

Add and subtract as indicated. [4.4]

43. $\frac{3}{5} - \frac{1}{10} + \frac{8}{15}$

44. $\frac{3}{4} - \frac{1}{8} + \frac{3}{2}$

45. $\dfrac{5}{x - 5} - \dfrac{x}{x - 5}$

46. $\dfrac{y}{x^2 - y^2} - \dfrac{x}{x^2 - y^2}$

47. $\dfrac{1}{x} + \dfrac{1}{x^2} + \dfrac{1}{x^3}$

48. $\dfrac{1}{x} - \dfrac{1}{x^2} - \dfrac{1}{x^3}$

49. $\dfrac{8}{y^2 - 16} - \dfrac{7}{y^2 - y - 12}$

REVIEW FOR CHAPTER 4

50. $\dfrac{6}{y^2 - 9} - \dfrac{5}{y^2 - y - 6}$

51. $\dfrac{x - 2}{x^2 + 5x + 4} - \dfrac{x - 4}{2x^2 + 12x + 16}$

52. $\dfrac{x - 1}{x^2 + 4x + 3} - \dfrac{x - 1}{2x^2 + 10x + 12}$

53. $3 + \dfrac{4}{5x - 2}$ **54.** $7 - \dfrac{2}{7x - 3}$

55. $\dfrac{-4}{x^2 + 5x + 6} - \dfrac{5}{x^2 + 7x + 12} + \dfrac{10}{x^2 + 6x + 8}$

56. $\dfrac{3}{x^2 + 8x + 15} - \dfrac{1}{x^2 + 7x + 12} - \dfrac{1}{x^2 + 9x + 20}$

57. $\dfrac{5}{3x^2 + x - 2} - \dfrac{1}{3x^2 + 5x + 2}$

58. $\dfrac{7}{4x^2 - x - 3} - \dfrac{1}{4x^2 - 7x + 3}$

Simplify each complex fraction. [4.5]

59. $\dfrac{1 + \frac{2}{3}}{1 - \frac{2}{3}}$ **60.** $\dfrac{1 - \frac{3}{4}}{1 + \frac{3}{4}}$

61. $\dfrac{\frac{4a}{2a^3 + 2}}{\frac{8a}{4a + 4}}$ **62.** $\dfrac{\frac{2a}{3a^3 - 3}}{\frac{4a}{6a - 6}}$

63. $1 + \dfrac{1}{x + \dfrac{1}{x}}$ **64.** $1 + \dfrac{x}{1 + \dfrac{1}{x}}$

65. $\dfrac{1 - \dfrac{9}{x^2}}{1 - \dfrac{1}{x} - \dfrac{6}{x^2}}$ **66.** $\dfrac{4 - \dfrac{1}{x^2}}{4 + \dfrac{4}{x} + \dfrac{1}{x^2}}$

Solve each equation. [4.6]

67. $\dfrac{3}{x - 1} = \dfrac{3}{5}$ **68.** $\dfrac{2}{x + 1} = \dfrac{4}{5}$

69. $\dfrac{x + 1}{3} + \dfrac{x - 3}{4} = \dfrac{1}{6}$

70. $\dfrac{x + 2}{3} + \dfrac{x - 5}{5} = -\dfrac{3}{5}$

71. $\dfrac{5}{y + 1} = \dfrac{4}{y + 2}$ **72.** $\dfrac{3}{y - 2} = \dfrac{2}{y - 3}$

73. $\dfrac{x + 6}{x + 3} - 2 = \dfrac{3}{x + 3}$

74. $\dfrac{x + 2}{x + 1} - 2 = \dfrac{1}{x + 1}$

75. $\dfrac{4}{x^2 - x - 12} + \dfrac{1}{x^2 - 9} = \dfrac{2}{x^2 - 7x + 12}$

76. $\dfrac{1}{x^2 + 3x - 4} + \dfrac{3}{x^2 - 1} = \dfrac{-1}{x^2 + 5x + 4}$

77. $\dfrac{a + 4}{a^2 + 5a} = \dfrac{-2}{a^2 - 25}$

78. $\dfrac{a}{2} + \dfrac{3}{a - 3} = \dfrac{a}{a - 3}$

79. $\dfrac{3x}{x - 5} - \dfrac{2x}{x + 1} = \dfrac{-42}{x^2 - 4x - 5}$

80. $\dfrac{2x}{x + 2} = \dfrac{x}{x + 3} - \dfrac{3}{x^2 + 5x + 6}$

81. $1 - \dfrac{1}{x} = \dfrac{6}{x^2}$ **82.** $2 - \dfrac{11}{x} = \dfrac{-12}{x^2}$

Solve each application. [4.7]

83. One number is twice another. The sum of their reciprocals is $\frac{1}{2}$. Find the two numbers.

84. One number is 4 times another. The sum of their reciprocals is $\frac{5}{8}$. Find the numbers.

85. A bathtub can be filled by the cold-water faucet in 10 minutes and by the hot-water faucet in 12 minutes. How long does it take to fill the tub if both faucets are left open?

86. A water faucet can fill a sink in 6 minutes, while the drain can empty it in 4 minutes. If the sink is full, how long will it take to empty if both the water faucet and the drain are open?

87. The sum of a number and twice its reciprocal is $\frac{9}{2}$. Find the number.

88. The sum of a number and three times its reciprocal is 4. Find the number.

89. A boat takes 1.5 hours to travel 6 miles downstream and the same distance back. If the boat travels at 9 miles/hour, what is the speed of the current?

90. A car makes a 120 mile trip 10 miles/hour faster than a truck. The truck takes 2 hours longer to make the trip. What are the speeds of the car and the truck?

91. A jogger covers 3.5 miles in 28 minutes. Find the average speed of the jogger in miles per hour.

92. The speed of sound is 1,088 feet per second. Convert the speed of sound to miles per hour. Round your answer to the nearest whole number.

Graph each rational function. [4.6, 4.7]

93. $f(x) = \dfrac{6}{x + 2}$

94. $g(x) = \dfrac{x - 3}{x + 2}$

Chapter 4 Test

Reduce to lowest terms. [4.1]

1. $\dfrac{x^2 - y^2}{x - y}$

2. $\dfrac{2x^2 - 5x + 3}{2x^2 - x - 3}$

Divide. [4.2]

3. $\dfrac{24x^3 y + 12x^2 y^2 - 16xy^3}{4xy}$

4. $\dfrac{2x^3 - 9x^2 + 10}{2x - 1}$

Multiply and divide as indicated. [4.3]

5. $\dfrac{a^2 - 16}{5a - 15} \cdot \dfrac{10(a - 3)^2}{a^2 - 7a + 12}$

6. $\dfrac{a^4 - 81}{a^2 + 9} \div \dfrac{a^2 - 8a + 15}{4a - 20}$

7. $\dfrac{x^3 - 8}{2x^2 - 9x + 10} \div \dfrac{x^2 + 2x + 4}{2x^2 + x - 15}$

Add and subtract as indicated. [4.4]

8. $\frac{4}{21} + \frac{6}{35}$

9. $\frac{3}{4} - \frac{1}{2} + \frac{5}{8}$

10. $\dfrac{a}{a^2 - 9} + \dfrac{3}{a^2 - 9}$

11. $\dfrac{1}{x} + \dfrac{2}{x - 3}$

12. $\dfrac{4x}{x^2 + 6x + 5} - \dfrac{3x}{x^2 + 5x + 4}$

13. $\dfrac{2x + 8}{x^2 + 4x + 3} - \dfrac{x + 4}{x^2 + 5x + 6}$

Simplify each complex fraction. [4.5]

14. $\dfrac{3 - \dfrac{1}{a + 3}}{3 + \dfrac{1}{a + 3}}$

15. $\dfrac{1 - \dfrac{9}{x^2}}{1 + \dfrac{1}{x} - \dfrac{6}{x^2}}$

Solve each of the following equations. [4.6]

16. $\dfrac{1}{x} + 3 = \dfrac{4}{3}$

17. $\dfrac{x}{x - 3} + 3 = \dfrac{3}{x - 3}$

18. $\dfrac{y + 3}{2y} + \dfrac{5}{y - 1} = \dfrac{1}{2}$

19. $1 - \dfrac{1}{x} = \dfrac{6}{x^2}$

Solve the following applications. Be sure to show the equation in each case. [4.7]

20. What number must be subtracted from the denominator of $\frac{10}{23}$ to make the result $\frac{1}{3}$?

21. The current of a river is 2 miles/hour. It takes a motorboat a total of 3 hours to travel 8 miles upstream and return 8 miles downstream. What is the speed of the boat in still water?

22. An inlet pipe can fill a pool in 10 hours while an outlet pipe can empty it in 15 hours. If the pool is half full and both pipes are left open, how long will it take to fill the pool the rest of the way?

23. The top of Mount Whitney, the highest point in California, is 14,494 feet above sea level. Give this height in miles to the nearest tenth of a mile.

24. A bullet fired from a gun travels a distance of 4,750 feet in 3.2 seconds. Find the average speed of the bullet in miles per hour. Round to the nearest whole number.

25. Graph $f(x) = \dfrac{x + 4}{x - 1}$.

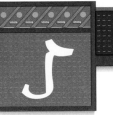

Rational Exponents and Roots

Introduction

Figure 1 shows a square in which each of the four sides is 1 inch long. To find the square of the length of the diagonal c we apply the Pythagorean Theorem:

$$c^2 = 1^2 + 1^2$$
$$c^2 = 2$$

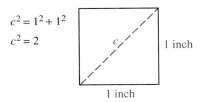

FIGURE 1

Because we know that c is positive and that its square is 2, we call c the *positive square root* of 2, and we write $c = \sqrt{2}$. The ability to associate numbers, such as $\sqrt{2}$, with the diagonal of a square (or a rectangle) allows us to analyze some interesting problems from geometry. Two particularly interesting geometric objects that we will study in this chapter are shown in Figures 2 and 3. Both are constructed from right triangles, and the numbers associated with the diagonals in each are found from the Pythagorean Theorem.

The Golden Rectangle

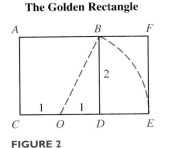

FIGURE 2

The Spiral of Roots

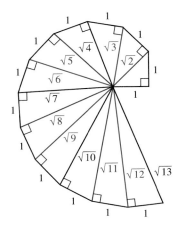

FIGURE 3

$\mathcal{O}$verview

The work we have done previously with exponents and polynomials will be very useful in understanding the concepts presented in this chapter. The square roots, cube roots, and higher roots we will study take us in the reverse direction from exponents. Although roots and expressions involving roots are new to you, you will find that they behave in much the same way as the polynomials we have dealt with in the past. That is, the rules we have developed for working with polynomials can be applied to these new expressions, so that finding the product $(\sqrt{5} + 3)(\sqrt{5} - 2)$ is very similar to finding the product $(x + 3)(x - 2)$.

Also in this chapter is our introduction to a larger set of numbers, the *complex numbers*. These numbers include the real numbers as well as numbers we have not worked with yet. Complex numbers also behave like polynomials in many situations, so the rules we have for polynomials can be applied to complex numbers as well.

If you intend to take more classes in mathematics, and you want to ensure your success in those classes, then you can work toward this goal: *Become the type of student who can learn mathematics on your own.* Most people who have degrees in mathematics were students who could learn mathematics outside the classroom. This doesn't mean that you have to learn it all on your own; it simply means that if you have to, you can do it. Attaining this goal gives you independence, and puts you in control of your success in any math class you take.

$\mathcal{S}$tudy $\mathcal{S}$kills

The study skills for this chapter are about how you approach new situations in mathematics. The first study skill is a reminder about your point of view toward your natural instincts for what does and doesn't work in mathematics. The second study skill gives you a way of testing your instincts.

1. **Remember: Be careful about trusting your intuition** Now that you have more experience in mathematics, you know that your intuition can get you into trouble if you are not careful. For example, I always have some students who think that $(x + 5)^2$ is the same as $x^2 + 25$, which is incorrect. When they get to this chapter, these same students will simplify $\sqrt{x^2 + 25}$ by writing it as $x + 5$, which is also incorrect. In both cases, their intuition is getting them into trouble.

2. **Test properties you are unsure of** From time to time you will be in a situation where you would like to apply a property or rule, but you are not sure it is true. You can always test a property or statement by substituting numbers for variables. For instance, to avoid the mistake of writing $(x + 5)^2$ as $x^2 + 25$,

substitute a number for x in both expressions and see if the results are the same:

When $x = 10$, the expression $(x + 5)^2$ becomes $(10 + 5)^2 = 15^2 = 225$.

When $x = 10$, the expression $x^2 + 25$ becomes $10^2 + 25 = 100 + 25 = 125$.

When you test the equivalence of expressions by substituting numbers for the variable, make it easy on yourself by choosing numbers that are easy to work with, such as 10. Don't try to verify the equivalence of expressions by substituting 0, 1, or 2 for the variable, since these numbers will occasionally give you false results.

It is not good practice to trust your intuition or instincts in every new situation in algebra. If you have any doubt about the generalizations you are making, test them by replacing variables with numbers and simplifying.

Rational Exponents

In Chapter 1, we developed notation (exponents) to give us the square, cube, or any power of a number. For instance, if we wanted the square of 3, we wrote $3^2 = 9$. If we wanted the cube of 3, we wrote $3^3 = 27$. In this section, we will develop notation that will take us in the reverse direction, that is, from the square of a number, say 25, back to the original number, 5.

DEFINITION

If x is a nonnegative real number, then the expression $\sqrt{x}$ is called the **positive square root** of x, and we say that

$$(\sqrt{x})^2 = x$$

In Words: $\sqrt{x}$ is the positive number we square to get x.

The negative square root of x, $-\sqrt{x}$, is defined in a similar manner.

EXAMPLE 1 The positive square root of 64 is 8 because 8 is the positive number with the property $8^2 = 64$. The negative square root of 64 is -8 since -8 is the negative number whose square is 64. We can summarize these facts by saying

$$\sqrt{64} = 8 \qquad \text{and} \qquad -\sqrt{64} = -8$$

Note: It is a common mistake to assume that an expression like $\sqrt{25}$ indicates both square roots, 5 and -5. The expression $\sqrt{25}$ indicates only the positive square root of 25, which is 5. If we want the negative square root, we must use a negative sign: $-\sqrt{25} = -5$.

The higher roots—cube roots, fourth roots, and so on—are defined in a similar way, as outlined at the top of the next page.

DEFINITION

If x is a real number and n is a positive integer, then:

Positive square root of x, $\sqrt{x}$, is such that $(\sqrt{x})^2 = x$ $x \geq 0$

Cube root of x, $\sqrt[3]{x}$, is such that $(\sqrt[3]{x})^3 = x$

Positive fourth root of x, $\sqrt[4]{x}$, is such that $(\sqrt[4]{x})^4 = x$ $x \geq 0$

Fifth root of x, $\sqrt[5]{x}$, is such that $(\sqrt[5]{x})^5 = x$

$\vdots$ $\vdots$ $\vdots$

The **nth root of x**, $\sqrt[n]{x}$, is such that $(\sqrt[n]{x})^n = x$ $x \geq 0$

                                                          if n is even

---

Note: We have restricted the even roots in this definition to nonnegative numbers. Even roots of negative numbers exist, but they are not represented by real numbers. That is, $\sqrt{-4}$ is not a real number since there is no real number whose square is -4.

Here is a table of the most common roots used in this book. Any of the roots that are unfamiliar should be memorized.

Square Roots		Cube Roots	Fourth Roots
$\sqrt{0} = 0$	$\sqrt{49} = 7$	$\sqrt[3]{0} = 0$	$\sqrt[4]{0} = 0$
$\sqrt{1} = 1$	$\sqrt{64} = 8$	$\sqrt[3]{1} = 1$	$\sqrt[4]{1} = 1$
$\sqrt{4} = 2$	$\sqrt{81} = 9$	$\sqrt[3]{8} = 2$	$\sqrt[4]{16} = 2$
$\sqrt{9} = 3$	$\sqrt{100} = 10$	$\sqrt[3]{27} = 3$	$\sqrt[4]{81} = 3$
$\sqrt{16} = 4$	$\sqrt{121} = 11$	$\sqrt[3]{64} = 4$	
$\sqrt{25} = 5$	$\sqrt{144} = 12$	$\sqrt[3]{125} = 5$	
$\sqrt{36} = 6$	$\sqrt{169} = 13$		

Notation: An expression like $\sqrt[3]{8}$ that involves a root is called a **radical expression.** In the expression $\sqrt[3]{8}$, the ^{3} is called the **index,** the $\sqrt{}$ is the **radical sign,** and 8 is called the **radicand.** The index of a radical must be a positive integer greater than 1. If no index is written, it is assumed to be 2.

Roots and Negative Numbers

When dealing with negative numbers and radicals, the only restriction concerns negative numbers under even roots. We can have negative signs in front of radicals and negative numbers under odd roots and still obtain real numbers. Here are some examples to help clarify this. In the last section of this chapter we will see how to deal with even roots of negative numbers.

EXAMPLES   Simplify each expression, if possible.

2. $\sqrt[3]{-8} = -2$ because $(-2)^3 = -8$

3. $\sqrt{-4}$ is not a real number since there is no real number whose square is -4

4. $-\sqrt{25} = -5$ is the negative square root of 25

5. $\sqrt[5]{-32} = -2$ because $(-2)^5 = -32$

6. $\sqrt[4]{-81}$ is not a real number since there is no real number we can raise to the fourth power to obtain -81

Variables under a Radical

From the preceding examples it is clear that we must be careful that we do not try to take an even root of a negative number. For this reason, we will assume that all variables appearing under a radical sign represent nonnegative numbers.

EXAMPLES   Assume all variables represent nonnegative numbers and simplify each expression as much as possible.

7. $\sqrt{25a^4b^6} = 5a^2b^3$ because $(5a^2b^3)^2 = 25a^4b^6$

8. $\sqrt[3]{x^6y^{12}} = x^2y^4$ because $(x^2y^4)^3 = x^6y^{12}$

9. $\sqrt[4]{81r^8s^{20}} = 3r^2s^5$ because $(3r^2s^5)^4 = 81r^8s^{20}$

Rational Numbers as Exponents

We will now develop a second kind of notation involving exponents that will allow us to designate square roots, cube roots, and so on in another way.

Consider the equation $x = 8^{1/3}$. Although we have not encountered fractional exponents before, let's assume that all the properties of exponents hold in this case. Cubing both sides of the equation, we have

$$x^3 = (8^{1/3})^3$$
$$x^3 = 8^{(1/3)(3)}$$
$$x^3 = 8^1$$
$$x^3 = 8$$

The last line tells us that x is the number whose cube is 8. It must be true, then, that x is the cube root of 8, $x = \sqrt[3]{8}$. Since we started with $x = 8^{1/3}$, it follows that

$$8^{1/3} = \sqrt[3]{8}$$

It seems reasonable, then, to define fractional exponents as indicating roots. The formal definition is given at the top of the next page.

Section 5.1   Rational Exponents

> ### DEFINITION
>
> If x is a real number and n is a positive integer greater than 1, then
>
> $$x^{1/n} = \sqrt[n]{x} \qquad x \geq 0 \text{ when } n \text{ is even}$$
>
> **In Words:** The quantity $x^{1/n}$ is the nth root of x.

With this definition we have a way of representing roots with exponents. Here are some examples.

EXAMPLES Write each expression as a root and then simplify, if possible.

10. $8^{1/3} = \sqrt[3]{8} = 2$

11. $36^{1/2} = \sqrt{36} = 6$

12. $-25^{1/2} = -\sqrt{25} = -5$

13. $(-25)^{1/2} = \sqrt{-25}$, which is not a real number

14. $\left(\frac{4}{9}\right)^{1/2} = \sqrt{\frac{4}{9}} = \frac{2}{3}$

Note: If we were using a scientific calculator to work Example 10, this is how the sequence of key strokes would look:

$$8 \;\boxed{y^x}\; 3 \;\boxed{1/x}\; \boxed{=} \qquad \text{or} \qquad 8 \;\boxed{y^x}\; \boxed{(}\; 1 \;\boxed{\div}\; 3 \;\boxed{)}\; \boxed{=}$$

The properties of exponents developed in Chapter 1 were applied to integer exponents only. We will now extend these properties to include rational exponents also. We do so without proof.

> ### PROPERTIES OF EXPONENTS
>
> If a and b are real numbers, r and s are rational numbers, and a and b are nonnegative whenever r and s indicate even roots, then
>
> **1.** $a^r \cdot a^s = a^{r+s}$ $\qquad$ **4.** $a^{-r} = \dfrac{1}{a^r} \quad (a \neq 0)$
>
> **2.** $(a^r)^s = a^{rs}$ $\qquad$ **5.** $\left(\dfrac{a}{b}\right)^r = \dfrac{a^r}{b^r} \quad (b \neq 0)$
>
> **3.** $(ab)^r = a^r b^r$ $\qquad$ **6.** $\dfrac{a^r}{a^s} = a^{r-s} \quad (a \neq 0)$

There are times when rational exponents can simplify our work with radicals. Here are Examples 8 and 9 again, but this time we will work them using rational exponents.

EXAMPLES Write each radical with a rational exponent and then simplify.

15. $\sqrt[3]{x^6y^{12}} = (x^6y^{12})^{1/3}$
$= (x^6)^{1/3}(y^{12})^{1/3}$
$= x^2y^4$

16. $\sqrt[4]{81r^8s^{20}} = (81r^8s^{20})^{1/4}$
$= 81^{1/4}(r^8)^{1/4}(s^{20})^{1/4}$
$= 3r^2s^5$

So far, the numerators of all the rational exponents we have encountered have been 1. The next theorem extends the work we can do with rational exponents to rational exponents with numerators other than 1.

THEOREM 5.1

If a is a nonnegative real number, m is an integer, and n is a positive integer, then

$$a^{m/n} = (a^{1/n})^m = (a^m)^{1/n}$$

Proof
We can prove Theorem 5.1 using the properties of exponents. Since $\frac{m}{n} = m(\frac{1}{n})$, we have

$a^{m/n} = a^{m(1/n)}$ $a^{m/n} = a^{(1/n)(m)}$
$= (a^m)^{1/n}$ $= (a^{1/n})^m$

Here are some examples that illustrate how we use this theorem.

EXAMPLES Simplify as much as possible.

17. $8^{2/3} = (8^{1/3})^2$ Theorem 5.1
$= 2^2$ Definition of fractional exponents
$= 4$ The square of 2 is 4.

Note: On a scientific calculator, Example 17 would look like this:

$8 \boxed{y^x} \boxed{(} 2 \boxed{\div} 3 \boxed{)} \boxed{=}$

18. $25^{3/2} = (25^{1/2})^3$ Theorem 5.1
$= 5^3$ Definition of fractional exponents
$= 125$ The cube of 5 is 125.

19. $9^{-3/2} = (9^{1/2})^{-3}$ Theorem 5.1
$= 3^{-3}$ Definition of fractional exponents
$= \frac{1}{3^3}$ Property 4 for exponents
$= \frac{1}{27}$ The cube of 3 is 27.

20. $\left(\dfrac{27}{8}\right)^{-4/3} = \left[\left(\dfrac{27}{8}\right)^{1/3}\right]^{-4}$ Theorem 5.1

$\qquad = \left(\dfrac{3}{2}\right)^{-4}$ Definition of fractional exponents

$\qquad = \left(\dfrac{2}{3}\right)^{4}$ Property 4 for exponents.

$\qquad = \dfrac{16}{81}$ $\left(\dfrac{2}{3}\right)^4 = \dfrac{16}{81}$

The following examples show the application of the properties of exponents to rational exponents.

EXAMPLES Assume all variables represent positive quantities and simplify the answers as much as possible.

21. $x^{1/3} \cdot x^{5/6} = x^{1/3+5/6}$ Property 1

$\qquad = x^{2/6+5/6}$ LCD is 6.

$\qquad = x^{7/6}$ Add fractions.

22. $(y^{2/3})^{3/4} = y^{(2/3)(3/4)}$ Property 2

$\qquad = y^{1/2}$ Multiply fractions: $\frac{2}{3} \cdot \frac{3}{4} = \frac{6}{12} = \frac{1}{2}$.

23. $\dfrac{z^{1/3}}{z^{1/4}} = z^{1/3-1/4}$ Property 6

$\qquad = z^{4/12-3/12}$ LCD is 12.

$\qquad = z^{1/12}$ Subtract fractions.

24. $\left(\dfrac{a^{-1/3}}{b^{1/2}}\right)^6 = \dfrac{(a^{-1/3})^6}{(b^{1/2})^6}$ Property 5

$\qquad = \dfrac{a^{-2}}{b^3}$ Property 2

$\qquad = \dfrac{1}{a^2b^3}$ Property 4

25. $\dfrac{(x^{-3}y^{1/2})^4}{x^{10}y^{3/2}} = \dfrac{(x^{-3})^4(y^{1/2})^4}{x^{10}y^{3/2}}$ Property 3

$\qquad = \dfrac{x^{-12}y^2}{x^{10}y^{3/2}}$ Property 2

$\qquad = x^{-22}y^{1/2}$ Property 6

$\qquad = \dfrac{y^{1/2}}{x^{22}}$ Property 4

Facts from Geometry: The Pythagorean Theorem (Again) and the Golden Rectangle

Now that we have had some experience working with square roots, we can rewrite the Pythagorean Theorem using a square root. If triangle ABC is a right triangle with $C = 90°$, then the length of the longest side is the *positive square root* of the sum of the squares of the other two sides, as shown in Figure 1.

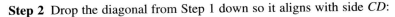

$$c = \sqrt{a^2 + b^2}$$

FIGURE 1

In the introduction to this chapter we mentioned the **golden rectangle.** Its origins can be traced back over 2,000 years to the Greek civilization that produced Pythagoras, Socrates, Plato, Aristotle, and Euclid. The most important mathematical work to come from that Greek civilization was Euclid's *Elements,* an elegantly written summary of all that was known about geometry at that time in history. Euclid's *Elements,* according to Howard Eves (an authority on the history of mathematics), exercised a greater influence on scientific thinking than any other work. Here is how we construct a golden rectangle from a square of side 2, using the same method that Euclid used in his *Elements.*

Constructing a Golden Rectangle from a Square of Side 2

Step 1 Draw a square with a side of length 2. Connect the midpoint of side CD to corner B, as shown in the margin. (Note that we have labeled the midpoint of segment CD with the letter O.)

Step 2 Drop the diagonal from Step 1 down so it aligns with side CD:

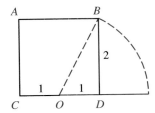

Step 3 Form rectangle $ACEF$. This is a golden rectangle.

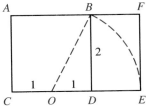

All golden rectangles are constructed from squares. Every golden rectangle, no matter how large or small it is, will have the same shape. To associate a number with the shape of the golden rectangle, we use the ratio of its length to its width. This ratio is called the **golden ratio.** To calculate the golden ratio we must first find the length of the diagonal we used to construct the golden rectangle. Figure 2 shows the golden rectangle that we constructed from a square of side 2. The length of the diagonal OB is found by applying the Pythagorean Theorem to triangle OBD.

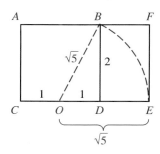

$$OB = \sqrt{1^2 + 2^2}$$
$$= \sqrt{1 + 4}$$
$$= \sqrt{5}$$

FIGURE 2

The length of segment OE is equal to the length of diagonal OB; both are $\sqrt{5}$. Since the distance from C to O is 1, the length CE of the golden rectangle is $1 + \sqrt{5}$. Now we can find the golden ratio:

$$\text{Golden ratio} = \frac{\text{Length}}{\text{Width}} = \frac{CE}{EF} = \frac{1 + \sqrt{5}}{2}$$

Using Technology: Graphing Calculators — A Word of Caution

Some graphing calculators give surprising results when evaluating expressions such as $(-8)^{2/3}$. As you know from reading this section, the expression $(-8)^{2/3}$ simplifies to 4, either by taking the cube root first and then squaring the result, or by squaring the base first and then taking the cube root of the result. Here are three different ways to evaluate this expression on your calculator:

1. $(-8)^\wedge(2/3)$ To evaluate $(-8)^{2/3}$
2. $((-8)^\wedge 2)^\wedge(1/3)$ To evaluate $((-8)^2)^{1/3}$
3. $((-8)^\wedge(1/3))^\wedge 2$ To evaluate $((-8)^{1/3})^2$

Note any differences in the results.
Next, graph each of the following functions, one at a time.

1. $Y_1 = X^{2/3}$ **2.** $Y_2 = (X^2)^{1/3}$ **3.** $Y_3 = (X^{1/3})^2$

The correct graph is shown in Figure 3. Note which of your graphs match the correct graph.

Different calculators evaluate exponential expressions in different ways. You should use the method (or methods) that gave you the correct graph.

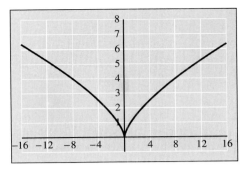

FIGURE 3

Problem Set

5.1

Unless indicated otherwise, the problems in this section are to be worked without a calculator. It is okay, however, to check your work with a calculator.

Find each of the following roots, if possible.

1. $\sqrt{144}$

2. $-\sqrt{144}$

3. $\sqrt{-144}$

4. $\sqrt{-49}$

5. $-\sqrt{49}$

6. $\sqrt{49}$

7. $\sqrt[3]{-27}$

8. $-\sqrt[3]{27}$

9. $\sqrt[4]{16}$

10. $-\sqrt[4]{16}$

11. $\sqrt[4]{-16}$

12. $-\sqrt[4]{-16}$

13. $\sqrt{0.04}$

14. $\sqrt{0.81}$

15. $\sqrt[3]{0.008}$

16. $\sqrt[3]{0.125}$

Simplify each expression. Assume all variables represent nonnegative numbers.

17. $\sqrt{36a^8}$

18. $\sqrt{49a^{10}}$

19. $\sqrt[3]{27a^{12}}$

20. $\sqrt[3]{8a^{15}}$

21. $\sqrt[3]{x^3y^6}$

22. $\sqrt[3]{x^6y^3}$

23. $\sqrt[5]{32x^{10}y^5}$

24. $\sqrt[5]{32x^5y^{10}}$

25. $\sqrt[4]{16a^{12}b^{20}}$

26. $\sqrt[4]{81a^{24}b^8}$

Use the definition of rational exponents to write each of the following with the appropriate root. Then simplify.

27. $36^{1/2}$

28. $49^{1/2}$

29. $-9^{1/2}$

30. $-16^{1/2}$

31. $8^{1/3}$

32. $-8^{1/3}$

33. $(-8)^{1/3}$

34. $-27^{1/3}$

35. $32^{1/5}$

36. $81^{1/4}$

37. $\left(\frac{81}{25}\right)^{1/2}$

38. $\left(\frac{9}{16}\right)^{1/2}$

39. $\left(\frac{64}{125}\right)^{1/3}$

40. $\left(\frac{8}{27}\right)^{1/3}$

Use Theorem 5.1 to simplify each of the following as much as possible.

41. $27^{2/3}$

42. $8^{4/3}$

43. $25^{3/2}$

44. $9^{3/2}$

45. $16^{3/4}$

46. $81^{3/4}$

Simplify each expression. Remember, negative exponents give reciprocals.

47. $27^{-1/3}$

48. $9^{-1/2}$

49. $81^{-3/4}$

50. $4^{-3/2}$

51. $\left(\frac{25}{36}\right)^{-1/2}$

52. $\left(\frac{16}{49}\right)^{-1/2}$

53. $\left(\frac{81}{16}\right)^{-3/4}$

54. $\left(\frac{27}{8}\right)^{-2/3}$

55. $16^{1/2} + 27^{1/3}$

56. $25^{1/2} + 100^{1/2}$

57. $8^{-2/3} + 4^{-1/2}$

58. $49^{-1/2} + 25^{-1/2}$

Use the properties of exponents to simplify each of the following as much as possible. Assume all bases are positive.

59. $x^{3/5} \cdot x^{1/5}$

60. $x^{3/4} \cdot x^{5/4}$

61. $(a^{3/4})^{4/3}$

62. $(a^{2/3})^{3/4}$

63. $\dfrac{x^{1/5}}{x^{3/5}}$

64. $\dfrac{x^{2/7}}{x^{5/7}}$

65. $\dfrac{x^{5/6}}{x^{2/3}}$

66. $\dfrac{x^{7/8}}{x^{8/7}}$

67. $(x^{3/5}y^{5/6}z^{1/3})^{3/5}$

68. $(x^{3/4}y^{1/8}z^{5/6})^{4/5}$

69. $\dfrac{a^{3/4}b^2}{a^{7/8}b^{1/4}}$

70. $\dfrac{a^{1/3}b^4}{a^{3/5}b^{1/3}}$

71. $\dfrac{(y^{2/3})^{3/4}}{(y^{1/3})^{3/5}}$

72. $\dfrac{(y^{5/4})^{2/5}}{(y^{1/4})^{4/3}}$

73. $\left(\dfrac{a^{-1/4}}{b^{1/2}}\right)^8$

74. $\left(\dfrac{a^{-1/5}}{b^{1/3}}\right)^{15}$

75. $\dfrac{(r^{-2}s^{1/3})^6}{r^8 s^{3/2}}$

76. $\dfrac{(r^{-5}s^{1/2})^4}{r^{12}s^{5/2}}$

77. $\dfrac{(25a^6 b^4)^{1/2}}{(8a^{-9}b^3)^{-1/3}}$

78. $\dfrac{(27a^3 b^6)^{1/3}}{(81a^8 b^{-4})^{1/4}}$

79. Show that the expression $(a^{1/2} + b^{1/2})^2$ is not equal to $a + b$ by replacing a with 9 and b with 4 in both expressions and then simplifying each.

80. Show that the statement $(a^2 + b^2)^{1/2} = a + b$ is not true, in general, by replacing a with 3 and b with 4 and then simplifying both sides.

81. You may have noticed, if you have been using a calculator to find roots, that you can find the fourth root of a number by pressing the square root key twice. Written in symbols, this fact looks like this:

$$\sqrt{\sqrt{a}} = \sqrt[4]{a} \qquad a \geq 0$$

Show that this statement is true by rewriting each side with exponents instead of radical notation and then simplifying the left side.

82. Show that the statement below is true by rewriting each side with exponents instead of radical notation and then simplifying the left side.

$$\sqrt[3]{\sqrt{a}} = \sqrt[6]{a} \qquad a \geq 0$$

Applying the Concepts

83. **Maximum Speed** The maximum speed $v(r)$ that an automobile can travel around a curve of

radius r without skidding is given by the equation

$$v(r) = \left(\dfrac{5r}{2}\right)^{1/2}$$

where $v(r)$ is in miles per hour and r is measured in feet. What is the maximum speed a car can travel around a curve with a radius of 250 feet without skidding?

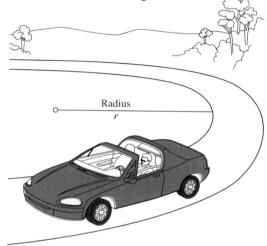

84. **Relativity** The equation

$$L = \left(1 - \dfrac{v^2}{c^2}\right)^{1/2}$$

gives the relativistic length of a 1 foot ruler traveling with velocity v. Find L if $\dfrac{v}{c} = \dfrac{3}{5}$.

85. **Golden Ratio** The golden ratio is the ratio of the length to the width in any golden rectangle. The exact value of this number is

$$\dfrac{1 + \sqrt{5}}{2}$$

Use a calculator to find a decimal approximation to this number and round it to the nearest thousandth.

86. **Golden Ratio** The reciprocal of the golden ratio is

$$\dfrac{2}{1 + \sqrt{5}}$$

Find a decimal approximation to this number that is accurate to the nearest thousandth.

Review Problems

 The problems that follow review material we covered in Sections 1.5 and 4.2. Reviewing these problems will help you understand the next section.

Multiply. [1.5]

87. $x^2(x^4 - x)$
88. $5x^2(2x^3 - x)$
89. $(x - 3)(x + 5)$
90. $(x - 2)(x + 2)$
91. $(x^2 - 5)^2$
92. $(x^2 + 5)^2$
93. $(x - 3)(x^2 + 3x + 9)$
94. $(x + 3)(x^2 - 3x + 9)$

Divide. [4.2]

95. $\dfrac{15x^2y - 20x^4y^2}{5xy}$
96. $\dfrac{12x^3y^2 - 24x^2y^3}{6xy}$

Using Technology

 If you have not read the word of caution at the end of Section 5.1, do so now.

Graph the equation $y = x^{3/4}$ and then use the Trace and Zoom features to approximate each of the following to the nearest tenth.

97. $2^{3/4}$
98. $3^{3/4}$
99. $10^{3/4}$
100. $16^{3/4}$

Graph $y = x^{3/4}$ and $y = x^{4/3}$ on the same coordinate system, using a window with X from -4 to 4 and Y from -1 to 4. Use the graphs to answer the following questions.

101. Where do the two graphs intersect?
102. For what values of x is $x^{3/4} \geq x^{4/3}$?

Carbon-14 dating is used extensively in science to find the age of fossils. If at one time a nonliving substance contains an amount A of carbon-14, then t years later the amount of carbon-14 it contains is given by the formula

$$A \cdot 2^{-t/5,600}$$

Use this formula and a calculator to solve the following problems.

103. A nonliving substance contains 3 micrograms of carbon-14. How much carbon-14 will be left at the end of:
(a) 5,000 years?   (b) 10,000 years?
(c) 56,000 years?   (d) 112,000 years?

104. A nonliving substance contains 5 micrograms of carbon-14. How much carbon-14 will be left at the end of:
(a) 500 years?   (b) 5,000 years?
(c) 56,000 years?   (d) 112,000 years?

More Expressions Involving Rational Exponents

In this section we will look at multiplication, division, factoring, and simplification of some expressions that resemble polynomials but contain rational exponents. The problems in this section will be of particular interest to you if you are planning to take either an engineering calculus class or a business calculus class. As was the case in the previous section, we will assume all variables represent nonnegative real numbers. That way, we will not have to worry about the possibility of introducing undefined terms (even roots of negative numbers) into any of our examples. Let's begin this section with a look at multiplication of expressions containing rational exponents.

EXAMPLE 1 Multiply: $x^{2/3}(x^{4/3} - x^{1/3})$.

SOLUTION Applying the distributive property and then simplifying, we have

$$
\begin{aligned}
x^{2/3}(x^{4/3} - x^{1/3}) &= x^{2/3}x^{4/3} - x^{2/3}x^{1/3} &&\text{Distributive property} \\
&= x^{6/3} - x^{3/3} &&\text{Add exponents.} \\
&= x^2 - x &&\text{Simplify.}
\end{aligned}
$$

◢

EXAMPLE 2 Multiply: $(x^{2/3} - 3)(x^{2/3} + 5)$.

> **SOLUTION** Applying the FOIL method, we multiply as if we were multiplying two binomials:

$$(x^{2/3} - 3)(x^{2/3} + 5) = x^{2/3}x^{2/3} + 5x^{2/3} - 3x^{2/3} - 15$$
$$= x^{4/3} + 2x^{2/3} - 15$$

EXAMPLE 3 Multiply: $(3a^{1/3} - 2b^{1/3})(4a^{1/3} - b^{1/3})$.

> **SOLUTION** Again, we use the FOIL method to multiply:

$$(3a^{1/3} - 2b^{1/3})(4a^{1/3} - b^{1/3}) = 3a^{1/3}4a^{1/3} - 3a^{1/3}b^{1/3} - 2b^{1/3}4a^{1/3} + 2b^{1/3}b^{1/3}$$
$$= 12a^{2/3} - 11a^{1/3}b^{1/3} + 2b^{2/3}$$

EXAMPLE 4 Expand: $(t^{1/2} - 5)^2$.

> **SOLUTION**
> *Method 1*   We can use the definition of exponents and the FOIL method:

$$(t^{1/2} - 5)^2 = (t^{1/2} - 5)(t^{1/2} - 5)$$
$$= t^{1/2}t^{1/2} - 5t^{1/2} - 5t^{1/2} + 25$$
$$= t - 10t^{1/2} + 25$$

> *Method 2*   We can obtain the same result by using the formula for the square of a binomial, $(a - b)^2 = a^2 - 2ab + b^2$:

$$(t^{1/2} - 5)^2 = (t^{1/2})^2 - 2t^{1/2} \cdot 5 + 5^2$$
$$= t - 10t^{1/2} + 25$$

EXAMPLE 5 Multiply: $(x^{3/2} - 2^{3/2})(x^{3/2} + 2^{3/2})$.

> **SOLUTION** This product has the form $(a - b)(a + b)$, which will result in the difference of two squares, $a^2 - b^2$:

$$(x^{3/2} - 2^{3/2})(x^{3/2} + 2^{3/2}) = (x^{3/2})^2 - (2^{3/2})^2$$
$$= x^3 - 2^3$$
$$= x^3 - 8$$

EXAMPLE 6 Multiply: $(a^{1/3} - b^{1/3})(a^{2/3} + a^{1/3}b^{1/3} + b^{2/3})$.

> **SOLUTION** We can find this product by multiplying in columns:

$$
\begin{array}{r}
a^{2/3} + a^{1/3}b^{1/3} + b^{2/3} \\
a^{1/3} - b^{1/3} \\
\hline
a \quad + a^{2/3}b^{1/3} + a^{1/3}b^{2/3} \\
- a^{2/3}b^{1/3} - a^{1/3}b^{2/3} - b \\
\hline
a \qquad\qquad\qquad\qquad - b
\end{array}
$$

The product is $a - b$.

Our next example involves division with expressions that contain rational exponents. As you will see, this kind of division is very similar to division of a polynomial by a monomial (as shown previously in Section 4.2).

EXAMPLE 7 Divide: $\dfrac{15x^{2/3}y^{1/3} - 20x^{4/3}y^{2/3}}{5x^{1/3}y^{1/3}}$.

SOLUTION We can approach this problem in the same way we approached division by a monomial. We simply divide each term in the numerator by the term in the denominator:

$$\frac{15x^{2/3}y^{1/3} - 20x^{4/3}y^{2/3}}{5x^{1/3}y^{1/3}} = \frac{15x^{2/3}y^{1/3}}{5x^{1/3}y^{1/3}} - \frac{20x^{4/3}y^{2/3}}{5x^{1/3}y^{1/3}}$$
$$= 3x^{1/3} - 4xy^{1/3}$$

The next three examples involve factoring. In the first example, we are told what to factor from each term of an expression.

EXAMPLE 8 Factor $3(x - 2)^{1/3}$ from $12(x - 2)^{4/3} - 9(x - 2)^{1/3}$ and then simplify, if possible.

SOLUTION This solution is similar to factoring out the greatest common factor:

$$12(x - 2)^{4/3} - 9(x - 2)^{1/3} = 3(x - 2)^{1/3}[4(x - 2) - 3]$$
$$= 3(x - 2)^{1/3}(4x - 11)$$

Although an expression containing rational exponents is not a polynomial—remember, a polynomial must have exponents that are whole numbers—we are going to treat the expressions that follow as if they were polynomials.

EXAMPLE 9 Factor $x^{2/3} - 3x^{1/3} - 10$ as if it were a trinomial.

SOLUTION We can think of $x^{2/3} - 3x^{1/3} - 10$ as a trinomial in which the variable is $x^{1/3}$. To see this, replace $x^{1/3}$ with y to get

$$y^2 - 3y - 10$$

Since this trinomial in y factors as $(y - 5)(y + 2)$, we can factor the original expression similarly:

$$x^{2/3} - 3x^{1/3} - 10 = (x^{1/3} - 5)(x^{1/3} + 2)$$

Remember, with factoring, we can always multiply the factors to check that we have factored correctly.

EXAMPLE 10 Factor $6x^{2/5} + 11x^{1/5} - 10$ as if it were a trinomial.

SOLUTION We can think of this expression as a trinomial in $x^{1/5}$:

$$6x^{2/5} + 11x^{1/5} - 10 = (3x^{1/5} - 2)(2x^{1/5} + 5)$$

In our next example, we combine two expressions by applying the methods we used to add and subtract fractions or rational expressions in Chapter 4.

EXAMPLE 11 Subtract: $(x^2 + 4)^{1/2} - \dfrac{x^2}{(x^2 + 4)^{1/2}}$.

SOLUTION In order to combine these two expressions, we need to find a least common denominator, change to equivalent fractions, and subtract numerators. The least common denominator is $(x^2 + 4)^{1/2}$.

$$(x^2 + 4)^{1/2} - \frac{x^2}{(x^2 + 4)^{1/2}} = \frac{(x^2 + 4)^{1/2}}{1} \cdot \frac{(x^2 + 4)^{1/2}}{(x^2 + 4)^{1/2}} - \frac{x^2}{(x^2 + 4)^{1/2}}$$

$$= \frac{x^2 + 4 - x^2}{(x^2 + 4)^{1/2}}$$

$$= \frac{4}{(x^2 + 4)^{1/2}}$$

EXAMPLE 12 If you purchase an investment for P dollars and t years later it is worth A dollars, then the annual rate of return, r, on that investment is given by the formula

$$r = \left(\frac{A}{P}\right)^{1/t} - 1$$

Find the annual rate of return on a coin collection that was purchased for $500 and sold 4 years later for $800.

SOLUTION Using $A = 800$, $P = 500$, and $t = 4$ in the formula, we have

$$r = \left(\frac{800}{500}\right)^{1/4} - 1$$

The easiest way to simplify this expression is by using a scientific calculator. If we first change $\frac{1}{4}$ to 0.25 and $\frac{800}{500}$ to 1.6, then pressing the keys in the following sequence will give us an approximation to $(\frac{800}{500})^{1/4}$:

$$1.6 \boxed{y^x} .25 \boxed{=}$$

To three decimal places, the result is 1.125. Using this result, we can complete the problem:

$$r = (1.6)^{0.25} - 1$$
$$= 1.125 - 1$$
$$= 0.125 \quad \text{or } 12.5\%$$

The annual return on the coin collection is approximately 12.5%. To do as well with a savings account, we would have to invest the original $500 in an account that paid 12.5%, compounded annually.

Problem Set

5.2

Multiply. (Assume all variables in this problem set represent nonnegative real numbers.)

1. $x^{2/3}(x^{1/3} + x^{4/3})$ **2.** $x^{2/5}(x^{3/5} - x^{8/5})$
3. $a^{1/2}(a^{3/2} - a^{1/2})$ **4.** $a^{1/4}(a^{3/4} + a^{7/4})$
5. $2x^{1/3}(3x^{8/3} - 4x^{5/3} + 5x^{2/3})$
6. $5x^{1/2}(4x^{5/2} + 3x^{3/2} + 2x^{1/2})$
7. $4x^{1/2}y^{3/5}(3x^{3/2}y^{-3/5} - 9x^{-1/2}y^{7/5})$
8. $3x^{4/5}y^{1/3}(4x^{6/5}y^{-1/3} - 12x^{-4/5}y^{5/3})$
9. $(x^{2/3} - 4)(x^{2/3} + 2)$
10. $(x^{2/3} - 5)(x^{2/3} + 2)$
11. $(a^{1/2} - 3)(a^{1/2} - 7)$
12. $(a^{1/2} - 6)(a^{1/2} - 2)$
13. $(4y^{1/3} - 3)(5y^{1/3} + 2)$
14. $(5y^{1/3} - 2)(4y^{1/3} + 3)$
15. $(5x^{2/3} + 3y^{1/2})(2x^{2/3} + 3y^{1/2})$
16. $(4x^{2/3} - 2y^{1/2})(5x^{2/3} - 3y^{1/2})$
17. $(t^{1/2} + 5)^2$ **18.** $(t^{1/2} - 3)^2$
19. $(x^{3/2} + 4)^2$ **20.** $(x^{3/2} - 6)^2$
21. $(a^{1/2} - b^{1/2})^2$ **22.** $(a^{1/2} + b^{1/2})^2$
23. $(2x^{1/2} - 3y^{1/2})^2$ **24.** $(5x^{1/2} + 4y^{1/2})^2$
25. $(a^{1/2} - 3^{1/2})(a^{1/2} + 3^{1/2})$
26. $(a^{1/2} - 5^{1/2})(a^{1/2} + 5^{1/2})$
27. $(x^{3/2} + y^{3/2})(x^{3/2} - y^{3/2})$
28. $(x^{5/2} + y^{5/2})(x^{5/2} - y^{5/2})$
29. $(t^{1/2} - 2^{3/2})(t^{1/2} + 2^{3/2})$
30. $(t^{1/2} - 5^{3/2})(t^{1/2} + 5^{3/2})$
31. $(2x^{3/2} + 3^{1/2})(2x^{3/2} - 3^{1/2})$
32. $(3x^{1/2} + 2^{3/2})(3x^{1/2} - 2^{3/2})$
33. $(x^{1/3} + y^{1/3})(x^{2/3} - x^{1/3}y^{1/3} + y^{2/3})$
34. $(x^{1/3} - y^{1/3})(x^{2/3} + x^{1/3}y^{1/3} + y^{2/3})$
35. $(a^{1/3} - 2)(a^{2/3} + 2a^{1/3} + 4)$
36. $(a^{1/3} + 3)(a^{2/3} - 3a^{1/3} + 9)$
37. $(2x^{1/3} + 1)(4x^{2/3} - 2x^{1/3} + 1)$
38. $(3x^{1/3} - 1)(9x^{2/3} + 3x^{1/3} + 1)$
39. $(t^{1/4} - 1)(t^{1/4} + 1)(t^{1/2} + 1)$
40. $(t^{1/4} - 2)(t^{1/4} + 2)(t^{1/2} + 4)$

Divide.

41. $\dfrac{18x^{3/4} + 27x^{1/4}}{9x^{1/4}}$ **42.** $\dfrac{25x^{1/4} + 30x^{3/4}}{5x^{1/4}}$

43. $\dfrac{12x^{2/3}y^{1/3} - 16x^{1/3}y^{2/3}}{4x^{1/3}y^{1/3}}$

44. $\dfrac{12x^{4/3}y^{1/3} - 18x^{1/3}y^{4/3}}{6x^{1/3}y^{1/3}}$

45. $\dfrac{21a^{7/5}b^{3/5} - 14a^{2/5}b^{8/5}}{7a^{2/5}b^{3/5}}$

46. $\dfrac{24a^{9/5}b^{3/5} - 16a^{4/5}b^{8/5}}{8a^{4/5}b^{3/5}}$

47. Factor $3(x - 2)^{1/2}$ from

$$12(x - 2)^{3/2} - 9(x - 2)^{1/2}$$

48. Factor $4(x + 1)^{1/3}$ from

$$4(x + 1)^{4/3} + 8(x + 1)^{1/3}$$

49. Factor $5(x - 3)^{7/5}$ from

$$5(x - 3)^{12/5} - 15(x - 3)^{7/5}$$

50. Factor $6(x + 3)^{8/7}$ from

$$6(x + 3)^{15/7} - 12(x + 3)^{8/7}$$

51. Factor $3(x + 1)^{1/2}$ from

$$9x(x + 1)^{3/2} + 6(x + 1)^{1/2}$$

52. Factor $4x(x + 1)^{1/2}$ from

$$4x^2(x + 1)^{1/2} + 8x(x + 1)^{3/2}$$

Factor each of the following as if it were a trinomial.

53. $x^{2/3} - 5x^{1/3} + 6$ **54.** $x^{2/3} - x^{1/3} - 6$
55. $a^{2/5} - 2a^{1/5} - 8$ **56.** $a^{2/5} + 2a^{1/5} - 8$
57. $2y^{2/3} - 5y^{1/3} - 3$ **58.** $3y^{2/3} + 5y^{1/3} - 2$
59. $9t^{2/5} - 25$ **60.** $16t^{2/5} - 49$
61. $4x^{2/7} + 20x^{1/7} + 25$
62. $25x^{2/7} - 20x^{1/7} + 4$

Simplify each of the following to a single fraction.

63. $\dfrac{3}{x^{1/2}} + x^{1/2}$

64. $\dfrac{2}{x^{1/2}} - x^{1/2}$

65. $x^{2/3} + \dfrac{5}{x^{1/3}}$

66. $x^{3/4} - \dfrac{7}{x^{1/4}}$

67. $\dfrac{3x^2}{(x^3 + 1)^{1/2}} + (x^3 + 1)^{1/2}$

68. $\dfrac{x^3}{(x^2 - 1)^{1/2}} + 2x(x^2 - 1)^{1/2}$

69. $\dfrac{x^2}{(x^2 + 4)^{1/2}} - (x^2 + 4)^{1/2}$

70. $\dfrac{x^5}{(x^2 - 2)^{1/2}} + 4x^3(x^2 - 2)^{1/2}$

Use a calculator to find approximations to each of the following. Round your answers for Problems 75 and 76 to the nearest thousandth.

71. $16^{0.25}$

72. $81^{0.25}$

73. $9^{1.5}$

74. $32^{0.4}$

75. $\left(\frac{1}{2}\right)^{1/5}$

76. $\left(\frac{1}{2}\right)^{1/10}$

Applying the Concepts

77. Return on Investment A coin collection is purchased as an investment for $500 and sold 4 years later for $900. Find the annual rate of return on the investment.

78. Return on Investment An investor buys stock in a company for $800. Five years later, the same stock is worth $1,600. Find the annual rate of return on the stocks.

79. Return on Investment Find the annual rate of return on a home that is purchased for $60,000 and sold 5 years later for $80,000.

80. Return on Investment Find the annual rate of return on a home that is purchased for $75,000 and sold 10 years later for $150,000.

Review Problems

 NOTE: The problems that follow review material we covered in Sections 4.1 and 4.2.

Reduce to lowest terms. [4.1]

81. $\dfrac{x^2 - 9}{x^4 - 81}$

82. $\dfrac{6 - a - a^2}{3 - 2a - a^2}$

Divide. [4.2]

83. $\dfrac{15x^2y - 20x^4y^2}{5xy}$

84. $\dfrac{12x^3y^2 - 24x^2y^3}{6xy}$

Divide using long division. [4.2]

85. $\dfrac{10x^2 + 7x - 12}{2x + 3}$

86. $\dfrac{6x^2 - x - 35}{2x - 5}$

87. $\dfrac{x^3 - 125}{x - 5}$

88. $\dfrac{x^3 + 64}{x + 4}$

SECTION 5.3

Simplified Form for Radicals

In this section we will use radical notation instead of rational exponents. We will begin by stating two properties of radicals. Following this, we will give a definition for the simplified form for radical expressions. The examples in this section show how we use the properties of radicals to write radical expressions in simplified form.

For our first two properties of radicals, we will assume a and b are nonnegative real numbers whenever n is an even number.

PROPERTY I FOR RADICALS

$$\sqrt[n]{ab} = \sqrt[n]{a}\,\sqrt[n]{b}$$

In Words: The nth root of a product is the product of the nth roots.

Proof of Property 1

$$\sqrt[n]{ab} = (ab)^{1/n} \quad \text{Definition of fractional exponents}$$
$$= a^{1/n}b^{1/n} \quad \text{Exponents distribute over products.}$$
$$= \sqrt[n]{a}\,\sqrt[n]{b} \quad \text{Definition of fractional exponents}$$

PROPERTY 2 FOR RADICALS

$$\sqrt[n]{\frac{a}{b}} = \frac{\sqrt[n]{a}}{\sqrt[n]{b}} \qquad b \neq 0$$

In Words: The nth root of a quotient is the quotient of the nth roots.

The proof of property 2 is similar to the proof of property 1.

Note: There is no property for radicals that says the nth root of a sum is the sum of the nth roots. That is,

$$\sqrt[n]{a + b} \neq \sqrt[n]{a} + \sqrt[n]{b}$$

Properties 1 and 2 allow us to change the form of radical expressions and simplify them without changing their value.

Simplified Form for Radical Expressions

A radical expression is in **simplified form** if all of the following are true:

1. None of the factors of the radicand (the quantity under the radical sign) can be written as powers greater than or equal to the index—that is, no perfect squares can be factors of the quantity under a square root sign, no perfect cubes can be factors of the quantity under a cube root sign, and so forth.
2. There are no fractions under the radical sign.
3. There are no radicals in the denominator.

Satisfying the first condition for simplified form actually amounts to taking as much out from under the radical sign as possible. The following examples illustrate the first condition for simplified form.

EXAMPLE 1 Write $\sqrt{50}$ in simplified form.

SOLUTION The largest perfect square that divides 50 is 25. We write 50 as $25 \cdot 2$ and apply property 1 for radicals:

$$\sqrt{50} = \sqrt{25 \cdot 2} \quad 50 = 25 \cdot 2$$
$$= \sqrt{25}\,\sqrt{2} \quad \text{Property 1}$$
$$= 5\sqrt{2} \quad \sqrt{25} = 5$$

We have taken as much as possible out from under the radical sign—in this case, factoring 25 from 50 and then writing $\sqrt{25}$ as 5.

EXAMPLE 2 Write in simplified form: $\sqrt{48x^4y^3}$, where $x, y \geq 0$.

SOLUTION The largest perfect square that is a factor of the radicand is $16x^4y^2$. Applying property 1 again, we have

$$\begin{aligned}\sqrt{48x^4y^3} &= \sqrt{16x^4y^2 \cdot 3y}\\ &= \sqrt{16x^4y^2}\,\sqrt{3y}\\ &= 4x^2y\sqrt{3y}\end{aligned}$$

EXAMPLE 3 Write $\sqrt[3]{40a^5b^4}$ in simplified form.

SOLUTION We now want to factor the largest perfect cube from the radicand. We write $40a^5b^4$ as $8a^3b^3 \cdot 5a^2b$ and proceed as we did in Examples 1 and 2.

$$\begin{aligned}\sqrt[3]{40a^5b^4} &= \sqrt[3]{8a^3b^3 \cdot 5a^2b}\\ &= \sqrt[3]{8a^3b^3}\,\sqrt[3]{5a^2b}\\ &= 2ab\sqrt[3]{5a^2b}\end{aligned}$$

Here are some further examples concerning the first condition for simplified form.

EXAMPLES Write each expression in simplified form.

4. $\begin{aligned}\sqrt{12x^7y^6} &= \sqrt{4x^6y^6 \cdot 3x}\\ &= \sqrt{4x^6y^6}\,\sqrt{3x}\\ &= 2x^3y^3\,\sqrt{3x}\end{aligned}$ **5.** $\begin{aligned}\sqrt[3]{54a^6b^2c^4} &= \sqrt[3]{27a^6c^3 \cdot 2b^2c}\\ &= \sqrt[3]{27a^6c^3}\,\sqrt[3]{2b^2c}\\ &= 3a^2c\,\sqrt[3]{2b^2c}\end{aligned}$

The second property of radicals is used to simplify a radical that contains a fraction.

EXAMPLE 6 Simplify: $\sqrt{\frac{3}{4}}$.

SOLUTION Applying property 2 for radicals, we have

$$\sqrt{\frac{3}{4}} = \frac{\sqrt{3}}{\sqrt{4}} \quad \text{Property 2}$$

$$= \frac{\sqrt{3}}{2} \quad \sqrt{4} = 2$$

The last expression is in simplified form because it satisfies all three conditions for simplified form.

EXAMPLE 7 Write $\sqrt{\frac{5}{6}}$ in simplified form.

SOLUTION Proceeding as in Example 6, we have

$$\sqrt{\frac{5}{6}} = \frac{\sqrt{5}}{\sqrt{6}}$$

The resulting expression satisfies the second condition for simplified form since neither radical contains a fraction. However, it violates condition 3 since it has a radical in the denominator. Getting rid of the radical in the denominator is called **rationalizing the denominator** and is accomplished, in this case, by multiplying the numerator and denominator by $\sqrt{6}$:

$$\frac{\sqrt{5}}{\sqrt{6}} = \frac{\sqrt{5}}{\sqrt{6}} \cdot \frac{\sqrt{6}}{\sqrt{6}}$$

$$= \frac{\sqrt{30}}{(\sqrt{6})^2}$$

$$= \frac{\sqrt{30}}{6}$$

EXAMPLES Rationalize the denominator.

8. $\dfrac{4}{\sqrt{3}} = \dfrac{4}{\sqrt{3}} \cdot \dfrac{\sqrt{3}}{\sqrt{3}}$

$\quad = \dfrac{4\sqrt{3}}{(\sqrt{3})^2}$

$\quad = \dfrac{4\sqrt{3}}{3}$

9. $\dfrac{2\sqrt{3x}}{\sqrt{5y}} = \dfrac{2\sqrt{3x}}{\sqrt{5y}} \cdot \dfrac{\sqrt{5y}}{\sqrt{5y}}$

$\quad = \dfrac{2\sqrt{15xy}}{(\sqrt{5y})^2}$

$\quad = \dfrac{2\sqrt{15xy}}{5y}$

When the denominator involves a cube root, we must multiply by a radical that will produce a perfect cube under the cube root sign in the denominator, as our next example illustrates.

EXAMPLE 10 Rationalize the denominator in $\dfrac{7}{\sqrt[3]{4}}$.

SOLUTION Since $4 = 2^2$, we can multiply both the numerator and denominator by $\sqrt[3]{2}$ and obtain $\sqrt[3]{2^3}$ in the denominator:

$$\frac{7}{\sqrt[3]{4}} = \frac{7}{\sqrt[3]{2^2}}$$

$$= \frac{7}{\sqrt[3]{2^2}} \cdot \frac{\sqrt[3]{2}}{\sqrt[3]{2}}$$

$$= \frac{7\sqrt[3]{2}}{\sqrt[3]{2^3}}$$

$$= \frac{7\sqrt[3]{2}}{2}$$

EXAMPLE 11 Simplify: $\sqrt{\dfrac{12x^5y^3}{5z}}$.

SOLUTION We use property 2 to write the numerator and denominator as two separate radicals:

$$\sqrt{\frac{12x^5y^3}{5z}} = \frac{\sqrt{12x^5y^3}}{\sqrt{5z}}$$

Simplifying the numerator, we have

$$\frac{\sqrt{12x^5y^3}}{\sqrt{5z}} = \frac{\sqrt{4x^4y^2}\,\sqrt{3xy}}{\sqrt{5z}}$$

$$= \frac{2x^2y\,\sqrt{3xy}}{\sqrt{5z}}$$

To rationalize the denominator we multiply the numerator and denominator by $\sqrt{5z}$:

$$\frac{2x^2y\,\sqrt{3xy}}{\sqrt{5z}} \cdot \frac{\sqrt{5z}}{\sqrt{5z}} = \frac{2x^2y\,\sqrt{15xyz}}{\sqrt{(5z)^2}}$$

$$= \frac{2x^2y\,\sqrt{15xyz}}{5z}$$

The Square Root of a Perfect Square

So far in this chapter, we have assumed that all our variables are nonnegative when they appear under a square root symbol. There are times, however, when this is not the case.

Consider the following two statements:

$$\sqrt{3^2} = \sqrt{9} = 3 \qquad \text{and} \qquad \sqrt{(-3)^2} = \sqrt{9} = 3$$

Whether we operate on 3 or -3, the result is the same: both expressions simplify to 3. The other operation we have worked with in the past that produces the same result is absolute value. That is,

$$|3| = 3 \qquad \text{and} \qquad |-3| = 3$$

This leads us to the next property of radicals.

PROPERTY 3 FOR RADICALS

If a is a real number, then $\sqrt{a^2} = |a|$.

The result of this discussion and property 3 is simply this:

> If we know a is positive, then $\sqrt{a^2} = a$.
> If we know a is negative, then $\sqrt{a^2} = |a|$.
> If we don't know whether a is positive or negative, then $\sqrt{a^2} = |a|$.

EXAMPLES Simplify each expression. Do *not* assume the variables represent positive numbers.

12. $\sqrt{9x^2} = 3|x|$

13. $\sqrt{x^3} = |x|\sqrt{x}$

14. $\sqrt{x^2 - 6x + 9} = \sqrt{(x-3)^2} = |x-3|$

15. $\sqrt{x^3 - 5x^2} = \sqrt{x^2(x-5)} = |x|\sqrt{x-5}$

As you can see, we must use absolute value symbols when we take a square root of a perfect square, unless we know that the base of the perfect square is a positive number. The same idea holds for higher even roots, but not for odd roots. With odd roots, no absolute value symbols are necessary.

EXAMPLES Simplify each expression.

16. $\sqrt[3]{(-2)^3} = \sqrt[3]{-8} = -2$ **17.** $\sqrt[3]{(-5)^3} = \sqrt[3]{-125} = -5$

We can extend the discussion above to all roots as follows:

EXTENDING PROPERTY 3 FOR RADICALS

If a is a real number, then

$$\sqrt[n]{a^n} = |a| \quad \text{if} \quad n \text{ is even}$$
$$\sqrt[n]{a^n} = a \quad \text{if} \quad n \text{ is odd}$$

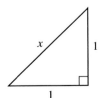

ℱacts from 𝒢eometry: 𝒯he 𝒮piral of ℛoots

In order to visualize the square roots of the positive integers, we can construct the spiral of roots that was illustrated in the introduction to this chapter. To begin, we draw two line segments, each of length 1, at right angles to each other. Then we use the Pythagorean Theorem to find the length of the diagonal. Figure 1 illustrates this procedure.

Next, we construct a second triangle by connecting a line segment of length 1 to the end of the first diagonal so that the angle formed is a right angle. We find the length of the second diagonal using the Pythagorean Theorem. Figure 2 (p. 306) illustrates this procedure. Continuing to draw new triangles by connecting line segments of length 1 to the end of each new diagonal so that the angle formed is a right angle, the spiral of roots begins to appear (Figure 3).

FIGURE I

$$x = \sqrt{1^2 + 1^2}$$
$$= \sqrt{2}$$

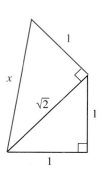

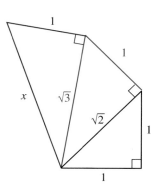

FIGURE 2

FIGURE 3

$x = \sqrt{(\sqrt{2})^2 + 1^2}$
$\quad = \sqrt{2 + 1}$
$\quad = \sqrt{3}$

$x = \sqrt{(\sqrt{3})^2 + 1^2}$
$\quad = \sqrt{3 + 1}$
$\quad = \sqrt{4}$

The Spiral of Roots and Function Notation

Looking over the diagrams and calculations in the discussion above, we see that each diagonal in the spiral of roots is found by using the length of the previous diagonal.

First diagonal: $\quad \sqrt{1^2 + 1^2} = \sqrt{2}$

Second diagonal: $\quad \sqrt{(\sqrt{2})^2 + 1^2} = \sqrt{3}$

Third diagonal: $\quad \sqrt{(\sqrt{3})^2 + 1^2} = \sqrt{4}$

Fourth diagonal: $\quad \sqrt{(\sqrt{4})^2 + 1^2} = \sqrt{5}$

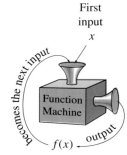

A function machine showing the recursive process.

A process like this one, in which the answer to one calculation is used to find the answer to the next calculation, is called a **recursive** process. In this particular case, we can use function notation to model the process. If we let x represent the length of any diagonal, then the length of the next diagonal is given by

$$f(x) = \sqrt{x^2 + 1}$$

To begin the process of finding the diagonals, we let $x = 1$:

$$f(1) = \sqrt{1^2 + 1} = \sqrt{2}$$

To find the next diagonal, we substitute $\sqrt{2}$ for x to obtain

$$f(f(1)) = f(\sqrt{2}) = \sqrt{(\sqrt{2})^2 + 1} = \sqrt{3}$$
$$f(f(f(1))) = f(\sqrt{3}) = \sqrt{(\sqrt{3})^2 + 1} = \sqrt{4}$$

We can describe the process of finding the diagonals of the spiral of roots very concisely this way:

$$f(1), \quad f(f(1)), \quad f(f(f(1))), \quad \ldots \qquad \text{where } f(x) = \sqrt{x^2 + 1}$$

This sequence of function values is a special case of a general category of similar sequences that are closely connected to *fractals* and *chaos,* two topics in mathematics that are currently receiving a good deal of attention.

 Using Technology: Graphing Calculators

As our discussion above indicates, the length of each diagonal in the spiral of roots is used to calculate the length of the next diagonal. The ANS key on a graphing calculator can be used very effectively in a situation like this. To begin, we store the number 1 in the variable ANS. Next, we key in the formula used to produce each diagonal using ANS for the variable. After that, it is simply a matter of pressing ENTER, as many times as we like, to produce the lengths of as many diagonals as we like. Here is a summary of what we do:

ENTER THIS	DISPLAY SHOWS
1 ENTER	1.000
$\sqrt{\ }$ (ANS^2 + 1) ENTER	1.414
ENTER	1.732
ENTER	2.000
ENTER	2.236

If you continue to press the ENTER key, you will produce decimal approximations for as many of the diagonals in the spiral of roots as you like.

Problem Set

5.3

Use property 1 for radicals to write each of the following expressions in simplified form. *(Assume all variables are nonnegative.)*

1. $\sqrt{8}$
2. $\sqrt{32}$
3. $\sqrt{98}$
4. $\sqrt{75}$
5. $\sqrt{288}$
6. $\sqrt{128}$
7. $\sqrt{80}$
8. $\sqrt{200}$
9. $\sqrt{48}$
10. $\sqrt{27}$

NOTE: Remember, simplified form for radicals does not always give us a simpler looking expression. Sometimes putting a radical expression in simplified form produces an expression that is more complicated than the original expression.

11. $\sqrt{675}$ **12.** $\sqrt{972}$

13. $\sqrt[3]{54}$ **14.** $\sqrt[3]{24}$

15. $\sqrt[3]{128}$ **16.** $\sqrt[3]{162}$

17. $\sqrt[3]{432}$ **18.** $\sqrt[3]{1,536}$

19. $\sqrt[5]{64}$ **20.** $\sqrt[4]{48}$

21. $\sqrt{18x^3}$ **22.** $\sqrt{27x^5}$

23. $\sqrt[4]{32y^7}$ **24.** $\sqrt[5]{32y^7}$

25. $\sqrt[3]{40x^4y^7}$ **26.** $\sqrt[3]{128x^6y^2}$

27. $\sqrt{48a^2b^3c^4}$ **28.** $\sqrt{72a^4b^3c^2}$

29. $\sqrt[3]{48a^2b^3c^4}$ **30.** $\sqrt[3]{72a^4b^3c^2}$

31. $\sqrt[5]{64x^8y^{12}}$ **32.** $\sqrt[4]{32x^9y^{10}}$

33. $\sqrt[5]{243x^7y^{10}z^5}$ **34.** $\sqrt[5]{64x^8y^4z^{11}}$

Substitute the given numbers into the expression $\sqrt{b^2 - 4ac}$ *and then simplify.*

35. $a = 2, b = -6, c = 3$

36. $a = 6, b = 7, c = -5$

37. $a = 1, b = 2, c = 6$

38. $a = 2, b = 5, c = 3$

39. $a = \frac{1}{2}, b = -\frac{1}{2}, c = -\frac{5}{4}$

40. $a = \frac{7}{4}, b = -\frac{3}{4}, c = -2$

Rationalize the denominator in each of the following expressions. (Assume all variables represent positive numbers.)

41. $\dfrac{2}{\sqrt{3}}$ **42.** $\dfrac{3}{\sqrt{2}}$

43. $\dfrac{5}{\sqrt{6}}$ **44.** $\dfrac{7}{\sqrt{5}}$

45. $\sqrt{\dfrac{1}{2}}$ **46.** $\sqrt{\dfrac{1}{3}}$

47. $\sqrt{\dfrac{1}{5}}$ **48.** $\sqrt{\dfrac{1}{6}}$

49. $\dfrac{4}{\sqrt[3]{2}}$ **50.** $\dfrac{5}{\sqrt[3]{3}}$

51. $\dfrac{2}{\sqrt[3]{9}}$ **52.** $\dfrac{3}{\sqrt[3]{4}}$

53. $\sqrt[4]{\dfrac{3}{2x^2}}$ **54.** $\sqrt[4]{\dfrac{5}{3x^2}}$

55. $\sqrt[4]{\dfrac{8}{y}}$ **56.** $\sqrt[4]{\dfrac{27}{y}}$

57. $\sqrt[3]{\dfrac{4x}{3y}}$ **58.** $\sqrt[3]{\dfrac{7x}{6y}}$

59. $\sqrt[3]{\dfrac{2x}{9y}}$ **60.** $\sqrt[3]{\dfrac{5x}{4y}}$

61. $\sqrt[4]{\dfrac{1}{8x^3}}$ **62.** $\sqrt[4]{\dfrac{8}{9x^3}}$

Write each of the following in simplified form. (Assume all variables represent positive numbers.)

63. $\sqrt{\dfrac{27x^3}{5y}}$ **64.** $\sqrt{\dfrac{12x^5}{7y}}$

65. $\sqrt{\dfrac{75x^3y^2}{2z}}$ **66.** $\sqrt{\dfrac{50x^2y^3}{3z}}$

67. $\sqrt[3]{\dfrac{16a^4b^3}{9c}}$ **68.** $\sqrt[3]{\dfrac{54a^5b^4}{25c^2}}$

69. $\sqrt[3]{\dfrac{8x^3y^6}{9z}}$ **70.** $\sqrt[3]{\dfrac{27x^6y^3}{2z^2}}$

Simplify each expression. Do not assume the variables represent positive numbers.

71. $\sqrt{25x^2}$ **72.** $\sqrt{49x^2}$

73. $\sqrt{27x^3y^2}$ **74.** $\sqrt{40x^3y^2}$

75. $\sqrt{x^2 - 10x + 25}$ **76.** $\sqrt{x^2 - 16x + 64}$

77. $\sqrt{4x^2 + 12x + 9}$

78. $\sqrt{16x^2 + 40x + 25}$

79. $\sqrt{4a^4 + 16a^3 + 16a^2}$

80. $\sqrt{9a^4 + 18a^3 + 9a^2}$

81. $\sqrt{4x^3 - 8x^2}$ **82.** $\sqrt{18x^3 - 9x^2}$

83. Show that the statement $\sqrt{a + b} = \sqrt{a} + \sqrt{b}$ is not true, in general, by replacing a with 9 and b with 16 and simplifying both sides.

84. Find a pair of values for a and b that will make the statement $\sqrt{a + b} = \sqrt{a} + \sqrt{b}$ true.

Applying the Concepts

85. Diagonal Distance The distance, d, between opposite corners of a rectangular room with length l and width w is given by

$$d = \sqrt{l^2 + w^2}$$

How far is it between opposite corners of a living room that measures 10 by 15 feet?

86. Radius of a Sphere The radius r of a sphere with volume V can be found by using the formula

$$r = \sqrt[3]{\frac{3V}{4\pi}}$$

Find the radius of a sphere with volume 9 cubic feet. Write your answer in simplified form. (Use $\frac{22}{7}$ for π.)

87. Spiral of Roots Construct your own spiral of roots by using a ruler. Draw the first triangle by using two 1 inch lines. The first diagonal will have a length of $\sqrt{2}$ inches. Each new triangle will be formed by drawing a 1 inch line segment at the end of the previous diagonal so that the angle formed is 90°.

88. Spiral of Roots Construct a spiral of roots by using line segments of length 2 inches. The length of the first diagonal will be $2\sqrt{2}$ inches. The length of the second diagonal will be $2\sqrt{3}$ inches.

89. Spiral of Roots If $f(x) = \sqrt{x^2 + 1}$, find the first six terms in the sequence below. Use your results to predict the value of the 10th term and the 100th term.

$$f(1), \quad f(f(1)), \quad f(f(f(1))), \quad \cdots$$

90. Spiral of Roots If $f(x) = \sqrt{x^2 + 4}$ find the first six terms in the sequence below. Use your results to predict the value of the 10th term and the 100th term. (The numbers in this sequence are the lengths of the diagonals of the spiral you drew in Problem 88.)

$$f(2), \quad f(f(2)), \quad f(f(f(2))), \quad \cdots$$

Review Problems

 The problems below review material we covered in Section 4.3.

Perform the indicated operations.

91. $\dfrac{8xy^3}{9x^2y} \div \dfrac{16x^2y^2}{18xy^3}$

92. $\dfrac{25x^2}{5y^4} \cdot \dfrac{30y^3}{2x^5}$

93. $\dfrac{12a^2 - 4a - 5}{2a + 1} \cdot \dfrac{7a + 3}{42a^2 - 17a - 15}$

94. $\dfrac{20a^2 - 7a - 3}{4a + 1} \cdot \dfrac{25a^2 - 5a - 6}{5a + 2}$

95. $\dfrac{8x^3 + 27}{27x^3 + 1} \div \dfrac{6x^2 + 7x - 3}{9x^2 - 1}$

96. $\dfrac{27x^3 + 8}{8x^3 + 1} \div \dfrac{6x^2 + x - 2}{4x^2 - 1}$

One Step Further

Factor each radicand into the product of prime factors. Then simplify each radical.

97. $\sqrt[3]{8,640}$ **98.** $\sqrt{8,640}$

99. $\sqrt[3]{10,584}$ **100.** $\sqrt{10,584}$

Assume a is a positive number and rationalize each denominator.

101. $\dfrac{1}{\sqrt[10]{a^3}}$

102. $\dfrac{1}{\sqrt[12]{a^7}}$

103. $\dfrac{1}{\sqrt[20]{a^{11}}}$

104. $\dfrac{1}{\sqrt[15]{a^{13}}}$

Using Technology

105. Show that the two expressions $\sqrt{x^2 + 1}$ and $x + 1$ are not, in general, equal to each other by graphing $y = \sqrt{x^2 + 1}$ and $y = x + 1$ in the same viewing window.

106. Show that the two expressions $\sqrt{x^2 + 9}$ and $x + 3$ are not, in general, equal to each other by graphing $y = \sqrt{x^2 + 9}$ and $y = x + 3$ in the same viewing window.

107. Approximately how far apart are the graphs in Problem 105 when $x = 2$?

108. Approximately how far apart are the graphs in Problem 106 when $x = 2$?

109. For what value of x are the expressions $\sqrt{x^2 + 1}$ and $x + 1$ equal?

110. For what value of x are the expressions $\sqrt{x^2 + 9}$ and $x + 3$ equal?

SECTION
5.4

Addition and Subtraction with Radical Expressions

In Chapter 1 we found that we could add similar terms when combining polynomials. The same idea applies to addition and subtraction of radical expressions.

> **DEFINITION**
>
> Two radicals are said to be **similar radicals** if they have the same index and the same radicand.

The expressions $5\sqrt[3]{7}$ and $-8\sqrt[3]{7}$ are similar since the index is 3 in both cases and the radicands are 7. The expressions $3\sqrt[4]{5}$ and $7\sqrt[3]{5}$ are not similar since they have different indexes, while the expressions $2\sqrt[5]{8}$ and $3\sqrt[5]{9}$ are not similar because the radicands are not the same.

We add and subtract radical expressions in the same way we add and subtract polynomials—by combining similar terms under the distributive property.

EXAMPLE 1 Combine: $5\sqrt{3} - 4\sqrt{3} + 6\sqrt{3}$.

SOLUTION All three radicals are similar. We apply the distributive property to get

$$5\sqrt{3} - 4\sqrt{3} + 6\sqrt{3} = (5 - 4 + 6)\sqrt{3}$$
$$= 7\sqrt{3}$$

EXAMPLE 2 Combine: $3\sqrt{8} + 5\sqrt{18}$.

SOLUTION The two radicals do not seem to be similar. We must rewrite each in simplified form before applying the distributive property.

$$3\sqrt{8} + 5\sqrt{18} = 3\sqrt{4 \cdot 2} + 5\sqrt{9 \cdot 2}$$
$$= 3\sqrt{4}\sqrt{2} + 5\sqrt{9}\sqrt{2}$$
$$= 3 \cdot 2\sqrt{2} + 5 \cdot 3\sqrt{2}$$
$$= 6\sqrt{2} + 15\sqrt{2}$$
$$= (6 + 15)\sqrt{2}$$
$$= 21\sqrt{2}$$

The result of Example 2 can be generalized to the following rule for sums and differences of radical expressions.

> ◢ **RULE**
>
> To add or subtract radical expressions, put each in simplified form and apply the distributive property if possible. We can add only similar radicals. (We must write each expression in simplified form before we can tell if the radicals are similar.)

EXAMPLE 3 Combine $7\sqrt{75xy^3} - 4y\sqrt{12xy}$, where $x, y \geq 0$.

SOLUTION We write each expression in simplified form and combine similar radicals:

$$\begin{aligned}
7\sqrt{75xy^3} - 4y\sqrt{12xy} &= 7\sqrt{25y^2}\sqrt{3xy} - 4y\sqrt{4}\sqrt{3xy} \\
&= 35y\sqrt{3xy} - 8y\sqrt{3xy} \\
&= (35y - 8y)\sqrt{3xy} \\
&= 27y\sqrt{3xy}
\end{aligned}$$

EXAMPLE 4 Combine: $10\sqrt[3]{8a^4b^2} + 11a\sqrt[3]{27ab^2}$.

SOLUTION Writing each radical in simplified form and combining similar terms, we have

$$\begin{aligned}
10\sqrt[3]{8a^4b^2} + 11a\sqrt[3]{27ab^2} &= 10\sqrt[3]{8a^3}\sqrt[3]{ab^2} + 11a\sqrt[3]{27}\sqrt[3]{ab^2} \\
&= 20a\sqrt[3]{ab^2} + 33a\sqrt[3]{ab^2} \\
&= 53a\sqrt[3]{ab^2}
\end{aligned}$$

EXAMPLE 5 Combine: $\dfrac{\sqrt{3}}{2} + \dfrac{1}{\sqrt{3}}$.

SOLUTION We begin by writing the second term in simplified form.

$$\begin{aligned}
\frac{\sqrt{3}}{2} + \frac{1}{\sqrt{3}} &= \frac{\sqrt{3}}{2} + \frac{1}{\sqrt{3}} \cdot \frac{\sqrt{3}}{\sqrt{3}} \\
&= \frac{\sqrt{3}}{2} + \frac{\sqrt{3}}{3} \\
&= \frac{1}{2}\sqrt{3} + \frac{1}{3}\sqrt{3} \\
&= \left(\frac{1}{2} + \frac{1}{3}\right)\sqrt{3} \\
&= \frac{5}{6}\sqrt{3} = \frac{5\sqrt{3}}{6}
\end{aligned}$$

EXAMPLE 6 Construct a golden rectangle from a square of side 4. Then show that the ratio of the length to the width is the golden ratio $\dfrac{1 + \sqrt{5}}{2}$.

SOLUTION Figure 1 shows the golden rectangle constructed from a square of side 4.

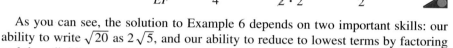

The length of the diagonal OB is found from the Pythagorean Theorem:

$$OB = \sqrt{2^2 + 4^2} = \sqrt{4 + 16} = \sqrt{20} = 2\sqrt{5}$$

The ratio of the length to the width for the rectangle is the golden ratio:

$$\text{Golden ratio} = \frac{CE}{EF} = \frac{2 + 2\sqrt{5}}{4} = \frac{2(1 + \sqrt{5})}{2 \cdot 2} = \frac{1 + \sqrt{5}}{2}$$

FIGURE 1

As you can see, the solution to Example 6 depends on two important skills: our ability to write $\sqrt{20}$ as $2\sqrt{5}$, and our ability to reduce to lowest terms by factoring and then dividing out the common factor 2 from the numerator and denominator.

Problem Set

5.4

Combine the following expressions. (Assume any variables under an even root are nonnegative.)

1. $3\sqrt{5} + 4\sqrt{5}$
2. $6\sqrt{3} - 5\sqrt{3}$
3. $3x\sqrt{7} - 4x\sqrt{7}$
4. $6y\sqrt{a} + 7y\sqrt{a}$
5. $5\sqrt[3]{10} - 4\sqrt[3]{10}$
6. $6\sqrt[4]{2} + 9\sqrt[4]{2}$
7. $8\sqrt[5]{6} - 2\sqrt[5]{6} + 3\sqrt[5]{6}$
8. $7\sqrt[6]{7} - \sqrt[6]{7} + 4\sqrt[6]{7}$
9. $3x\sqrt{2} - 4x\sqrt{2} + x\sqrt{2}$
10. $5x\sqrt{6} - 3x\sqrt{6} - 2x\sqrt{6}$
11. $\sqrt{20} - \sqrt{80} + \sqrt{45}$
12. $\sqrt{8} - \sqrt{32} - \sqrt{18}$
13. $4\sqrt{8} - 2\sqrt{50} - 5\sqrt{72}$
14. $\sqrt{48} - 3\sqrt{27} + 2\sqrt{75}$
15. $5x\sqrt{8} + 3\sqrt{32x^2} - 5\sqrt{50x^2}$
16. $2\sqrt{50x^2} - 8x\sqrt{18} - 3\sqrt{72x^2}$

NOTE: You can check your work by substituting values for the variables in both the original expression and in the simplified expression. If your results match, you have worked the problem correctly.

17. $5\sqrt[3]{16} - 4\sqrt[3]{54}$
18. $\sqrt[3]{81} + 3\sqrt[3]{24}$
19. $\sqrt[3]{x^4y^2} + 7x\sqrt[3]{xy^2}$
20. $2\sqrt[3]{x^8y^6} - 3y^2\sqrt[3]{8x^8}$
21. $5a^2\sqrt{27ab^3} - 6b\sqrt{12a^5b}$
22. $9a\sqrt{20a^3b^2} + 7b\sqrt{45a^5}$
23. $b\sqrt[3]{24a^5b} + 3a\sqrt[3]{81a^2b^4}$
24. $7\sqrt[3]{a^4b^3c^2} - 6ab\sqrt[3]{ac^2}$
25. $5x\sqrt[4]{3y^5} + y\sqrt[4]{243x^4y} + \sqrt[4]{48x^4y^5}$
26. $x\sqrt[4]{5xy^8} + y\sqrt[4]{405x^5y^4} + y^2\sqrt[4]{80x^5}$

27. $\dfrac{\sqrt{2}}{2} + \dfrac{1}{\sqrt{2}}$
28. $\dfrac{\sqrt{3}}{3} + \dfrac{1}{\sqrt{3}}$

29. $\dfrac{\sqrt{5}}{3} + \dfrac{1}{\sqrt{5}}$
30. $\dfrac{\sqrt{6}}{2} + \dfrac{1}{\sqrt{6}}$

31. $\sqrt{x} - \dfrac{1}{\sqrt{x}}$
32. $\sqrt{x} + \dfrac{1}{\sqrt{x}}$

33. $\dfrac{\sqrt{18}}{6} + \sqrt{\dfrac{1}{2}} + \dfrac{\sqrt{2}}{2}$ **34.** $\dfrac{\sqrt{12}}{6} + \sqrt{\dfrac{1}{3}} + \dfrac{\sqrt{3}}{3}$

35. $\sqrt{6} - \sqrt{\tfrac{2}{3}} + \sqrt{\tfrac{1}{6}}$ **36.** $\sqrt{15} - \sqrt{\tfrac{3}{5}} + \sqrt{\tfrac{5}{3}}$

37. $\sqrt[3]{25} + \dfrac{3}{\sqrt[3]{5}}$ **38.** $\sqrt[4]{8} + \dfrac{1}{\sqrt[4]{2}}$

39. Use a calculator to find a decimal approximation for $\sqrt{12}$ and for $2\sqrt{3}$.

40. Use a calculator to find decimal approximations for $\sqrt{50}$ and $5\sqrt{2}$.

41. Use a calculator to find a decimal approximation for $\sqrt{8} + \sqrt{18}$. Is it equal to the decimal approximation for $\sqrt{26}$ or $\sqrt{50}$?

42. Use a calculator to find a decimal approximation for $\sqrt{3} + \sqrt{12}$. Is it equal to the decimal approximation for $\sqrt{15}$ or $\sqrt{27}$?

Each statement below is false. Correct the right side of each one.

43. $3\sqrt{2x} + 5\sqrt{2x} = 8\sqrt{4x}$

44. $5\sqrt{3} - 7\sqrt{3} = -2\sqrt{9}$

45. $\sqrt{9 + 16} = 3 + 4$

46. $\sqrt{36 + 64} = 6 + 8$

Applying the Concepts

47. **Golden Rectangle** Construct a golden rectangle from a square of side 8. Then show that the ratio of the length to the width is the golden ratio $\dfrac{1 + \sqrt{5}}{2}$.

48. **Golden Rectangle** Construct a golden rectangle from a square of side 10. Then show that the ratio of the length to the width is the golden ratio $\dfrac{1 + \sqrt{5}}{2}$.

49. **Golden Rectangle** Use a ruler to construct a golden rectangle from a square of side 1 inch. Then show that the ratio of the length to the width is the golden ratio.

50. **Golden Rectangle** Use a ruler to construct a golden rectangle from a square of side $\tfrac{2}{3}$ inch.

Then show that the ratio of the length to the width is the golden ratio.

51. **Golden Rectangle** To show that all golden rectangles have the same ratio of length to width, construct a golden rectangle from a square of side $2x$. Then show that the ratio of the length to the width is the golden ratio.

52. **Golden Rectangle** To show that all golden rectangles have the same ratio of length to width, construct a golden rectangle from a square of side x. Then show that the ratio of the length to the width is the golden ratio.

Review Problems

 The problems below review material we covered in Section 4.4.

Add and subtract as indicated.

53. $\dfrac{2a - 4}{a + 2} - \dfrac{a - 6}{a + 2}$ **54.** $\dfrac{2a - 3}{a - 2} - \dfrac{a - 1}{a - 2}$

55. $3 + \dfrac{4}{3 - t}$ **56.** $6 + \dfrac{2}{5 - t}$

57. $\dfrac{3}{2x - 5} - \dfrac{39}{8x^2 - 14x - 15}$

58. $\dfrac{2}{4x - 5} + \dfrac{9}{8x^2 - 38x + 35}$

59. $\dfrac{1}{x - y} - \dfrac{3xy}{x^3 - y^3}$ **60.** $\dfrac{1}{x + y} + \dfrac{3xy}{x^3 + y^3}$

One Step Further

Simplify as much as possible. Assume all variables represent positive numbers.

61. $5\sqrt{x + 3} + 3\sqrt{4x + 12}$

62. $3x\sqrt{x + 5} + 2\sqrt{x^3 + 5x^2}$

63. $5\sqrt{x^3 + 4x^2} - x\sqrt{25x + 100}$

64. $x\sqrt{9x + 18} - 3\sqrt{x^3 + 2x^2}$

65. $2x\sqrt{x^2 + 10x + 25} + \sqrt{4x^4 + 40x^3 + 100x^2}$

66. $x\sqrt{x^2y^2 + 6xy^2 + 9y^2} + y\sqrt{x^4 + 6x^3 + 9x^2}$

Multiplication and Division with Radical Expressions

In this section we will look at multiplication and division of expressions that contain radicals. As you will see, multiplication of expressions that contain radicals is very similar to multiplication of polynomials. The division problems in this section are just an extension of the work we did previously when we rationalized denominators.

EXAMPLE 1 Multiply: $(3\sqrt{5})(2\sqrt{7})$.

SOLUTION We can rearrange the order and grouping of the numbers in this product by applying the commutative and associative properties. Then we apply property 1 for radicals and multiply:

$$
\begin{aligned}
(3\sqrt{5})(2\sqrt{7}) &= (3 \cdot 2)(\sqrt{5}\sqrt{7}) && \text{Commutative and associative properties} \\
&= (3 \cdot 2)(\sqrt{5 \cdot 7}) && \text{Property 1 for radicals} \\
&= 6\sqrt{35} && \text{Multiplication}
\end{aligned}
$$

In actual practice, it is not necessary to show either of the first two steps.

EXAMPLE 2 Multiply: $\sqrt{3}(2\sqrt{6} - 5\sqrt{12})$.

SOLUTION Applying the distributive property, we have

$$
\begin{aligned}
\sqrt{3}(2\sqrt{6} - 5\sqrt{12}) &= \sqrt{3} \cdot 2\sqrt{6} - \sqrt{3} \cdot 5\sqrt{12} \\
&= 2\sqrt{18} - 5\sqrt{36}
\end{aligned}
$$

Writing each radical in simplified form gives

$$
\begin{aligned}
2\sqrt{18} - 5\sqrt{36} &= 2\sqrt{9}\sqrt{2} - 5\sqrt{36} \\
&= 6\sqrt{2} - 30
\end{aligned}
$$

EXAMPLE 3 Multiply: $(\sqrt{3} + \sqrt{5})(4\sqrt{3} - \sqrt{5})$.

SOLUTION The same principle that applies when multiplying two binomials applies to this product. We must multiply each term in the first expression by each term in the second one. Any convenient method can be used. Let's use the FOIL method.

$$
\begin{aligned}
(\sqrt{3} + \sqrt{5})(4\sqrt{3} - \sqrt{5}) &= \overset{\text{F}}{\sqrt{3} \cdot 4\sqrt{3}} - \overset{\text{O}}{\sqrt{3}\sqrt{5}} + \overset{\text{I}}{\sqrt{5} \cdot 4\sqrt{3}} - \overset{\text{L}}{\sqrt{5}\sqrt{5}} \\
&= 4 \cdot 3 - \sqrt{15} + 4\sqrt{15} - 5 \\
&= 12 + 3\sqrt{15} - 5 \\
&= 7 + 3\sqrt{15}
\end{aligned}
$$

EXAMPLE 4 Expand and simplify: $(\sqrt{x} + 3)^2$.

SOLUTION

Method 1 We can write this problem as a multiplication problem and proceed as we did in Example 3:

$$(\sqrt{x} + 3)^2 = (\sqrt{x} + 3)(\sqrt{x} + 3)$$

$$\begin{array}{cccc} \text{F} & \text{O} & \text{I} & \text{L} \end{array}$$
$$= \sqrt{x} \cdot \sqrt{x} + 3\sqrt{x} + 3\sqrt{x} + 3 \cdot 3$$
$$= x + 6\sqrt{x} + 9$$

Method 2 We can obtain the same result by applying the formula for the square of a sum: $(a + b)^2 = a^2 + 2ab + b^2$.

$$(\sqrt{x} + 3)^2 = (\sqrt{x})^2 + 2(\sqrt{x})(3) + 3^2$$
$$= x + 6\sqrt{x} + 9$$

EXAMPLE 5 Expand $(3\sqrt{x} - 2\sqrt{y})^2$ and simplify the result.

SOLUTION Let's apply the formula for the square of a difference: $(a - b)^2 = a^2 - 2ab + b^2$.

$$(3\sqrt{x} - 2\sqrt{y})^2 = (3\sqrt{x})^2 - 2(3\sqrt{x})(2\sqrt{y}) + (2\sqrt{y})^2$$
$$= 9x - 12\sqrt{xy} + 4y$$

EXAMPLE 6 Expand and simplify: $(\sqrt{x + 2} - 1)^2$.

SOLUTION Applying the formula $(a - b)^2 = a^2 - 2ab + b^2$, we have

$$(\sqrt{x + 2} - 1)^2 = (\sqrt{x + 2})^2 - 2\sqrt{x + 2}\,(1) + 1^2$$
$$= x + 2 - 2\sqrt{x + 2} + 1$$
$$= x + 3 - 2\sqrt{x + 2}$$

EXAMPLE 7 Multiply: $(\sqrt{6} + \sqrt{2})(\sqrt{6} - \sqrt{2})$.

SOLUTION We notice that the product is of the form $(a + b)(a - b)$, which always gives the difference of two squares, $a^2 - b^2$:

$$(\sqrt{6} + \sqrt{2})(\sqrt{6} - \sqrt{2}) = (\sqrt{6})^2 - (\sqrt{2})^2$$
$$= 6 - 2 = 4$$

In Example 7 the two expressions $(\sqrt{6} + \sqrt{2})$ and $(\sqrt{6} - \sqrt{2})$ are called **conjugates.** In general, the conjugate of $\sqrt{a} + \sqrt{b}$ is $\sqrt{a} - \sqrt{b}$. If a and b are positive integers, multiplying conjugates of this form always produces a rational number. That is, if a and b are positive integers, then

$$(\sqrt{a} + \sqrt{b})(\sqrt{a} - \sqrt{b}) = \sqrt{a}\sqrt{a} - \sqrt{a}\sqrt{b} + \sqrt{a}\sqrt{b} - \sqrt{b}\sqrt{b}$$
$$= a - \sqrt{ab} + \sqrt{ab} - b$$
$$= a - b$$

The result, $a - b$, is an integer (because a and b are integers) and therefore is a rational number.

Division with radical expressions is the same as rationalizing the denominator. In Section 5.3 we were able to divide $\sqrt{3}$ by $\sqrt{2}$ by rationalizing the denominator, as shown at the top of the next page.

$$\frac{\sqrt{3}}{\sqrt{2}} = \frac{\sqrt{3}}{\sqrt{2}} \cdot \frac{\sqrt{2}}{\sqrt{2}} = \frac{\sqrt{6}}{2}$$

We can accomplish the same result with expressions such as $\dfrac{6}{\sqrt{5} - \sqrt{3}}$ by multiplying the numerator and denominator by the conjugate of the denominator.

EXAMPLE 8 Divide: $\dfrac{6}{\sqrt{5} - \sqrt{3}}$. (Rationalize the denominator.)

SOLUTION Since the product of two conjugates is a rational number, we multiply the numerator and denominator by the conjugate of the denominator.

$$\frac{6}{\sqrt{5} - \sqrt{3}} = \frac{6}{\sqrt{5} - \sqrt{3}} \cdot \frac{(\sqrt{5} + \sqrt{3})}{(\sqrt{5} + \sqrt{3})}$$

$$= \frac{6\sqrt{5} + 6\sqrt{3}}{(\sqrt{5})^2 - (\sqrt{3})^2}$$

$$= \frac{6\sqrt{5} + 6\sqrt{3}}{5 - 3}$$

$$= \frac{6\sqrt{5} + 6\sqrt{3}}{2}$$

The numerator and denominator of this last expression have a factor of 2 in common. We can reduce to lowest terms by factoring 2 from the numerator and then dividing both the numerator and denominator by 2:

$$= \frac{2(3\sqrt{5} + 3\sqrt{3})}{2} = 3\sqrt{5} + 3\sqrt{3}$$

EXAMPLE 9 Rationalize the denominator: $\dfrac{\sqrt{5} - 2}{\sqrt{5} + 2}$.

SOLUTION To rationalize the denominator, we multiply the numerator and denominator by the conjugate of the denominator:

$$\frac{\sqrt{5} - 2}{\sqrt{5} + 2} = \frac{\sqrt{5} - 2}{\sqrt{5} + 2} \cdot \frac{(\sqrt{5} - 2)}{(\sqrt{5} - 2)}$$

$$= \frac{5 - 2\sqrt{5} - 2\sqrt{5} + 4}{(\sqrt{5})^2 - 2^2}$$

$$= \frac{9 - 4\sqrt{5}}{5 - 4}$$

$$= \frac{9 - 4\sqrt{5}}{1} = 9 - 4\sqrt{5}$$

EXAMPLE 10 A golden rectangle constructed from a square of side 2 is shown in Figure 1. Show that the smaller rectangle *BDEF* is also a golden rectangle by finding the ratio of its length to its width.

SOLUTION First, we find expressions for the length and width of the smaller rectangle.

$$\text{Length} = EF = 2 \qquad \text{Width} = DE = \sqrt{5} - 1$$

Next, we find the ratio of length to width:

$$\text{Ratio of length to width} = \frac{EF}{DE} = \frac{2}{\sqrt{5} - 1}$$

To show that the small rectangle is a golden rectangle, we must show that the ratio of length to width is the golden ratio. We do so by rationalizing the denominator, as we did in the preceding examples.

$$\frac{2}{\sqrt{5} - 1} = \frac{2}{\sqrt{5} - 1} \cdot \frac{\sqrt{5} + 1}{\sqrt{5} + 1}$$

$$= \frac{2(\sqrt{5} + 1)}{5 - 1}$$

$$= \frac{2(\sqrt{5} + 1)}{4}$$

$$= \frac{\sqrt{5} + 1}{2} \qquad \text{Divide out common factor 2.}$$

Since addition is commutative, this last expression is the golden ratio. Therefore, the small rectangle in Figure 1 is a golden rectangle.

FIGURE 1

Problem Set

5.5

Assume all expressions appearing under a square root symbol represent nonnegative numbers throughout this problem set.

Multiply.

1. $\sqrt{6}\sqrt{3}$

2. $\sqrt{6}\sqrt{2}$

3. $(2\sqrt{3})(5\sqrt{7})$

4. $(3\sqrt{5})(2\sqrt{7})$

5. $(4\sqrt{6})(2\sqrt{15})(3\sqrt{10})$

6. $(4\sqrt{35})(2\sqrt{21})(5\sqrt{15})$

7. $(3\sqrt[3]{3})(6\sqrt[3]{9})$

8. $(2\sqrt[3]{2})(6\sqrt[3]{4})$

9. $\sqrt{3}(\sqrt{2} - 3\sqrt{3})$

10. $\sqrt{2}(5\sqrt{3} + 4\sqrt{2})$

11. $6\sqrt[3]{4}(2\sqrt[3]{2} + 1)$

12. $7\sqrt[3]{5}(3\sqrt[3]{25} - 2)$

13. $(\sqrt{3} + \sqrt{2})(3\sqrt{3} - \sqrt{2})$

14. $(\sqrt{5} - \sqrt{2})(3\sqrt{5} + 2\sqrt{2})$

15. $(\sqrt{x} + 5)(\sqrt{x} - 3)$ **16.** $(\sqrt{x} + 4)(\sqrt{x} + 2)$

17. $(3\sqrt{6} + 4\sqrt{2})(\sqrt{6} + 2\sqrt{2})$

18. $(\sqrt{7} - 3\sqrt{3})(2\sqrt{7} - 4\sqrt{3})$

19. $(\sqrt{3} + 4)^2$ **20.** $(\sqrt{5} - 2)^2$

21. $(\sqrt{x} - 3)^2$ **22.** $(\sqrt{x} + 4)^2$

23. $(2\sqrt{a} - 3\sqrt{b})^2$ **24.** $(5\sqrt{a} - 2\sqrt{b})^2$

25. $(\sqrt{x - 4} + 2)^2$ **26.** $(\sqrt{x - 3} + 2)^2$

27. $(\sqrt{x - 5} - 3)^2$ **28.** $(\sqrt{x - 3} - 4)^2$

29. $(\sqrt{3} - \sqrt{2})(\sqrt{3} + \sqrt{2})$

30. $(\sqrt{5} - \sqrt{2})(\sqrt{5} + \sqrt{2})$

31. $(\sqrt{a} + 7)(\sqrt{a} - 7)$ **32.** $(\sqrt{a} + 5)(\sqrt{a} - 5)$

33. $(5 - \sqrt{x})(5 + \sqrt{x})$ **34.** $(3 - \sqrt{x})(3 + \sqrt{x})$

35. $(\sqrt{x - 4} + 2)(\sqrt{x - 4} - 2)$

36. $(\sqrt{x + 3} + 5)(\sqrt{x + 3} - 5)$

37. $(\sqrt{3} + 1)^3$ **38.** $(\sqrt{5} - 2)^3$

Rationalize the denominator in each of the following.

39. $\dfrac{\sqrt{2}}{\sqrt{6} - \sqrt{2}}$ **40.** $\dfrac{\sqrt{5}}{\sqrt{5} + \sqrt{3}}$

41. $\dfrac{\sqrt{5}}{\sqrt{5} + 1}$ **42.** $\dfrac{\sqrt{7}}{\sqrt{7} - 1}$

43. $\dfrac{\sqrt{x}}{\sqrt{x} - 3}$ **44.** $\dfrac{\sqrt{x}}{\sqrt{x} + 2}$

45. $\dfrac{\sqrt{5}}{2\sqrt{5} - 3}$ **46.** $\dfrac{\sqrt{7}}{3\sqrt{7} - 2}$

47. $\dfrac{3}{\sqrt{x} - \sqrt{y}}$ **48.** $\dfrac{2}{\sqrt{x} + \sqrt{y}}$

49. $\dfrac{\sqrt{6} + \sqrt{2}}{\sqrt{6} - \sqrt{2}}$ **50.** $\dfrac{\sqrt{5} - \sqrt{3}}{\sqrt{5} + \sqrt{3}}$

51. $\dfrac{\sqrt{7} - 2}{\sqrt{7} + 2}$ **52.** $\dfrac{\sqrt{11} + 3}{\sqrt{11} - 3}$

53. $\dfrac{\sqrt{a} + \sqrt{b}}{\sqrt{a} - \sqrt{b}}$ **54.** $\dfrac{\sqrt{a} - \sqrt{b}}{\sqrt{a} + \sqrt{b}}$

55. $\dfrac{\sqrt{x} + 2}{\sqrt{x} - 2}$ **56.** $\dfrac{\sqrt{x} - 3}{\sqrt{x} + 3}$

57. $\dfrac{2\sqrt{3} - \sqrt{7}}{3\sqrt{3} + \sqrt{7}}$ **58.** $\dfrac{5\sqrt{6} + 2\sqrt{2}}{\sqrt{6} - \sqrt{2}}$

59. $\dfrac{3\sqrt{x} + 2}{1 + \sqrt{x}}$ **60.** $\dfrac{5\sqrt{x} - 1}{2 + \sqrt{x}}$

61. Show that the following product is 5.

$$(\sqrt[3]{2} + \sqrt[3]{3})(\sqrt[3]{4} - \sqrt[3]{6} + \sqrt[3]{9})$$

62. Show that the following product is $x + 8$.

$$(\sqrt[3]{x} + 2)(\sqrt[3]{x^2} - 2\sqrt[3]{x} + 4)$$

Each statement below is false. Correct the right side of each one.

63. $5(2\sqrt{3}) = 10\sqrt{15}$ **64.** $3(2\sqrt{x}) = 6\sqrt{3x}$

65. $(\sqrt{x} + 3)^2 = x + 9$ **66.** $(\sqrt{x} - 7)^2 = x - 49$

67. $(5\sqrt{3})^2 = 15$ **68.** $(3\sqrt{5})^2 = 15$

Applying the Concepts

69. **Gravity** If an object is dropped from the top of a 100 foot building, the amount of time t (in seconds) that it takes for the object to be h feet from the ground is given by the formula

$$t = \frac{\sqrt{100 - h}}{4}$$

How long does it take before the object is 50 feet from the ground? How long does it take to reach the ground? (When it is on the ground, h is 0.)

70. **Gravity** Use the formula given in Problem 69 to determine h if t is 1.25 seconds.

71. **Golden Rectangle** Rectangle *ACEF* in Figure 2 is a golden rectangle. If side *AC* is 6 inches, show that the smaller rectangle *BDEF* is also a golden rectangle.

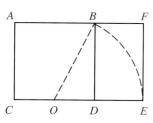

FIGURE 2

72. Golden Rectangle Rectangle *ACEF* in Figure 2 is a golden rectangle. If side *AC* is 1 inch, show that the smaller rectangle *BDEF* is also a golden rectangle.

73. Golden Rectangle If side *AC* in Figure 2 is 2*x*, show that rectangle *BDEF* is a golden rectangle.

74. Golden Rectangle If side *AC* in Figure 2 is *x*, show that rectangle *BDEF* is a golden rectangle.

Review Problems

NOTE: The problems that follow review material we covered in Section 4.5.

Simplify each complex fraction.

75. $\dfrac{\frac{1}{4} - \frac{1}{3}}{\frac{1}{2} + \frac{1}{6}}$

76. $\dfrac{\frac{1}{8} - \frac{1}{3}}{\frac{1}{4} - \frac{1}{3}}$

77. $\dfrac{1 - \dfrac{2}{y}}{1 + \dfrac{2}{y}}$

78. $\dfrac{1 + \dfrac{3}{y}}{1 - \dfrac{3}{y}}$

79. $\dfrac{4 + \dfrac{4}{x} + \dfrac{1}{x^2}}{4 - \dfrac{1}{x^2}}$

80. $\dfrac{1 - \dfrac{1}{x} - \dfrac{6}{x^2}}{1 - \dfrac{9}{x^2}}$

One Step Further

Rationalize the denominator.

81. $\dfrac{\sqrt{x - 4}}{\sqrt{x - 4} + 2}$

82. $\dfrac{\sqrt{x + 3}}{\sqrt{x + 3} + 5}$

83. $\dfrac{\sqrt{x + 3} + \sqrt{x - 3}}{\sqrt{x + 3} - \sqrt{x - 3}}$

84. $\dfrac{\sqrt{x + 5} + \sqrt{x - 5}}{\sqrt{x + 5} - \sqrt{x - 5}}$

85. $\dfrac{1}{\sqrt[3]{x} + 2}$

86. $\dfrac{1}{\sqrt[3]{x} - 2}$

87. $\dfrac{1}{\sqrt[3]{3} + \sqrt[3]{2}}$

88. $\dfrac{1}{\sqrt[3]{3} - \sqrt[3]{2}}$

SECTION
5.6

Equations with Radicals

This section is concerned with solving equations that involve one or more radicals. The first step in solving an equation that contains a radical is to eliminate the radical from the equation. To do so we need an additional property.

> ● **SQUARING PROPERTY OF EQUALITY**
>
> If both sides of an equation are squared, the solutions to the original equation are solutions to the resulting equation.

We will never lose solutions to our equations by squaring both sides. We may, however, introduce **extraneous solutions.** These extraneous solutions satisfy the equation obtained by squaring both sides of the original equation, but do not satisfy the original equation.

We know that if two real numbers a and b are equal, then so are their squares:

$$\text{If} \quad a = b$$
$$\text{then} \quad a^2 = b^2$$

On the other hand, extraneous solutions are introduced when we square opposites. That is, even though opposites are not equal, their squares are. For example,

$$5 = -5 \qquad \text{A false statement}$$
$$(5)^2 = (-5)^2 \quad \text{Square both sides.}$$
$$25 = 25 \qquad \text{A true statement}$$

We are free to square both sides of an equation any time it is convenient. We must be aware, however, that doing so may introduce extraneous solutions. We must, therefore, check all our solutions in the original equation if at any time we square both sides of the original equation.

EXAMPLE 1 Solve for x: $\sqrt{3x + 4} = 5$.

SOLUTION We square both sides and proceed as usual:

$$\sqrt{3x + 4} = 5$$
$$(\sqrt{3x + 4})^2 = 5^2$$
$$3x + 4 = 25$$
$$3x = 21$$
$$x = 7$$

Check Checking $x = 7$ in the original equation, we have

$$\sqrt{3(7) + 4} \overset{?}{=} 5$$
$$\sqrt{21 + 4} \overset{?}{=} 5$$
$$\sqrt{25} \overset{?}{=} 5$$
$$5 = 5$$

The solution $x = 7$ satisfies the original equation.

EXAMPLE 2 Solve: $\sqrt{4x - 7} = -3$.

SOLUTION Squaring both sides, we have

$$\sqrt{4x - 7} = -3$$
$$(\sqrt{4x - 7})^2 = (-3)^2$$
$$4x - 7 = 9$$
$$4x = 16$$
$$x = 4$$

Check   Checking $x = 4$ in the original equation gives

$$\sqrt{4(4) - 7} \overset{?}{=} -3$$
$$\sqrt{16 - 7} \overset{?}{=} -3$$
$$\sqrt{9} \overset{?}{=} -3$$
$$3 = -3 \quad \text{A false statement}$$

The solution $x = 4$ produces a false statement when checked in the original equation. Since $x = 4$ was the only possible solution, there is no solution to the original equation. The possible solution $x = 4$ is an extraneous solution. It satisfies the equation obtained by squaring both sides of the original equation, but does not satisfy the original equation.

Note: The fact that there is no solution to the equation in Example 2 was obvious to begin with. Notice that the left side of the equation is the *positive* square root of $4x - 7$, which must be a positive number or 0. The right side of the equation is -3. Since we cannot have a number that is either positive or 0 equal to a negative number, there is no solution to the equation.

EXAMPLE 3   Solve: $\sqrt{5x - 1} + 3 = 7$.

SOLUTION   We must isolate the radical on the left side of the equation. If we attempt to square both sides without doing so, the resulting equation will also contain a radical. Adding -3 to both sides, we have

$$\sqrt{5x - 1} + 3 = 7$$
$$\sqrt{5x - 1} = 4$$

We can now square both sides and proceed as usual:

$$(\sqrt{5x - 1})^2 = 4^2$$
$$5x - 1 = 16$$
$$5x = 17$$
$$x = \frac{17}{5}$$

Check   Checking $x = \frac{17}{5}$, we have

$$\sqrt{5\left(\frac{17}{5}\right) - 1} + 3 \overset{?}{=} 7$$
$$\sqrt{17 - 1} + 3 \overset{?}{=} 7$$
$$\sqrt{16} + 3 \overset{?}{=} 7$$
$$4 + 3 \overset{?}{=} 7$$
$$7 = 7$$

EXAMPLE 4 Solve: $t + 5 = \sqrt{t + 7}$.

SOLUTION This time, squaring both sides of the equation results in a quadratic equation:

$$(t + 5)^2 = (\sqrt{t + 7})^2 \quad \text{Square both sides.}$$
$$t^2 + 10t + 25 = t + 7$$
$$t^2 + 9t + 18 = 0 \quad \text{Standard form}$$
$$(t + 3)(t + 6) = 0 \quad \text{Factor the left side.}$$
$$t + 3 = 0 \quad \text{or} \quad t + 6 = 0 \quad \text{Set factors equal to 0.}$$
$$t = -3 \quad \text{or} \quad t = -6$$

Check We must check each solution in the original equation:

Check $t = -3$:
$$-3 + 5 \overset{?}{=} \sqrt{-3 + 7}$$
$$2 \overset{?}{=} \sqrt{4}$$
$$2 = 2$$
A true statement

Check $t = -6$:
$$-6 + 5 \overset{?}{=} \sqrt{-6 + 7}$$
$$-1 \overset{?}{=} \sqrt{1}$$
$$-1 = 1$$
A false statement

Since $t = -6$ does not check, our only solution is $t = -3$.

EXAMPLE 5 Solve: $\sqrt{x - 3} = \sqrt{x} - 3$.

The note that follows this example will help if you are having trouble convincing yourself that what is written below is true.

SOLUTION We begin by squaring both sides. Note carefully what happens when we square the right side of the equation, and compare the square of the right side with the square of the left side. You must convince yourself that these results are correct.

$$(\sqrt{x - 3})^2 = (\sqrt{x} - 3)^2$$
$$x - 3 = x - 6\sqrt{x} + 9$$

Now we still have a radical in our equation, so we will have to square both sides again. Before we do, though, let's isolate the remaining radical.

$$x - 3 = x - 6\sqrt{x} + 9$$
$$-3 = -6\sqrt{x} + 9 \quad \text{Add } -x \text{ to each side.}$$
$$-12 = -6\sqrt{x} \quad \text{Add } -9 \text{ to each side.}$$
$$2 = \sqrt{x} \quad \text{Divide each side by } -6.$$
$$4 = x \quad \text{Square each side.}$$

Check Our only possible solution is $x = 4$, which we check in our original equation as follows:

$$\sqrt{4 - 3} \overset{?}{=} \sqrt{4} - 3$$
$$\sqrt{1} \overset{?}{=} 2 - 3$$
$$1 = -1 \quad \text{A false statement}$$

Substituting 4 for x in the original equation yields a false statement. Since 4 was our only possible solution, there is no solution to the equation.

Note: In reading through Example 5, it is very important that you realize that the square of $(\sqrt{x} - 3)$ is not $x + 9$. Remember, when we square a difference with two terms, we use the formula

$$(a - b)^2 = a^2 - 2ab + b^2$$

Applying this formula to $(\sqrt{x} - 3)^2$, we have

$$(\sqrt{x} - 3)^2 = (\sqrt{x})^2 - 2(\sqrt{x})(3) + 3^2$$
$$= x - 6\sqrt{x} + 9$$

Here is another example of an equation for which we must apply the squaring property twice before all radicals are eliminated.

EXAMPLE 6   Solve: $\sqrt{x + 1} = 1 - \sqrt{2x}$.

SOLUTION   This equation has two separate terms involving radical signs. Squaring both sides gives

$$x + 1 = 1 - 2\sqrt{2x} + 2x$$
$$-x = -2\sqrt{2x} \qquad \text{Add } -2x \text{ and } -1 \text{ to both sides.}$$
$$x^2 = 4(2x) \qquad \text{Square both sides.}$$
$$x^2 - 8x = 0 \qquad \text{Standard form}$$

Our equation is a quadratic equation in standard form. To solve for x, we factor the left side and set each factor equal to 0:

$$x(x - 8) = 0 \quad \text{Factor left side.}$$
$$x = 0 \quad \text{or} \quad x - 8 = 0 \quad \text{Set factors equal to 0.}$$
$$x = 8$$

Check   Since we squared both sides of our equation, we have the possibility that one or both of the solutions are extraneous. We must check each one in the original equation:

Check $x = 8$:
$$\sqrt{8 + 1} \overset{?}{=} 1 - \sqrt{2 \cdot 8}$$
$$\sqrt{9} \overset{?}{=} 1 - \sqrt{16}$$
$$3 \overset{?}{=} 1 - 4$$
$$3 = -3$$
A false statement

Check $x = 0$:
$$\sqrt{0 + 1} \overset{?}{=} 1 - \sqrt{2 \cdot 0}$$
$$\sqrt{1} \overset{?}{=} 1 - \sqrt{0}$$
$$1 \overset{?}{=} 1 - 0$$
$$1 = 1$$
A true statement

Since $x = 8$ does not check, it is an extraneous solution. Our only solution is $x = 0$.

EXAMPLE 7   Solve: $\sqrt{x + 1} = \sqrt{x + 2} - 1$.

SOLUTION   Squaring both sides, we have

$$(\sqrt{x + 1})^2 = (\sqrt{x + 2} - 1)^2$$
$$x + 1 = x + 2 - 2\sqrt{x + 2} + 1$$

Once again we are left with a radical in the equation. Before we square each side again, we must isolate the radical on the right side of the equation.

$$
\begin{array}{ll}
x + 1 = x + 3 - 2\sqrt{x + 2} & \text{Simplify the right side.} \\
1 = 3 - 2\sqrt{x + 2} & \text{Add } -x \text{ to each side.} \\
-2 = -2\sqrt{x + 2} & \text{Add } -3 \text{ to each side.} \\
1 = \sqrt{x + 2} & \text{Divide each side by } -2. \\
1 = x + 2 & \text{Square both sides.} \\
-1 = x & \text{Add } -2 \text{ to each side.}
\end{array}
$$

Check   Checking our only possible solution, $x = -1$, in the original equation, we have

$$\sqrt{-1 + 1} \overset{?}{=} \sqrt{-1 + 2} - 1$$
$$\sqrt{0} \overset{?}{=} \sqrt{1} - 1$$
$$0 \overset{?}{=} 1 - 1$$
$$0 = 0 \qquad \text{A true statement}$$

Our solution checks.   ◢

It is also possible to raise both sides of an equation to powers greater than 2. We only need to check for extraneous solutions when we raise both sides of an equation to an even power. Raising both sides of an equation to an odd power will not produce extraneous solutions.

EXAMPLE 8   Solve: $\sqrt[3]{4x + 5} = 3$.

SOLUTION   Cubing both sides, we have

$$(\sqrt[3]{4x + 5})^3 = 3^3$$
$$4x + 5 = 27$$
$$4x = 22$$
$$x = \frac{22}{4}$$
$$x = \frac{11}{2}$$

We do not need to check for extraneous solutions, since we raised both sides to an odd power.   ◢

We end this section by looking at the graphs of some functions that contain radicals.

EXAMPLE 9   Graph $f(x) = \sqrt{x}$ and $g(x) = \sqrt[3]{x}$.

SOLUTION   The graphs are shown in Figures 1 and 2. Notice that the graph of $y = \sqrt{x}$ appears in the first quadrant only, because in the equation $y = \sqrt{x}$, x and y cannot be negative.

The graph of $y = \sqrt[3]{x}$ appears in quadrants I and III since the cube root of a positive number is also a positive number, and the cube root of a negative number is a negative number. That is, when x is positive, y will be positive and when x is negative, y will be negative.

The graphs of both equations contain the origin, since $y = 0$ when $x = 0$ in both equations.

x	y
-4	Undefined
-1	Undefined
0	0
1	1
4	2
9	3
16	4

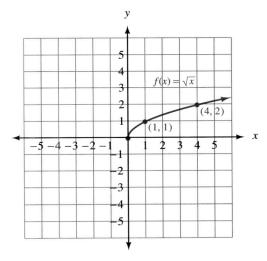

FIGURE 1

x	y
-27	-3
-8	-2
-1	-1
0	0
1	1
8	2
27	3

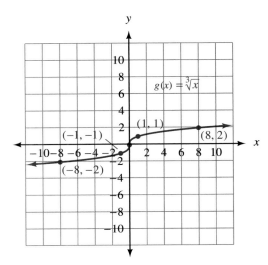

FIGURE 2

Problem Set

5·6

Solve each of the following equations.

1. $\sqrt{2x + 1} = 3$
2. $\sqrt{3x + 1} = 4$
3. $\sqrt{4x + 1} = -5$
4. $\sqrt{6x + 1} = -5$
5. $\sqrt{2y - 1} = 3$
6. $\sqrt{3y - 1} = 2$
7. $\sqrt{5x - 7} = -1$
8. $\sqrt{8x + 3} = -6$
9. $\sqrt{2x - 3} - 2 = 4$
10. $\sqrt{3x + 1} - 4 = 1$
11. $\sqrt{4a + 1} + 3 = 2$
12. $\sqrt{5a - 3} + 6 = 2$
13. $\sqrt[4]{3x + 1} = 2$
14. $\sqrt[4]{4x + 1} = 3$
15. $\sqrt[3]{2x - 5} = 1$
16. $\sqrt[3]{5x + 7} = 2$
17. $\sqrt[3]{3a + 5} = -3$
18. $\sqrt[3]{2a + 7} = -2$
19. $\sqrt{y - 3} = y - 3$
20. $\sqrt{y + 3} = y - 3$
21. $\sqrt{a + 2} = a + 2$
22. $\sqrt{a + 10} = a - 2$
23. $\sqrt{2x + 4} = \sqrt{1 - x}$
24. $\sqrt{3x + 4} = -\sqrt{2x + 3}$
25. $\sqrt{4a + 7} = -\sqrt{a + 2}$
26. $\sqrt{7a - 1} = \sqrt{2a + 4}$
27. $\sqrt[4]{5x - 8} = \sqrt[4]{4x - 1}$
28. $\sqrt[4]{6x + 7} = \sqrt[4]{x + 2}$
29. $x + 1 = \sqrt{5x + 1}$
30. $x - 1 = \sqrt{6x + 1}$
31. $t + 5 = \sqrt{2t + 9}$
32. $t + 7 = \sqrt{2t + 13}$
33. $\sqrt{y - 8} = \sqrt{8 - y}$
34. $\sqrt{2y + 5} = \sqrt{5y + 2}$
35. $\sqrt[3]{3x + 5} = \sqrt[3]{5 - 2x}$
36. $\sqrt[3]{4x + 9} = \sqrt[3]{3 - 2x}$

NOTE: The following equations will require that you square both sides twice before all the radicals are eliminated.

Solve each equation using the methods shown in Examples 5, 6, and 7.

37. $\sqrt{x - 8} = \sqrt{x} - 2$
38. $\sqrt{x + 3} = \sqrt{x} - 3$
39. $\sqrt{x + 1} = \sqrt{x} + 1$
40. $\sqrt{x - 1} = \sqrt{x} - 1$
41. $\sqrt{x + 8} = \sqrt{x - 4} + 2$

42. $\sqrt{x + 5} = \sqrt{x - 3} + 2$
43. $\sqrt{x - 5} - 3 = \sqrt{x - 8}$
44. $\sqrt{x - 3} - 4 = \sqrt{x - 3}$
45. $\sqrt{x + 4} = 2 - \sqrt{2x}$
46. $\sqrt{5x + 1} = 1 + \sqrt{5x}$
47. $\sqrt{2x + 4} = \sqrt{x + 3} + 1$
48. $\sqrt{2x - 1} = \sqrt{x - 4} + 2$

Applying the Concepts

49. **Solving a Formula** Solve the following formula for h:
$$t = \frac{\sqrt{100 - h}}{4}$$

50. **Solving a Formula** Solve the following formula for h:
$$t = \sqrt{\frac{2h - 40t}{g}}$$

51. **Pendulum Clock** The length of time T (in seconds) it takes the pendulum of a grandfather clock to swing through one complete cycle is given by the formula
$$T = 2\pi\sqrt{\frac{L}{32}}$$
where L is the length (in feet) of the pendulum and π is approximately $\frac{22}{7}$. How long must the pendulum be if one complete cycle takes 2 seconds?

52. **Pendulum Clock** Solve the formula in Problem 51 for L.

Graph each equation.

53. $y = 2\sqrt{x}$
54. $y = -2\sqrt{x}$
55. $y = \sqrt{x} - 2$
56. $y = \sqrt{x} + 2$
57. $f(x) = \sqrt{x - 2}$
58. $f(x) = \sqrt{x + 2}$
59. $f(x) = 3\sqrt[3]{x}$
60. $f(x) = -3\sqrt[3]{x}$

61. $g(x) = \sqrt[3]{x} + 3$

62. $g(x) = \sqrt[3]{x} - 3$

63. $g(x) = \sqrt[3]{x + 3}$

64. $g(x) = \sqrt[3]{x - 3}$

Review Problems

NOTE: The problems that follow review material we covered in Section 5.5. Reviewing these problems will help you understand the next section.

Multiply.

65. $\sqrt{2}(\sqrt{3} - \sqrt{2})$

66. $(\sqrt{x} - 4)(\sqrt{x} + 5)$

67. $(\sqrt{x} + 5)^2$

68. $(\sqrt{5} + \sqrt{3})(\sqrt{5} - \sqrt{3})$

Rationalize the denominator.

69. $\dfrac{\sqrt{x}}{\sqrt{x} + 3}$

70. $\dfrac{\sqrt{5} - \sqrt{3}}{\sqrt{5} + \sqrt{3}}$

One Step Further

Solve each equation.

71. $\dfrac{x}{3\sqrt{2x - 3}} - \dfrac{1}{\sqrt{2x - 3}} = \dfrac{1}{3}$

72. $\dfrac{x}{5\sqrt{2x + 10}} + \dfrac{1}{\sqrt{2x + 10}} = \dfrac{1}{5}$

73. $x + 1 = \sqrt[3]{4x + 4}$ **74.** $x - 1 = \sqrt[3]{4x - 4}$

Solve for y in terms of x.

75. $y + 2 = \sqrt{x^2 + (y - 2)^2}$

76. $y + \frac{1}{2} = \sqrt{x^2 + (y - \frac{1}{2})^2}$

Using Technology

77. Use your Y variables list, or write a program, to graph the family of curves $Y = \sqrt{X} + B$ for $B = -3, -2, -1, 0, 1, 2,$ and 3.

78. Use your Y variables list, or write a program, to graph the family of curves $Y = \sqrt{X + B}$ for $B = -3, -2, -1, 0, 1, 2,$ and 3.

79. Summarize the results of Problem 77 by giving a written description of the effect of b on the graph of $y = \sqrt{x} + b$.

80. Summarize the results of Problem 78 by giving a written description of the effect of b on the graph of $y = \sqrt{x + b}$.

81. Use your Y variables list, or write a program, to graph the family of curves $Y = \sqrt[3]{X} + B$ for $B = -3, -2, -1, 0, 1, 2,$ and 3.

82. Use your Y variables list, or write a program, to graph the family of curves $Y = \sqrt[3]{X + B}$ for $B = -3, -2, -1, 0, 1, 2,$ and 3.

83. Summarize the results of Problem 81 by giving a written description of the effect of b on the graph of $y = \sqrt[3]{x} + b$.

84. Summarize the results of Problem 82 by giving a written description of the effect of b on the graph of $y = \sqrt[3]{x + b}$.

85. Use your Y variables list, or write a program, to graph the family of curves $Y = A\sqrt{X}$ for $A = -3, -2, -1, 0, 1, 2,$ and 3.

86. Use your Y variables list, or write a program, to graph the family of curves $Y = A\sqrt{X}$ for $A = \frac{1}{4}, \frac{1}{3}, \frac{1}{2}, 1, 2,$ and 3.

87. Summarize the results of Problems 85 and 86 by giving a written description of the effect of a on the graph of $y = a\sqrt{x}$.

SECTION
5.7

Complex Numbers

The equation $x^2 = -9$ has no real number solutions since the square of a real number is always positive. We have been unable to work with square roots of negative numbers like $\sqrt{-25}$ and $\sqrt{-16}$ for the same reason. *Complex numbers* allow us to expand our work with radicals to include square roots of negative numbers and to solve equations like $x^2 = -9$ and $x^2 = -64$. Our work with complex numbers is based on the definition given at the top of the next page.

DEFINITION

The **number** i is such that $i = \sqrt{-1}$ (which is the same as saying $i^2 = -1$).

The number i, as we have defined it here, is not a real number. Because of the way we have defined i, we can use it to simplify square roots of negative numbers.

SQUARE ROOTS OF NEGATIVE NUMBERS

If a is a positive number, then $\sqrt{-a}$ can always be written as $i\sqrt{a}$. That is,

$$\sqrt{-a} = i\sqrt{a} \qquad \text{if } a \text{ is a positive number}$$

To justify our rule we simply square the quantity $i\sqrt{a}$ to obtain $-a$:

$$(i\sqrt{a})^2 = i^2 \cdot (\sqrt{a})^2$$
$$= -1 \cdot a$$
$$= -a$$

Here are some examples that illustrate the use of our new rule.

EXAMPLES Write each square root in terms of the number i.

1. $\sqrt{-25} = i\sqrt{25} = i \cdot 5 = 5i$ **2.** $\sqrt{-49} = i\sqrt{49} = i \cdot 7 = 7i$

3. $\sqrt{-12} = i\sqrt{12} = i \cdot 2\sqrt{3} = 2i\sqrt{3}$ **4.** $\sqrt{-17} = i\sqrt{17}$

Note: In Examples 3 and 4 we wrote i before the radical simply to avoid confusion. If we were to write the answer to Example 3 as $2\sqrt{3}\, i$, some people might think the i is under the radical sign, and, of course, it is not.

If we assume all the properties of exponents hold when the base is i, we can write any power of i as i, -1, $-i$, or 1. Using the fact that $i^2 = -1$, we have

$$i^1 = i$$
$$i^2 = -1$$
$$i^3 = i^2 \cdot i = -1(i) = -i$$
$$i^4 = i^2 \cdot i^2 = -1(-1) = 1$$

Since $i^4 = 1$, i^5 will simplify to i, and we will repeat the sequence i, -1, $-i$, 1 as we simplify higher powers of i. Any power of i simplifies to i, -1, $-i$, or 1. The easiest way to simplify higher powers of i is to write them in terms of i^2. For instance, to simplify i^{21}, we would write

$$i^{21} = (i^2)^{10} \cdot i \quad \text{Because } 2 \cdot 10 + 1 = 21$$

Then, since $i^2 = -1$, we have

$$(-1)^{10} \cdot i = 1 \cdot i = i$$

EXAMPLES Simplify as much as possible.

5. $i^{30} = (i^2)^{15} = (-1)^{15} = -1$

6. $i^{11} = (i^2)^5 \cdot i = (-1)^5 \cdot i = (-1)i = -i$

7. $i^{40} = (i^2)^{20} = (-1)^{20} = 1$

DEFINITION

A **complex number** is any number that can be put in the form

$$a + bi$$

where a and b are real numbers and $i = \sqrt{-1}$. The form $a + bi$ is called **standard form** for complex numbers. The number a is called the **real part** of the complex number. The number b is called the **imaginary part** of the complex number.

Every real number is also a complex number. The real number 8, for example, can be written as $8 + 0i$; therefore, 8 is also considered a complex number.

Equality for Complex Numbers

Two complex numbers are equal if and only if their real parts are equal and their imaginary parts are equal. That is, for real numbers, $a, b, c,$ and d,

$$a + bi = c + di \quad \text{if and only if} \quad a = c \quad \text{and} \quad b = d$$

EXAMPLE 8 Find x and y if $3x + 4i = 12 - 8yi$.

SOLUTION Since the two complex numbers are equal, their real parts are equal and their imaginary parts are equal:

$$3x = 12 \quad \text{and} \quad 4 = -8y$$
$$x = 4 \qquad\qquad y = -\frac{1}{2}$$

EXAMPLE 9 Find x and y if $(4x - 3) + 7i = 5 + (2y - 1)i$.

SOLUTION The real parts are $4x - 3$ and 5. The imaginary parts are 7 and $2y - 1$:

$$4x - 3 = 5 \quad \text{and} \quad 7 = 2y - 1$$
$$4x = 8 \qquad\qquad 8 = 2y$$
$$x = 2 \qquad\qquad y = 4$$

Addition and Subtraction of Complex Numbers

To add two complex numbers, add their real parts and add their imaginary parts. That is, if $a, b, c,$ and d are real numbers, then

$$(a + bi) + (c + di) = (a + c) + (b + d)i$$

If we assume that the commutative, associative, and distributive properties hold for the number i, then the definition of addition is simply an extension of these properties.

We define subtraction in a similar manner. If a, b, c, and d are real numbers, then

$$(a + bi) - (c + di) = (a - c) + (b - d)i$$

EXAMPLES   Add or subtract as indicated.

10. $(3 + 4i) + (7 - 6i) = (3 + 7) + (4 - 6)i = 10 - 2i$

11. $(7 + 3i) - (5 + 6i) = (7 - 5) + (3 - 6)i = 2 - 3i$

12. $(5 - 2i) - (9 - 4i) = (5 - 9) + (-2 + 4)i = -4 + 2i$

Multiplication with Complex Numbers

Since complex numbers have the same form as binomials, we find the product of two complex numbers the same way we found the product of two binomials.

EXAMPLE 13   Multiply: $(3 - 4i)(2 + 5i)$.

SOLUTION   Multiplying each term in the second complex number by each term in the first, we have

$$\overset{\text{F}\quad\;\text{O}\quad\;\text{I}\quad\;\text{L}}{(3 - 4i)(2 + 5i) = 3 \cdot 2 + 3 \cdot 5i - 2 \cdot 4i - 5i(4i)}$$
$$= 6 + 15i - 8i - 20i^2$$

Combining similar terms and using the fact that $i^2 = -1$, we can simplify this expression as follows:

$$6 + 15i - 8i - 20i^2 = 6 + 7i - 20(-1)$$
$$= 6 + 7i + 20$$
$$= 26 + 7i$$

The product of the complex numbers $3 - 4i$ and $2 + 5i$ is the complex number $26 + 7i$.

EXAMPLE 14   Multiply: $2i(4 - 6i)$.

SOLUTION   Applying the distributive property gives us

$$2i(4 - 6i) = 2i \cdot 4 - 2i \cdot 6i$$
$$= 8i - 12i^2$$
$$= 12 + 8i$$

EXAMPLE 15   Expand: $(3 + 5i)^2$.

SOLUTION   We treat this like the square of a binomial. *Remember:* $(a + b)^2 = a^2 + 2ab + b^2$.

$$(3 + 5i)^2 = 3^2 + 2(3)(5i) + (5i)^2$$
$$= 9 + 30i + 25i^2$$
$$= 9 + 30i - 25$$
$$= -16 + 30i$$

EXAMPLE 16   Multiply: $(2 - 3i)(2 + 3i)$.

SOLUTION   This product has the form $(a - b)(a + b)$, which we know results in the difference of two squares, $a^2 - b^2$:

$$(2 - 3i)(2 + 3i) = 2^2 - (3i)^2$$
$$= 4 - 9i^2$$
$$= 4 + 9$$
$$= 13$$

In Example 16 the product of the two complex numbers $2 - 3i$ and $2 + 3i$ is the real number 13. The two complex numbers $2 - 3i$ and $2 + 3i$ are called **complex conjugates.** The fact that their product is a real number is very useful.

DEFINITION

The complex numbers $a + bi$ and $a - bi$ are called **complex conjugates.** One important property they have is that their product is the real number $a^2 + b^2$. Here's why:

$$(a + bi)(a - bi) = a^2 - (bi)^2$$
$$= a^2 - b^2i^2$$
$$= a^2 - b^2(-1)$$
$$= a^2 + b^2$$

Division with Complex Numbers

The fact that the product of two complex conjugates is a real number is the key to division with complex numbers.

EXAMPLE 17   Divide: $\dfrac{2 + i}{3 - 2i}$.

SOLUTION   We want a complex number in standard form that is equivalent to the quotient $(2 + i)/(3 - 2i)$. We need to eliminate i from the denominator. Multiplying the numerator and denominator by $3 + 2i$ will give us what we want, as shown on the next page.

$$\frac{2 + i}{3 - 2i} = \frac{2 + i}{3 - 2i} \cdot \frac{(3 + 2i)}{(3 + 2i)}$$

$$= \frac{6 + 4i + 3i + 2i^2}{9 - 4i^2}$$

$$= \frac{6 + 7i - 2}{9 + 4}$$

$$= \frac{4 + 7i}{13}$$

$$= \frac{4}{13} + \frac{7}{13}i$$

So, dividing the complex number $2 + i$ by $3 - 2i$ gives the complex number $\frac{4}{13} + \frac{7}{13}i$.

EXAMPLE 18 Divide: $\dfrac{7 - 4i}{i}$.

SOLUTION The conjugate of the denominator is $-i$. Multiplying the numerator and denominator by this number, we have

$$\frac{7 - 4i}{i} = \frac{7 - 4i}{i} \cdot \frac{-i}{-i}$$

$$= \frac{-7i + 4i^2}{-i^2}$$

$$= \frac{-7i + 4(-1)}{-(-1)}$$

$$= -4 - 7i$$

Problem Set

5.7

Write the following in terms of i and simplify as much as possible.

1. $\sqrt{-36}$
2. $\sqrt{-49}$
3. $-\sqrt{-25}$
4. $-\sqrt{-81}$
5. $\sqrt{-72}$
6. $\sqrt{-48}$
7. $-\sqrt{-12}$
8. $-\sqrt{-75}$

Write each of the following as i, -1, $-i$, or 1.

9. i^{28}
10. i^{31}
11. i^{26}
12. i^{37}
13. i^{75}
14. i^{42}

Find x and y so that each of the following equations is true.

15. $2x + 3yi = 6 - 3i$
16. $4x - 2yi = 4 + 8i$
17. $2 - 5i = -x + 10yi$
18. $4 + 7i = 6x - 14yi$
19. $2x + 10i = -16 - 2yi$
20. $4x - 5i = -2 + 3yi$
21. $(2x - 4) - 3i = 10 - 6yi$
22. $(4x - 3) - 2i = 8 + yi$
23. $(7x - 1) + 4i = 2 + (5y + 2)i$
24. $(5x + 2) - 7i = 4 + (2y + 1)i$

Combine the following complex numbers:

25. $(2 + 3i) + (3 + 6i)$
26. $(4 + i) + (3 + 2i)$
27. $(3 - 5i) + (2 + 4i)$
28. $(7 + 2i) + (3 - 4i)$
29. $(5 + 2i) - (3 + 6i)$
30. $(6 + 7i) - (4 + i)$
31. $(3 - 5i) - (2 + i)$
32. $(7 - 3i) - (4 + 10i)$
33. $[(3 + 2i) - (6 + i)] + (5 + i)$
34. $[(4 - 5i) - (2 + i)] + (2 + 5i)$
35. $[(7 - i) - (2 + 4i)] - (6 + 2i)$
36. $[(3 - i) - (4 + 7i)] - (3 - 4i)$
37. $(3 + 2i) - [(3 - 4i) - (6 + 2i)]$
38. $(7 - 4i) - [(-2 + i) - (3 + 7i)]$
39. $(4 - 9i) + [(2 - 7i) - (4 + 8i)]$
40. $(10 - 2i) - [(2 + i) - (3 - i)]$

Find the following products:

41. $3i(4 + 5i)$ **42.** $2i(3 + 4i)$
43. $6i(4 - 3i)$ **44.** $11i(2 - i)$
45. $(3 + 2i)(4 + i)$ **46.** $(2 - 4i)(3 + i)$
47. $(4 + 9i)(3 - i)$ **48.** $(5 - 2i)(1 + i)$
49. $(1 + i)^3$ **50.** $(1 - i)^3$
51. $(2 - i)^3$ **52.** $(2 + i)^3$
53. $(2 + 5i)^2$ **54.** $(3 + 2i)^2$
55. $(1 - i)^2$ **56.** $(1 + i)^2$
57. $(3 - 4i)^2$ **58.** $(6 - 5i)^2$
59. $(2 + i)(2 - i)$ **60.** $(3 + i)(3 - i)$
61. $(6 - 2i)(6 + 2i)$ **62.** $(5 + 4i)(5 - 4i)$
63. $(2 + 3i)(2 - 3i)$ **64.** $(2 - 7i)(2 + 7i)$
65. $(10 + 8i)(10 - 8i)$ **66.** $(11 - 7i)(11 + 7i)$

Find the following quotients. Write all answers in standard form for complex numbers.

67. $\dfrac{2 - 3i}{i}$ **68.** $\dfrac{3 + 4i}{i}$

69. $\dfrac{5 + 2i}{-i}$ **70.** $\dfrac{4 - 3i}{-i}$

71. $\dfrac{4}{2 - 3i}$ **72.** $\dfrac{3}{4 - 5i}$

73. $\dfrac{6}{-3 + 2i}$ **74.** $\dfrac{-1}{-2 - 5i}$

75. $\dfrac{2 + 3i}{2 - 3i}$ **76.** $\dfrac{4 - 7i}{4 + 7i}$

77. $\dfrac{5 + 4i}{3 + 6i}$ **78.** $\dfrac{2 + i}{5 - 6i}$

79. $\dfrac{3 - 7i}{9 - 5i}$ **80.** $\dfrac{4 + 10i}{3 + 6i}$

Review Problems

NOTE: The following problems review material we covered in Sections 4.6 and 4.7.

Solve each equation. [4.6]

81. $\dfrac{t}{3} - \dfrac{1}{2} = -1$

82. $\dfrac{x}{x - 2} + \dfrac{2}{3} = \dfrac{2}{x - 2}$

83. $2 + \dfrac{5}{y} = \dfrac{3}{y^2}$ **84.** $1 - \dfrac{1}{y} = \dfrac{12}{y^2}$

Solve each application problem. [4.7]

85. The sum of a number and its reciprocal is $\frac{41}{20}$. Find the number.

86. It takes an inlet pipe 8 hours to fill a tank. The drain can empty the tank in 6 hours. If the tank is full and both the inlet pipe and drain are open, how long will it take to drain the tank?

One Step Further

87. Show that $-i$ and $\dfrac{1}{i}$ (the opposite and the reciprocal of i) are the same number.

88. Show that i^{2n+1} is the same as i for all positive even integers n.

89. Show that $x = 1 + i$ is a solution to the equation $x^2 - 2x + 2 = 0$.

90. Show that $x = 1 - i$ is a solution to the equation $x^2 - 2x + 2 = 0$.

91. Show that $x = 2 + i$ is a solution to the equation $x^3 - 11x + 20 = 0$.

92. Show that $x = 2 - i$ is a solution to the equation $x^3 - 11x + 20 = 0$.

REVIEW FOR CHAPTER 5

Examples

Examples

1. The number 49 has two square roots, 7 and -7. They are written like this:

$$\sqrt{49} = 7 \qquad -\sqrt{49} = -7$$

2. $\quad \sqrt[3]{8} = 2$

$\quad \sqrt[3]{-27} = -3$

3. $25^{1/2} = \sqrt{25} = 5$

$\quad 8^{2/3} = (\sqrt[3]{8})^2 = 2^2 = 4$

$\quad 9^{3/2} = (\sqrt{9})^3 = 3^3 = 27$

4. $\sqrt{4 \cdot 5} = \sqrt{4}\sqrt{5}$

$\qquad = 2\sqrt{5}$

$\sqrt{\dfrac{7}{9}} = \dfrac{\sqrt{7}}{\sqrt{9}} = \dfrac{\sqrt{7}}{3}$

5. $\sqrt{\dfrac{4}{5}} = \dfrac{\sqrt{4}}{\sqrt{5}}$

$\qquad = \dfrac{2}{\sqrt{5}} \cdot \dfrac{\sqrt{5}}{\sqrt{5}}$

$\qquad = \dfrac{2\sqrt{5}}{5}$

6. $5\sqrt{3} - 7\sqrt{3}$

$\qquad = (5 - 7)\sqrt{3}$

$\qquad = -2\sqrt{3}$

$\sqrt{20} + \sqrt{45}$

$\qquad = 2\sqrt{5} + 3\sqrt{5}$

$\qquad = (2 + 3)\sqrt{5}$

$\qquad = 5\sqrt{5}$

Chapter 5 Summary

Square Roots [5.1]

Every positive real number x has two square roots. The **positive square root** of x is written $\sqrt{x}$, while the **negative square root** of x is written $-\sqrt{x}$. Both the positive and the negative square roots of x are numbers we square to get x. That is,

$$(\sqrt{x})^2 = x \qquad \text{and} \qquad (-\sqrt{x})^2 = x \qquad \text{for } x \geq 0$$

Higher Roots [5.1]

In the expression $\sqrt[n]{a}$, n is the **index,** a is the **radicand,** and $\sqrt{}$ is the **radical sign.** The expression $\sqrt[n]{a}$ is such that

$$(\sqrt[n]{a})^n = a \qquad a \geq 0 \text{ when } n \text{ is even}$$

Rational Exponents [5.1, 5.2]

Rational exponents are used to indicate roots. The relationship between rational exponents and roots is as follows:

$$a^{1/n} = \sqrt[n]{a} \qquad \text{and} \qquad a^{m/n} = (a^{1/n})^m = (a^m)^{1/n} \qquad a \geq 0 \text{ when } n \text{ is even}$$

Properties of Radicals [5.3]

If a and b are nonnegative real numbers whenever n is even, then

1. $\sqrt[n]{ab} = \sqrt[n]{a}\,\sqrt[n]{b}$ **2.** $\sqrt[n]{\dfrac{a}{b}} = \dfrac{\sqrt[n]{a}}{\sqrt[n]{b}}$ $(b \neq 0)$

Simplified Form for Radicals [5.3]

A radical expression is said to be in **simplified form** if all of the following are true:

1. There is no factor of the radicand that can be written as a power greater than or equal to the index.
2. There are no fractions under the radical sign.
3. There are no radicals in the denominator.

Addition and Subtraction of Radical Expressions [5.4]

We add and subtract radical expressions by using the distributive property to combine similar radicals. **Similar radicals** are radicals with the same index and the same radicand.

7. $(\sqrt{x} + 2)(\sqrt{x} + 3)$

$\quad = \sqrt{x}\,\sqrt{x} + 3\sqrt{x}$
$\qquad + 2\sqrt{x} + 2 \cdot 3$

$\quad = x + 5\sqrt{x} + 6$

Multiplication of Radical Expressions [5.5]

We multiply radical expressions in the same way that we multiply polynomials. We can use the distributive property and the FOIL method.

8. $\dfrac{3}{\sqrt{2}} = \dfrac{3}{\sqrt{2}} \cdot \dfrac{\sqrt{2}}{\sqrt{2}} = \dfrac{3\sqrt{2}}{2}$

$\dfrac{3}{\sqrt{5} - \sqrt{3}}$

$\quad = \dfrac{3}{\sqrt{5} - \sqrt{3}} \cdot \dfrac{\sqrt{5} + \sqrt{3}}{\sqrt{5} + \sqrt{3}}$

$\quad = \dfrac{3\sqrt{5} + 3\sqrt{3}}{5 - 3}$

$\quad = \dfrac{3\sqrt{5} + 3\sqrt{3}}{2}$

Rationalizing the Denominator [5.3, 5.5]

When a fraction contains a square root in the denominator, we **rationalize the denominator** by multiplying the numerator and denominator by one of the following:

1. The square root itself if there is only one term in the denominator.
2. The **conjugate** of the denominator if there are two terms in the denominator.

Rationalizing the denominator is also called *division* of radical expressions.

9. $\quad \sqrt{2x + 1} = 3$

$\quad (\sqrt{2x + 1})^2 = 3^2$

$\qquad\quad 2x + 1 = 9$

$\qquad\qquad\quad x = 4$

Squaring Property of Equality [5.6]

We may square both sides of an equation any time it is convenient to do so, as long as we check all resulting solutions in the original equation.

10. $3 + 4i$ is a complex number.

Addition

$(3 + 4i) + (2 - 5i) = 5 - i$

Multiplication

$(3 + 4i)(2 - 5i)$

$\quad = 6 - 15i + 8i - 20i^2$

$\quad = 6 - 7i + 20$

$\quad = 26 - 7i$

Division

$\dfrac{2}{3 + 4i} = \dfrac{2}{3 + 4i} \cdot \dfrac{3 - 4i}{3 - 4i}$

$\quad = \dfrac{6 - 8i}{9 + 16}$

$\quad = \dfrac{6}{25} - \dfrac{8}{25}\,i$

Complex Numbers [5.7]

A **complex number** is any number than can be put in the **standard form**

$$a + bi$$

where a and b are real numbers and $i = \sqrt{-1}$. The **real part** of the complex number is a, and b is the **imaginary part.**

If a, b, c, and d are real numbers, then we have the following definitions associated with complex numbers:

1. Equality

$$a + bi = c + di \qquad \text{if and only if} \qquad a = c \quad \text{and} \quad b = d$$

2. Addition and subtraction

$$(a + bi) + (c + di) = (a + c) + (b + d)i$$
$$(a + bi) - (c + di) = (a - c) + (b - d)i$$

3. Multiplication with complex numbers is similar to multiplication with two binomials.

4. Division is similar to rationalizing the denominator.

REVIEW FOR CHAPTER 5

Common Mistakes

1. The most common mistake when working with radicals is to assume that the square root of a sum is the sum of the square roots:

$$\sqrt{x + y} = \sqrt{x} + \sqrt{y} \quad \text{Mistake}$$

The problem with this is that it just isn't true. If we try it with 16 and 9, the mistake becomes obvious:

$$\sqrt{16 + 9} \overset{?}{=} \sqrt{16} + \sqrt{9}$$
$$\sqrt{25} \overset{?}{=} 4 + 3$$
$$5 \neq 7$$

2. A common mistake when working with complex numbers is to mistake i for -1. The letter i is not -1; it is the *square root* of -1. That is, $i = \sqrt{-1}$.

3. When both sides of an equation are squared in the process of solving the equation, a common mistake occurs when the resulting solutions are not checked in the original equation. Remember, every time we square both sides of an equation, there is the possibility that we have introduced an extraneous root.

4. When squaring a quantity that has two terms involving radicals, it is a common mistake to omit the middle term in the result. For example,

$$(\sqrt{x + 3} + \sqrt{2x})^2 = (x + 3) + (2x) \quad \text{Mistake}$$

It should look like this:

$$(\sqrt{x + 3} + \sqrt{2x})^2 = (x + 3) + 2\sqrt{2x}\sqrt{x + 3} + (2x)$$

Remember: $(a + b)^2 = a^2 + 2ab + b^2$.

Chapter 5 Review Problems

Simplify each expression as much as possible. [5.1]

1. $\sqrt{49}$

2. $\sqrt[3]{8}$

3. $(-27)^{1/3}$

4. $27^{1/3}$

5. $16^{1/4}$

6. $\left(\dfrac{27}{64}\right)^{1/3}$

7. $9^{3/2}$

8. $8^{2/3}$

9. $\sqrt[5]{32x^{15}y^{10}}$

10. $\sqrt[3]{125x^9y^{12}}$

11. $8^{-4/3}$

12. $8^{-2/3} + 25^{-1/2}$

Use the properties of exponents to simplify each expression. Assume all bases represent positive numbers. [5.1]

13. $x^{2/3} \cdot x^{4/3}$

14. $(y^{3/4})^{4/3}$

15. $(a^{2/3}b^{4/3})^3$

16. $x^{2/3} \cdot x^{3/4}$

17. $\dfrac{a^{3/5}}{a^{1/4}}$

18. $(x^{1/3}y^{1/2}z^{5/6})^{6/5}$

19. $\dfrac{a^{2/3}b^3}{a^{1/4}b^{1/3}}$

20. $\dfrac{(y^{3/4})^{8/3}}{(y^{1/3})^{3/2}}$

Multiply. [5.2]

21. $(3x^{1/2} + 5y^{1/2})(4x^{1/2} - 3y^{1/2})$

22. $(4x^{1/2} - 3)(5x^{1/2} + 2)$

23. $(a^{1/3} - 5)^2$

24. $(2t^{1/3} - 1)(4t^{2/3} + 2t^{1/3} + 1)$

Divide. (Assume $x > 0$.) [5.2]

25. $\dfrac{28x^{5/6} + 14x^{7/6}}{7x^{1/3}}$

26. $\dfrac{39a^{5/7}b^{4/5} - 26a^{3/7}b^{6/5}}{13a^{2/7}b^{3/5}}$

27. Factor $2(x - 3)^{1/4}$ from the following expression. [5.2]

$$8(x - 3)^{5/4} - 2(x - 3)^{1/4}$$

28. Factor $6x^{2/5} - 11x^{1/5} - 10$ as if it were a trinomial. [5.2]

Simplify into a single fraction. (Assume $x > 0$.) [5.2]

29. $x^{3/4} + \dfrac{5}{x^{1/4}}$

30. $\dfrac{4x^3}{(x^2 + 1)^{1/2}} + (x^2 + 1)^{1/2}$

Write each expression in simplified form for radicals. (Assume all variables represent nonnegative numbers.) [5.3]

31. $\sqrt{12}$

32. $\sqrt{27}$

33. $\sqrt{50}$

34. $\sqrt{20}$

35. $\sqrt[3]{16}$

36. $\sqrt[3]{32}$

37. $\sqrt{18x^2}$

38. $\sqrt{72y^5}$

39. $\sqrt{80a^3b^4c^2}$

40. $\sqrt[3]{27x^4y^3}$

41. $\sqrt[4]{32a^4b^5c^6}$

42. $\sqrt[4]{162a^6b^5c^4}$

Rationalize the denominator in each expression. [5.3]

43. $\dfrac{3}{\sqrt{2}}$

44. $\sqrt{\dfrac{2}{5}}$

45. $\dfrac{6}{\sqrt[3]{2}}$

46. $\dfrac{7}{\sqrt[3]{9}}$

Write each expression in simplified form. (Assume all variables represent positive numbers.) [5.3]

47. $\sqrt{\dfrac{48x^3}{7y}}$

48. $\sqrt{\dfrac{75x^2y^3}{2z}}$

49. $\sqrt[3]{\dfrac{40x^2y^3}{3z}}$

50. $\sqrt[3]{\dfrac{54x^4y^3}{5z^2}}$

Combine the following expressions. (Assume all variables represent positive numbers.) [5.4]

51. $5x\sqrt{6} + 2x\sqrt{6} - 9x\sqrt{6}$

52. $3x\sqrt{7} + 7x\sqrt{7} - 2x\sqrt{7}$

53. $\sqrt{12} + \sqrt{3}$

54. $\sqrt{18} + \sqrt{8}$

55. $\dfrac{3}{\sqrt{5}} + \sqrt{5}$

56. $\sqrt{15} - \sqrt{\dfrac{3}{5}}$

57. $3\sqrt{8} - 4\sqrt{72} + 5\sqrt{50}$

58. $3\sqrt{48} - 3\sqrt{75} + 2\sqrt{27}$

59. $3b\sqrt{27a^5b} + 2a\sqrt{3a^3b^3}$

60. $4a\sqrt{18a^3b^2} - 2b\sqrt{50a^5}$

61. $2x\sqrt[3]{xy^3z^2} - 6y\sqrt[3]{x^4z^2}$

62. $7x\sqrt[3]{81x^2y^4} - 3y\sqrt[3]{24x^5y}$

Multiply. [5.5]

63. $\sqrt{2}(\sqrt{3} - 2\sqrt{2})$

64. $4\sqrt{5}(2\sqrt{10} - 3\sqrt{5})$

65. $(\sqrt{x} - 2)(\sqrt{x} - 3)$

66. $(\sqrt{6} + 3\sqrt{2})(2\sqrt{6} + \sqrt{2})$

67. $(5\sqrt{6} - 2\sqrt{3})(2\sqrt{6} + \sqrt{3})$

68. $(\sqrt{x} - 2)^2$

69. $(\sqrt{8} - \sqrt{2})(\sqrt{8} + \sqrt{2})$

70. $(3\sqrt{5} + 1)(3\sqrt{5} - 1)$

REVIEW FOR CHAPTER 5

Rationalize the denominator. (Assume x, y > 0.) [5.5]

71. $\dfrac{3}{\sqrt{5} - 2}$ **72.** $\dfrac{3}{\sqrt{6} - \sqrt{3}}$

73. $\dfrac{\sqrt{7} + \sqrt{5}}{\sqrt{7} - \sqrt{5}}$ **74.** $\dfrac{\sqrt{x} + \sqrt{y}}{\sqrt{x} - \sqrt{y}}$

75. $\dfrac{3\sqrt{7}}{3\sqrt{7} - 4}$ **76.** $\dfrac{5\sqrt{6}}{2\sqrt{6} + 7}$

Solve each equation. [5.6]

77. $\sqrt{4a + 1} = 1$ **78.** $\sqrt{7x - 4} = -2$
79. $\sqrt[3]{3x - 8} = 1$ **80.** $\sqrt[3]{8 - 3x} = -1$
81. $\sqrt{3x + 1} - 3 = 1$ **82.** $\sqrt{4x + 8} - 2 = 2$
83. $\sqrt{x + 4} = \sqrt{x} - 2$
84. $\sqrt{x + 11} = \sqrt{x - 5} + 2$
85. $\sqrt{2y - 8} = y - 4$
86. $\sqrt{y + 3} = y + 3$

Graph each equation. [5.6]

87. $y = 3\sqrt{x}$ **88.** $y = \sqrt{x} + 3$
89. $f(x) = \sqrt[3]{x} + 2$ **90.** $g(x) = 4\sqrt[3]{x}$

Write in terms of i and then simplify. [5.7]
91. $\sqrt{-49}$ **92.** $\sqrt{-80}$

Write each of the following as i, -1, $-i$, or 1. [5.7]
93. i^{24} **94.** i^{27}

Find x and y so that each of the following equations is true. [5.7]

95. $3 - 4i = -2x + 8yi$
96. $(3x + 2) - 8i = -4 + 2yi$

Combine the following complex numbers. [5.7]
97. $(3 + 5i) + (6 - 2i)$
98. $(3 + 5i) - (6 - 2i)$
99. $(2 + 5i) - [(3 + 2i) + (6 - i)]$
100. $[(6 + 2i) - (3 - 4i)] - (5 - i)$

Multiply. [5.7]

101. $3i(4 + 2i)$ **102.** $-5i(6 - i)$
103. $(2 + 3i)(4 + i)$ **104.** $(6 - 3i)(2 + 4i)$
105. $(4 + 2i)^2$ **106.** $(1 + i)^2$
107. $(4 + 3i)(4 - 3i)$ **108.** $(3 + i)(3 - i)$

Divide. Write all answers in standard form for complex numbers. [5.7]

109. $\dfrac{3 + i}{i}$ **110.** $\dfrac{2 - i}{-i}$

111. $\dfrac{-3}{2 + i}$ **112.** $\dfrac{3 + 2i}{3 - 2i}$

113. $\dfrac{4 - 3i}{4 + 3i}$ **114.** $\dfrac{7 - i}{3 - 2i}$

Chapter 5 Test

Assume all variable bases are positive integers and all variable exponents are positive real numbers throughout this test.

Simplify each of the following. [5.1]

1. $27^{-2/3}$ **2.** $\left(\frac{25}{49}\right)^{-1/2}$

3. $a^{3/4} \cdot a^{-1/3}$ **4.** $\dfrac{(x^{2/3}y^{-3})^{1/2}}{(x^{3/4}y^{1/2})^{-1}}$

5. $\sqrt{49x^8y^{10}}$ **6.** $\sqrt[5]{32x^{10}y^{20}}$

7. $\dfrac{(36a^8b^4)^{1/2}}{(27a^9b^6)^{1/3}}$ **8.** $\dfrac{(x^ny^{1/n})^n}{(x^{1/n}y^n)^{n^2}}$

Multiply. [5.2]

9. $2a^{1/2}(3a^{3/2} - 5a^{1/2})$ **10.** $(4a^{3/2} - 5)^2$

Factor. [5.2]

11. $3x^{2/3} + 5x^{1/3} - 2$ **12.** $9x^{2/3} - 49$

Combine. [5.2]

13. $\dfrac{4}{x^{1/2}} + x^{1/2}$

14. $\dfrac{x^2}{(x^2 - 3)^{1/2}} - (x^2 - 3)^{1/2}$

Write in simplified form. [5.3]

15. $\sqrt{125x^3y^5}$

16. $\sqrt[3]{40x^7y^8}$

17. $\sqrt{\dfrac{2}{3}}$

18. $\sqrt{\dfrac{12a^4b^3}{5c}}$

Combine. [5.4]

19. $3\sqrt{12} - 4\sqrt{27}$

20. $2\sqrt[3]{24a^3b^3} - 5a\sqrt[3]{3b^3}$

Multiply. [5.5]

21. $(\sqrt{x} + 7)(\sqrt{x} - 4)$

22. $(3\sqrt{2} - \sqrt{3})^2$

Rationalize the denominator. [5.5]

23. $\dfrac{5}{\sqrt{3} - 1}$

24. $\dfrac{\sqrt{x} - \sqrt{2}}{\sqrt{x} + \sqrt{2}}$

Solve for x. [5.6]

25. $\sqrt{3x + 1} = x - 3$

26. $\sqrt[3]{2x + 7} = -1$

27. $\sqrt{x + 3} = \sqrt{x + 4} - 1$

Graph. [5.6]

28. $f(x) = \sqrt{x - 2}$

29. $g(x) = \sqrt[3]{x} + 3$

30. Solve for x and y so that the following equation is true: [5.7]

$$(2x + 5) - 4i = 6 - (y - 3)i$$

Perform the indicated operations. [5.7]

31. $(3 + 2i) - [(7 - i) - (4 + 3i)]$

32. $(2 - 3i)(4 + 3i)$

33. $(5 - 4i)^2$

34. $\dfrac{2 - 3i}{2 + 3i}$

35. Show that i^{38} can be written as -1. [5.7]

Quadratic Equations

6

Contents

Introduction

Table 1 is taken from the trail map given to skiers at Northstar at Tahoe ski resort in Lake Tahoe, California. The table gives the length of each chair lift at Northstar, along with the change in elevation from the beginning of the lift to the end of the lift.

TABLE I		
From the Trail Map for Northstar at Tahoe Ski Resort		
Lift Information		
Lift	Vertical Rise	Length
Big Springs Gondola	480′	4,100′
Bear Paw Double	120′	790′
Echo Triple	710′	4,890′
Aspen Express Quad	900′	5,100′
Forest Double	1,170′	5,750′
Lookout Double	960′	4,330′
Comstock Express Quad	1,250′	5,900′
Rendezvous Triple	650′	2,900′
Schaffer Camp Triple	1,860′	6,150′
Chipmunk Tow Lift	28′	280′
Bear Cub Tow Lift	120′	750′

Right triangles are good mathematical models for chair lifts. In this chapter we will use our knowledge of right triangles, along with the new material developed in the chapter, to solve problems involving chair lifts and a variety of others.

342

Overview

We begin this chapter by extending the work we started in Chapter 2 with quadratic equations. In Chapter 2 we solved quadratic equations by factoring. Up to this point, factoring is the only method we have had for solving quadratic equations. The new methods we will develop in this chapter will allow us to solve any quadratic equation, regardless of whether it is factorable.

Once we have presented the new methods of solving quadratic equations, we will take a closer look at the relationship between solutions to equations and the equations themselves. The results will give us a better overall view of equations and their solutions, and will allow us to build equations from their solutions.

Then we will encounter new graphs called *parabolas,* which are connected very closely to the quadratic equations presented in this chapter.

In the last section of the chapter, we will turn our attention to quadratic inequalities. Our ability to solve a wider variety of equations will increase the types of inequalities we can solve.

To be successful in this chapter, you should have a working knowledge of factoring, binomial squares, square roots, and complex numbers. Also important for your success in this chapter is your ability to graph equations in two variables by constructing tables of ordered pairs and by using intercepts.

Study Skills

This is the last chapter where we will mention study skills. You should know by now what works best for you and what you have to do to achieve your goals for this course. From now on it is simply a matter of sticking with the things that work for you and avoiding the things that do not. It seems simple, but as with anything that takes effort, it is up to you to see that you maintain the skills that get you where you want to be in the course.

SECTION 6.1

Completing the Square

In this section we will develop the first of our new methods for solving quadratic equations. The new method is called **completing the square.** Completing the square on a quadratic equation allows us to obtain solutions, regardless of whether the equation can be factored. Before we solve equations by completing the square, we need to know how to solve equations by taking square roots of both sides.

Consider the equation

$$x^2 = 16$$

We could solve it by writing it in standard form, factoring the left side, and proceeding as we did in Chapter 2. However, we can shorten our work considerably if we simply notice that x must be either the positive square root of 16 or the negative square root of 16. That is,

$$\text{If } x^2 = 16:$$
$$\text{then } \quad x = \sqrt{16} \quad \text{ or } \quad x = -\sqrt{16}$$
$$x = 4 \quad \text{ or } \quad x = -4$$

We can generalize this result into a theorem as follows:

THEOREM 6.1

If $a^2 = b$, where b is a real number, then $a = \sqrt{b}$ or $a = -\sqrt{b}$.

Notation: The expression $a = \sqrt{b}$ or $a = -\sqrt{b}$ can be written in shorthand form as $a = \pm\sqrt{b}$. The symbol $\pm$ is read "plus or minus."

We can apply Theorem 6.1 to some fairly complicated quadratic equations.

EXAMPLE 1 Solve: $(2x - 3)^2 = 25$.

SOLUTION

$$(2x - 3)^2 = 25$$
$$2x - 3 = \pm\sqrt{25} \quad \text{Theorem 6.1}$$
$$2x - 3 = \pm 5 \quad \sqrt{25} = 5$$
$$2x = 3 \pm 5 \quad \text{Add 3 to both sides.}$$
$$x = \frac{3 \pm 5}{2} \quad \text{Divide both sides by 2.}$$

The last equation can be written as two separate statements:

$$x = \frac{3 + 5}{2} \quad \text{or} \quad x = \frac{3 - 5}{2}$$
$$= \frac{8}{2} \quad \text{or} \quad = \frac{-2}{2}$$
$$= 4 \quad \text{or} \quad = -1$$

The solution set is $\{4, -1\}$.

Notice that we could have solved the equation in Example 1 by expanding the left side, writing the resulting equation in standard form, and then factoring. The problem would look like this:

$$(2x - 3)^2 = 25 \quad \text{Original equation}$$
$$4x^2 - 12x + 9 = 25 \quad \text{Expand the left side.}$$
$$4x^2 - 12x - 16 = 0 \quad \text{Add } -25 \text{ to each side.}$$
$$4(x^2 - 3x - 4) = 0 \quad \text{Begin factoring.}$$

$$4(x - 4)(x + 1) = 0 \qquad \text{Factor completely.}$$

$$x - 4 = 0 \quad \text{or} \quad x + 1 = 0 \qquad \text{Set variable factors equal to 0.}$$

$$x = 4 \quad \text{or} \qquad x = -1$$

As you can see, solving the equation by factoring leads to the same two solutions.

EXAMPLE 2   Solve for x: $(3x - 1)^2 = -12$.

SOLUTION
$$(3x - 1)^2 = -12$$

$$3x - 1 = \pm\sqrt{-12} \qquad \text{Theorem 6.1}$$

$$3x - 1 = \pm 2i\sqrt{3} \qquad \sqrt{-12} = 2i\sqrt{3}$$

$$3x = 1 \pm 2i\sqrt{3} \qquad \text{Add 1 to both sides.}$$

$$x = \frac{1 \pm 2i\sqrt{3}}{3} \qquad \text{Divide both sides by 3.}$$

The solution set is

$$\left\{ \frac{1 + 2i\sqrt{3}}{3}, \frac{1 - 2i\sqrt{3}}{3} \right\}$$

Check   Both solutions are complex. Here is a check of the first solution:

$$\text{When} \quad x = \frac{1 + 2i\sqrt{3}}{3}:$$

$$\text{the equation} \qquad (3x - 1)^2 = -12$$

$$\text{becomes} \qquad \left(3 \cdot \frac{\mathbf{1 + 2i\sqrt{3}}}{\mathbf{3}} - 1\right)^2 \overset{?}{=} -12$$

$$\text{or} \qquad (1 + 2i\sqrt{3} - 1)^2 \overset{?}{=} -12$$

$$(2i\sqrt{3})^2 \overset{?}{=} -12$$

$$4 \cdot i^2 \cdot 3 \overset{?}{=} -12$$

$$12(-1) \overset{?}{=} -12$$

$$-12 = -12$$

Note: We cannot solve the equation in Example 2 by factoring. If we expand the left side and write the resulting equation in standard form, we are left with a quadratic equation that does not factor:

$$(3x - 1)^2 = -12 \qquad \text{Equation from Example 2}$$

$$9x^2 - 6x + 1 = -12 \qquad \text{Expand the left side.}$$

$$9x^2 - 6x + 13 = 0 \qquad \text{Standard form, but not factorable}$$

EXAMPLE 3   Solve: $x^2 + 6x + 9 = 12$.

SOLUTION   We can solve this equation as we have the equations in Examples 1 and 2 if we first write the left side as $(x + 3)^2$.

$$x^2 + 6x + 9 = 12 \qquad \text{Original equation}$$
$$(x + 3)^2 = 12 \qquad \text{Write } x^2 + 6x + 9 \text{ as } (x + 3)^2.$$
$$x + 3 = \pm 2\sqrt{3} \qquad \text{Theorem 6.1}$$
$$x = -3 \pm 2\sqrt{3} \qquad \text{Add } -3 \text{ to each side.}$$

We have two irrational solutions: $-3 + 2\sqrt{3}$ and $-3 - 2\sqrt{3}$. What is important about this problem, however, is the fact that the equation was easy to solve because the left side was a perfect square trinomial.

Completing the Square

The method of **completing the square** is simply a way of transforming any quadratic equation into an equation of the form found in the preceding three examples.

The key to understanding the method of completing the square lies in recognizing the relationship between the last two terms of any perfect square trinomial whose leading coefficient is 1.

Consider the following list of perfect square trinomials and their corresponding binomial squares:

$$x^2 - 6x + 9 = (x - 3)^2$$
$$x^2 + 8x + 16 = (x + 4)^2$$
$$x^2 - 10x + 25 = (x - 5)^2$$
$$x^2 + 12x + 36 = (x + 6)^2$$

In each case the leading coefficient is 1. A more important observation comes from noticing the relationship between the linear and constant terms (middle and last terms) in each trinomial. Observe that the constant term in each case is the square of half the coefficient of x in the middle term. For example, in the last expression, the constant term, 36, is the square of half of 12, where 12 is the coefficient of x in the middle term. (Notice also that the second terms in all the binomials on the right side are half the coefficients of the middle terms of the trinomials on the left side.) We can use these observations to build our own perfect square trinomials and, in doing so, solve some quadratic equations. Consider the following equation:

$$x^2 + 6x = 3$$

We can think of the left side as having the first two terms of a perfect square trinomial. We need only add the correct constant term. If we take half the coefficient of x, we get 3. If we then square this quantity, we have 9. Adding 9 to both sides, the equation becomes

$$x^2 + 6x + 9 = 3 + 9$$

The left side is the perfect square $(x + 3)^2$; the right side is 12:

$$(x + 3)^2 = 12$$

The equation is now in the correct form. We can apply Theorem 6.1 and finish the solution, as we did in Example 3.

This method is called **completing the square,** since we complete the square on the left side of the original equation by adding the appropriate constant term.

EXAMPLE 4   Solve by completing the square: $x^2 + 5x - 2 = 0$.

SOLUTION   We must begin by adding 2 to both sides. (The left side of the equation, as it is, is not a perfect square, because it does not have the correct constant term. We will simply "move" that term to the other side and use our own constant term.)

$$x^2 + 5x = 2 \qquad \text{Add 2 to each side.}$$

We complete the square by adding the square of half the coefficient of the linear term to both sides:

$$x^2 + 5x + \frac{25}{4} = 2 + \frac{25}{4} \qquad \text{Half of 5 is } \tfrac{5}{2}; \text{ the square of } \tfrac{5}{2} \text{ is } \tfrac{25}{4}.$$

$$\left(x + \frac{5}{2}\right)^2 = \frac{33}{4} \qquad 2 + \tfrac{25}{4} = \tfrac{8}{4} + \tfrac{25}{4} = \tfrac{33}{4}$$

$$x + \frac{5}{2} = \pm\sqrt{\frac{33}{4}} \qquad \text{Theorem 6.1}$$

$$x + \frac{5}{2} = \pm\frac{\sqrt{33}}{2} \qquad \text{Simplify the radical.}$$

$$x = -\frac{5}{2} \pm \frac{\sqrt{33}}{2} \qquad \text{Add } -\tfrac{5}{2} \text{ to both sides.}$$

$$x = \frac{-5 \pm \sqrt{33}}{2}$$

The solution set is

$$\left\{\frac{-5 + \sqrt{33}}{2}, \frac{-5 - \sqrt{33}}{2}\right\}$$

We can use a calculator to get decimal approximations to these solutions. If $\sqrt{33} \approx 5.74$, then

$$\frac{-5 + 5.74}{2} = 0.37 \qquad \frac{-5 - 5.74}{2} = -5.37$$

EXAMPLE 5   Solve for x: $3x^2 - 8x + 7 = 0$.

SOLUTION
$$3x^2 - 8x + 7 = 0$$

$$3x^2 - 8x = -7 \qquad \text{Add } -7 \text{ to both sides.}$$

We cannot complete the square on the left side because the leading coefficient is not 1. We take an extra step and divide both sides by 3:

$$\frac{3x^2}{3} - \frac{8x}{3} = -\frac{7}{3}$$

$$x^2 - \frac{8}{3}x = -\frac{7}{3}$$

Half of $\frac{8}{3}$ is $\frac{4}{3}$, and the square of $\frac{4}{3}$ is $\frac{16}{9}$:

$$x^2 - \frac{8}{3}x + \frac{16}{9} = -\frac{7}{3} + \frac{16}{9} \qquad \text{Add } \tfrac{16}{9} \text{ to both sides.}$$

$$\left(x - \frac{4}{3}\right)^2 = -\frac{5}{9} \qquad \text{Simplify right side.}$$

$$x - \frac{4}{3} = \pm\sqrt{-\frac{5}{9}} \qquad \text{Theorem 6.1}$$

$$x - \frac{4}{3} = \pm\frac{i\sqrt{5}}{3} \qquad \sqrt{-\frac{5}{9}} = \frac{\sqrt{-5}}{3} = \frac{i\sqrt{5}}{3}$$

$$x = \frac{4}{3} \pm \frac{i\sqrt{5}}{3} \qquad \text{Add } \tfrac{4}{3} \text{ to both sides.}$$

$$x = \frac{4 \pm i\sqrt{5}}{3}$$

The solution set is

$$\left\{ \frac{4 + i\sqrt{5}}{3}, \frac{4 - i\sqrt{5}}{3} \right\}$$

Strategy for Solving a Quadratic Equation by Completing the Square

To summarize the method used in the preceding two examples, we list the following steps:

Step 1 Write the equation in the form $ax^2 + bx = c$.

Step 2 If the leading coefficient is not 1, divide both sides by the coefficient so that the resulting equation has a leading coefficient of 1. That is, if $a \neq 1$, then divide both sides by a.

Step 3 Add the square of half the coefficient of the linear term to both sides of the equation.

Step 4 Write the left side of the equation as the square of a binomial and simplify the right side, if possible.

Step 5 Apply Theorem 6.1 and solve as usual.

Facts from Geometry: More Special Triangles

The triangles shown in Figures 1 and 2 (at the top of the next page) occur frequently in mathematics.

Note that both triangles are right triangles. We refer to the triangle in Figure 1 as a 30°–60°–90° triangle, and the triangle in Figure 2 as a 45°–45°–90° triangle.

348

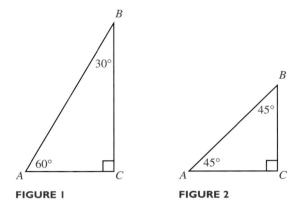

FIGURE I   **FIGURE 2**

EXAMPLE 6 If the shortest side in a 30°–60°–90° triangle is 1 inch, find the lengths of the other two sides.

SOLUTION In Figure 3, triangle ABC is a 30°–60°–90° triangle in which the shortest side, AC, is 1 inch long. Triangle DBC is also a 30°–60°–90° triangle in which the shortest side, DC, is 1 inch long.

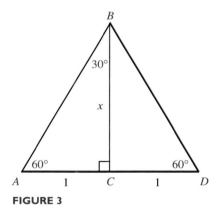

FIGURE 3

Notice that the large triangle ABD is an equilateral triangle because each of its interior angles is 60°. Each side of triangle ABD is 2 inches long. Therefore, side AB in triangle ABC is 2 inches long. To find the length of side BC, we use the Pythagorean Theorem:

$$BC^2 + AC^2 = AB^2$$
$$x^2 + 1^2 = 2^2$$
$$x^2 + 1 = 4$$
$$x^2 = 3$$
$$x = \sqrt{3} \text{ inches}$$

Note that we write only the positive square root because x is the length of a side in a triangle and is therefore a positive number.

EXAMPLE 7   Table 1 in the introduction to this chapter gives the vertical rise of the Forest Double chair lift as 1,170 feet and the length of the chair lift as 5,750 feet. To the nearest foot, find the horizontal distance covered by a person riding this lift.

SOLUTION   The figure below is a model of the Forest Double chair lift. A rider gets on the lift at point A and exits at point B. The length of the lift is AB.

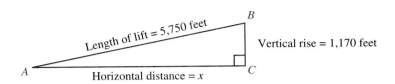

To find the horizontal distance covered by a person riding the chair lift we use the Pythagorean Theorem:

$$5{,}750^2 = x^2 + 1{,}170^2 \qquad \text{Pythagorean Theorem}$$
$$33{,}062{,}500 = x^2 + 1{,}368{,}900 \qquad \text{Simplify squares.}$$
$$x^2 = 33{,}062{,}500 - 1{,}368{,}900 \qquad \text{Solve for } x^2.$$
$$x^2 = 31{,}693{,}600 \qquad \text{Simplify the right side.}$$
$$x = \sqrt{31{,}693{,}600} \qquad \text{Theorem 6.1}$$
$$x = 5{,}630 \text{ feet} \qquad \text{To the nearest foot}$$

A rider getting on the lift at point A and riding to point B will cover a horizontal distance of approximately 5,630 feet.

Problem Set

6.1

Solve the following equations.

1. $x^2 = 25$
2. $x^2 = 16$
3. $a^2 = -9$
4. $a^2 = -49$
5. $y^2 = \frac{3}{4}$
6. $y^2 = \frac{5}{9}$
7. $x^2 + 12 = 0$
8. $x^2 + 8 = 0$
9. $4a^2 - 45 = 0$
10. $9a^2 - 20 = 0$
11. $(2y - 1)^2 = 25$
12. $(3y + 7)^2 = 1$
13. $(2a + 3)^2 = -9$
14. $(3a - 5)^2 = -49$
15. $(5x + 2)^2 = -8$
16. $(6x - 7)^2 = -75$
17. $x^2 + 8x + 16 = -27$
18. $x^2 - 12x + 36 = -8$
19. $4a^2 - 12a + 9 = -4$
20. $9a^2 - 12a + 4 = -9$

Simplify the left side of each equation and then solve for x.

21. $(x + 5)^2 + (x - 5)^2 = 52$
22. $(2x + 1)^2 + (2x - 1)^2 = 10$
23. $(2x + 3)^2 + (2x - 3)^2 = 26$
24. $(3x + 2)^2 + (3x - 2)^2 = 26$
25. $(3x + 4)(3x - 4) - (x + 2)(x - 2) = -4$
26. $(5x + 2)(5x - 2) - (x + 3)(x - 3) = 29$

Copy each of the following and fill in the blanks so that the left side of each is a perfect square trinomial. That is, complete the square.

27. $x^2 + 12x + \underline{\quad} = (x + \underline{\quad})^2$

28. $x^2 + 6x +$ _____ $= (x +$ _____ $)^2$
29. $x^2 - 4x +$ _____ $= (x -$ _____ $)^2$
30. $x^2 - 2x +$ _____ $= (x -$ _____ $)^2$
31. $a^2 - 10a +$ _____ $= (a -$ _____ $)^2$
32. $a^2 - 8a +$ _____ $= (a -$ _____ $)^2$
33. $x^2 + 5x +$ _____ $= (x +$ _____ $)^2$
34. $x^2 + 3x +$ _____ $= (x +$ _____ $)^2$
35. $y^2 - 7y +$ _____ $= (y -$ _____ $)^2$
36. $y^2 - y +$ _____ $= (y -$ _____ $)^2$

Solve each of the following quadratic equations by completing the square.

37. $x^2 + 4x = 12$ **38.** $x^2 - 2x = 8$
39. $x^2 + 12x = -27$ **40.** $x^2 - 6x = 16$
41. $a^2 - 2a + 5 = 0$
42. $a^2 + 10a + 22 = 0$
43. $y^2 - 8y + 1 = 0$ **44.** $y^2 + 6y - 1 = 0$
45. $x^2 - 5x - 3 = 0$ **46.** $x^2 - 5x - 2 = 0$
47. $2x^2 - 4x - 8 = 0$
48. $3x^2 - 9x - 12 = 0$
49. $3t^2 - 8t + 1 = 0$
50. $5t^2 + 12t - 1 = 0$
51. $4x^2 - 3x + 5 = 0$
52. $7x^2 - 5x + 2 = 0$

Applying the Concepts

53. If the shortest side in a $30°-60°-90°$ triangle is $\frac{1}{2}$ inch long, find the lengths of the other two sides.

54. If the shortest side in a $30°-60°-90°$ triangle is 3 feet long, find the lengths of the other two sides.

55. If the shortest side of a $30°-60°-90°$ triangle is x in length, find the lengths of the other two sides in terms of x.

56. If the longest side of a $30°-60°-90°$ triangle is x in length, find the lengths of the other two sides in terms of x.

57. If the shorter sides of a $45°-45°-90°$ triangle are each 1 inch long, find the length of the hypotenuse.

58. If the shorter sides of a $45°-45°-90°$ triangle are each 3 feet long, find the length of the hypotenuse.

59. If the hypotenuse of a $45°-45°-90°$ triangle is 1 inch long, find the length of the shorter sides.

60. If the hypotenuse of a $45°-45°-90°$ triangle is 2 feet long, find the length of the shorter sides.

61. If the shorter sides of a $45°-45°-90°$ triangle are each x in length, find the length of the hypotenuse in terms of x.

62. If the hypotenuse of a $45°-45°-90°$ triangle is x in length, find the length of the shorter sides in terms of x.

63. **Chair Lift** Use Table 1 from the introduction to this chapter to find the horizontal distance covered by a person riding the Bear Paw Double chair lift. Round your answer to the nearest foot.

64. **Chair Lift** Use Table 1 from the introduction to this chapter to find the horizontal distance covered by a person riding the Big Springs Gondola lift. Round your answer to the nearest foot.

65. **Chair Lift** Using a right triangle to model the Forest Double chair lift, find the slope of the lift to the nearest hundredth.

66. **Chair Lift** Using a right triangle to model the Echo Triple chair lift, find the slope of the lift to the nearest hundredth.

Review Problems

| NOTE | The problems that follow review material we covered in Section 5.3. Reviewing these problems will help you with the next section. |

Write each of the following in simplified form for radicals.

67. $\sqrt{45}$ **68.** $\sqrt{24}$ **69.** $\sqrt{27y^5}$ **70.** $\sqrt{8y^3}$
71. $\sqrt[3]{54x^6y^5}$ **72.** $\sqrt[3]{16x^9y^7}$

73. Simplify $\sqrt{b^2 - 4ac}$ when $a = 6$, $b = 7$, and $c = -5$.

74. Simplify $\sqrt{b^2 - 4ac}$ when $a = 2$, $b = -6$, and $c = 3$.

Rationalize the denominator.

75. $\dfrac{3}{\sqrt{2}}$ **76.** $\dfrac{5}{\sqrt{3}}$ **77.** $\dfrac{2}{\sqrt[3]{4}}$ **78.** $\dfrac{3}{\sqrt[3]{2}}$

One Step Further

Solve for x.

79. $(x + a)^2 + (x - a)^2 = 10a^2$
80. $(ax + 1)^2 + (ax - 1)^2 = 10$

Assume a is a positive number and solve for x by completing the square on x.

81. $x^2 + 2ax = -a^2$

82. $x^2 + 2ax = -4a^2$
83. $x^2 + 2ax = 0$
84. $x^2 + ax = 0$

Assume p and q are positive numbers and solve for x by completing the square on x.

85. $x^2 + px + q = 0$
86. $x^2 - px + q = 0$

The Quadratic Formula

In this section we will use the method of completing the square from the preceding section to derive the *quadratic formula*. The quadratic formula is a very useful tool in mathematics. It allows us to solve all types of quadratic equations.

> ◣ **THE QUADRATIC FORMULA**
>
> For any quadratic equation in the form $ax^2 + bx + c = 0$, where $a \neq 0$, the two solutions are
>
> $$x = \frac{-b + \sqrt{b^2 - 4ac}}{2a} \quad \text{and} \quad x = \frac{-b - \sqrt{b^2 - 4ac}}{2a}$$

Proof
We prove the quadratic formula by completing the square on $ax^2 + bx + c = 0$:

$$ax^2 + bx + c = 0$$
$$ax^2 + bx = -c \qquad \text{Add } -c \text{ to both sides.}$$
$$x^2 + \frac{b}{a}x = -\frac{c}{a} \qquad \text{Divide both sides by } a.$$

To complete the square on the left side, we add the square of one-half of $\dfrac{b}{a}$ to both sides:

$$x^2 + \frac{b}{a}x + \left(\frac{b}{2a}\right)^2 = -\frac{c}{a} + \left(\frac{b}{2a}\right)^2 \qquad \frac{1}{2} \text{ of } \frac{b}{a} \text{ is } \frac{b}{2a}.$$

We now simplify the right side as a separate step. We square the second term and combine the two terms by writing each with the least common denominator $4a^2$:

$$-\frac{c}{a} + \left(\frac{b}{2a}\right)^2 = -\frac{c}{a} + \frac{b^2}{4a^2} = \frac{4a}{4a}\left(\frac{-c}{a}\right) + \frac{b^2}{4a^2} = \frac{-4ac + b^2}{4a^2}$$

It is convenient to write this last expression as

$$\frac{b^2 - 4ac}{4a^2}$$

Continuing with the proof, we have

$$x^2 + \frac{b}{a}x + \left(\frac{b}{2a}\right)^2 = \frac{b^2 - 4ac}{4a^2}$$

$$\left(x + \frac{b}{2a}\right)^2 = \frac{b^2 - 4ac}{4a^2} \qquad \text{Write left side as a binomial square.}$$

$$x + \frac{b}{2a} = \pm\frac{\sqrt{b^2 - 4ac}}{2a} \qquad \text{Theorem 6.1}$$

$$x = -\frac{b}{2a} \pm \frac{\sqrt{b^2 - 4ac}}{2a} \qquad \text{Add } -\frac{b}{2a} \text{ to both sides.}$$

$$x = \frac{-b \pm \sqrt{b^2 - 4ac}}{2a}$$

Our proof is now complete. What we have is this: If our equation is in the form $ax^2 + bx + c = 0$ (standard form), where $a \neq 0$, the two solutions are always given by the formula

$$x = \frac{-b \pm \sqrt{b^2 - 4ac}}{2a}$$

This formula is known as the **quadratic formula.** If we substitute the coefficients a, b, and c of any quadratic equation in standard form into the formula, we need only perform some basic arithmetic to arrive at the solution set.

EXAMPLE 1 Use the quadratic formula to solve: $6x^2 + 7x - 5 = 0$.

SOLUTION Using $a = 6$, $b = 7$, and $c = -5$ in the formula

$$x = \frac{-b \pm \sqrt{b^2 - 4ac}}{2a}$$

we have

$$x = \frac{-7 \pm \sqrt{49 - 4(6)(-5)}}{2(6)}$$

or

$$x = \frac{-7 \pm \sqrt{49 + 120}}{12}$$

$$= \frac{-7 \pm \sqrt{169}}{12}$$

$$= \frac{-7 \pm 13}{12}$$

We separate the last equation into the two statements

$$x = \frac{-7 + 13}{12} \quad \text{or} \quad x = \frac{-7 - 13}{12}$$

$$x = \frac{1}{2} \quad \text{or} \quad x = -\frac{5}{3}$$

The solution set is $\{\frac{1}{2}, -\frac{5}{3}\}$.

Whenever the solutions to a quadratic equation are rational numbers, as they are in Example 1, it means that the original equation was solvable by factoring. To illustrate, let's solve the equation from Example 1 again, but this time by factoring:

$$6x^2 + 7x - 5 = 0 \quad \text{Equation in standard form}$$
$$(3x + 5)(2x - 1) = 0 \quad \text{Factor the left side.}$$
$$3x + 5 = 0 \quad \text{or} \quad 2x - 1 = 0 \quad \text{Set factors equal to 0.}$$
$$x = -\frac{5}{3} \quad \text{or} \quad x = \frac{1}{2}$$

When an equation can be solved by factoring, then factoring is usually the faster method of solution. It is best to try to factor first, and then if you have trouble factoring, go to the quadratic formula. It always works.

EXAMPLE 2 Solve: $\frac{x^2}{3} - x = -\frac{1}{2}$.

SOLUTION Multiplying through by 6 and writing the result in standard form, we have

$$2x^2 - 6x + 3 = 0$$

The left side of this equation is not factorable. Therefore, we use the quadratic formula with $a = 2$, $b = -6$, and $c = 3$. The two solutions are given by

$$x = \frac{-(-6) \pm \sqrt{36 - 4(2)(3)}}{2(2)}$$
$$= \frac{6 \pm \sqrt{12}}{4}$$
$$= \frac{6 \pm 2\sqrt{3}}{4} \qquad \sqrt{12} = \sqrt{4 \cdot 3} = \sqrt{4}\sqrt{3} = 2\sqrt{3}$$

We can reduce this last expression to lowest terms by factoring 2 from the numerator and denominator and then dividing the numerator and denominator by 2:

$$x = \frac{2(3 \pm \sqrt{3})}{2 \cdot 2} = \frac{3 \pm \sqrt{3}}{2}$$

EXAMPLE 3   Solve: $\dfrac{1}{x+2} - \dfrac{1}{x} = \dfrac{1}{3}$.

SOLUTION   To solve this equation, we must first put it in standard form. To do so, we must clear the equation of fractions by multiplying each side by the LCD for all the denominators, which is $3x(x+2)$.

$$3x(x+2)\left(\dfrac{1}{x+2} - \dfrac{1}{x}\right) = \dfrac{1}{3} \cdot 3x(x+2) \quad \text{Multiply each side by the LCD.}$$

$$3x(x+2) \cdot \dfrac{1}{x+2} - 3x(x+2) \cdot \dfrac{1}{x} = \dfrac{1}{3} \cdot 3x(x+2)$$

$$3x - 3(x+2) = x(x+2)$$

$$3x - 3x - 6 = x^2 + 2x \quad \text{Multiplication}$$

$$-6 = x^2 + 2x \quad \text{Simplify left side.}$$

$$0 = x^2 + 2x + 6 \quad \text{Add 6 to each side.}$$

Since the right side of the last equation is not factorable, we use the quadratic formula. From the last equation, we have $a = 1$, $b = 2$, and $c = 6$. Using these numbers for a, b, and c in the quadratic formula gives

$$x = \dfrac{-2 \pm \sqrt{4 - 4(1)(6)}}{2(1)}$$

$$= \dfrac{-2 \pm \sqrt{4 - 24}}{2} \quad \text{Simplify inside the radical.}$$

$$= \dfrac{-2 \pm \sqrt{-20}}{2}$$

$$= \dfrac{-2 \pm 2i\sqrt{5}}{2} \quad \sqrt{-20} = i\sqrt{20} = i\sqrt{4}\sqrt{5} = 2i\sqrt{5}$$

$$= \dfrac{2(-1 \pm i\sqrt{5})}{2} \quad \text{Factor 2 from the numerator.}$$

$$= -1 \pm i\sqrt{5} \quad \text{Divide the numerator and denominator by 2.}$$

Since neither of the two solutions, $-1 + i\sqrt{5}$ nor $-1 - i\sqrt{5}$, will make any of the denominators in our original equation 0, they are both solutions. ◢

Although the equation in our next example is not a quadratic equation, we can solve it by using both factoring and the quadratic formula.

EXAMPLE 4   Solve: $27t^3 - 8 = 0$.

SOLUTION   It would be a mistake to add 8 to each side of this equation and then take the cube root of each side because we would lose two of the solutions. Instead, we factor the left side and then set the factors equal to 0:

$$27t^3 - 8 = 0 \quad \text{Equation in standard form}$$

$$(3t - 2)(9t^2 + 6t + 4) = 0 \quad \text{Factor as the difference of two cubes.}$$

$$3t - 2 = 0 \quad \text{or} \quad 9t^2 + 6t + 4 = 0 \quad \text{Set each factor equal to 0.}$$

The first equation leads to a solution of $t = \frac{2}{3}$. The second equation does not factor, so we use the quadratic formula with $a = 9$, $b = 6$, and $c = 4$:

$$t = \frac{-6 \pm \sqrt{36 - 4(9)(4)}}{2(9)}$$

$$= \frac{-6 \pm \sqrt{36 - 144}}{18}$$

$$= \frac{-6 \pm \sqrt{-108}}{18}$$

$$= \frac{-6 \pm 6i\sqrt{3}}{18} \qquad \sqrt{-108} = i\sqrt{36 \cdot 3} = 6i\sqrt{3}$$

$$= \frac{6(-1 \pm i\sqrt{3})}{6 \cdot 3} \qquad \text{Factor 6 from the numerator and denominator.}$$

$$= \frac{-1 \pm i\sqrt{3}}{3} \qquad \text{Divide out common factor 6.}$$

The three solutions to the original equation are

$$\frac{2}{3}, \quad \frac{-1 + i\sqrt{3}}{3}, \quad \text{and} \quad \frac{-1 - i\sqrt{3}}{3}$$

EXAMPLE 5 If an object is thrown downward with an initial velocity of 20 feet/second, the distance $s(t)$, in feet, it travels in t seconds is given by the function $s(t) = 20t + 16t^2$. How long does it take the object to fall 40 feet?

SOLUTION We let $s(t) = 40$ and solve for t:

When $s(t) = 40$:

the function $s(t) = 20t + 16t^2$

becomes $40 = 20t + 16t^2$

or $16t^2 + 20t - 40 = 0$

$4t^2 + 5t - 10 = 0 \qquad \text{Divide by 4.}$

Using the quadratic formula, we have

$$t = \frac{-5 \pm \sqrt{25 - 4(4)(-10)}}{2(4)}$$

$$t = \frac{-5 \pm \sqrt{185}}{8}$$

$$t = \frac{-5 + \sqrt{185}}{8} \quad \text{or} \quad t = \frac{-5 - \sqrt{185}}{8}$$

The second solution is impossible since it is a negative number and time t must be positive. It takes

$$t = \frac{-5 + \sqrt{185}}{8} \quad \text{or approximately} \quad \frac{-5 + 13.60}{8} \approx 1.08 \text{ seconds}$$

for the object to fall 40 feet.

Recall from Chapter 3 that the relationship between profit, revenue, and cost is given by the formula

$$P(x) = R(x) - C(x)$$

where $P(x)$ is the profit, $R(x)$ is the total revenue, and $C(x)$ is the total cost of producing and selling x items.

EXAMPLE 6 A company produces and sells copies of an accounting program for home computers. The total weekly cost (in dollars) to produce x copies of the program is $C(x) = 8x + 500$, while the weekly revenue for selling all x copies of the program is $R(x) = 35x - 0.1x^2$. How many programs must be sold each week for the weekly profit to be $1,200?

SOLUTION Substituting the given expressions for $R(x)$ and $C(x)$ in the equation $P(x) = R(x) - C(x)$, we have a polynomial in x that represents the weekly profit $P(x)$:

$$\begin{aligned} P(x) &= R(x) - C(x) \\ &= 35x - 0.1x^2 - (8x + 500) \\ &= 35x - 0.1x^2 - 8x - 500 \\ &= -500 + 27x - 0.1x^2 \end{aligned}$$

Setting this expression equal to 1,200, we have a quadratic equation to solve that will give us the number of programs, x, that need to be sold each week to bring in a profit of $1,200:

$$1,200 = -500 + 27x - 0.1x^2$$

We can write this equation in standard form by adding the opposite of each term on the right side of the equation to both sides of the equation. Doing so produces the following equation:

$$0.1x^2 - 27x + 1,700 = 0$$

Applying the quadratic formula to this equation with $a = 0.1$, $b = -27$, and $c = 1,700$, we have

$$x = \frac{27 \pm \sqrt{(-27)^2 - 4(0.1)(1,700)}}{2(0.1)}$$

$$= \frac{27 \pm \sqrt{729 - 680}}{0.2}$$

$$= \frac{27 \pm \sqrt{49}}{0.2}$$

$$= \frac{27 \pm 7}{0.2}$$

Writing this last expression as two separate expressions, we have our two solutions:

$$x = \frac{27 + 7}{0.2} \qquad \text{or} \qquad x = \frac{27 - 7}{0.2}$$

$$= \frac{34}{0.2} \qquad\qquad\qquad = \frac{20}{0.2}$$

$$= 170 \qquad\qquad\qquad = 100$$

The weekly profit will be $1,200 if the company produces and sells 100 programs or 170 programs.

What is interesting about the equation we solved in Example 6 is that it has rational solutions, meaning it could have been solved by factoring. But looking back at the equation, factoring does not seem like a reasonable method of solution because the coefficients are either very large or very small. So, there are times when using the quadratic formula is a faster method of solution, even though the equation you are solving is factorable.

 Using Technology: Graphing Calculators

More about Example 5

We can solve the problem discussed in Example 5 by graphing the function $Y_1 = 20X + 16X^2$ in a window with X from 0 to 2 (because X is taking the place of t and we know t is a positive quantity) and Y from 0 to 50 (because we are looking for X when Y_1 is 40). Graphing Y_1 gives a graph similar to the graph in Figure 1 (at the top of the next page). Using the Zoom and Trace features at $Y_1 = 40$ gives us X = 1.08 to the nearest hundredth, matching the results we obtained by solving the original equation algebraically.

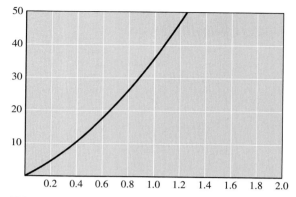

FIGURE 1

More about Example 6

To visualize the functions in Example 6, we set up our calculator this way:

$$Y_1 = 35X - .1X^2 \quad \text{Revenue function}$$
$$Y_2 = 8X + 500 \quad \text{Cost function}$$
$$Y_3 = Y_1 - Y_2 \quad \text{Profit function}$$

Window: X from 0 to 350, Y from 0 to 3500

Graphing these functions produces graphs similar to the ones shown in Figure 2. The lower graph is the graph of the profit function. Using the Zoom and Trace features on the lower graph at $Y_3 = 1{,}200$ produces two corresponding values of X, 170 and 100, which match the results in Example 6.

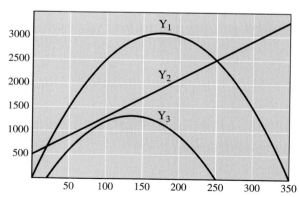

FIGURE 2

We will continue this discussion of the relationship between graphs of functions and solutions to equations in the Using Technology material in the next section.

Problem Set

6.2

Solve each equation. Use factoring or the quadratic formula, whichever is appropriate.

> **NOTE:** Try factoring first. If you have any difficulty factoring, then go right to the quadratic formula.

1. $x^2 + 5x + 6 = 0$

2. $x^2 + 5x - 6 = 0$

3. $a^2 - 4a + 1 = 0$

4. $a^2 + 4a + 1 = 0$

5. $\frac{1}{6}x^2 - \frac{1}{2}x + \frac{1}{3} = 0$

6. $\frac{1}{4}x^2 + \frac{1}{4}x - \frac{1}{2} = 0$

7. $\frac{x^2}{2} + 1 = \frac{2x}{3}$

8. $\frac{x^2}{2} + \frac{2}{3} = -\frac{2x}{3}$

9. $y^2 - 5y = 0$

10. $2y^2 + 10y = 0$

11. $30x^2 + 40x = 0$

12. $50x^2 - 20x = 0$

13. $\frac{2t^2}{3} - t = -\frac{1}{6}$

14. $\frac{t^2}{3} - \frac{t}{2} = -\frac{3}{2}$

15. $0.01x^2 + 0.06x - 0.08 = 0$

16. $0.02x^2 - 0.03x + 0.05 = 0$

17. $2x + 3 = -2x^2$

18. $2x - 3 = 3x^2$

19. $100x^2 - 200x + 100 = 0$

20. $100x^2 - 600x + 900 = 0$

21. $\frac{1}{2}r^2 = \frac{1}{6}r - \frac{2}{3}$

22. $\frac{1}{4}r^2 = \frac{2}{5}r + \frac{1}{10}$

23. $(x - 3)(x - 5) = 1$

24. $(x - 3)(x + 1) = -6$

25. $(x + 3)^2 + (x - 8)(x - 1) = 16$

26. $(x - 4)^2 + (x + 2)(x + 1) = 9$

27. $\frac{x^2}{3} - \frac{5x}{6} = \frac{1}{2}$

28. $\frac{x^2}{6} + \frac{5}{6} = -\frac{x}{3}$

Multiply both sides of each equation by its LCD. Then solve the resulting equation.

29. $\frac{1}{x + 1} - \frac{1}{x} = \frac{1}{2}$

30. $\frac{1}{x + 1} + \frac{1}{x} = \frac{1}{3}$

31. $\frac{1}{y - 1} + \frac{1}{y + 1} = 1$

32. $\frac{2}{y + 2} + \frac{3}{y - 2} = 1$

33. $\frac{1}{x + 2} + \frac{1}{x + 3} = 1$

34. $\frac{1}{x + 3} + \frac{1}{x + 4} = 1$

35. $\frac{6}{r^2 - 1} - \frac{1}{2} = \frac{1}{r + 1}$

36. $2 + \frac{5}{r - 1} = \frac{12}{(r - 1)^2}$

Solve each equation. In each case you will have three solutions.

37. $x^3 - 8 = 0$

38. $x^3 - 27 = 0$

39. $8a^3 + 27 = 0$

40. $27a^3 + 8 = 0$

41. $125t^3 - 1 = 0$

42. $64t^3 + 1 = 0$

Each of the following equations has three solutions. Look first for the greatest common factor, then use the quadratic formula to find all solutions.

43. $2x^3 + 2x^2 + 3x = 0$

44. $6x^3 - 4x^2 + 6x = 0$

45. $3y^4 = 6y^3 - 6y^2$

46. $4y^4 = 16y^3 - 20y^2$

47. $6t^5 + 4t^4 = -2t^3$

48. $8t^5 + 2t^4 = -10t^3$

49. One solution to a quadratic equation is $\frac{-3 + 2i}{5}$. What is the other solution?

50. One solution to a quadratic equation is $\frac{-2 + 3i\sqrt{2}}{5}$. What is the other solution?

360

Applying the Concepts

Problems 51–58 may be solved using a graphing calculator. For Problems 51–54, assume all distances are in feet and all times are in seconds.

51. Falling Object An object is thrown downward with an initial velocity of 5 feet/second. The relationship between the distance s it travels and time t is given by $s(t) = 5t + 16t^2$. How long does it take the object to fall 74 feet?

52. Falling Object The distance an object falls from rest is given by the equation $s(t) = 16t^2$, where s = distance and t = time. How long does it take an object dropped from a 100 foot cliff to hit the ground?

53. Ball Toss A ball is thrown upward with an initial velocity of 20 feet/second. The equation that gives the height h of the ball at any time t is $h(t) = 20t - 16t^2$. At what times will the ball be 4 feet off the ground?

54. Coin Toss A coin is propelled upward with an initial velocity of 32 feet/second from a height of 16 feet above the ground. The equation giving the coin's height h at any time t is $h(t) = 16 + 32t - 16t^2$. Does the coin ever reach a height of 32 feet?

55. Profit The total cost (in dollars) for a company to manufacture and sell x items per week is $C(x) = 60x + 300$, while the revenue brought in by selling all x items is $R(x) = 100x - 0.5x^2$. How many items must be sold to obtain a weekly profit of $300?

56. Profit The total cost (in dollars) for a company to produce and sell x items per week is $C(x) = 200x + 1,600$, while the revenue brought in by selling all x items is $R(x) = 300x - 0.5x^2$. How many items must be sold to make a weekly profit of $2,150?

57. Profit Suppose it costs a company selling patterns $C(x) = 800 + 6.5x$ dollars to produce and sell x patterns a month. If the revenue obtained by selling x patterns is $R(x) = 10x - 0.002x^2$, how many patterns must be sold each month if the company wants a monthly profit of $700?

58. Profit Suppose a company manufactures and sells x picture frames each month with a total cost of $C(x) = 1,200 + 3.5x$ dollars. If the revenue obtained by selling x frames is $R(x) = 9x - 0.002x^2$, find the number of frames the company must sell each month if its monthly profit is to be $2,300.

Review Problems

NOTE: The problems that follow review material we covered in Sections 4.2 and 5.1. Reviewing the problems from Section 4.2 will help you with the next section.

Divide, using long division. [4.2]

59. $\dfrac{8y^2 - 26y - 9}{2y - 7}$ **60.** $\dfrac{6y^2 + 7y - 18}{3y - 4}$

61. $\dfrac{x^3 + 9x^2 + 26x + 24}{x + 2}$

62. $\dfrac{x^3 + 6x^2 + 11x + 6}{x + 3}$

Simplify each expression. (Assume x, $y > 0$.) [5.1]

63. $25^{1/2}$ **64.** $8^{1/3}$

65. $(\frac{9}{25})^{3/2}$ **66.** $(\frac{16}{81})^{3/4}$

67. $8^{-2/3}$ **68.** $4^{-3/2}$

69. $\dfrac{(49x^8y^{-4})^{1/2}}{(27x^{-3}y^9)^{-1/3}}$ **70.** $\dfrac{(x^{-2}y^{1/3})^6}{x^{-10}y^{3/2}}$

One Step Further

So far, all the equations we have solved have had coefficients that were rational numbers. Here are some equations that have irrational coefficients and some that have complex coefficients. Solve each equation. (Remember: $i^2 = -1$.)

71. $x^2 + \sqrt{3}x - 6 = 0$
72. $x^2 - \sqrt{5}x - 5 = 0$
73. $\sqrt{2}x^2 + 2x - \sqrt{2} = 0$
74. $\sqrt{7}x^2 + 2\sqrt{2}x - \sqrt{7} = 0$
75. $x^2 + ix + 2 = 0$
76. $x^2 + 3ix - 2 = 0$
77. $ix^2 + 3x + 4i = 0$
78. $4ix^2 + 5x + 9i = 0$

Additional Items Involving Solutions to Equations

In this section we will do two things. First, we will define the discriminant and use it to determine the kind of solutions a quadratic equation has without solving the equation. Second, we will use the zero-factor property to build equations from their solutions.

The Discriminant

The quadratic formula

$$x = \frac{-b \pm \sqrt{b^2 - 4ac}}{2a}$$

gives the solutions to any quadratic equation in standard form. There are times, when working with quadratic equations, when it is important only to know what kind of solutions the equation has. The **discriminant** helps us find out.

> **DEFINITION**
>
> The expression under the radical in the quadratic formula is called the **discriminant:**
>
> $$\text{Discriminant} = D = b^2 - 4ac$$

When the original equation has integer coefficients, the discriminant indicates the number and type of solutions to a quadratic equation. For example, if we were to use the quadratic formula to solve the equation $2x^2 + 2x + 3 = 0$, we would find the discriminant to be

$$b^2 - 4ac = 2^2 - 4(2)(3) = -20$$

Since the discriminant appears under a square root symbol, we have the square root of a negative number in the quadratic formula. Our solutions would therefore be complex numbers. Similarly, if the discriminant were 0, the quadratic formula would yield

$$x = \frac{-b \pm \sqrt{0}}{2a} = \frac{-b \pm 0}{2a} = \frac{-b}{2a}$$

and the equation would have one rational solution: the number $\frac{-b}{2a}$.

The following table gives the relationship between the discriminant and the type of solutions to the equation.

For the equation $ax^2 + bx + c = 0$, where a, b, and c are integers and $a \neq 0$:

IF THE DISCRIMINANT $b^2 - 4ac$ IS:	THEN THE EQUATION WILL HAVE:
Negative	Two complex solutions containing i
Zero	One rational solution
A positive number that is also a perfect square	Two rational solutions
A positive number that is not a perfect square	Two irrational solutions

In the second and third cases, when the discriminant is 0 or a positive perfect square, the solutions are rational numbers. The quadratic equations in these two cases are the ones that can be factored.

EXAMPLES For each equation, give the number and kind of solutions.

1. $x^2 - 3x - 40 = 0$

Using $a = 1$, $b = -3$, and $c = -40$ in $b^2 - 4ac$, we have

$$(-3)^2 - 4(1)(-40) = 9 + 160 = 169$$

The discriminant is a perfect square. Therefore, the equation has two rational solutions.

2. $2x^2 - 3x + 4 = 0$

Using $a = 2$, $b = -3$, and $c = 4$, we have

$$b^2 - 4ac = (-3)^2 - 4(2)(4) = 9 - 32 = -23$$

The discriminant is negative, implying the equation has two complex solutions that contain i.

3. $4x^2 - 12x + 9 = 0$

Using $a = 4$, $b = -12$, and $c = 9$, the discriminant is

$$b^2 - 4ac = (-12)^2 - 4(4)(9) = 144 - 144 = 0$$

Since the discriminant is 0, the equation will have one rational solution.

4. $x^2 + 6x = 8$

We must first put the equation in standard form by adding -8 to each side. If we do so, the resulting equation is

$$x^2 + 6x - 8 = 0$$

Now we identify a, b, and c as 1, 6, and -8, respectively:

$$b^2 - 4ac = 6^2 - 4(1)(-8) = 36 + 32 = 68$$

The discriminant is a positive number, but not a perfect square. Therefore, the equation will have two irrational solutions.

EXAMPLE 5   Find an appropriate k so that the equation $4x^2 - kx = -9$ has exactly one rational solution.

SOLUTION   We begin by writing the equation in standard form:

$$4x^2 - kx + 9 = 0$$

Using $a = 4$, $b = -k$, and $c = 9$, we have

$$b^2 - 4ac = (-k)^2 - 4(4)(9)$$
$$= k^2 - 144$$

An equation has exactly one rational solution when the discriminant is 0. We set the discriminant equal to 0 and solve:

$$k^2 - 144 = 0$$
$$k^2 = 144$$
$$k = \pm 12$$

Choosing k to be 12 or -12 will result in an equation with one rational solution.

Building Equations from Their Solutions

Suppose we know that the solutions to an equation are $x = 3$ and $x = -2$. We can find equations with these solutions by using the zero-factor property. First, let's write our solutions as equations with 0 on the right side:

If	$x = 3$	First solution		If	$x = -2$	Second solution
then	$x - 3 = 0$	Add -3 to each side.		then	$x + 2 = 0$	Add 2 to each side.

Now, since both $x - 3$ and $x + 2$ are 0, their product must be 0 also. Therefore, we can write

$$(x - 3)(x + 2) = 0 \quad \text{Zero-factor property}$$
$$x^2 - x - 6 = 0 \quad \text{Multiply out the left side.}$$

There are many other equations that have 3 and -2 as solutions. For example, any constant multiple of $x^2 - x - 6 = 0$, such as $5x^2 - 5x - 30 = 0$, also has 3 and -2 as solutions. Similarly, any equation built from positive integer powers of the factors $x - 3$ and $x + 2$ will also have 3 and -2 as solutions. One such equation is

$$(x - 3)^2(x + 2) = 0$$
$$(x^2 - 6x + 9)(x + 2) = 0$$
$$x^3 - 4x^2 - 3x + 18 = 0$$

In mathematics, we distinguish between the solutions to this last equation and those to the equation $x^2 - x - 6 = 0$ by saying $x = 3$ is a solution of **multiplicity 2** in the equation $x^3 - 4x^2 - 3x + 18 = 0$, and a solution of **multiplicity 1** in the equation $x^2 - x - 6 = 0$.

EXAMPLE 6 Find an equation that has solutions $t = 5$, $t = -5$, and $t = 3$.

SOLUTION First, we use the given solutions to write equations that have 0 on the right side.

$$\text{If} \quad t = 5 \qquad t = -5 \qquad t = 3$$
$$\text{then} \quad t - 5 = 0 \qquad t + 5 = 0 \qquad t - 3 = 0$$

Since $t - 5$, $t + 5$, and $t - 3$ are all 0, their product is also 0 by the zero-factor property. An equation with solutions of 5, -5, and 3 is

$$(t - 5)(t + 5)(t - 3) = 0 \quad \text{Zero-factor property}$$
$$(t^2 - 25)(t - 3) = 0 \quad \text{Multiply first two binomials.}$$
$$t^3 - 3t^2 - 25t + 75 = 0 \quad \text{Complete the multiplication.}$$

Remember, there are many other equations with these same solutions.

The last line gives us an equation with solutions of 5, -5, and 3.

EXAMPLE 7 Find an equation with solutions $x = -\frac{2}{3}$ and $x = \frac{4}{5}$.

SOLUTION The solution $x = -\frac{2}{3}$ can be rewritten as $3x + 2 = 0$ as follows:

$$x = -\frac{2}{3} \quad \text{The first solution}$$
$$3x = -2 \quad \text{Multiply each side by 3.}$$
$$3x + 2 = 0 \quad \text{Add 2 to each side.}$$

Similarly, the solution $x = \frac{4}{5}$ can be rewritten as $5x - 4 = 0$:

$$x = \frac{4}{5} \quad \text{The second solution}$$
$$5x = 4 \quad \text{Multiply each side by 5.}$$
$$5x - 4 = 0 \quad \text{Add } -4 \text{ to each side.}$$

Since both $3x + 2$ and $5x - 4$ are 0, their product is 0 also, giving us the equation we are looking for:

$$(3x + 2)(5x - 4) = 0 \quad \text{Zero-factor property}$$
$$15x^2 - 2x - 8 = 0 \quad \text{Multiplication}$$

From Example 6 and the discussion that preceded it, we know that if $x = a$ is a solution to an equation, then $x - a$ must be a factor of the same equation. We can use this fact to solve equations that we would otherwise be unable to solve.

EXAMPLE 8 Find all solutions to the equation

$$x^3 + 9x^2 + 26x + 24 = 0$$

if $x = -2$ is one of its solutions.

SOLUTION Since $x = -2$ is a solution to the equation, $x + 2$ must be a factor of the left side of the equation—meaning $x + 2$ divides the left side evenly. In Example

You may want to look back
at Section 4.2 to see what
we are talking about.

10 of Section 4.2, we used long division to divide $x^3 + 9x^2 + 26x + 24$ by $x + 2$
and found the quotient to be $x^2 + 7x + 12$. Without showing that division problem
again, here is how we use it to find all solutions to the equation in question:

$$x^3 + 9x^2 + 26x + 24 = 0$$

$$(x + 2)(x^2 + 7x + 12) = 0 \qquad \text{From long division by } x + 2$$

$$(x + 2)(x + 3)(x + 4) = 0 \qquad \text{Factoring the trinomial}$$

$$x + 2 = 0 \quad \text{or} \quad x + 3 = 0 \quad \text{or} \quad x + 4 = 0 \qquad \text{Setting each factor to 0}$$

$$x = -2 \quad \text{or} \quad x = -3 \quad \text{or} \quad x = -4 \qquad \text{The three solutions}$$

Using Technology: Graphing Calculators

Solving Equations

Now that we have explored the relationship between equations and their solutions,
we can look at how a graphing calculator can be used in the solution process. To
begin, let's solve the equation $x^2 = x + 2$ using techniques from algebra: writing it
in standard form, factoring, and then setting each factor equal to 0.

$$x^2 - x - 2 = 0 \qquad \text{Standard form}$$

$$(x - 2)(x + 1) = 0 \qquad \text{Factor.}$$

$$x - 2 = 0 \quad \text{or} \quad x + 1 = 0 \qquad \text{Set each factor equal to 0.}$$

$$x = 2 \quad \text{or} \quad x = -1 \qquad \text{Solve.}$$

Our original equation, $x^2 = x + 2$, has two solutions, $x = 2$ and $x = -1$. To
solve the equation using a graphing calculator, we need to associate it with an equation (or equations) in two variables. One way to do this is to associate the left side
with the equation $y = x^2$ and the right side of the equation with $y = x + 2$. To do
so, we set up the functions list in our calculator this way:

$$Y_1 = X^2$$
$$Y_2 = X + 2$$
Window: X from -5 to 5, Y from -5 to 5

Graphing these functions in this window will produce a graph similar to the one
shown in Figure 1 (at the top of the next page).

If we use the Trace feature to find the coordinates of the points of intersection,
we find that the two curves intersect at $(-1, 1)$ and $(2, 4)$. We note that the x-
coordinates of these two points match the solutions to the equation $x^2 = x + 2$,
which we found using algebraic techniques. This makes sense because if two graphs
intersect at a point (x, y), then the coordinates of that point satisfy both equations.
If a point (x, y) satisfies both $y = x^2$ and $y = x + 2$, then, for that particular point,
$x^2 = x + 2$. From this we conclude that the x-coordinates of the points of intersec-

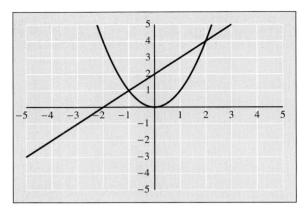

FIGURE 1

tion are solutions to our original equation. Here is a summary of what we have discovered:

Conclusion 1 If the graphs of two functions $y = f(x)$ and $y = g(x)$ intersect in the coordinate plane, then the x-coordinates of the points of intersection are solutions to the equation $f(x) = g(x)$.

A second method of solving our original equation $x^2 = x + 2$ graphically requires the use of one function instead of two. To begin, we write the equation in standard form as $x^2 - x - 2 = 0$. Next, we graph the function $y = x^2 - x - 2$. The x-intercepts of the graph are the points with y-coordinates of 0. Therefore they satisfy the equation $0 = x^2 - x - 2$, which is equivalent to our original equation. The graph in Figure 2 shows $Y_1 = X^2 - X - 2$ in a window with X from -5 to 5 and Y from -5 to 5.

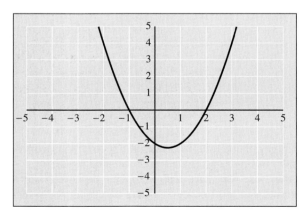

FIGURE 2

Using the Trace feature, we find that the x-intercepts of the graph are $x = -1$ and $x = 2$, which match the solutions to our original equation $x^2 = x + 2$. We can summarize the relationship between solutions to an equation and the intercepts of its associated graph this way:

Conclusion 2 If $y = f(x)$ is a function, then any x-intercept on the graph of $y = f(x)$ is a solution to the equation $f(x) = 0$.

Solving Equations with Complex Solutions

There are limitations to using a graph to solve an equation. To illustrate, suppose we solve the equation $4x^3 - 6x^2 + 2x - 3 = 0$ by factoring:

$$4x^3 - 6x^2 + 2x - 3 = 0 \quad \text{Original equation}$$
$$2x^2(2x - 3) + (2x - 3) = 0$$
$$(2x^2 + 1)(2x - 3) = 0 \quad \text{Factor by grouping.}$$

$$2x^2 + 1 = 0 \qquad \text{or} \quad 2x - 3 = 0 \quad \text{Set factors equal to 0.}$$
$$2x^2 = -1 \qquad \text{or} \qquad 2x = 3$$
$$x^2 = -\frac{1}{2} \qquad \text{or} \qquad x = \frac{3}{2} \quad \text{Solve the resulting equations.}$$

$$x = \pm\frac{1}{\sqrt{2}}\, i$$
$$= \pm\frac{\sqrt{2}}{2}\, i$$

We have three solutions:

$$-\frac{\sqrt{2}}{2}\, i, \quad \frac{\sqrt{2}}{2}\, i, \quad \frac{3}{2}$$

Figure 3 shows the graph of $y = 4x^3 - 6x^2 + 2x - 3$. As you can see, the graph crosses the x-axis exactly once at $x = \frac{3}{2}$, which we expect. The rectangular coordinate system consists of ordered pairs of *real numbers,* so our complex solutions cannot appear on the graph.

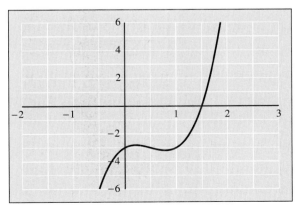

FIGURE 3

Every cubic equation will have at least one real number solution. If we are interested in exact values for all solutions to our equation, including complex solutions as well as irrational solutions, we can take our one real solution and, with the aid of long division and factoring or the quadratic formula, find the other solutions, just as we did in Example 8. But how do we find the one real solution in the first place? We use a graphing calculator.

Suppose we want to solve the equation $x^3 = 15x + 4$ completely. We can use a graphing calculator to graph the function $y = x^3 - 15x - 4$, and note that it crosses the x-axis at $x = 4$. The graph is shown in Figure 4.

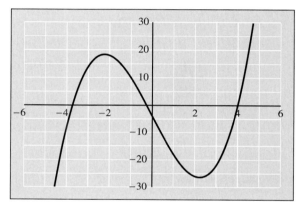

FIGURE 4

If we zoom and trace on the other two solutions, we find they are approximated by -3.732 and -0.268 to the nearest thousandth. To find exact values for these last two solutions, we reason that if $x = 4$ is a solution, then $x - 4$ must be a factor of the equation $x^3 - 15x - 4$ without a remainder. Here is the division problem:

$$
\begin{array}{r}
x^2 + 4x + 1 \\
x - 4 \overline{)x^3 + 0x^2 - 15x - 4} \\
\underline{x^3 - 4x^2} \\
4x^2 - 15x \\
\underline{4x^2 - 16x} \\
x - 4 \\
\underline{x - 4} \\
0
\end{array}
$$

We find that $x^3 - 15x - 4$ factors into $(x - 4)(x^2 + 4x + 1)$. This allows us to find all solutions to our original equation:

$$x^3 - 15x - 4 = 0$$
$$(x - 4)(x^2 + 4x + 1) = 0$$
$$x - 4 = 0 \quad \text{or} \quad x^2 + 4x + 1 = 0$$
$$x = 4 \quad \text{or} \quad x = \frac{-4 \pm \sqrt{16 - 4(1)(1)}}{2(1)}$$
$$= \frac{-4 \pm \sqrt{12}}{2}$$
$$= \frac{-4 \pm 2\sqrt{3}}{2}$$
$$= -2 \pm \sqrt{3}$$

The three exact solutions to our equation are 4, $-2 + \sqrt{3}$, and $-2 - \sqrt{3}$.

Problem Set 6.3

Use the discriminant to find the number and kind of solutions for each of the following equations.

1. $x^2 - 6x + 5 = 0$
2. $x^2 - x - 12 = 0$
3. $4x^2 - 4x = -1$
4. $9x^2 + 12x = -4$
5. $x^2 + x - 1 = 0$
6. $x^2 - 2x + 3 = 0$
7. $2y^2 = 3y + 1$
8. $3y^2 = 4y - 2$
9. $x^2 - 9 = 0$
10. $4x^2 - 81 = 0$
11. $5a^2 - 4a = 5$
12. $3a = 4a^2 - 5$

Determine k so that each of the following has exactly one real solution.

13. $x^2 - kx + 25 = 0$
14. $x^2 + kx + 25 = 0$
15. $x^2 = kx - 36$
16. $x^2 = kx - 49$
17. $4x^2 - 12x + k = 0$
18. $9x^2 + 30x + k = 0$
19. $kx^2 - 40x = 25$
20. $kx^2 - 2x = -1$
21. $3x^2 - kx + 2 = 0$
22. $5x^2 + kx + 1 = 0$

For each of the following problems, find an equation that has the given solutions.

23. $x = 5, x = 2$
24. $x = -5, x = -2$
25. $t = -3, t = 6$
26. $t = -4, t = 2$
27. $y = 2, y = -2, y = 4$
28. $y = 1, y = -1, y = 3$
29. $x = \frac{1}{2}, x = 3$
30. $x = \frac{1}{3}, x = 5$

31. $t = -\frac{3}{4}, t = 3$
32. $t = -\frac{4}{5}, t = 2$
33. $x = 3, x = -3, x = \frac{5}{6}$
34. $x = 5, x = -5, x = \frac{2}{3}$
35. $a = -\frac{1}{2}, a = \frac{3}{5}$
36. $a = -\frac{1}{3}, a = \frac{4}{7}$
37. $x = -\frac{2}{3}, x = \frac{2}{3}, x = 1$
38. $x = -\frac{4}{5}, x = \frac{4}{5}, x = -1$
39. $x = 2, x = -2, x = 3, x = -3$
40. $x = 1, x = -1, x = 5, x = -5$

41. Find an equation that has a solution of $x = 3$ of multiplicity 1 and a solution $x = -5$ of multiplicity 2.

42. Find an equation that has a solution of $x = 5$ of multiplicity 1 and a solution $x = -3$ of multiplicity 2.

43. Find an equation that has solutions $x = 3$ and $x = -3$, both of multiplicity 2.

44. Find an equation that has solutions $x = 4$ and $x = -4$, both of multiplicity 2.

45. Find all solutions to $x^3 + 6x^2 + 11x + 6 = 0$ if $x = -3$ is one of its solutions.

46. Find all solutions to $x^3 + 10x^2 + 29x + 20 = 0$ if $x = -4$ is one of its solutions.

47. One solution to $y^3 + 5y^2 - 2y - 24 = 0$ is $y = -3$. Find all solutions.

48. One solution to $y^3 + 3y^2 - 10y - 24 = 0$ is $y = -2$. Find all solutions.

49. If $x = 3$ is one solution to $x^3 - 5x^2 + 8x = 6$, find the other solutions.

50. If $x = 2$ is one solution to $x^3 - 6x^2 + 13x = 10$, find the other solutions.

51. Find all solutions to $t^3 = 13t^2 - 65t + 125$ if $t = 5$ is one of the solutions.

52. Find all solutions to $t^3 = 8t^2 - 25t + 26$ if $t = 2$ is one of the solutions.

Review Problems

 NOTE: The problems that follow review material we covered in Section 5.2. Reviewing these problems will help you with the next section.

Multiply.

53. $a^4(a^{3/2} - a^{1/2})$

54. $(a^{1/2} - 5)(a^{1/2} + 3)$

55. $(x^{3/2} - 3)^2$

56. $(x^{1/2} - 8)(x^{1/2} + 8)$

Divide.

57. $\dfrac{30x^{3/4} - 25x^{5/4}}{5x^{1/4}}$

58. $\dfrac{45x^{5/3}\, y^{7/3} - 36x^{8/3}y^{4/3}}{9x^{2/3}\, y^{1/3}}$

59. Factor $5(x - 3)^{1/2}$ from
$$10(x - 3)^{3/2} - 15(x - 3)^{1/2}$$

60. Factor $2(x + 1)^{1/3}$ from
$$8(x + 1)^{4/3} - 2(x + 1)^{1/3}$$

Factor each of the following as if they were trinomials.

61. $2x^{2/3} - 11x^{1/3} + 12$

62. $9x^{2/3} + 12x^{1/3} + 4$

One Step Further

63. Find all solutions to $x^4 + x^3 - x^2 + x - 2 = 0$ if $x = -2$ is one solution.

64. Find all solutions to $x^4 - x^3 + 2x^2 - 4x - 8 = 0$ if $x = 2$ is one solution.

65. Find all solutions to $x^3 + 3ax^2 + 3a^2x + a^3 = 0$ if $x = -a$ is one solution.

66. If $x = -2a$ is one solution to $x^3 + 6ax^2 + 12a^2x + 8a^3 = 0$, find the other solutions.

Using Technology

Find all solutions to the following equations. Solve using algebra and by graphing. If rounding is necessary, round to the nearest hundredth.

67. $x^2 = 4x + 5$

68. $4x^2 = 8x + 5$

69. $x^2 - 1 = 2x$

70. $4x^2 - 1 = 4x$

Find all solutions to each equation. If rounding is necessary, round to the nearest hundredth.

71. $2x^3 - x^2 - 2x + 1 = 0$

72. $3x^3 - 2x^2 - 3x + 2 = 0$

73. $2x^3 + 2 = x^2 + 4x$

74. $3x^3 - 9x = 2x^2 - 6$

Each equation below has only one real solution. Find it by graphing. Then use long division or the quadratic formula to find the remaining solutions.

75. $3x^3 - 8x^2 + 10x - 4 = 0$

76. $10x^3 + 6x^2 + x - 2 = 0$

SECTION 6.4

Equations That Are Quadratic in Form

We are now in a position to put our knowledge of quadratic equations to work to solve a variety of equations.

EXAMPLE 1   Solve: $(x + 3)^2 - 2(x + 3) - 8 = 0$.

SOLUTION

Method 1   We can see that this equation is quadratic in form by replacing $x + 3$ with another variable, say y. Replacing $x + 3$ with y, we have

$$y^2 - 2y - 8 = 0$$

We can solve this equation by factoring the left side and then setting each factor equal to 0:

$$y^2 - 2y - 8 = 0$$
$$(y - 4)(y + 2) = 0 \qquad \text{Factor.}$$
$$y - 4 = 0 \quad \text{or} \quad y + 2 = 0 \qquad \text{Set factors to 0.}$$
$$y = 4 \quad \text{or} \qquad y = -2$$

Since our original equation was written in terms of the variable x, we would like our solutions in terms of x also. Replacing y with $x + 3$ and then solving for x, we have

$$x + 3 = 4 \qquad \text{or} \qquad x + 3 = -2$$
$$x = 1 \qquad \text{or} \qquad x = -5$$

The solutions to our original equation are 1 and -5.

   This method lends itself well to other types of equations that are quadratic in form, as we will see.

Method 2   In this example, there is another method that works just as well as the method used above. Let's solve our original equation again, but this time, let's begin by expanding $(x + 3)^2$ and $2(x + 3)$:

$$(x + 3)^2 - 2(x + 3) - 8 = 0$$
$$x^2 + 6x + 9 - 2x - 6 - 8 = 0 \qquad \text{Multiply.}$$
$$x^2 + 4x - 5 = 0 \qquad \text{Combine similar terms.}$$
$$(x - 1)(x + 5) = 0 \qquad \text{Factor.}$$
$$x - 1 = 0 \quad \text{or} \quad x + 5 = 0 \qquad \text{Set factors to 0.}$$
$$x = 1 \quad \text{or} \qquad x = -5$$

   As you can see, either method produces the same results.

EXAMPLE 2   Solve: $4x^4 + 7x^2 = 2$.

The choice of the letter y is arbitrary. We could just as easily use the substitution $m = x^2$.

SOLUTION   This equation is quadratic in x^2. We can make it easier to see this by using the substitution $y = x^2$. Making the substitution $y = x^2$ and then solving the resulting equation, we have

$$4y^2 + 7y = 2$$

$$4y^2 + 7y - 2 = 0 \qquad \text{Standard form}$$

$$(4y - 1)(y + 2) = 0 \qquad \text{Factor.}$$

$$4y - 1 = 0 \quad \text{or} \quad y + 2 = 0 \qquad \text{Set factors to 0.}$$

$$y = \tfrac{1}{4} \quad \text{or} \qquad y = -2$$

Now we replace y with x^2 in order to solve for x:

$$x^2 = \tfrac{1}{4} \qquad\qquad \text{or} \qquad x^2 = -2$$

$$x = \pm\sqrt{\tfrac{1}{4}} \qquad \text{or} \qquad x = \pm\sqrt{-2} \quad \text{Theorem 6.1}$$

$$x = \pm\tfrac{1}{2} \qquad \text{or} \qquad x = \pm i\sqrt{2}$$

The solution set is $\{-\tfrac{1}{2}, \tfrac{1}{2}, i\sqrt{2}, -i\sqrt{2}\}$.

EXAMPLE 3 Solve for x: $x + \sqrt{x} - 6 = 0$.

SOLUTION To see that this equation is quadratic in form, we have to notice that $(\sqrt{x})^2 = x$. That is, the equation can be rewritten as

$$(\sqrt{x})^2 + \sqrt{x} - 6 = 0$$

Replacing $\sqrt{x}$ with y and solving as usual, we have

$$y^2 + y - 6 = 0$$

$$(y + 3)(y - 2) = 0$$

$$y + 3 = 0 \quad \text{or} \quad y - 2 = 0$$

$$y = -3 \quad \text{or} \qquad y = 2$$

Again, to find x we replace y with $\sqrt{x}$ and solve:

$$\sqrt{x} = -3 \qquad \text{or} \qquad \sqrt{x} = 2$$

$$x = 9 \qquad\qquad\qquad x = 4 \quad \text{Square both sides of each equation.}$$

Since we squared both sides of each equation, we have the possibility of obtaining extraneous solutions. We have to check both solutions in our original equation.

Check

When $x = 9$:

the equation $x + \sqrt{x} - 6 = 0$

becomes $9 + \sqrt{9} - 6 \overset{?}{=} 0$

$9 + 3 - 6 \overset{?}{=} 0$

$6 \neq 0$

This means 9 is extraneous.

When $x = 4$:

the equation $x + \sqrt{x} - 6 = 0$

becomes $4 + \sqrt{4} - 6 \overset{?}{=} 0$

$4 + 2 - 6 \overset{?}{=} 0$

$0 = 0$

This means 4 is a solution.

The only solution to the equation $x + \sqrt{x} - 6 = 0$ is $x = 4$.

We should note here that the two possible solutions, 9 and 4, to the equation in Example 3 can be obtained by another method. Instead of substituting for $\sqrt{x}$, we can isolate it on one side of the equation and then square both sides to clear the equation of radicals:

$$x + \sqrt{x} - 6 = 0$$

$$\sqrt{x} = -x + 6 \qquad \text{Isolate } \sqrt{x}.$$

$$x = x^2 - 12x + 36 \qquad \text{Square both sides.}$$

$$0 = x^2 - 13x + 36 \qquad \text{Add } -x \text{ to both sides.}$$

$$0 = (x - 4)(x - 9) \qquad \text{Factor.}$$

$$x - 4 = 0 \quad \text{or} \quad x - 9 = 0$$

$$x = 4 \qquad\qquad x = 9$$

We obtain the same two possible solutions. Since we squared both sides of the equation to find them, we would have to check each one in the original equation. As was the case in Example 3, only $x = 4$ is a solution; $x = 9$ is extraneous.

EXAMPLE 4 If an object is tossed into the air with an upward velocity of 12 feet/second from the top of a building h feet high, the time (in seconds) it takes for the object to hit the ground below is given by the formula

$$16t^2 - 12t - h = 0$$

Solve this formula for t.

SOLUTION The formula is in standard form and is quadratic in t. The coefficients a, b, and c that we need to apply to the quadratic formula are $a = 16$, $b = -12$, and $c = -h$. Substituting these quantities into the quadratic formula, we have

$$t = \frac{12 \pm \sqrt{144 - 4(16)(-h)}}{2(16)}$$

$$t = \frac{12 \pm \sqrt{144 + 64h}}{32}$$

We can factor the perfect square 16 from the two terms under the radical and simplify our radical somewhat:

$$t = \frac{12 \pm \sqrt{16(9 + 4h)}}{32}$$

$$t = \frac{12 \pm 4\sqrt{9 + 4h}}{32}$$

Now we can reduce to lowest terms by factoring a 4 from the numerator and denominator:

$$t = \frac{4(3 \pm \sqrt{9 + 4h})}{4 \cdot 8}$$

$$t = \frac{3 \pm \sqrt{9 + 4h}}{8}$$

If we were given a value of h, we would find that one of the solutions to this last formula would be a negative number. Since time is always measured in positive units, we wouldn't use that solution.

 Facts from Geometry: More about the Golden Ratio

In Section 5.1 we derived the golden ratio $\dfrac{1 + \sqrt{5}}{2}$ by finding the ratio of length to width for a golden rectangle. The golden ratio was actually discovered before the golden rectangle by the Greeks who lived before Euclid. The early Greeks found the golden ratio by dividing a line segment into two parts so that the ratio of the shorter part to the longer part was the same as the ratio of the longer part to the whole segment. When they divided a line segment in this manner they said it was divided in "extreme and mean ratio." Figure 1 illustrates a line segment divided this way.

FIGURE 1

If point B divides segment AC in "extreme and mean ratio," then

$$\frac{\text{Length of shorter segment}}{\text{Length of longer segment}} = \frac{\text{Length of longer segment}}{\text{Length of whole segment}}$$

$$\frac{AB}{BC} = \frac{BC}{AC}$$

EXAMPLE 5 If the length of segment AB in Figure 1 is 1 inch, find the length of BC so that the whole segment AC is divided in "extreme and mean ratio."

SOLUTION Using Figure 1 as a guide, if we let $x =$ the length of segment BC, then the length of AC is $x + 1$. If B divides AC into "extreme and mean ratio," then the ratio of AB to BC must equal the ratio of BC to AC. Writing this relationship using the variable x, we have

$$\frac{1}{x} = \frac{x}{x + 1}$$

If we multiply both sides of this equation by the LCD $x(x + 1)$, we have

$$x + 1 = x^2$$
$$0 = x^2 - x - 1 \quad \text{Write equation in standard form.}$$

Since this last equation is not factorable, we apply the quadratic formula:

$$x = \frac{1 \pm \sqrt{(-1)^2 - 4(1)(-1)}}{2}$$

$$= \frac{1 \pm \sqrt{5}}{2}$$

Our equation has two solutions, which we approximate using decimals:

$$\frac{1 + \sqrt{5}}{2} \approx 1.618 \qquad \frac{1 - \sqrt{5}}{2} \approx -0.618$$

Since we originally let x equal the length of segment BC, we use only the positive solution to our equation. As you can see, the positive solution is the golden ratio.

Using Technology: Graphing Calculators

More about Example 1

As we have mentioned before, algebraic expressions entered into a graphing calculator do not have to be simplified in order to be evaluated. This fact applies to equations as well. We can graph the equation $y = (x + 3)^2 - 2(x + 3) - 8$ to assist us in solving the equation in Example 1. The graph is shown in Figure 2. Using the Zoom and Trace features at the x-intercepts gives us $x = 1$ and $x = -5$ as the solutions to the equation $0 = (x + 3)^2 - 2(x + 3) - 8$.

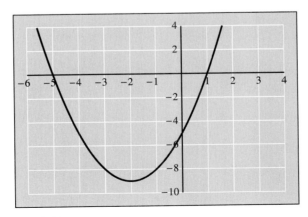

FIGURE 2

More about Example 2

Figure 3 shows the graph of $y = 4x^4 + 7x^2 - 2$. As we expect, the x-intercepts give the real number solutions to the equation $0 = 4x^4 + 7x^2 - 2$. The complex solutions do not appear on the graph.

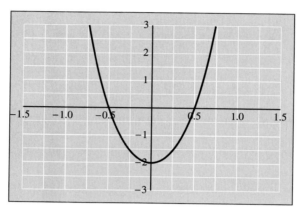

FIGURE 3

More about Example 3

In solving the equation in Example 3, we found that one of the possible solutions was an extraneous solution. If we solve the equation $x + \sqrt{x} - 6 = 0$ by graphing the function $y = x + \sqrt{x} - 6$, we find that the extraneous solution, 9, is not an x-intercept. Figure 4 shows that the only solution to the equation occurs at the x-intercept 4.

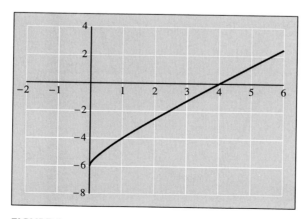

FIGURE 4

Problem Set
6.4

Solve each equation.

1. $(x - 3)^2 + 3(x - 3) + 2 = 0$
2. $(x + 4)^2 - (x + 4) - 6 = 0$
3. $2(x + 4)^2 + 5(x + 4) - 12 = 0$
4. $3(x - 5)^2 + 14(x - 5) - 5 = 0$
5. $x^4 - 6x^2 - 27 = 0$
6. $x^4 + 2x^2 - 8 = 0$
7. $x^4 + 9x^2 = -20$
8. $x^4 - 11x^2 = -30$
9. $(2a - 3)^2 - 9(2a - 3) = -20$
10. $(3a - 2)^2 + 2(3a - 2) = 3$
11. $2(4a + 2)^2 = 3(4a + 2) + 20$
12. $6(2a + 4)^2 = (2a + 4) + 2$
13. $6t^4 = -t^2 + 5$ 14. $3t^4 = -2t^2 + 8$
15. $9x^4 - 49 = 0$ 16. $25x^4 - 9 = 0$

Solve each of the following equations.

NOTE: Remember, if you square both sides of an equation in the process of solving it, you have to check all solutions in the original equation.

17. $x - 7\sqrt{x} + 10 = 0$ 18. $x - 6\sqrt{x} + 8 = 0$
19. $t - 2\sqrt{t} - 15 = 0$ 20. $t - 3\sqrt{t} - 10 = 0$
21. $6x + 11\sqrt{x} = 35$ 22. $2x + \sqrt{x} = 15$
23. $(a - 2) - 11\sqrt{a - 2} + 30 = 0$
24. $(a - 3) - 9\sqrt{a - 3} + 20 = 0$
25. $(2x + 1) - 8\sqrt{2x + 1} + 15 = 0$
26. $(2x - 3) - 7\sqrt{2x - 3} + 12 = 0$

Applying the Concepts

For Problems 27–30 t is in seconds.

27. **Falling Object** An object is tossed into the air with an upward velocity of 8 feet/second from the top of a building h feet high. The time it takes for the object to hit the ground below is given by the formula $16t^2 - 8t - h = 0$. Solve this formula for t.

28. **Falling Object** An object is tossed into the air with an upward velocity of 6 feet/second from the top of a building h feet high. The time it takes for the object to hit the ground below is given by the formula $16t^2 - 6t - h = 0$. Solve this formula for t.

29. **Falling Object** An object is tossed into the air with an upward velocity of v feet/second from the top of a building 20 feet high. The time it takes for the object to hit the ground below is given by the formula $16t^2 - vt - 20 = 0$. Solve this formula for t.

30. **Falling Object** An object is tossed into the air with an upward velocity of v feet/second from the top of a building 40 feet high. The time it takes for the object to hit the ground below is given by the formula $16t^2 - vt - 40 = 0$. Solve this formula for t.

Golden Ratio *Use Figure 1 from this section as a guide to working Problems 31–34.*

31. If AB in Figure 1 is 4 inches, and B divides AC in "extreme and mean ratio," find BC and then show that BC is four times the golden ratio.

32. If AB in Figure 1 is $\frac{1}{2}$ inch, and B divides AC in "extreme and mean ratio," find BC and then show that BC is half the golden ratio.

33. If AB in Figure 1 is 2 inches, and B divides AC in "extreme and mean ratio," find BC and then show that the ratio of BC to AB is the golden ratio.

34. If AB in Figure 1 is $\frac{1}{2}$ inch, and B divides AC in "extreme and mean ratio," find BC and then show that the ratio of BC to AB is the golden ratio.

35. Solve the formula $16t^2 - vt - h = 0$ for t.
36. Solve the formula $16t^2 + vt + h = 0$ for t.
37. Solve the formula $kx^2 + 8x + 4 = 0$ for x.
38. Solve the formula $k^2x^2 + kx + 4 = 0$ for x.

39. Solve $x^2 + 2xy + y^2 = 0$ for x by using the quadratic formula with $a = 1$, $b = 2y$, and $c = y^2$.

40. Solve $x^2 - 2xy + y^2 = 0$ for x.

Review Problems

 NOTE: The problems that follow review material we covered in Sections 5.4 and 5.5.

Combine, if possible. [5.4]

41. $5\sqrt{7} - 2\sqrt{7}$ **42.** $6\sqrt{2} - 9\sqrt{2}$

43. $\sqrt{18} - \sqrt{8} + \sqrt{32}$

44. $\sqrt{50} + \sqrt{72} - \sqrt{8}$

45. $9x\sqrt{20x^3y^2} + 7y\sqrt{45x^5}$

46. $5x^2\sqrt{27xy^3} - 6y\sqrt{12x^5y}$

Multiply. [5.5]

47. $(\sqrt{5} - 2)(\sqrt{5} + 8)$

48. $(2\sqrt{3} - 7)(2\sqrt{3} + 7)$

49. $(\sqrt{x} + 2)^2$

50. $(3 - \sqrt{x})(3 + \sqrt{x})$

Rationalize the denominator. [5.5]

51. $\dfrac{\sqrt{7}}{\sqrt{7} - 2}$ **52.** $\dfrac{\sqrt{5} - \sqrt{2}}{\sqrt{5} + \sqrt{2}}$

One Step Further

Find the x- and y-intercepts.

53. $y = x^3 - 4x$

54. $y = x^4 - 10x^2 + 9$

55. $y = 3x^3 + x^2 - 27x - 9$

56. $y = 2x^3 + x^2 - 8x - 4$

57. The graph of $y = 2x^3 - 7x^2 - 5x + 4$ crosses the x-axis at $x = 4$. Where else does it cross the x-axis?

58. The graph of $y = 6x^3 + x^2 - 12x + 5$ crosses the x-axis at $x = 1$. Where else does it cross the x-axis?

SECTION 6.5

Graphing Parabolas

Recall from Section 3.5 that the solution set to equations such as

$$y = x^2 - 3$$

will consist of ordered pairs. One method of graphing the solution set is to find a number of ordered pairs that satisfy the equation and to graph them. We can obtain some ordered pairs that are solutions to $y = x^2 - 3$ by use of a table as follows:

x	$y = x^2 - 3$	y	Solutions
-3	$y = (-3)^2 - 3 = 9 - 3 = 6$	6	$(-3, 6)$
-2	$y = (-2)^2 - 3 = 4 - 3 = 1$	1	$(-2, 1)$
-1	$y = (-1)^2 - 3 = 1 - 3 = -2$	-2	$(-1, -2)$
0	$y = 0^2 - 3 = 0 - 3 = -3$	-3	$(0, -3)$
1	$y = 1^2 - 3 = 1 - 3 = -2$	-2	$(1, -2)$
2	$y = 2^2 - 3 = 4 - 3 = 1$	1	$(2, 1)$
3	$y = 3^2 - 3 = 9 - 3 = 6$	6	$(3, 6)$

Graphing these solutions and then connecting them with a smooth curve, we have the graph of $y = x^2 - 3$. (See Figure 1.)

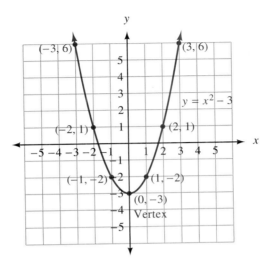

FIGURE I

The graph in Figure 1 is an example of a **parabola.** All equations of the form $y = ax^2 + bx + c$, $a \neq 0$, have parabolas for graphs.

Although it is always possible to graph parabolas by making a table of values of x and y that satisfy the equation, there are other methods that are faster and, in some cases, more accurate.

The important points associated with the graph of a parabola are the highest (or lowest) point on the graph and the x-intercepts. The y-intercepts can also be useful.

Intercepts for Parabolas

The graph of the equation $y = ax^2 + bx + c$ will cross the y-axis at $y = c$, since substituting $x = 0$ into $y = ax^2 + bx + c$ yields $y = c$.

Since the graph will cross the x-axis when $y = 0$, the x-intercepts are those values of x that are solutions to the quadratic equation $0 = ax^2 + bx + c$.

The Vertex of a Parabola

The highest or lowest point on a parabola is called the **vertex.** The vertex for the graph of $y = ax^2 + bx + c$ will always occur when

$$x = \frac{-b}{2a}$$

To see this, we must transform the right side of $y = ax^2 + bx + c$ into an expression that contains x in just one of its terms. This is accomplished by completing the square on the first two terms. Here is what it looks like:

$$y = ax^2 + bx + c$$
$$y = a\left(x^2 + \frac{b}{a}x\right) + c$$
$$y = a\left[x^2 + \frac{b}{a}x + \left(\frac{b}{2a}\right)^2\right] + c - a\left(\frac{b}{2a}\right)^2$$
$$y = a\left(x + \frac{b}{2a}\right)^2 + \frac{4a}{4a}\cdot c - a\cdot\frac{b^2}{4a^2}$$
$$y = a\left(x + \frac{b}{2a}\right)^2 + \frac{4ac - b^2}{4a}$$

It may not look like it, but this last line indicates that the vertex of the graph of $y = ax^2 + bx + c$ has an x-coordinate of $\frac{-b}{2a}$. Since a, b, and c are constants, the only quantity that is varying in the last expression is the x in $\left(x + \frac{b}{2a}\right)^2$. Since the quantity $\left(x + \frac{b}{2a}\right)^2$ is the square of $x + \frac{b}{2a}$, the smallest it will ever be is 0, and that will happen when $x = \frac{-b}{2a}$.

We can use the vertex point along with the x- and y-intercepts to sketch the graph of any equation of the form $y = ax^2 + bx + c$. Here is a summary of what we know about parabolas:

Graphing Parabolas

The graph of $y = ax^2 + bx + c$, $a \neq 0$, will have:

1. A y-intercept at $y = c$
2. x-intercepts (if they exist) at

$$x = \frac{-b \pm \sqrt{b^2 - 4ac}}{2a}$$

3. A vertex when $x = \frac{-b}{2a}$

EXAMPLE 1   Sketch the graph of $y = x^2 - 6x + 5$.

SOLUTION   To find the x-intercepts we let $y = 0$ and solve for x:

$$0 = x^2 - 6x + 5$$
$$0 = (x - 5)(x - 1)$$
$$x = 5 \quad \text{or} \quad x = 1$$

To find the coordinates of the vertex we first find

$$x = \frac{-b}{2a} = \frac{-(-6)}{2(1)} = 3$$

The x-coordinate of the vertex is 3. To find the y-coordinate we substitute 3 for x in our original equation:

$$y = 3^2 - 6(3) + 5 = 9 - 18 + 5 = -4$$

The graph crosses the x-axis at 1 and 5 and has its vertex at $(3, -4)$.

Plotting these points and connecting them with a smooth curve, we have the graph shown in Figure 2. The graph is a parabola that opens up, so we say the graph is **concave up.** The vertex is the lowest point on the graph. (Note that the graph crosses the y-axis at 5, which is the value of y we obtain when we let $x = 0$.)

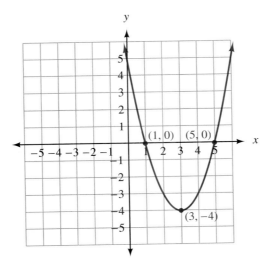

FIGURE 2

Finding the Vertex by Completing the Square

Another way to locate the vertex of the parabola in Example 1 is by completing the square on the first two terms on the right side of the equation $y = x^2 - 6x + 5$. In this case, we would add 9 and subtract 9 on the right side of the equation. (This amounts to adding 0 to the equation, so we know we haven't changed its solutions.)

This is what it looks like:

$$y = (x^2 - 6x \qquad) + 5$$
$$y = (x^2 - 6x \mathbf{+ 9}) + 5 \mathbf{- 9}$$
$$y = (x - 3)^2 - 4$$

You may have to look at this last equation awhile to see this, but when $x = 3$, then $y = (x - 3)^2 - 4 = 0^2 - 4 = -4$ is the smallest y will ever be. And that is why the vertex is at $(3, -4)$. As a matter of fact, this is the same kind of reasoning we used when we derived the formula $x = \dfrac{-b}{2a}$ for the x-coordinate of the vertex.

EXAMPLE 2   Graph: $f(x) = -x^2 - 2x + 3$.

SOLUTION   To find the x-intercepts, we let $f(x) = 0$:

$$0 = -x^2 - 2x + 3$$
$$0 = x^2 + 2x - 3 \quad \text{Multiply each side by } -1.$$
$$0 = (x + 3)(x - 1)$$
$$x = -3 \quad \text{or} \quad x = 1$$

The x-coordinate of the vertex is given by

$$x = \frac{-b}{2a} = \frac{-(-2)}{2(-1)} = \frac{2}{-2} = -1$$

To find the y-coordinate of the vertex, we substitute -1 for x in our original equation to get

$$y = -(-1)^2 - 2(-1) + 3 = -1 + 2 + 3 = 4$$

Our parabola has x-intercepts at -3 and 1, and a vertex at $(-1, 4)$. Figure 3 shows the graph. We say the graph is **concave down** since it opens downward.

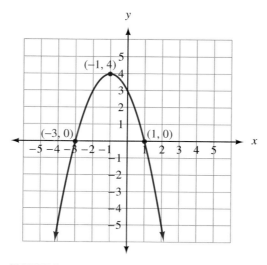

FIGURE 3

Again, we could have obtained the coordinates of the vertex by completing the square on the first two terms on the right side of our equation. To do so we must first factor -1 from the first two terms. (Remember, the leading coefficient must be 1 in order to complete the square.) When we complete the square, we add 1 inside the parentheses, which actually decreases the right side of the equation by -1 since everything in the parentheses is multiplied by -1. To make up for it, we add 1 outside the parentheses.

$$y = -1(x^2 + 2x \qquad) + 3$$
$$y = -1(x^2 + 2x + 1) + 3 + 1$$
$$y = -1(x + 1)^2 + 4$$

The last line tells us that the *largest* value of y will be 4, and that will occur when $x = -1$.

EXAMPLE 3 Graph: $y = 3x^2 - 6x + 1$.

SOLUTION To find the x-intercepts, we let $y = 0$ and solve for x:

$$0 = 3x^2 - 6x + 1$$

Since the right side of this equation does not factor, we can look at the discriminant to see what kind of solutions are possible. The discriminant for this equation is

$$b^2 - 4ac = 36 - 4(3)(1) = 24$$

Since the discriminant is a positive number but not a perfect square, the equation will have irrational solutions. This means that the x-intercepts are irrational numbers and will have to be approximated with decimals using the quadratic formula. Rather than use the quadratic formula, we will find some other points on the graph, but first let's find the vertex.

Here are both methods of finding the vertex:

Using the formula that gives us the x-coordinate of the vertex, we have

$$x = \frac{-b}{2a} = \frac{-(-6)}{2(3)} = 1$$

Substituting 1 for x in the equation gives us the y-coordinate of the vertex:

$$y = 3 \cdot 1^2 - 6 \cdot 1 + 1 = -2$$

To complete the square on the right side of the equation, we factor 3 from the first two terms, add 1 inside the parentheses, and add -3 outside the parentheses (this amounts to adding 0 to the right side):

$$y = 3(x^2 - 2x \qquad) + 1$$
$$y = 3(x^2 - 2x + 1) + 1 - 3$$
$$y = 3(x - 1)^2 - 2$$

In either case, the vertex is $(1, -2)$.

If we can find two points, one on each side of the vertex, we can sketch the graph. Let's let $x = 0$ and $x = 2$, since each of these numbers is the same distance from $x = 1$, and $x = 0$ will give us the y-intercept.

$$\text{When} \quad x = 0: \quad y = 3(0)^2 - 6(0) + 1$$
$$= 0 - 0 + 1$$
$$= 1$$
$$\text{When} \quad x = 2: \quad y = 3(2)^2 - 6(2) + 1$$
$$= 12 - 12 + 1$$
$$= 1$$

The two points just found are $(0, 1)$ and $(2, 1)$. Plotting these two points along with the vertex $(1, -2)$, we have the graph shown in Figure 4.

Note: Although we chose not to find the irrational x-intercepts, we can see they exist by observing where the parabola crosses the x-axis.

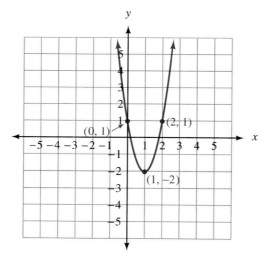

FIGURE 4

EXAMPLE 4   Graph: $y = -2x^2 + 6x - 5$.

SOLUTION   Letting $y = 0$, we have

$$0 = -2x^2 + 6x - 5$$

Again, the right side of this equation does not factor. The discriminant is $b^2 - 4ac = 36 - 4(-2)(-5) = -4$, which indicates that the solutions are complex numbers. This means that our original equation does not have x-intercepts; that is, the graph does not cross the x-axis.

Let's find the vertex.

Using our formula for the x-coordinate of the vertex, we have

$$x = \frac{-b}{2a} = \frac{-6}{2(-2)} = \frac{6}{4} = \frac{3}{2}$$

To find the y-coordinate, we let $x = \frac{3}{2}$:

$$y = -2\left(\frac{3}{2}\right)^2 + 6\left(\frac{3}{2}\right) - 5$$

$$= \frac{-18}{4} + \frac{18}{2} - 5$$

$$= \frac{-18 + 36 - 20}{4}$$

$$= -\frac{1}{2}$$

Finding the vertex by completing the square is a more complicated matter. In order to make the coefficient of x^2 a 1, we must factor -2 from the first two terms. To complete the square inside the parentheses, we add $\frac{9}{4}$. Since each term inside the parentheses is multiplied by -2, we add $\frac{9}{2}$ outside the parentheses so that the net result is the same as adding 0 to the right side:

$$y = -2(x^2 - 3x \qquad) - 5$$

$$y = -2\left(x^2 - 3x + \frac{9}{4}\right) - 5 + \frac{9}{2}$$

$$y = -2\left(x - \frac{3}{2}\right)^2 - \frac{1}{2}$$

The vertex is $(\frac{3}{2}, -\frac{1}{2})$. Since this is the only point we have so far, we must find two others. Let's let $x = 3$ and $x = 0$, since each of these points is the same distance from $x = \frac{3}{2}$ and on either side:

When $x = 3$:
$$y = -2(3)^2 + 6(3) - 5$$
$$= -18 + 18 - 5$$
$$= -5$$

When $x = 0$:
$$y = -2(0)^2 + 6(0) - 5$$
$$= 0 + 0 - 5$$
$$= -5$$

The two additional points on the graph are $(3, -5)$ and $(0, -5)$. Figure 5 shows the graph. The graph is concave down. The vertex is the highest point on the graph.

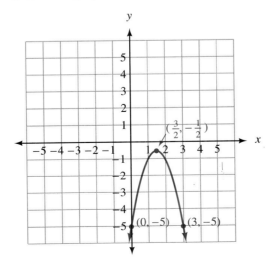

FIGURE 5

By looking at the equations and graphs in Examples 1–4, we can conclude that the graph of $y = ax^2 + bx + c$ will be concave up when a is positive, and concave down when a is negative. Taking this even further, if $a > 0$, then the vertex is the lowest point on the graph, and if $a < 0$, the vertex is the highest point on the graph. We can use this information to solve some problems in which we are interested in finding the largest or smallest value of a variable.

EXAMPLE 5 A company selling copies of an accounting program for home computers finds that it will make a weekly profit of P dollars from selling x copies of the program, according to the equation

$$P(x) = -0.1x^2 + 27x - 500$$

How many copies of the program should it sell to make the largest possible profit, and what is the largest possible profit?

SOLUTION Since the coefficient of x^2 is negative, we know the graph of this parabola will be concave down, meaning that the vertex is the highest point of the curve. We find the vertex by first finding its x-coordinate:

$$x = \frac{-b}{2a} = \frac{-27}{2(-0.1)} = \frac{27}{0.2} = 135$$

This represents the number of programs the company needs to sell each week in order to make a maximum profit. To find the maximum profit, we substitute 135 for x in the original equation. (A calculator is helpful for these kinds of calculations.)

$$\begin{aligned}
P(135) &= -0.1(135)^2 + 27(135) - 500 \\
&= -0.1(18{,}225) + 3{,}645 - 500 \\
&= -1{,}822.5 + 3{,}645 - 500 \\
&= 1{,}322.5
\end{aligned}$$

The maximum weekly profit is $1,322.50 and is obtained by selling 135 programs a week.

EXAMPLE 6 An art supply store finds that they can sell x sketch pads each week at p dollars each, according to the equation $x = 900 - 300p$. Graph the revenue equation $R = xp$. Then use the graph to find the price p that will bring in the maximum revenue. Finally, find the maximum revenue.

SOLUTION As it stands, the revenue equation contains three variables. Since we are asked to find the value of p that gives us the maximum value of R, we rewrite the equation using just the variables R and p. Since $x = 900 - 300p$, we have

$$R = xp = (900 - 300p)p$$

The graph of this equation is shown in Figure 6. The graph appears in the first quadrant only, since R and p are both positive quantities. From the graph we see that the maximum value of R occurs when $p = \$1.50$. We can calculate the maximum value of R from the equation.

When $p = 1.5$:

the equation $R = (900 - 300p)p$

becomes

$$R = (900 - 300 \cdot 1.5)1.5$$
$$= (900 - 450)1.5$$
$$= 450 \cdot 1.5$$
$$= 675$$

The maximum revenue is $675. It is obtained by setting the price of each sketch pad at $p = \$1.50$.

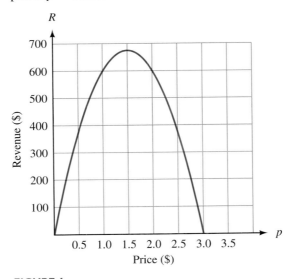

FIGURE 6

Problem Set
6.5

If you have been using a graphing calculator for some of the material in this course, you are well aware that your calculator can draw all the graphs in this section very easily. However, it is important that you are able to recognize and sketch the graph of any parabola by hand. It is a skill that all successful intermediate algebra students should possess, even if they are proficient in the use of a graphing calculator. My suggestion is that you work the problems in this section and problem set without your calculator. Then use your calculator to check your results.

For each of the following equations, give the x-intercepts and the coordinates of the vertex, and sketch the graph.

1. $y = x^2 + 2x - 3$
2. $y = x^2 - 2x - 3$
3. $y = -x^2 - 4x + 5$
4. $y = x^2 + 4x - 5$
5. $f(x) = x^2 - 1$
6. $f(x) = x^2 - 4$
7. $f(x) = -x^2 + 9$
8. $f(x) = -x^2 + 1$
9. $y = 2x^2 - 4x - 6$
10. $y = 2x^2 + 4x - 6$
11. $y = x^2 - 2x - 4$
12. $y = x^2 - 2x - 2$

Find the vertex and any two convenient points to sketch the graphs of the following:

13. $g(x) = x^2 - 4x - 4$
14. $g(x) = x^2 - 2x + 3$
15. $g(x) = -x^2 + 2x - 5$
16. $g(x) = -x^2 + 4x - 2$
17. $y = x^2 + 1$ **18.** $y = x^2 + 4$
19. $y = -x^2 - 3$ **20.** $y = -x^2 - 2$
21. $y = 3x^2 + 4x + 1$ **22.** $y = 2x^2 + 4x + 3$

For each of the following equations, find the coordinates of the vertex and indicate whether the vertex is the highest point on the graph or the lowest point on the graph. (Do not graph.)

23. $y = x^2 - 6x + 5$
24. $y = -x^2 + 6x - 5$
25. $y = -x^2 + 2x + 8$
26. $y = x^2 - 2x - 8$
27. $y = 12 + 4x - x^2$
28. $y = -12 - 4x + x^2$
29. $y = -x^2 - 8x$
30. $y = x^2 + 8x$

Applying the Concepts

31. **Maximum Profit** A company earns a weekly profit of P dollars by selling x items, according to the equation $P(x) = -0.5x^2 + 40x - 300$. Find the number of items the company must sell each week in order to obtain the largest possible profit. Then, find the largest possible profit.

32. **Maximum Profit** A company earns a weekly profit of P dollars by selling x items, according to the equation $P(x) = -0.5x^2 + 100x - 1,600$. Find the number of items the company must sell each week in order to obtain the largest possible profit. Then, find the largest possible profit.

33. **Maximum Profit** A company finds that it can make a profit of P dollars each month by selling x patterns, according to the formula $P(x) = -0.002x^2 + 3.5x - 800$. How many patterns must it sell each month in order to have a maximum profit? What is the maximum profit?

34. **Maximum Profit** A company selling picture frames finds that it can make a profit of P dollars each month by selling x frames, according to the formula $P(x) = -0.002x^2 + 5.5x - 1,200$. How many frames must it sell each month in order to have a maximum profit? What is the maximum profit?

35. **Maximum Height** Chaudra is tossing a softball into the air with an underhand motion. The distance of the ball above her hand at any time is given by the function

$$h(t) = 32t - 16t^2 \qquad \text{for } 0 \le t \le 2$$

where $h(t)$ is the height of the ball (in feet) and t is the time (in seconds). Find the times at which the ball is in her hand, and the maximum height of the ball.

36. **Maximum Height** Hali is tossing a quarter into the air with an underhand motion. The distance of the quarter above her hand at any time is given by the function

$$h(t) = 16t - 16t^2 \qquad \text{for } 0 \le t \le 1$$

where $h(t)$ is the height of the quarter (in feet) and t is the time (in seconds). Find the times at which the quarter is in her hand, and the maximum height of the quarter.

37. **Maximum Height** An arrow is shot straight up into the air with an initial velocity of 128 feet/second. If h represents the height (in feet) of the arrow at any time t (in seconds), then the equation that gives h in terms of t is $h(t) = 128t - 16t^2$. Find the maximum height attained by the arrow.

38. **Maximum Height** A ball is projected into the air with an upward velocity of 64 feet/second. The equation that gives the height h (in feet) of the ball at any time t (in seconds) is $h(t) = 64t - 16t^2$. Find the maximum height attained by the ball.

39. **Maximum Revenue** A company that manufactures typewriter ribbons knows that the number of ribbons, x, it can sell each week is related to the price, p, of each ribbon by the equation $x = 1,200 - 100p$. Graph the revenue equation $R = xp$. Then use the graph to find the price p that will bring in the maximum revenue. Finally, find the maximum revenue.

40. **Maximum Revenue** A company that manufactures diskettes for home computers finds that it can sell x diskettes each day at p dollars per diskette, according to the equation $x = 800 - 100p$. Graph the revenue equation $R = xp$. Then use the graph to find the price p that will bring in the maximum revenue. Finally, find the maximum revenue.

41. **Maximum Revenue** The relationship between the number of calculators, x, a company sells each day and the price, p, of each calculator is given by the equation $x = 1,700 - 100p$. Graph the revenue equation $R = xp$, and use the graph to find the price p that will bring in the maximum revenue. Then find the maximum revenue.

42. **Maximum Revenue** The relationship between the number, x, of pencil sharpeners a company sells each week and the price, p, of each sharpener is given by the equation $x = 1,800 - 100p$. Graph the revenue equation $R = xp$, and use the graph to find the price p that will bring in the maximum revenue. Then find the maximum revenue.

Review Problems

 The problems that follow review material we covered in Section 5.7.

Perform the indicated operations.

43. $(3 - 5i) - (2 - 4i)$ 44. $2i(5 - 6i)$

45. $(3 + 2i)(7 - 3i)$ 46. $(4 + 5i)^2$

47. $\dfrac{i}{3 + i}$ 48. $\dfrac{2 + 3i}{2 - 3i}$

Quadratic Inequalities

Quadratic inequalities in one variable are inequalities of the form

$$ax^2 + bx + c < 0 \qquad ax^2 + bx + c > 0$$
$$ax^2 + bx + c \leq 0 \qquad ax^2 + bx + c \geq 0$$

where a, b, and c are constants, with $a \neq 0$. The technique we will use to solve inequalities of this type involves graphing. Suppose, for example, we wish to find the solution set for the inequality $x^2 - x - 6 > 0$. We begin by factoring the left side to obtain

$$(x - 3)(x + 2) > 0$$

We have two real numbers $x - 3$ and $x + 2$ whose product $(x - 3)(x + 2)$ is greater than zero. That is, their product is positive. The only way the product can be positive is either if both factors, $(x - 3)$ and $(x + 2)$, are positive or if they are both negative. To help visualize where $x - 3$ is positive and where it is negative, we draw a real number line and label it accordingly:

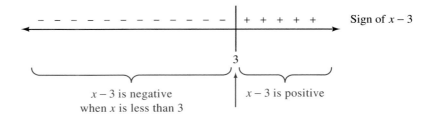

Here is a similar diagram showing where the factor $x + 2$ is positive and where it is negative:

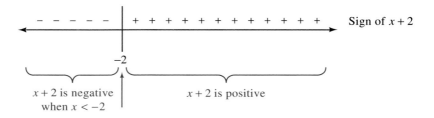

Drawing the two number lines together, we have

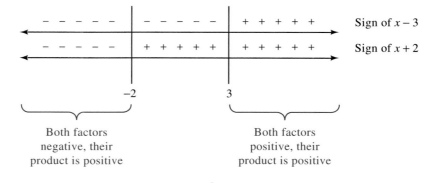

We can see from this diagram that the graph of the solution to $x^2 - x - 6 > 0$ is

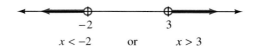

Using Technology

We can solve the preceding problem by using a graphing calculator to visualize where the product $(x - 3)(x + 2)$ is positive. First, we graph the function $y = (x - 3)(x + 2)$ as shown in Figure 1.

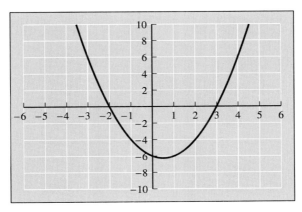

FIGURE 1

Next, we observe where the graph is above the x-axis. As you can see, the graph is above the x-axis to the right of 3 and to the left of -2, as shown in Figure 2.

Graph is above the x-axis when x is here.

Graph is above the x-axis when x is here.

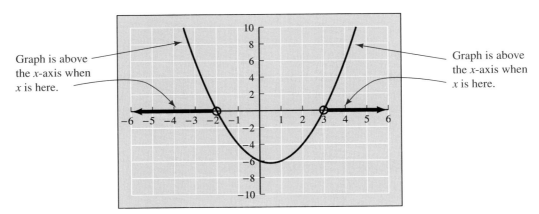

FIGURE 2

When the graph is above the x-axis, we have points whose y-coordinates are positive. Since these y-coordinates are the same as the expression $(x - 3)(x + 2)$, the values of x for which the graph of $y = (x - 3)(x + 2)$ is above the x-axis are the values of x for which the inequality $(x - 3)(x + 2) > 0$ is true. Therefore, our solution set is

$$x < -2 \qquad \text{or} \qquad x > 3$$

EXAMPLE 1   Solve for x: $x^2 - 2x - 8 \le 0$.

ALGEBRAIC SOLUTION   We begin by factoring:

$$x^2 - 2x - 8 \le 0$$
$$(x - 4)(x + 2) \le 0$$

The product $(x - 4)(x + 2)$ is negative or zero. The factors must have opposite signs. We draw a diagram showing where each factor is positive and where each factor is negative:

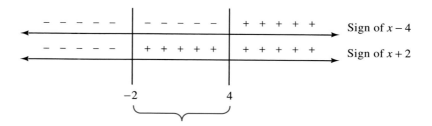

From the diagram we have the graph of the solution set:

GRAPHICAL SOLUTION   To solve this inequality with a graphing calculator, we graph the function $y = (x - 4)(x + 2)$ and observe where the graph is below the x-axis. These are the points that have negative y-coordinates, which means that the product $(x - 4)(x + 2)$ is negative for these points. Figure 3 shows the graph of

When x is here, the graph is on or below the x-axis.

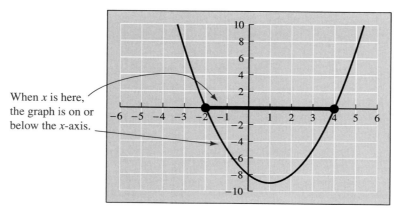

FIGURE 3

$y = (x - 4)(x + 2)$, along with the region on the x-axis where the graph contains points with negative y-coordinates.

As you can see, the graph is below the x-axis when x is between -2 and 4. Since our original inequality includes the possibility that $(x - 4)(x + 2)$ is 0, we include the endpoints, -2 and 4, with our solution set.

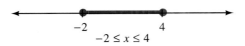

EXAMPLE 2   Solve for x: $6x^2 - x \geq 2$.

ALGEBRAIC SOLUTION
$$6x^2 - x \geq 2$$
$$6x^2 - x - 2 \geq 0 \leftarrow \text{Standard form}$$
$$(3x - 2)(2x + 1) \geq 0$$

The product is positive, so the factors must agree in sign. Here is the diagram showing where that occurs:

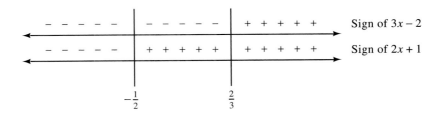

Since the factors agree in sign below $-\frac{1}{2}$ and above $\frac{2}{3}$, the graph of the solution set is

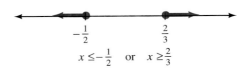

GRAPHICAL SOLUTION   To solve this inequality with a graphing calculator, we graph the function $y = (3x - 2)(2x + 1)$ and observe where the graph is above the x-axis. These are the points that have positive y-coordinates, which means that the

product $(3x - 2)(2x + 1)$ is positive for these points. Figure 4 shows the graph of $y = (3x - 2)(2x + 1)$, along with the regions on the x-axis where the graph is on or above the x-axis.

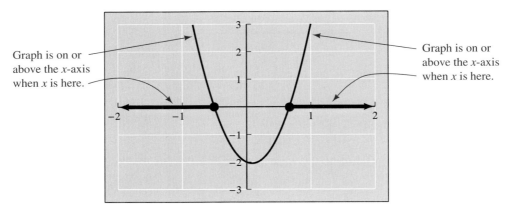

Graph is on or above the x-axis when x is here.

Graph is on or above the x-axis when x is here.

FIGURE 4

To find the points where the graph crosses the x-axis, we need to use either the Trace and Zoom features to zoom in on each point, or the calculator function that finds the intercepts automatically (on the TI-82/83 this is the root function under the CALC key). Whichever method we use, we will obtain the following result:

$$x \leq -0.5 \quad \text{or} \quad x \geq 0.67$$

EXAMPLE 3   Solve: $x^2 - 6x + 9 \geq 0$.

ALGEBRAIC SOLUTION
$$x^2 - 6x + 9 \geq 0$$
$$(x - 3)^2 \geq 0$$

This is a special case in which both factors are the same. Since $(x - 3)^2$ is always positive or 0, the solution set is all real numbers. That is, any real number that is used in place of x in the original inequality will produce a true statement.

GRAPHICAL SOLUTION   The graph of $y = (x - 3)^2$ is shown in Figure 5. Notice that it touches the x-axis at 3 and is above the x-axis everywhere else. This means that every point on the graph has a y-coordinate greater than or equal to 0, no matter what the value of x. The conclusion that we draw from the graph is that the inequality $(x - 3)^2 \geq 0$ is true for all values of x.

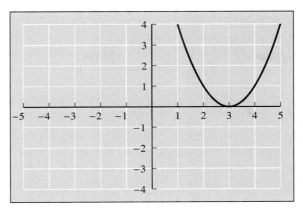

FIGURE 5

Our next two examples involve inequalities that contain rational expressions.

EXAMPLE 4   Solve: $\dfrac{x - 4}{x + 1} \le 0$.

SOLUTION   The inequality indicates that the quotient of $(x - 4)$ and $(x + 1)$ is negative or 0 (less than or equal to 0). We can use the same reasoning we used to solve the first three examples, because quotients are positive or negative under the same conditions that products are positive or negative. Here is the diagram that shows where each factor is positive and where each factor is negative:

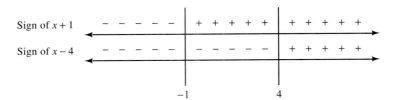

Between -1 and 4 the factors have opposite signs, making the quotient negative. Thus, the region between -1 and 4 is where the solutions lie, since the original inequality indicates the quotient $\dfrac{x - 4}{x + 1}$ is negative. The solution set and its graph are shown below:

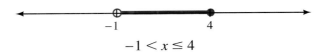

$$-1 < x \le 4$$

Notice that the left endpoint is open—that is, it is not included in the solution set—because $x = -1$ would make the denominator in the original inequality 0. It is

396

important to check all endpoints of solution sets to inequalities that involve rational expressions.

EXAMPLE 5   Solve: $\dfrac{3}{x-2} - \dfrac{2}{x-3} > 0$.

SOLUTION   We begin by adding the two rational expressions on the left side. The common denominator is $(x-2)(x-3)$:

$$\frac{3}{x-2} \cdot \frac{(x-3)}{(x-3)} - \frac{2}{x-3} \cdot \frac{(x-2)}{(x-2)} > 0$$

$$\frac{3x - 9 - 2x + 4}{(x-2)(x-3)} > 0$$

$$\frac{x-5}{(x-2)(x-3)} > 0$$

This time the quotient involves three factors. Here is the diagram that shows the signs of the three factors:

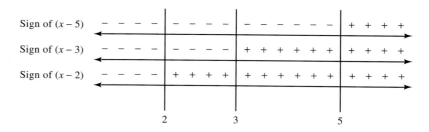

The original inequality indicates that the quotient is positive. In order for this to happen, either all three factors must be positive, or exactly two factors must be negative. Looking back at the diagram, we see the regions that satisfy these conditions are between 2 and 3 or above 5. Here is our solution set:

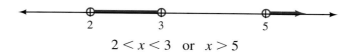

$$2 < x < 3 \quad \text{or} \quad x > 5$$

Problem Set

6.6

Solve each of the following inequalities and graph the solution set.

1. $x^2 + x - 6 > 0$

2. $x^2 + x - 6 < 0$

3. $x^2 - x - 12 \le 0$

4. $x^2 - x - 12 \ge 0$

5. $x^2 + 5x \ge -6$

6. $x^2 - 5x > 6$

7. $6x^2 < 5x - 1$

8. $4x^2 \ge -5x + 6$

9. $x^2 - 9 < 0$
10. $x^2 - 16 \geq 0$
11. $4x^2 - 9 \geq 0$
12. $9x^2 - 4 < 0$
13. $2x^2 - x - 3 < 0$
14. $3x^2 + x - 10 \geq 0$
15. $x^2 - 4x + 4 \geq 0$
16. $x^2 - 4x + 4 < 0$
17. $x^2 - 10x + 25 < 0$
18. $x^2 - 10x + 25 > 0$
19. $(x - 2)(x - 3)(x - 4) > 0$
20. $(x - 2)(x - 3)(x - 4) < 0$
21. $(x + 1)(x + 2)(x + 3) \leq 0$
22. $(x + 1)(x + 2)(x + 3) \geq 0$

23. $\dfrac{x - 1}{x + 4} \leq 0$

24. $\dfrac{x + 4}{x - 1} \leq 0$

25. $\dfrac{3x}{x + 6} - \dfrac{8}{x + 6} < 0$

26. $\dfrac{5x}{x + 1} - \dfrac{3}{x + 1} < 0$

27. $\dfrac{4}{x - 6} + 1 > 0$

28. $\dfrac{2}{x - 3} + 1 \geq 0$

29. $\dfrac{x - 2}{(x + 3)(x - 4)} < 0$

30. $\dfrac{x - 1}{(x + 2)(x - 5)} < 0$

31. $\dfrac{2}{x - 4} - \dfrac{1}{x - 3} > 0$

32. $\dfrac{4}{x + 3} - \dfrac{3}{x + 2} > 0$

33. $\dfrac{x + 7}{2x + 12} + \dfrac{6}{x^2 - 36} \leq 0$

34. $\dfrac{x + 1}{2x - 2} - \dfrac{2}{x^2 - 1} \leq 0$

35. The graph of $y = x^2 - 4$ is shown in Figure 6. Use the graph to write the solution set for each of the following:
    (a) $x^2 - 4 < 0$
    (b) $x^2 - 4 > 0$
    (c) $x^2 - 4 = 0$

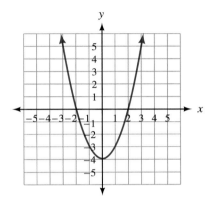

FIGURE 6

36. The graph of $y = 4 - x^2$ is shown in Figure 7. Use the graph to write the solution set for each of the following:
    (a) $4 - x^2 < 0$
    (b) $4 - x^2 > 0$
    (c) $4 - x^2 = 0$

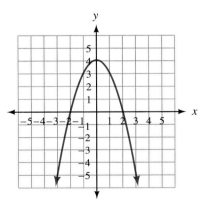

FIGURE 7

37. The graph of $y = x^2 - 3x - 10$ is shown in Figure 8. Use the graph to write the solution set for each of the following:

(a) $x^2 - 3x - 10 < 0$

(b) $x^2 - 3x - 10 > 0$

(c) $x^2 - 3x - 10 = 0$

39. The graph of $y = x^3 - 3x^2 - x + 3$ is shown in Figure 10. Use the graph to write the solution set for each of the following:

(a) $x^3 - 3x^2 - x + 3 < 0$

(b) $x^3 - 3x^2 - x + 3 > 0$

(c) $x^3 - 3x^2 - x + 3 = 0$

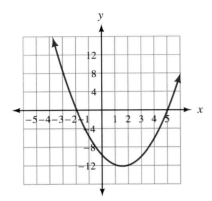

FIGURE 8

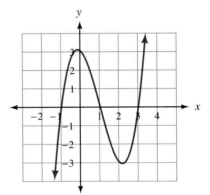

FIGURE 10

38. The graph of $y = x^2 + x - 12$ is shown in Figure 9. Use the graph to write the solution set for each of the following:

(a) $x^2 + x - 12 < 0$

(b) $x^2 + x - 12 > 0$

(c) $x^2 + x - 12 = 0$

40. The graph of $y = x^3 + 4x^2 - 4x - 16$ is shown in Figure 11. Use the graph to write the solution set for each of the following:

(a) $x^3 + 4x^2 - 4x - 16 < 0$

(b) $x^3 + 4x^2 - 4x - 16 > 0$

(c) $x^3 + 4x^2 - 4x - 16 = 0$

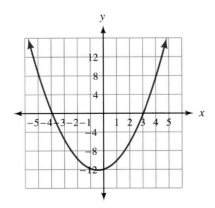

FIGURE 9

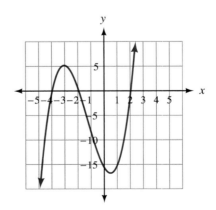

FIGURE 11

Applying the Concepts

41. Dimensions of a Rectangle The length of a rectangle is 3 inches more than twice the width. If the area is to be at least 44 square inches, what are the possibilities for the width?

42. Dimensions of a Rectangle The length of a rectangle is 5 inches less than three times the width. If the area is to be less than 12 square inches, what are the possibilities for the width?

43. Revenue A manufacturer of portable radios knows that the weekly revenue produced by selling x radios is given by the equation $R = 1,300p - 100p^2$, where p is the price of each radio (in dollars). What price should be charged for each radio if the weekly revenue is to be at least \$4,000?

44. Revenue A manufacturer of small calculators knows that the weekly revenue produced by selling x calculators is given by the equation $R = 1,700p - 100p^2$, where p is the price of each calculator (in dollars). What price should be charged for each calculator if the revenue is to be at least \$7,000 each week?

Review Problems

NOTE: The problems that follow review material we covered in Sections 3.1 and 5.6. Reviewing these problems will help you understand parts of the next section.

Graph each line by finding the x- and y-intercepts. [3.1]

45. $5x - 3y = 15$ **46.** $-2x + 4y = 6$

47. $\frac{1}{2}x + \frac{1}{3}y = 1$ **48.** $\frac{2}{3}x - \frac{4}{3}y = 4$

Solve each equation. [5.6]

49. $\sqrt{3t - 1} = 2$ **50.** $\sqrt{4t + 5} + 7 = 3$

51. $\sqrt{x + 3} = x - 3$ **52.** $\sqrt{x + 3} = \sqrt{x} - 3$

Graph each equation. [5.6]

53. $y = \sqrt[3]{x - 1}$ **54.** $y = \sqrt[3]{x} - 1$

One Step Further

Graph the solution set for each inequality.

55. $x^2 - 2x - 1 < 0$ **56.** $x^2 - 6x + 7 < 0$

57. $x^2 - 8x + 13 > 0$

58. $x^2 - 10x + 18 > 0$

400

██ REVIEW FOR CHAPTER 6 ██

Examples

1. If $(x - 3)^2 = 25$

then $x - 3 = \pm 5$

$x = 3 \pm 5$

$x = 8$ or $x = -2$

Chapter 6 Summary

Theorem 6.1 [6.1]

If $a^2 = b$, where b is a real number, then

$$a = \sqrt{b} \quad \text{or} \quad a = -\sqrt{b} \quad (\text{or} \quad a = \pm\sqrt{b})$$

2. Solve $x^2 - 6x - 6 = 0$.

$x^2 - 6x = 6$

$x^2 - 6x + 9 = 6 + 9$

$(x - 3)^2 = 15$

$x - 3 = \pm\sqrt{15}$

$x = 3 \pm \sqrt{15}$

To Solve a Quadratic Equation by Completing the Square [6.1]

Step 1 Write the equation in the form $ax^2 + bx = c$.

Step 2 If $a \neq 1$, divide through by the constant a so the coefficient of x^2 is 1.

Step 3 Complete the square on the left side by adding the square of $\frac{1}{2}$ the coefficient of x to both sides.

Step 4 Write the left side of the equation as the square of a binomial. Simplify the right side, if possible.

Step 5 Apply Theorem 6.1 and solve as usual.

3. If $2x^2 + 3x - 4 = 0$, then

$x = \dfrac{-3 \pm \sqrt{9 - 4(2)(-4)}}{2(2)}$

$= \dfrac{-3 \pm \sqrt{41}}{4}$

The Quadratic Formula [6.2]

For any quadratic equation in the form $ax^2 + bx + c = 0$, $a \neq 0$, the two solutions are

$$x = \frac{-b \pm \sqrt{b^2 - 4ac}}{2a}$$

This last expression is known as the **quadratic formula.**

4. The discriminant for

$x^2 + 6x + 9 = 0$

is

$D = 36 - 4(1)(9) = 0$

which means the equation has one rational solution.

The Discriminant [6.3]

The expression $b^2 - 4ac$ that appears under the radical sign in the quadratic formula is known as the **discriminant.**

We can classify the solutions to $ax^2 + bx + c = 0$:

THE SOLUTIONS ARE:	WHEN THE DISCRIMINANT IS:
Two complex numbers containing i	Negative
One rational number	Zero
Two rational numbers	A positive perfect square
Two irrational numbers	A positive number, but not a perfect square

Chapter 6   Quadratic Equations

5. $x^4 - x^2 - 12 = 0$ is a quadratic equation in x^2. Letting $y = x^2$, we have

$$y^2 - y - 12 = 0$$
$$(y - 4)(y + 3) = 0$$
$$y = 4 \quad \text{or} \quad y = -3$$

Resubstituting x^2 for y, we have

$$x^2 = 4 \quad \text{or} \quad x^2 = -3$$
$$x = \pm 2 \quad \text{or} \quad x = \pm i\sqrt{3}$$

6. The graph of $y = x^2 - 4$ will be a parabola. It will cross the x-axis at 2 and -2, and the vertex will be $(0, -4)$.

7. Solve $x^2 - 2x - 8 > 0$. We factor and draw the sign diagram:

$(x-4)(x+2) > 0$

$$- - \;|- - - - -|+ + + +\; (x-4)$$
$$- - |+ + + +|+ + + +\; (x+2)$$

$-2 \qquad 4$

The solution is $x < -2$ or $x > 4$.

Equations That Are Quadratic in Form [6.4]

There are a variety of equations whose form is quadratic. We solve most of them by making a substitution so the equation becomes quadratic, and then solving that equation by factoring or by the quadratic formula. For example,

THE EQUATION	IS QUADRATIC IN
$(2x - 3)^2 + 5(2x - 3) - 6 = 0$	$2x - 3$
$4x^4 - 7x^2 - 2 = 0$	x^2
$2x - 7\sqrt{x} + 3 = 0$	$\sqrt{x}$

Graphing Parabolas [6.5]

The graph of any equation of the form

$$y = ax^2 + bx + c \qquad a \neq 0$$

is a **parabola.** The graph is **concave up** if $a > 0$, and **concave down** if $a < 0$. The highest or lowest point on the graph is called the **vertex** and will always have an x-coordinate of $x = \dfrac{-b}{2a}$.

Quadratic Inequalities [6.6]

We solve quadratic inequalities by manipulating the inequality to get 0 on the right side and then factoring the left side. We then make a diagram that indicates where the factors are positive and where they are negative. From this sign diagram and the original inequality we graph the appropriate solution set.

Chapter 6 Review Problems

Solve each equation. [6.1]

1. $(2t - 5)^2 = 25$
2. $(3t - 2)^2 = 4$
3. $(3y - 4)^2 = -49$
4. $(8y + 1)^2 = -36$
5. $(2x + 6)^2 = 12$
6. $(3x - 4)^2 = 18$

Solve by completing the square. [6.1]

7. $2x^2 + 6x - 20 = 0$
8. $3x^2 + 15x = -18$

9. $a^2 + 9 = 6a$
10. $a^2 + 4 = 4a$
11. $2y^2 + 6y = -3$
12. $3y^2 + 3 = 9y$

Solve each equation. [6.2]

13. $\frac{1}{6}x^2 + \frac{1}{2}x - \frac{5}{3} = 0$
14. $\frac{2}{15}x^2 + \frac{1}{3}x + \frac{1}{5} = 0$
15. $8x^2 - 18x = 0$
16. $8x^2 - 18 = 0$
17. $4t^2 - 8t + 19 = 0$
18. $4t^2 - 4t + 9 = 0$

19. $100x^2 - 200x = 100$
20. $10x^2 - 60x = -70$
21. $0.06a^2 + 0.05a = 0.04$
22. $0.06a^2 - 0.01a = 0.02$
23. $9 - 6x = -x^2$ 24. $2x - 1 = x^2$
25. $(2x + 1)(x - 5) - (x + 3)(x - 2) = -17$
26. $(2x - 3)^2 - (3x + 2)(x - 2) = 2x + 6$
27. $2y^3 + 2y = 10y^2$ 28. $3y^3 - y = 5y^2$
29. $5x^2 = -2x + 3$ 30. $2x^2 = -3x + 7$
31. $x^3 - 27 = 0$ 32. $8a^3 + 27 = 0$

33. $3 - \dfrac{2}{x} + \dfrac{1}{x^2} = 0$ 34. $\dfrac{2}{x^2} + 1 = \dfrac{4}{x}$

35. $\dfrac{1}{x - 3} + \dfrac{1}{x + 2} = 1$

36. $\dfrac{1}{x + 3} + \dfrac{1}{x - 2} = 1$

37. The total cost (in dollars) for a company to produce x items per week is $C(x) = 7x + 400$. The revenue for selling all x items is $R(x) = 34x - 0.1x^2$. How many items must it produce and sell each week for its weekly profit to be $1,300? [6.2]

38. The total cost (in dollars) for a company to produce x items per week is $C(x) = 70x + 300$. The revenue for selling all x items is $R(x) = 110x - 0.5x^2$. How many items must it produce and sell each week for its weekly profit to be $300? [6.2]

Use the discriminant to find the number and kind of solutions for each equation. [6.3]

39. $2x^2 - 8x = -8$ 40. $4x^2 - 8x = -4$
41. $2x^2 + x - 3 = 0$ 42. $5x^2 + 11x = 12$
43. $x^2 - x = 1$ 44. $x^2 - 5x = -5$
45. $3x^2 + 5x = -4$ 46. $4x^2 - 3x = -6$

Determine k so that each equation has exactly one real solution. [6.3]

47. $25x^2 - kx + 4 = 0$
48. $4x^2 + kx + 25 = 0$

49. $kx^2 + 12x + 9 = 0$
50. $kx^2 - 16x + 16 = 0$
51. $9x^2 + 30x + k = 0$
52. $4x^2 + 28x + k = 0$

For each of the following problems, find an equation that has the given solutions. [6.3]

53. $x = 3, x = 5$ 54. $y = \frac{1}{2}, y = -4$
55. $t = 3, t = -3, t = 5$
56. $t = 2, t = -2, t = 6$

57. Find all solutions to $x^3 - 4x^2 + x + 6 = 0$ if $x = 2$ is one solution. [6.3]
58. Find all solutions to $x^3 - 3x^2 - 6x + 8 = 0$ if $x = 1$ is one solution. [6.3]

Find all solutions. [6.4]

59. $(x - 2)^2 - 4(x - 2) - 60 = 0$
60. $(x + 3)^2 - 3(x + 3) - 70 = 0$
61. $6(2y + 1)^2 - (2y + 1) - 2 = 0$
62. $3(4y - 1)^2 + (4y - 1) - 10 = 0$
63. $x^4 - x^2 = 12$ 64. $x^4 - 7x^2 = 18$
65. $x - \sqrt{x} - 2 = 0$ 66. $x - 3\sqrt{x} + 2 = 0$
67. $2x - 11\sqrt{x} = -12$ 68. $3x + \sqrt{x} = 2$
69. $\sqrt{x + 5} = \sqrt{x} + 1$ 70. $\sqrt{x - 2} = 2 - \sqrt{x}$
71. $\sqrt{y + 21} + \sqrt{y} = 7$
72. $\sqrt{y - 3} - \sqrt{y} = -1$
73. $\sqrt{y + 9} - \sqrt{y - 6} = 3$
74. $\sqrt{y + 7} - \sqrt{y + 2} = 1$

75. An object is tossed into the air with an upward velocity of 10 feet/second from the top of a building h feet high. The time (in seconds) it takes for the object to hit the ground below is given by the formula $16t^2 - 10t - h = 0$. Solve this formula for t. [6.4]

76. An object is tossed into the air with an upward velocity of v feet/second from the top of a 10 foot wall. The time (in seconds) it takes for the object to hit the ground below is given by the formula $16t^2 - vt - 10 = 0$. Solve this formula for t. [6.4]

Find the x-intercepts, if they exist, and the vertex for each parabola. Then use them to sketch the graph. [6.5]

77. $y = x^2 - 6x + 8$ **78.** $y = x^2 - x - 2$

Solve each inequality and graph the solution set. [6.6]

79. $x^2 - x - 2 < 0$ **80.** $x^2 - 4x + 4 < 0$
81. $(x + 2)(x - 3)(x + 4) > 0$
82. $x^3 + 2x^2 - 9x - 18 < 0$

83. $\dfrac{x}{x - 3} + \dfrac{2}{x - 3} > 0$

84. $\dfrac{x}{x + 4} - \dfrac{2}{x + 4} > 0$

85. $\dfrac{x - 3}{(x - 2)(x - 4)} > 0$

86. $\dfrac{x + 2}{(x + 1)(x - 1)} > 0$

Chapter 6 Test

Solve each equation. [6.1, 6.2]

1. $(2x + 4)^2 = 25$
2. $(2x - 6)^2 = -8$
3. $y^2 - 10y + 25 = -4$
4. $(y + 1)(y - 3) = -6$

5. $8t^3 - 125 = 0$ **6.** $\dfrac{1}{a + 2} - \dfrac{1}{3} = \dfrac{1}{a}$

7. Solve the formula $64(1 + r)^2 = A$ for r. [6.1]

8. Solve $x^2 - 4x = -2$ by completing the square. [6.1]

9. An object projected upward with an initial velocity of 32 feet/second will rise and fall according to the equation $s(t) = 32t - 16t^2$, where s is the distance (in feet) above the ground at time t (in seconds). At what times will the object be 12 feet above the ground? [6.2]

10. The total weekly cost for a company to make x ceramic coffee cups is given by the formula $C(x) = 2x + 100$. If the weekly revenue from selling all x cups is $R(x) = 25x - 0.2x^2$, how many cups must it sell a week to make a profit of $200 a week? [6.2]

11. Find k so that $kx^2 = 12x - 4$ has one rational solution. [6.3]

12. Use the discriminant to identify the number and kind of solutions to $2x^2 - 5x = 7$. [6.3]

Find equations that have the given solutions. [6.3]

13. $x = 5, x = -\frac{2}{3}$ **14.** $x = 2, x = -2, x = 7$

15. Find all the solutions to the equation $4x^3 - 16x^2 + 17x - 2 = 0$ if $x = 2$ is one solution. [6.3]

Solve each equation. [6.4]

16. $4x^4 - 7x^2 - 2 = 0$
17. $(2t + 1)^2 - 5(2t + 1) + 6 = 0$
18. $2t - 7\sqrt{t} + 3 = 0$

19. An object is tossed into the air with an upward velocity of 14 feet/second from the top of a building h feet high. The time (in seconds) it takes for the object to hit the ground below is given by the formula $16t^2 - 14t - h = 0$. Solve this formula for t. [6.4]

Sketch the graph of each of the following. Give the coordinates of the vertex in each case. [6.5]

20. $y = x^2 - 2x - 3$ **21.** $f(x) = -x^2 + 2x + 8$

22. Find the maximum weekly profit for a company with weekly costs of $C(x) = 5x + 100$ and weekly revenue of $R(x) = 25x - 0.1x^2$. [6.5]

Graph each of the following inequalities. [6.6]

23. $x^2 - x - 6 \le 0$ **24.** $2x^2 + 5x > 3$

25. $\dfrac{x + 2}{x - 1} \ge 0$

Systems of Equations

7

Introduction

Table 1 gives some corresponding temperatures on the Fahrenheit and Celsius temperature scales. Each pair of numbers is obtained by measuring the temperature of water using two different thermometers. The initial reading of 25°C and 77°F is the water at room temperature. The last reading, 100°C and 212°F, is boiling water. The numbers in between these pairs are obtained as the water is heated from room temperature to boiling.

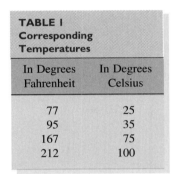

TABLE I
Corresponding Temperatures

In Degrees Fahrenheit	In Degrees Celsius
77	25
95	35
167	75
212	100

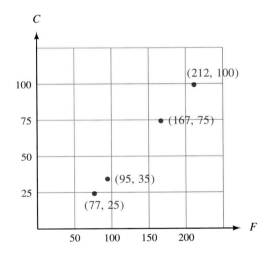

If we graph these pairs of numbers as ordered pairs (F, C), we find that they lie in a straight line—that is, the relationship between the variables C and F is a linear relationship. Although we could use the methods developed in Chapter 3 to find the formula that gives the exact relationship between C and F, the relationship can also be found as a *system of two equations in two variables,* which is one of the main topics of this chapter.

Systems of linear equations are used extensively in many different disciplines. They can be used to solve multiple-loop circuit problems in electronics, kinship patterns in anthropology, genetics problems in biology, and profit-and-cost problems in economics. There are many other applications as well.

406

Overview

In Chapter 3 we worked with linear equations in two variables. In this chapter we will extend our work with linear equations to include systems of linear equations in two variables. A **system of linear equations in two variables** is simply two linear equations considered at the same time. The solution to this type of system of equations consists of all the points that lie on both lines. That is, to solve a system of equations we look for all the points the individual equations have in common.

After we have introduced systems of equations in two variables, we will turn our attention to systems of equations in three variables. One of the ways in which we solve systems of equations in three variables is by reducing them to systems of equations in two variables.

Also in this chapter is our introduction to *determinants*. There are many places in mathematics where determinants are found. For our introduction to them we will see how they can be used to solve systems of linear equations.

To be successful in this chapter you should be familiar with the concepts in Chapter 3 that are concerned with linear equations in two variables.

Systems of Linear Equations in Two Variables

In Chapter 3 we found that the graph of an equation of the form $ax + by = c$ is a straight line. Since the graph is a straight line, the equation is said to be a linear equation. Two linear equations considered together form a **linear system of equations.** For example,

$$3x - 2y = 6$$
$$2x + 4y = 20$$

is a linear system. The solution set to the system is the set of all ordered pairs that satisfy both equations. If we graph each equation on the same set of axes, we can see the solution set (see Figure 1).

The point (4, 3) lies on both lines and therefore must satisfy both equations. It is obvious from the graph that this is the only point that does so. The solution set for the system is {(4, 3)}.

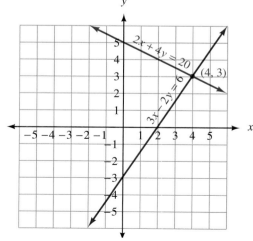

FIGURE I

More generally, if $a_1x + b_1y = c_1$ and $a_2x + b_2y = c_2$ are linear equations, then the solution set for the system

$$a_1x + b_1y = c_1$$
$$a_2x + b_2y = c_2$$

can be illustrated through one of the graphs shown in Figure 2.

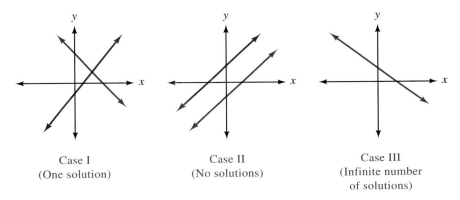

Case I	Case II	Case III
(One solution)	(No solutions)	(Infinite number of solutions)

FIGURE 2

Case I The two lines intersect at one and only one point. The coordinates of the point give the solution to the system. This is what usually happens.

Case II The lines are parallel and therefore have no points in common. The solution set to the system is the empty set, $\varnothing$. In this case, we say the equations are **inconsistent.**

Case III The lines coincide. That is, their graphs represent the same line. The solution set consists of all ordered pairs that satisfy either equation. In this case, the equations are said to be **dependent.**

In the beginning of this section we found the solution set for the system

$$3x - 2y = 6$$
$$2x + 4y = 20$$

by graphing each equation and then reading the solution set from the graph. Solving a system of linear equations by graphing is the least accurate method. If the coordinates of the point of intersection are not integers, it can be very difficult to read the solution set from the graph. There is another method of solving a linear system that does not depend on the graph. It is called the **addition method.**

EXAMPLE 1   Solve the system:

$$4x + 3y = 10$$
$$2x + \ y = 4$$

SOLUTION   If we multiply the bottom equation by -3, the coefficients of y in the resulting equation and the top equation will be opposites:

$$4x + 3y = 10 \xrightarrow{\text{No change}} 4x + 3y = 10$$
$$2x + \ y = 4 \xrightarrow[\text{Multiply by } -3]{} -6x - 3y = -12$$

Adding the left and right sides of the resulting equations, we have

$$
\begin{array}{r}
4x + 3y = 10 \\
-6x - 3y = -12 \\
\hline
-2x = -2
\end{array}
$$

The result is a linear equation in one variable. We have eliminated the variable y from the equations by addition. (It is for this reason we call this method of solving a linear system the *addition method*.) Solving $-2x = -2$ for x, we have

$$x = 1$$

This is the x-coordinate of the solution to our system. To find the y-coordinate, we substitute $x = 1$ into any of the equations containing both the variables x and y. Let's try the second equation in our original system:

$$2(1) + y = 4$$
$$2 + y = 4$$
$$y = 2$$

This is the y-coordinate of the solution to our system. The ordered pair $(1, 2)$ is the solution to the system.   ◢

EXAMPLE 2   Solve the system:

$$3x - 5y = -2$$
$$2x - 3y = 1$$

SOLUTION   We can eliminate either variable. Let's decide to eliminate the variable x. We can do so by multiplying the top equation by 2 and the bottom equation by -3, and then adding the left and right sides of the resulting equations:

$$
\begin{array}{r}
3x - 5y = -2 \xrightarrow{\text{Multiply by } 2} 6x - 10y = -4 \\
2x - 3y = 1 \xrightarrow[\text{Multiply by } -3]{} -6x + 9y = -3 \\
\hline
-y = -7 \\
y = 7
\end{array}
$$

The y-coordinate of the solution to the system is 7. Substituting this value of y into any of the equations with both x and y variables gives $x = 11$. The solution to the system is $(11, 7)$. This is the only ordered pair that satisfies both equations. ◢

EXAMPLE 3   Solve the system:

$$2x - 3y = 4$$
$$4x + 5y = 3$$

SOLUTION   We can eliminate x by multiplying the top equation by -2 and adding it to the bottom equation:

$$
\begin{array}{l}
2x - 3y = 4 \xrightarrow{\text{Multiply by } -2} -4x + 6y = -8 \\
4x + 5y = 3 \xrightarrow[\text{No change}]{} \underline{4x + 5y = 3} \\
\phantom{4x + 5y = 3 \xrightarrow[\text{No change}]{}} 11y = -5 \\
\phantom{4x + 5y = 3 \xrightarrow[\text{No change}]{}} y = -\tfrac{5}{11}
\end{array}
$$

The y-coordinate of our solution is $-\frac{5}{11}$. If we substitute this value of y back into either of the original equations, we find the arithmetic necessary to solve for x is cumbersome. For this reason, it is probably best to go back to the original system and solve it a second time — this time for x instead of y. Here is how we do that:

$$
\begin{array}{l}
2x - 3y = 4 \xrightarrow{\text{Multiply by } 5} 10x - 15y = 20 \\
4x + 5y = 3 \xrightarrow[\text{Multiply by } 3]{} \underline{12x + 15y = 9} \\
\phantom{4x + 5y = 3 \xrightarrow[\text{Multiply by } 3]{}} 22x = 29 \\
\phantom{4x + 5y = 3 \xrightarrow[\text{Multiply by } 3]{}} x = \tfrac{29}{22}
\end{array}
$$

The solution to our system is $\left(\frac{29}{22}, -\frac{5}{11}\right)$. ◢

EXAMPLE 4   Solve the system:

$$5x - 2y = 1$$
$$-10x + 4y = 3$$

SOLUTION   We can eliminate y by multiplying the first equation by 2 and adding the result to the second equation:

$$
\begin{array}{l}
5x - 2y = 1 \xrightarrow{\text{Multiply by } 2} 10x - 4y = 2 \\
-10x + 4y = 3 \xrightarrow[\text{No change}]{} \underline{-10x + 4y = 3} \\
\phantom{-10x + 4y = 3 \xrightarrow[\text{No change}]{}} 0 = 5
\end{array}
$$

The result is the false statement $0 = 5$, which indicates there is no solution to the system. If we were to graph the two lines, we would find that they are parallel. In a case like this, we say the system is **inconsistent.** Whenever both variables have been eliminated and the resulting statement is false, the solution set for the system will be the empty set, $\varnothing$. ◢

EXAMPLE 5 Solve the system:

$$4x - 3y = 2$$
$$8x - 6y = 4$$

SOLUTION Multiplying the top equation by -2 and adding, we can eliminate the variable x:

$$
\begin{array}{ll}
4x - 3y = 2 & \xrightarrow{\text{Multiply by } -2} & -8x + 6y = -4 \\
8x - 6y = 4 & \xrightarrow[\text{No change}]{} & \underline{8x - 6y = 4} \\
& & 0 = 0
\end{array}
$$

Both variables have been eliminated, and the resulting statement $0 = 0$ is true. In this case the lines coincide and the system is said to be **dependent.** The solution set consists of all ordered pairs that satisfy either equation. We can write the solution set as $\{(x, y) \mid 4x - 3y = 2\}$ or $\{(x, y) \mid 8x - 6y = 4\}$. ◢

The previous two examples illustrate the two special cases in which the graphs of the equations in the system either coincide or are parallel. In both cases the left-hand sides of the equations were multiples of one another. In the case of the dependent equations, the right-hand sides were also multiples. We can generalize these observations as follows:

The equations in the system

$$a_1 x + b_1 y = c_1$$
$$a_2 x + b_2 y = c_2$$

will be inconsistent (their graphs are parallel lines) if

$$\frac{a_1}{a_2} = \frac{b_1}{b_2} \neq \frac{c_1}{c_2}$$

and will be dependent (their graphs will coincide) if

$$\frac{a_1}{a_2} = \frac{b_1}{b_2} = \frac{c_1}{c_2}$$

EXAMPLE 6 Solve the system:

$$\tfrac{1}{2}x - \tfrac{1}{3}y = 2$$
$$\tfrac{1}{4}x + \tfrac{2}{3}y = 6$$

SOLUTION Although we could solve this system without clearing the equations of fractions, there is probably less chance for error if we have only integer coefficients to work with. So let's begin by multiplying both sides of the top equation by 6, and both sides of the bottom equation by 12, to clear each equation of fractions:

$$
\begin{array}{ll}
\tfrac{1}{2}x - \tfrac{1}{3}y = 2 & \xrightarrow{\text{Times } 6} & 3x - 2y = 12 \\
\tfrac{1}{4}x + \tfrac{2}{3}y = 6 & \xrightarrow[\text{Times } 12]{} & 3x + 8y = 72
\end{array}
$$

Now we can eliminate x by multiplying the top equation by -1 and leaving the bottom equation unchanged:

$$3x - 2y = 12 \xrightarrow{\text{Times } -1} -3x + 2y = -12$$
$$3x + 8y = 72 \xrightarrow{\text{No change}} \underline{3x + 8y = 72}$$
$$10y = 60$$
$$y = 6$$

We can substitute $y = 6$ into any equation that contains both x and y. Let's use $3x - 2y = 12$:

$$3x - 2(6) = 12$$
$$3x - 12 = 12$$
$$3x = 24$$
$$x = 8$$

The solution to the system is $(8, 6)$.

We end this section by considering another method of solving a linear system. The method is called the **substitution method** and is shown in the following examples.

EXAMPLE 7 Solve the system:

$$2x - 3y = -6$$
$$y = 3x - 5$$

SOLUTION The second equation tells us y is $3x - 5$. Substituting the expression $3x - 5$ for y in the first equation, we have

$$2x - 3(3x - 5) = -6$$

The result of the substitution is the elimination of the variable y. Solving the resulting linear equation in x as usual, we have

$$2x - 9x + 15 = -6$$
$$-7x + 15 = -6$$
$$-7x = -21$$
$$x = 3$$

Putting $x = 3$ into the second equation in the original system, we have

$$y = 3(3) - 5$$
$$= 9 - 5$$
$$= 4$$

The solution to the system is $(3, 4)$.

EXAMPLE 8 Solve by substitution:

$$2x + 3y = 5$$
$$x - 2y = 6$$

SOLUTION In order to use the substitution method we must solve one of the two equations for x or y. We can solve for x in the second equation by adding $2y$ to both sides:

$$x - 2y = 6$$
$$x = 2y + 6 \quad \text{Add } 2y \text{ to both sides.}$$

Substituting the expression $2y + 6$ for x in the first equation of our system, we have

$$2(2y + 6) + 3y = 5$$
$$4y + 12 + 3y = 5$$
$$7y + 12 = 5$$
$$7y = -7$$
$$y = -1$$

Using $y = -1$ in either equation in the original system, we find that $x = 4$. The solution is $(4, -1)$.

Both the substitution method and the addition method can be used to solve any system of linear equations in two variables. However, systems like the one in Example 7 are easier to solve using the substitution method, since one of the variables is already written in terms of the other. A system like the one in Example 2 is easier to solve using the addition method, since solving for one of the variables would lead to an expression involving fractions. The system in Example 8 could be solved easily by either method, since solving the second equation for x is a one-step process.

 Using Technology: Graphing Calculators

Solving Systems That Intersect in Exactly One Point

A graphing calculator can be used to solve a system of equations in two variables if the equations intersect in exactly one point. To solve the system shown in Example 3, we first solve each equation for y. Here is the result:

$$2x - 3y = 4 \qquad \text{becomes} \qquad y = \frac{4 - 2x}{-3}$$

$$4x + 5y = 3 \qquad \text{becomes} \qquad y = \frac{3 - 4x}{5}$$

Note that we do not have to write the equations in simplest form since it does not matter to the calculator whether there is a negative number in the denominator or not. Graphing these two functions on the calculator gives a diagram similar to the one in Figure 3.

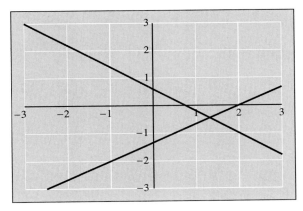

FIGURE 3

Using the Trace and Zoom features, we find that the two lines intersect at $x = 1.32$ and $y = -0.45$, which are the decimal equivalents (accurate to the nearest hundredth) of the fractions found in Example 3.

Special Cases

We cannot assume that two lines that look parallel in a calculator window are in fact parallel. If you graph the functions $y = x - 5$ and $y = 0.99x + 2$ in a window where x and y range from -10 to 10, the lines look parallel. However, we know this is not the case, since their slopes are different. As we zoom out repeatedly, the lines begin to look as if they coincide. We know this is not the case, because the two lines have different y-intercepts. To summarize: If we graph two lines on a calculator and the graphs look as if they are parallel or coincide, we should use algebraic methods, not the calculator, to determine the solution to the system.

Problem Set

7.1

Solve each system by graphing both equations on the same set of axes and then reading the solution from the graph.

1. $3x - 2y = 6$
$\quad\ x - y = 1$

2. $5x - 2y = 10$
$\quad\ x - y = -1$

3. $\quad\ y = \frac{3}{5}x - 3$
$\quad 2x - y = -4$

4. $\quad\ y = \frac{1}{2}x - 2$
$\quad 2x - y = -1$

5. $y = \frac{1}{2}x$
$\quad y = -\frac{3}{4}x + 5$

6. $y = \frac{2}{3}x$
$\quad y = -\frac{1}{3}x + 6$

7. $3x + 3y = -2$
$\quad\quad\quad y = -x + 4$

8. $2x - 2y = 6$
$\quad\quad\quad y = x - 3$

9. $2x - y = 5$
$\quad\quad\quad y = 2x - 5$

10. $x + 2y = 5$
$\quad\quad\quad y = -\frac{1}{2}x + 3$

Solve each of the following systems by the addition method.

11. $x + y = 5$
$\quad\;\; 3x - y = 3$

12. $x - y = 4$
$\quad\;\; -x + 2y = -3$

13. $3x + y = 4$
$\quad\;\; 4x + y = 5$

14. $6x - 2y = -10$
$\quad\;\; 6x + 3y = -15$

15. $3x - 2y = 6$
$\quad\;\; 6x - 4y = 12$

16. $4x + 5y = -3$
$\quad\;\; -8x - 10y = 3$

17. $x + 2y = 0$
$\quad\;\; 2x - 6y = 5$

18. $x + 3y = 3$
$\quad\;\; 2x - 9y = 1$

19. $2x - 5y = 16$
$\quad\;\; 4x - 3y = 11$

20. $5x - 3y = -11$
$\quad\;\; 7x + 6y = -12$

21. $6x + 3y = -1$
$\quad\;\; 9x + 5y = 1$

22. $5x + 4y = -1$
$\quad\;\; 7x + 6y = -2$

23. $4x + 3y = 14$
$\quad\;\; 9x - 2y = 14$

24. $7x - 6y = 13$
$\quad\;\; 6x - 5y = 11$

25. $2x - 5y = 3$
$\quad\;\; -4x + 10y = 3$

26. $3x - 2y = 1$
$\quad\;\; -6x + 4y = -2$

27. $\frac{1}{4}x - \frac{1}{6}y = -2$
$\quad\;\; -\frac{1}{6}x + \frac{1}{5}y = 4$

28. $-\frac{1}{3}x + \frac{1}{4}y = 0$
$\quad\;\; \frac{1}{5}x - \frac{1}{10}y = 1$

29. $\frac{1}{2}x + \frac{1}{3}y = 13$
$\quad\;\; \frac{2}{5}x + \frac{1}{4}y = 10$

30. $\frac{1}{2}x + \frac{1}{3}y = \frac{2}{3}$
$\quad\;\; \frac{2}{3}x + \frac{2}{5}y = \frac{14}{15}$

31. $\frac{2}{3}x + \frac{2}{5}y = 4$
$\quad\;\; \frac{1}{3}x - \frac{1}{2}y = -\frac{1}{3}$

32. $\frac{1}{2}x - \frac{1}{3}y = \frac{5}{6}$
$\quad\;\; -\frac{2}{5}x + \frac{1}{2}y = -\frac{9}{10}$

Solve each of the following systems by the substitution method.

33. $7x - y = 24$
$\quad\;\; x = 2y + 9$

34. $3x - y = -8$
$\quad\;\; y = 6x + 3$

35. $6x - y = 10$
$\quad\;\; y = -\frac{3}{4}x - 1$

36. $2x - y = 6$
$\quad\;\; y = -\frac{4}{3}x + 1$

37. $x - y = 4$
$\quad\;\; 2x - 3y = 6$

38. $x + y = 3$
$\quad\;\; 2x + 3y = -4$

39. $y = 3x - 2$
$\quad\;\; y = 4x - 4$

40. $y = 5x - 2$
$\quad\;\; y = -2x + 5$

41. $2x - y = 5$
$\quad\;\; 4x - 2y = 10$

42. $-10x + 8y = -6$
$\quad\;\; y = \frac{5}{4}x$

43. $\frac{1}{3}x - \frac{1}{2}y = 0$
$\quad\;\; x = \frac{3}{2}y$

44. $\frac{2}{5}x - \frac{2}{3}y = 0$
$\quad\;\; y = \frac{3}{5}x$

You may want to read Example 3 again before solving the systems that follow.

45. $4x - 7y = 3$
$\quad\;\; 5x + 2y = -3$

46. $3x - 4y = 7$
$\quad\;\; 6x - 3y = 5$

47. $9x - 8y = 4$
$\quad\;\; 2x + 3y = 6$

48. $4x - 7y = 10$
$\quad\;\; -3x + 2y = -9$

49. $3x - 5y = 2$
$\quad\;\; 7x + 2y = 1$

50. $4x - 3y = -1$
$\quad\;\; 5x + 8y = 2$

51. Multiply both sides of the second equation in the following system by 100 and then solve as usual.

$$x + y = 10{,}000$$
$$0.06x + 0.05y = 560$$

52. Multiply both sides of the second equation in the following system by 10 and then solve as usual.

$$x + y = 12$$
$$0.20x + 0.50y = 0.30(12)$$

53. What value of c will make the following system a dependent system (one in which the lines coincide)?

$$6x - 9y = 3$$
$$4x - 6y = c$$

54. What value of c will make the following system a dependent system?

$$5x - 7y = c$$
$$-15x + 21y = 9$$

Applying the Concepts

Problems 55 and 56 may be solved using a graphing calculator.

55. Cost of a Phone Call One telephone company charges 41 cents for the first minute and 32 cents for each additional minute for a certain long-distance phone call. If the number of additional minutes after the first minute is x and the cost (in cents) for the call is y, then the equation that gives the total cost (in cents) for the call is $y = 32x + 41$.

(a) How much does it cost to make a 10 minute long-distance call under these conditions?

(b) If a second phone company charges 45 cents for the first minute and 30 cents for each additional minute, write the equation that gives the total cost y of a call in terms of the number of additional minutes x.

(c) After how many additional minutes will the two companies charge an equal amount? (What is the x-coordinate of the point of intersection of the two lines?)

56. Cost of a Taxi Ride In a certain city, a taxi ride costs 75¢ for the first $\frac{1}{7}$ of a mile and 10¢ for every additional $\frac{1}{7}$ of a mile after the first $\frac{1}{7}$. If x is the number of additional $\frac{1}{7}$'s of a mile, then the total cost y (in cents) of a taxi ride is

$$y = 10x + 75$$

San Francisco AIRPORT

(a) How much does it cost to ride a taxi for 10 miles in this city?

(b) Suppose a taxi ride in another city costs 50¢ for the first $\frac{1}{7}$ of a mile, and 15¢ for each additional $\frac{1}{7}$ of a mile. Write an equation that gives the total cost y (in cents) to ride x $\frac{1}{7}$'s of a mile past the first $\frac{1}{7}$ in this city.

(c) Solve the two equations from parts (a) and (b) simultaneously (as a system of equations), and explain in words what your solution represents.

Review Problems

NOTE: The problems that follow review material we covered in Section 6.1.

Solve each equation.

57. $(2x - 1)^2 = 25$

58. $(3x + 5)^2 = -12$

59. What number would you add to $x^2 - 10x$ to make it a perfect square trinomial?

60. What number would you add to $x^2 - 5x$ to make it a perfect square trinomial?

Solve by completing the square.

61. $x^2 - 10x + 8 = 0$

62. $x^2 - 5x + 4 = 0$

63. $3x^2 - 6x + 6 = 0$

64. $4x^2 - 16x - 8 = 0$

One Step Further

65. Find a and b so that the line $ax + by = 7$ passes through the points $(1, -2)$ and $(3, 1)$.

66. Find a and b so that the line $ax + by = 2$ passes through the points $(2, 2)$ and $(6, 7)$.

67. Find a and b so the parabola $y = ax^2 + bx$ will pass through the points $(-1, 3)$ and $(3, 3)$. Then find the vertex of the parabola.

68. Find a and b so the parabola $y = ax^2 + bx$ will pass through the points $(1, 3)$ and $(-3, 3)$. Then find the vertex of the parabola.

Systems of Linear Equations in Three Variables

A solution to an equation in three variables such as

$$2x + y - 3z = 6$$

is an **ordered triple** of numbers (x, y, z). For example, the solutions to the equation $2x + y - 3z = 6$ are the ordered triples $(0, 0, -2)$, $(2, 2, 0)$, and $(0, 9, 1)$, since they produce a true statement when their coordinates are replaced for x, y, and z in the equation.

> **DEFINITION**
>
> The **solution set** for a system of three linear equations in three variables is the set of **ordered triples** that satisfy all three equations.

EXAMPLE 1 Solve the system:

$$
\begin{aligned}
x + y + z &= 6 &\quad (1)\\
2x - y + z &= 3 &\quad (2)\\
x + 2y - 3z &= -4 &\quad (3)
\end{aligned}
$$

SOLUTION We want to find the ordered triple (x, y, z) that satisfies all three equations. We have numbered the equations so it will be easier to keep track of where they are and what we are doing.

There are many ways to proceed. The main idea is to take two different pairs of equations and eliminate the same variable from each pair. We begin by adding equations (1) and (2) to eliminate the y variable. The resulting equation is numbered (4):

$$
\begin{aligned}
x + y + z &= 6 \quad (1)\\
\underline{2x - y + z} &= \underline{3} \quad (2)\\
3x \qquad + 2z &= 9 \quad (4)
\end{aligned}
$$

Adding twice equation (2) to equation (3) will also eliminate the variable y. The resulting equation is numbered (5):

$$
\begin{aligned}
4x - 2y + 2z &= 6 \quad \text{Twice (2)}\\
\underline{x + 2y - 3z} &= \underline{-4} \quad (3)\\
5x \qquad - z &= 2 \quad (5)
\end{aligned}
$$

Equations (4) and (5) form a linear system in two variables. By multiplying equation (5) by 2 and adding the result to equation (4), we will succeed in eliminating the variable z from the new pair of equations:

$$
\begin{aligned}
3x + 2z &= 9 \quad (4)\\
\underline{10x - 2z} &= \underline{4} \quad \text{Twice (5)}\\
13x \qquad &= 13\\
x &= 1
\end{aligned}
$$

Substituting $x = 1$ into equation (4), we have

$$3(1) + 2z = 9$$
$$2z = 6$$
$$z = 3$$

Using $x = 1$ and $z = 3$ in equation (1) gives us

$$1 + y + 3 = 6$$
$$y + 4 = 6$$
$$y = 2$$

The solution is the ordered triple $(1, 2, 3)$.

EXAMPLE 2   Solve the system:

$$2x + y - z = 3 \quad (1)$$
$$3x + 4y + z = 6 \quad (2)$$
$$2x - 3y + z = 1 \quad (3)$$

SOLUTION   It is easiest to eliminate z from the equations. The equation produced by adding (1) and (2) is

$$5x + 5y = 9 \quad (4)$$

The equation that results from adding (1) and (3) is

$$4x - 2y = 4 \quad (5)$$

Equations (4) and (5) form a linear system in two variables. We can eliminate the variable y from this system as follows:

$$5x + 5y = 9 \xrightarrow{\text{Multiply by 2}} 10x + 10y = 18$$
$$4x - 2y = 4 \xrightarrow[\text{Multiply by 5}]{} \underline{20x - 10y = 20}$$
$$30x \qquad\quad = 38$$
$$x = \frac{38}{30}$$
$$x = \frac{19}{15}$$

Substituting $x = \frac{19}{15}$ into equation (5) or equation (4) and solving for y gives

$$y = \frac{8}{15}$$

Using $x = \frac{19}{15}$ and $y = \frac{8}{15}$ in equation (1), (2), or (3) and solving for z results in

$$z = \frac{1}{15}$$

The ordered triple that satisfies all three equations is $(\frac{19}{15}, \frac{8}{15}, \frac{1}{15})$.

EXAMPLE 3   Solve the system:

$$2x + 3y - z = 5 \quad (1)$$
$$4x + 6y - 2z = 10 \quad (2)$$
$$x - 4y + 3z = 5 \quad (3)$$

SOLUTION Multiplying equation (1) by -2 and adding the result to equation (2) looks like this:

$$\begin{array}{rl} -4x - 6y + 2z = -10 & -2 \text{ times (1)} \\ \underline{4x + 6y - 2z = 10} & \text{(2)} \\ 0 = 0 & \end{array}$$

All three variables have been eliminated, and we are left with a true statement. As was the case in Section 7.1, this finding implies that the two equations are dependent.

  With a system of three equations in three variables, however, a dependent system can have no solution or an infinite number of solutions. After we have concluded the examples in this section, we will discuss the geometry behind these systems. Doing so will give you some additional insight into dependent systems. ◢

EXAMPLE 4 Solve the system:

$$\begin{array}{rl} x - 5y + 4z = 8 & \text{(1)} \\ 3x + y - 2z = 7 & \text{(2)} \\ -9x - 3y + 6z = 5 & \text{(3)} \end{array}$$

SOLUTION Multiplying equation (2) by 3 and adding the result to equation (3) produces

$$\begin{array}{rl} 9x + 3y - 6z = 21 & 3 \text{ times (2)} \\ \underline{-9x - 3y + 6z = 5} & \text{(3)} \\ 0 = 26 & \end{array}$$

In this case all three variables have been eliminated, and we are left with a false statement. The two equations are inconsistent; there are no ordered triples that satisfy both equations. The solution set for the system is the empty set, $\varnothing$. If equations (2) and (3) have no ordered triples in common, then certainly (1), (2), and (3) do not either. ◢

EXAMPLE 5 Solve the system:

$$\begin{array}{rl} x + 3y = 5 & \text{(1)} \\ 6y + z = 12 & \text{(2)} \\ x - 2z = -10 & \text{(3)} \end{array}$$

SOLUTION It may be helpful to rewrite the system as

$$\begin{array}{rl} x + 3y = 5 & \text{(1)} \\ 6y + z = 12 & \text{(2)} \\ x - 2z = -10 & \text{(3)} \end{array}$$

Equation (2) does not contain the variable x. If we multiply equation (3) by -1 and add the result to equation (1), we will be left with another equation that does not contain the variable x:

$$\begin{array}{rl} x + 3y = 5 & \text{(1)} \\ \underline{-x + 2z = 10} & -1 \text{ times (3)} \\ 3y + 2z = 15 & \text{(4)} \end{array}$$

Equations (2) and (4) form a linear system in two variables. Multiplying equation (2) by -2 and adding the result to equation (4) eliminates the variable z:

$$6y + z = 12 \xrightarrow{\text{Multiply by } -2} -12y - 2z = -24$$
$$3y + 2z = 15 \xrightarrow{\text{No change}} \underline{3y + 2z = 15}$$
$$-9y = -9$$
$$y = 1$$

Using $y = 1$ in equation (4) and solving for z, we have

$$z = 6$$

Substituting $y = 1$ into equation (1) gives

$$x = 2$$

The ordered triple that satisfies all three equations is $(2, 1, 6)$.

The Geometry Behind Equations in Three Variables

We can graph an ordered triple on a coordinate system with three axes. The graph will be a point in space. The coordinate system is drawn in perspective; you have to imagine that the x-axis comes out of the paper and is perpendicular to both the y-axis and the z-axis. To graph the point $(3, 4, 5)$ we move 3 units in the x direction, 4 units in the y direction, and then 5 units in the z direction, as shown in Figure 1.

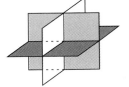

Case 1 The three planes have exactly one point in common, as in Examples 1, 2 and 5.

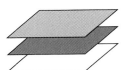

Case 2 The three planes have no points in common. The system they represent is an inconsistent system.

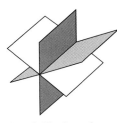

Case 3 The three planes intersect in a line. Any point on the line is a solution to the system of equations represented by the planes, so there is an infinite number of solutions to the system. This is an example of a dependent system.

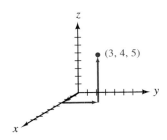

FIGURE 1

Although in actual practice it is sometimes difficult to graph equations in three variables, if we were to graph a linear equation in three variables, we would find that the graph was a plane in space. A system of three equations in three variables is represented by three planes in space.

There are a number of possible ways in which these three planes can intersect, some of which are shown in the margin. There are still other possibilities that are not among those shown.

In Example 3, we found that equations (1) and (2) were dependent equations. They represent the same plane. That is, they have all their points in common. But the system of equations that they came from has either no solution or an infinite number of solutions. It all depends on the third plane. If the third plane coincides

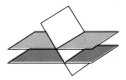

Case 4 Two of the planes are parallel, while the third plane intersects each of the parallel planes. There is no solution to the system; it is an inconsistent system.

with the first two, then the solution to the system is a plane. If the third plane is parallel to the first two, then there is no solution to the system. And finally, if the third plane intersects the first two, but does not coincide with them, then the solution to the system is that line of intersection.

In Example 4 we found that trying to eliminate a variable from the second and third equations resulted in a false statement. This means that the two planes represented by these equations are parallel. It makes no difference where the third plane is; there is no solution to the system in Example 4. (If we were to graph the three planes from Example 4, we would obtain a diagram similar to Case 2 or Case 4 in the margin.)

If, in the process of solving a system of linear equations in three variables, we eliminate all the variables from a pair of equations and are left with a false statement, we will say the system is inconsistent. If we eliminate all the variables and are left with a true statement, then we will say the system is dependent.

Problem Set
7.2

Solve the following systems.

1. $x + y + z = 4$
$x - y + 2z = 1$
$x - y - 3z = -4$

2. $x - y - 2z = -1$
$x + y + z = 6$
$x + y - z = 4$

3. $x + y + z = 6$
$x - y + 2z = 7$
$2x - y - 4z = -9$

4. $x + y + z = 0$
$x + y - z = 6$
$x - y + 2z = -7$

5. $x + 2y + z = 3$
$2x - y + 2z = 6$
$3x + y - z = 5$

6. $2x + y - 3z = -14$
$x - 3y + 4z = 22$
$3x + 2y + z = 0$

7. $2x + 3y - 2z = 4$
$x + 3y - 3z = 4$
$3x - 6y + z = -3$

8. $4x + y - 2z = 0$
$2x - 3y + 3z = 9$
$-6x - 2y + z = 0$

9. $-x + 4y - 3z = 2$
$2x - 8y + 6z = 1$
$3x - y + z = 0$

10. $4x + 6y - 8z = 1$
$-6x - 9y + 12z = 0$
$x - 2y - 2z = 3$

11. $\frac{1}{2}x - y + z = 0$
$2x + \frac{1}{3}y + z = 2$
$x + y + z = -4$

12. $\frac{1}{3}x + \frac{1}{2}y + z = -1$
$x - y + \frac{1}{5}z = 1$
$x + y + z = 5$

13. $2x - y - 3z = 1$
$x + 2y + 4z = 3$
$4x - 2y - 6z = 2$

14. $3x + 2y + z = 3$
$x - 3y + z = 4$
$-6x - 4y - 2z = 1$

15. $2x - y + 3z = 4$
$x + 2y - z = -3$
$4x + 3y + 2z = -5$

16. $6x - 2y + z = 5$
$3x + y + 3z = 7$
$x + 4y - z = 4$

17. $x + y = 9$
$y + z = 7$
$x - z = 2$

18. $x - y = -3$
$x + z = 2$
$y - z = 7$

19. $2x + y = 2$
$y + z = 3$
$4x - z = 0$

20. $2x + y = 6$
$3y - 2z = -8$
$x + z = 5$

21. $2x - 3y = 0$
$6y - 4z = 1$
$x + 2z = 1$

22. $3x + 2y = 3$
$y + 2z = 2$
$6x - 4z = 1$

23. $\frac{1}{2}x + \frac{2}{3}y = \frac{5}{2}$
$\frac{1}{5}x - \frac{1}{2}z = -\frac{3}{10}$
$\frac{1}{3}y - \frac{1}{4}z = \frac{3}{4}$

24. $\frac{1}{2}x - \frac{1}{3}y = \frac{1}{6}$
$\frac{1}{3}y - \frac{1}{3}z = 1$
$\frac{1}{5}x - \frac{1}{2}z = -\frac{4}{5}$

25. $\frac{1}{2}x - \frac{1}{4}y + \frac{1}{2}z = -2$
$\frac{1}{4}x - \frac{1}{12}y - \frac{1}{3}z = \frac{1}{4}$
$\frac{1}{6}x + \frac{1}{3}y - \frac{1}{2}z = \frac{3}{2}$

26. $\frac{1}{2}x + \frac{1}{2}y + z = \frac{1}{2}$
$\frac{1}{2}x - \frac{1}{4}y - \frac{1}{4}z = 0$
$\frac{1}{4}x + \frac{1}{12}y + \frac{1}{6}z = \frac{1}{6}$

Applying the Concepts

27. **Electric Current** In the following diagram of an electrical circuit, x, y, and z represent the amount of current (in amperes) flowing across the 5 ohm, 20 ohm, and 10 ohm resistors, respectively. (In circuit diagrams resistors are represented by $-W$ and potential differences by $-|{\vdash}\cdot$)

The system of equations used to find the three currents x, y, and z is given at the top of the next column.

$x - y - z = 0$
$5x + 20y = 80$
$20y - 10z = 50$

Solve the system for all variables.

28. **Cost of a Rental Car** If a car rental company charges \$10 a day and 8¢ a mile to rent one of its cars, then the cost z (in dollars) to rent a car for x days and drive y miles can be found from the equation

$$z = 10x + 0.08y$$

(a) How much does it cost to rent a car for 2 days and drive it 200 miles under these conditions?

(b) A second company charges \$12 a day and 6¢ a mile for the same car. Write an equation that gives the cost z (in dollars) to rent a car from this company for x days and drive it y miles.

(c) A car is rented from each of the companies mentioned above for 2 days. To find the mileage at which the cost of renting the cars from each of the two companies will be equal, solve the following system for y:

$$z = 10x + 0.08y$$
$$z = 12x + 0.06y$$
$$x = 2$$

Review Problems

NOTE ▶ The problems below review material we covered in Sections 2.2 and 6.2. Reviewing the problems from Section 2.2 will help you with some of the next section.

Solve each equation for y. [2.2]

29. $2x - 3y = 6$
30. $3x + 2y = 6$
31. $2x - 3y = 5$
32. $3x - 2y = 5$

Solve. [6.2]

33. $2x^2 + 4x - 3 = 0$
34. $3x^2 + 4x - 2 = 0$
35. $(2y - 3)(2y - 1) = -4$

36. $(y - 1)(3y - 3) = 10$

37. $t^3 - 125 = 0$ **38.** $8t^3 + 1 = 0$

39. $4x^5 - 16x^4 = 20x^3$ **40.** $3x^4 + 6x^2 = 6x^3$

41. $\dfrac{1}{x - 3} + \dfrac{1}{x + 2} = 1$

42. $\dfrac{1}{x + 3} + \dfrac{1}{x - 2} = 1$

One Step Further

43. Find a, b, and c so that the parabola

$$y = ax^2 + bx + c$$

passes through the points $(1, 0)$, $(3, -4)$, and $(5, 0)$.

44. Find a, b, and c so that the parabola

$$y = ax^2 + bx + c$$

passes through the points $(0, 3)$, $(1, 0)$, and $(-1, 4)$.

Solve each system for the solution (x, y, z, w).

45.
$$\begin{aligned}
x + y + z + w &= 10 \\
x + 2y - z + w &= 6 \\
x - y - z + 2w &= 4 \\
x - 2y + z - 3w &= -12
\end{aligned}$$

46.
$$\begin{aligned}
x + y + z + w &= 16 \\
x - y + 2z - w &= 1 \\
x + 3y - z - w &= -2 \\
x - 3y - 2z + 2w &= -4
\end{aligned}$$

SECTION 7.3

Introduction to Matrices and Determinants

We begin this section by defining what are called *matrices*. Then we will do some simple addition, subtraction, and multiplication with matrices. And, finally, we will expand and evaluate *determinants*. Later in this chapter we will see how determinants can be used to solve systems of equations. If you go on to take a probability or finite mathematics class, you will see how addition and multiplication of matrices can be applied to probability and business problems.

An **$m \times n$** (*m* by *n*) **matrix** is an array of numbers with **m rows** and **n columns,** enclosed with brackets [   ]. For example, the following is a 3×4 matrix:

$$\begin{bmatrix} 3 & 6 & -2 & 4 \\ 1 & -2 & 5 & 0 \\ 0 & 1 & 0 & -5 \end{bmatrix} \quad \text{3 rows}$$

4 columns

The numbers that make up the matrix are called **elements** of the matrix. The **dimensions** of the matrix are the numbers of rows and columns. If a matrix has an equal number of rows and columns, it is called a **square matrix.**

We add (or subtract) two matrices by adding (or subtracting) elements in corresponding positions in each matrix.

EXAMPLE 1   If $A = \begin{bmatrix} 3 & 4 \\ -2 & 7 \end{bmatrix}$ and $B = \begin{bmatrix} -8 & 1 \\ 0 & 2 \end{bmatrix}$, find $A + B$ and $A - B$.

SOLUTION   We add and subtract elements in corresponding positions to find the sum and difference of A and B:

$$A + B = \begin{bmatrix} 3 & 4 \\ -2 & 7 \end{bmatrix} + \begin{bmatrix} -8 & 1 \\ 0 & 2 \end{bmatrix} = \begin{bmatrix} 3 + (-8) & 4 + 1 \\ -2 + 0 & 7 + 2 \end{bmatrix}$$

$$= \begin{bmatrix} -5 & 5 \\ -2 & 9 \end{bmatrix}$$

$$A - B = \begin{bmatrix} 3 & 4 \\ -2 & 7 \end{bmatrix} - \begin{bmatrix} -8 & 1 \\ 0 & 2 \end{bmatrix} = \begin{bmatrix} 3 - (-8) & 4 - 1 \\ -2 - 0 & 7 - 2 \end{bmatrix}$$

$$= \begin{bmatrix} 11 & 3 \\ -2 & 5 \end{bmatrix}$$

Multiplying a matrix by a constant is called **scalar multiplication.** When we multiply a matrix by a constant, we multiply each element in the matrix by that constant.

EXAMPLE 2   If $A = \begin{bmatrix} 3 & 0 & 2 \\ 5 & -1 & 4 \end{bmatrix}$, find $5A$.

SOLUTION $5A = 5 \begin{bmatrix} 3 & 0 & 2 \\ 5 & -1 & 4 \end{bmatrix} = \begin{bmatrix} 5(3) & 5(0) & 5(2) \\ 5(5) & 5(-1) & 5(4) \end{bmatrix}$

$$= \begin{bmatrix} 15 & 0 & 10 \\ 25 & -5 & 20 \end{bmatrix}$$

A matrix with only one row is called a **row vector.** Similarly, a matrix with only one column is called a **column vector.** The product of a row vector and a column vector is the real number found by adding the products of first elements, second elements, third elements, and so on, in each of the vectors. Example 3 illustrates this.

EXAMPLE 3   If $A = [2 \quad 3 \quad 4]$ and $B = \begin{bmatrix} 5 \\ 6 \\ 7 \end{bmatrix}$, find AB.

SOLUTION $AB = [2 \quad 3 \quad 4] \begin{bmatrix} 5 \\ 6 \\ 7 \end{bmatrix} = 2(5) + 3(6) + 4(7) = 56$

If we extend this type of multiplication to matrices, we can think of each matrix as being composed of so many row and column vectors. For example, the product

$$[2 \quad 3 \quad 4] \begin{bmatrix} 5 & 0 \\ 6 & -2 \\ 7 & 3 \end{bmatrix}$$

is the row vector whose first element comes from the product

$$[2 \quad 3 \quad 4] \begin{bmatrix} 5 \\ 6 \\ 7 \end{bmatrix} = 56$$

and whose second element comes from the product

$$[2 \quad 3 \quad 4] \begin{bmatrix} 0 \\ -2 \\ 3 \end{bmatrix} = 2(0) + 3(-2) + 4(3) = 6$$

Showing all this at once looks like this:

$$[2 \quad 3 \quad 4] \begin{bmatrix} 5 & 0 \\ 6 & -2 \\ 7 & 3 \end{bmatrix} = [2(5) + 3(6) + 4(7) \quad 2(0) + 3(-2) + 4(3)]$$

$$= [56 \quad 6]$$

EXAMPLE 4 Let $A = \begin{bmatrix} 1 & 2 \\ 3 & 4 \end{bmatrix}$ and $B = \begin{bmatrix} -5 & 6 \\ 7 & -8 \end{bmatrix}$, and find AB.

SOLUTION
$$AB = \begin{bmatrix} 1 & 2 \\ 3 & 4 \end{bmatrix} \begin{bmatrix} -5 & 6 \\ 7 & -8 \end{bmatrix}$$

$$= \begin{bmatrix} 1(-5) + 2(7) & 1(6) + 2(-8) \\ 3(-5) + 4(7) & 3(6) + 4(-8) \end{bmatrix}$$

$$= \begin{bmatrix} 9 & -10 \\ 13 & -14 \end{bmatrix}$$

To summarize, the element in row i and column j of the matrix AB will come from the product of row i in matrix A with column j of matrix B.

EXAMPLE 5 If $A = \begin{bmatrix} 2 & 3 & 5 \\ -3 & 0 & 1 \\ 2 & -4 & 0 \end{bmatrix}$ and $B = \begin{bmatrix} -2 & 0 & 1 \\ 0 & -1 & 0 \\ 2 & 0 & 4 \end{bmatrix}$, find AB and BA.

SOLUTION

$$AB = \begin{bmatrix} 2 & 3 & 5 \\ -3 & 0 & 1 \\ 2 & -4 & 0 \end{bmatrix} \begin{bmatrix} -2 & 0 & 1 \\ 0 & -1 & 0 \\ 2 & 0 & 4 \end{bmatrix}$$

$$= \begin{bmatrix} -4 + 0 + 10 & 0 + (-3) + 0 & 2 + 0 + 20 \\ 6 + 0 + 2 & 0 + 0 + 0 & -3 + 0 + 4 \\ -4 + 0 + 0 & 0 + 4 + 0 & 2 + 0 + 0 \end{bmatrix}$$

$$= \begin{bmatrix} 6 & -3 & 22 \\ 8 & 0 & 1 \\ -4 & 4 & 2 \end{bmatrix}$$

The element in the 2nd row and 3rd column of AB comes from multiplying the 2nd row of A times the 3rd column of B.

$$BA = \begin{bmatrix} -2 & 0 & 1 \\ 0 & -1 & 0 \\ 2 & 0 & 4 \end{bmatrix} \begin{bmatrix} 2 & 3 & 5 \\ -3 & 0 & 1 \\ 2 & -4 & 0 \end{bmatrix}$$

$$= \begin{bmatrix} -4+0+2 & -6+0+(-4) & -10+0+0 \\ 0+3+0 & 0+0+0 & 0+(-1)+0 \\ 4+0+8 & 6+0+(-16) & 10+0+0 \end{bmatrix}$$

$$= \begin{bmatrix} -2 & -10 & -10 \\ 3 & 0 & -1 \\ 12 & -10 & 10 \end{bmatrix}$$

The element in the 3rd row and 1st column of BA comes from multiplying the 3rd row of matrix B times the 1st column of matrix A.

Notice that $AB \neq BA$.

Introduction to Determinants

Associated with every square matrix is a **determinant.** As you will see in the next section, determinants can be used to solve systems of equations. To distinguish between a matrix and its determinant, we use vertical lines to enclose the numbers in the determinant. (Remember, a matrix is enclosed in brackets.)

Although every square matrix has an associated determinant, we will consider only 2×2 and 3×3 determinants. Here is the definition for a 2×2 determinant.

DEFINITION

The value of the **2 × 2** (2 by 2) **determinant**

$$\begin{vmatrix} a & c \\ b & d \end{vmatrix}$$

is given by

$$\begin{vmatrix} a & c \\ b & d \end{vmatrix} = ad - bc$$

From the preceding definition we see that a determinant is simply a square array of numbers with two vertical lines enclosing it. The value of a 2×2 determinant is found by cross-multiplying on the diagonals, as indicated in the diagram:

$$\begin{vmatrix} a & c \\ b & d \end{vmatrix} = ad - bc$$

EXAMPLES Find the value of the following 2×2 determinants:

6. $\begin{vmatrix} 1 & 2 \\ 3 & 4 \end{vmatrix} = 1(4) - 3(2) = 4 - 6 = -2$

7. $\begin{vmatrix} 3 & 5 \\ -2 & 7 \end{vmatrix} = 3(7) - (-2)5 = 21 + 10 = 31$

EXAMPLE 8 Solve for x if

$$\begin{vmatrix} x^2 & 2 \\ x & 1 \end{vmatrix} = 8$$

SOLUTION We expand the determinant on the left side to get:

$$x^2(1) - x(2) = 8$$
$$x^2 - 2x = 8$$
$$x^2 - 2x - 8 = 0$$
$$(x - 4)(x + 2) = 0$$
$$x - 4 = 0 \quad \text{or} \quad x + 2 = 0$$
$$x = 4 \quad \text{or} \quad x = -2$$

We now turn our attention to 3×3 determinants. A 3×3 determinant is also a square array of numbers; its value is given by the following definition.

DEFINITION

The value of the **3 × 3 determinant**

$$\begin{vmatrix} a_1 & b_1 & c_1 \\ a_2 & b_2 & c_2 \\ a_3 & b_3 & c_3 \end{vmatrix}$$

is given by

$$\begin{vmatrix} a_1 & b_1 & c_1 \\ a_2 & b_2 & c_2 \\ a_3 & b_3 & c_3 \end{vmatrix} = a_1 b_2 c_3 + a_3 b_1 c_2 + a_2 b_3 c_1 - a_3 b_2 c_1 - a_1 b_3 c_2 - a_2 b_1 c_3$$

At first glance, the expansion of a 3×3 determinant looks a little complicated. There are actually two different methods that can be used to find the six products given above.

Method 1 We begin by writing the determinant with the first two columns repeated on the right:

$$\begin{vmatrix} a_1 & b_1 & c_1 \\ a_2 & b_2 & c_2 \\ a_3 & b_3 & c_3 \end{vmatrix} \begin{matrix} a_1 & b_1 \\ a_2 & b_2 \\ a_3 & b_3 \end{matrix}$$

The positive products in the definition come from multiplying down the three full diagonals:

$$\begin{vmatrix} a_1 & b_1 & c_1 \\ a_2 & b_2 & c_2 \\ a_3 & b_3 & c_3 \end{vmatrix} \begin{matrix} a_1 & b_1 \\ a_2 & b_2 \\ a_3 & b_3 \end{matrix}$$
$$+ \quad + \quad +$$

The negative products come from multiplying up the three full diagonals:

$$\begin{vmatrix} a_1 & b_1 & c_1 \\ a_2 & b_2 & c_2 \\ a_3 & b_3 & c_3 \end{vmatrix} \begin{matrix} a_1 & b_1 \\ a_2 & b_2 \\ a_3 & b_3 \end{matrix}$$

EXAMPLE 9   Find the value of

$$\begin{vmatrix} 1 & 3 & -2 \\ 2 & 0 & 1 \\ 4 & -1 & 1 \end{vmatrix}$$

SOLUTION   Repeating the first two columns and then finding the products up the diagonals and the products down the diagonals as given in Method 1, we have

$$\begin{vmatrix} 1 & 3 & -2 \\ 2 & 0 & 1 \\ 4 & -1 & 1 \end{vmatrix} \begin{matrix} 1 & 3 \\ 2 & 0 \\ 4 & -1 \end{matrix}$$

$$= 1(0)(1) + 3(1)(4) + (-2)(2)(-1)$$
$$- 4(0)(-2) - (-1)(1)(1) - 1(2)(3)$$
$$= 0 + 12 + 4 - 0 - (-1) - 6$$
$$= 11$$

Method 2   The second method of evaluating a 3×3 determinant is called **expansion by minors.**

DEFINITION

The **minor** for an element in a 3×3 determinant is the determinant consisting of the elements remaining when the row and column to which the element belongs are deleted. For example, in the determinant

$$\begin{vmatrix} a_1 & b_1 & c_1 \\ a_2 & b_2 & c_2 \\ a_3 & b_3 & c_3 \end{vmatrix}$$

$$\text{Minor for element } a_1 = \begin{vmatrix} b_2 & c_2 \\ b_3 & c_3 \end{vmatrix}$$

$$\text{Minor for element } b_2 = \begin{vmatrix} a_1 & c_1 \\ a_3 & c_3 \end{vmatrix}$$

$$\text{Minor for element } c_3 = \begin{vmatrix} a_1 & b_1 \\ a_2 & b_2 \end{vmatrix}$$

Before we can evaluate a 3×3 determinant by Method 2, we first define what is known as the **sign array** for a 3×3 determinant at the top of the next page.

DEFINITION

The **sign array** for a 3 × 3 determinant is a 3 × 3 array of signs in the following pattern:

$$\begin{vmatrix} + & - & + \\ - & + & - \\ + & - & + \end{vmatrix}$$

The sign array begins with a + sign in the upper left-hand corner. The signs then alternate between + and − across every row and down every column.

Strategy for Evaluating a 3 × 3 Determinant by Expansion of Minors

We can evaluate a 3 × 3 determinant by expanding across any row or down any column as follows:

Step 1 Choose a row or column to expand about.
Step 2 Write the product of each element in the row or column chosen in Step 1 with its minor.
Step 3 Connect the three products in Step 2 with the signs in the corresponding row or column in the sign array.

We will use the same determinant as in Example 9 to illustrate the procedure.

EXAMPLE 10   Expand across the first row:

$$\begin{vmatrix} 1 & 3 & -2 \\ 2 & 0 & 1 \\ 4 & -1 & 1 \end{vmatrix}$$

SOLUTION   The products of the three elements in row 1 with their minors are:

$$1\begin{vmatrix} 0 & 1 \\ -1 & 1 \end{vmatrix} \qquad 3\begin{vmatrix} 2 & 1 \\ 4 & 1 \end{vmatrix} \qquad (-2)\begin{vmatrix} 2 & 0 \\ 4 & -1 \end{vmatrix}$$

Connecting these three products with the signs from the first row of the sign array, we have

$$+1\begin{vmatrix} 0 & 1 \\ -1 & 1 \end{vmatrix} - 3\begin{vmatrix} 2 & 1 \\ 4 & 1 \end{vmatrix} + (-2)\begin{vmatrix} 2 & 0 \\ 4 & -1 \end{vmatrix}$$

We complete the problem by evaluating each of the three 2 × 2 determinants and then simplifying the resulting expression:

$$+1[0 - (-1)] - 3(2 - 4) + (-2)(-2 - 0) = 1(1) - 3(-2) + (-2)(-2)$$
$$= 1 + 6 + 4$$
$$= 11$$

The results of Examples 9 and 10 match. It makes no difference which method we use—the value of a 3×3 determinant is unique.

Note: The method shown in Example 10 is actually more valuable than our first method, because it will work with any size determinant from 3×3 to 4×4 to any higher-order determinant. Method 1 only works on 3×3 determinants. It cannot be used on a 4×4 determinant.

EXAMPLE II Expand down column 2:

$$\begin{vmatrix} 2 & 3 & -2 \\ 1 & 4 & 1 \\ 1 & 5 & -1 \end{vmatrix}$$

SOLUTION We connect the products of elements in column 2 and their minors with the signs from the second column in the sign array:

$$\begin{vmatrix} 2 & 3 & -2 \\ 1 & 4 & 1 \\ 1 & 5 & -1 \end{vmatrix} = -3 \begin{vmatrix} 1 & 1 \\ 1 & -1 \end{vmatrix} + 4 \begin{vmatrix} 2 & -2 \\ 1 & -1 \end{vmatrix} - 5 \begin{vmatrix} 2 & -2 \\ 1 & 1 \end{vmatrix}$$

$$= -3(-1-1) + 4[-2-(-2)] - 5[2-(-2)]$$
$$= -3(-2) + 4(0) - 5(4)$$
$$= 6 + 0 - 20$$
$$= -14$$

Problem Set 7.3

$Let\ A = \begin{bmatrix} 3 & -2 & 0 \\ 0 & 1 & 4 \\ 2 & 0 & 5 \end{bmatrix}, B = \begin{bmatrix} 1 & 2 & 0 \\ 3 & 4 & -5 \\ 0 & 0 & 3 \end{bmatrix},$

$C = \begin{bmatrix} 2 & -3 \\ -4 & 5 \end{bmatrix}, D = \begin{bmatrix} 1 & 0 \\ 2 & 1 \end{bmatrix}, E = [2 \quad 4 \quad 6],$

$F = [1 \quad 3 \quad 5],\ and\ G = \begin{bmatrix} 1 \\ 2 \\ 3 \end{bmatrix}. Find:$

1. $A + B$	**2.** $B + A$	**3.** $A - B$
4. $B - A$	**5.** $4A$	**6.** $-5B$
7. $2A - 3B$	**8.** $3B - 4A$	**9.** EG
10. FG	**11.** $2(EG)$	**12.** $3(FG)$
13. EA	**14.** FB	**15.** CD
16. DC	**17.** AB	**18.** BA
19. $E(A + B)$	**20.** $EA + EB$	

$Let\ A = \begin{bmatrix} 3 & -5 \\ -1 & 2 \end{bmatrix}, B = \begin{bmatrix} 2 & 5 \\ 1 & 3 \end{bmatrix},$

$and\ I = \begin{bmatrix} 1 & 0 \\ 0 & 1 \end{bmatrix}. Find:$

21. AI **22.** IB **23.** AB **24.** BA

25. Is matrix addition a commutative operation?

26. Is matrix multiplication a commutative operation?

Find the value of the following 2×2 determinants.

27. $\begin{vmatrix} 1 & 0 \\ 2 & 3 \end{vmatrix}$ **28.** $\begin{vmatrix} 5 & 4 \\ 3 & 2 \end{vmatrix}$

29. $\begin{vmatrix} 2 & 1 \\ 3 & 4 \end{vmatrix}$ **30.** $\begin{vmatrix} 4 & 1 \\ 5 & 2 \end{vmatrix}$

31. $\begin{vmatrix} 0 & 1 \\ 1 & 0 \end{vmatrix}$

32. $\begin{vmatrix} 1 & 0 \\ 0 & 1 \end{vmatrix}$

33. $\begin{vmatrix} -3 & 2 \\ 6 & -4 \end{vmatrix}$

34. $\begin{vmatrix} 8 & -3 \\ -2 & 5 \end{vmatrix}$

35. $\begin{vmatrix} -3 & -1 \\ 4 & -2 \end{vmatrix}$

36. $\begin{vmatrix} 5 & 3 \\ 7 & -6 \end{vmatrix}$

Solve each of the following for x.

37. $\begin{vmatrix} 2x & 1 \\ x & 3 \end{vmatrix} = 10$

38. $\begin{vmatrix} 3x & -2 \\ 2x & 3 \end{vmatrix} = 26$

39. $\begin{vmatrix} 1 & 2x \\ 2 & -3x \end{vmatrix} = 21$

40. $\begin{vmatrix} -5 & 4x \\ 1 & -x \end{vmatrix} = 27$

41. $\begin{vmatrix} 2x & -4 \\ 2 & x \end{vmatrix} = -8x$

42. $\begin{vmatrix} 3x & 2 \\ 2 & x \end{vmatrix} = -11x$

43. $\begin{vmatrix} x^2 & 3 \\ x & 1 \end{vmatrix} = 10$

44. $\begin{vmatrix} x^2 & -2 \\ x & 1 \end{vmatrix} = 35$

Find the value of the following 3 × 3 determinants by using Method 1 of this section.

45. $\begin{vmatrix} 1 & 2 & 0 \\ 0 & 2 & 1 \\ 1 & 1 & 1 \end{vmatrix}$

46. $\begin{vmatrix} -1 & 0 & 2 \\ 3 & 0 & 1 \\ 0 & 1 & 3 \end{vmatrix}$

47. $\begin{vmatrix} 1 & 2 & 3 \\ 3 & 2 & 1 \\ 1 & 1 & 1 \end{vmatrix}$

48. $\begin{vmatrix} -1 & 2 & 0 \\ 3 & -2 & 1 \\ 0 & 5 & 4 \end{vmatrix}$

Find the value of each determinant by using Method 2 and expanding across the first row.

49. $\begin{vmatrix} 0 & 1 & 2 \\ 1 & 0 & 1 \\ -1 & 2 & 0 \end{vmatrix}$

50. $\begin{vmatrix} 3 & -2 & 1 \\ 0 & -1 & 0 \\ 2 & 0 & 1 \end{vmatrix}$

51. $\begin{vmatrix} 3 & 0 & 2 \\ 0 & -1 & -1 \\ 4 & 0 & 0 \end{vmatrix}$

52. $\begin{vmatrix} 1 & 1 & 1 \\ 1 & -1 & 1 \\ 1 & 1 & -1 \end{vmatrix}$

Find the value of each of the following determinants.

53. $\begin{vmatrix} 2 & -1 & 0 \\ 1 & 0 & -2 \\ 0 & 1 & 2 \end{vmatrix}$

54. $\begin{vmatrix} 5 & 0 & -4 \\ 0 & 1 & 3 \\ -1 & 2 & -1 \end{vmatrix}$

55. $\begin{vmatrix} 1 & 3 & 7 \\ -2 & 6 & 4 \\ 3 & 7 & -1 \end{vmatrix}$

56. $\begin{vmatrix} 2 & 1 & 5 \\ 6 & -3 & 4 \\ 8 & 9 & -2 \end{vmatrix}$

In Words

57. Show that the determinant equation below is another way to write the slope-intercept form of the equation of a line.

$$\begin{vmatrix} y & x \\ m & 1 \end{vmatrix} = b$$

58. Show that the determinant equation below is another way to write the equation $F = \frac{9}{5}C + 32$.

$$\begin{vmatrix} C & F & 1 \\ 5 & 41 & 1 \\ -10 & 14 & 1 \end{vmatrix} = 0$$

Review Problems

 NOTE: The problems that follow review material we covered in Section 6.3.

Use the discriminant to find the number and kind of solutions to the following equations.

59. $2x^2 - 5x + 4 = 0$

60. $4x^2 - 12x = -9$

For each of the following problems, find an equation with the given solutions.

61. $x = -3, x = 5$

62. $x = 2, x = -2, x = 1$

63. $y = \frac{2}{3}, y = 3$

64. $y = -\frac{3}{5}, y = 2$

65. Find all solutions to $x^3 - 8x^2 + 21x - 18 = 0$ if $x = 3$ is one solution.

66. Find all solutions to $3x^3 - 2x^2 - 10x + 4 = 0$ if $x = 2$ is one solution.

One Step Further

A 4 × 4 determinant can be evaluated only by using Method 2, expansion by minors; Method 1 will not work. Below is a 4 × 4 determinant and its associated sign array.

$$\begin{vmatrix} 2 & 0 & 1 & -3 \\ -1 & 2 & 0 & 1 \\ -3 & 0 & 1 & 0 \\ 1 & 1 & 0 & 0 \end{vmatrix} \qquad \begin{vmatrix} + & - & + & - \\ - & + & - & + \\ + & - & + & - \\ - & + & - & + \end{vmatrix}$$

4 × 4 determinant   4 × 4 sign array

67. Use expansion by minors to evaluate this 4×4 determinant by expanding it across row 1.

68. Evaluate the preceding determinant by expanding it down column 4.

69. Use expansion by minors down column 3 to evaluate the preceding determinant.

70. Evaluate the preceding determinant by expanding it across row 4.

SECTION 7.4

Cramer's Rule

We begin this section with a look at how determinants can be used to solve a system of linear equations in two variables. The method we use is called **Cramer's rule.** We state it here as a theorem without proof.

THEOREM 7.1: CRAMER'S RULE

The solution to the system
$$a_1 x + b_1 y = c_1$$
$$a_2 x + b_2 y = c_2$$

is given by
$$x = \frac{D_x}{D} \qquad y = \frac{D_y}{D} \qquad (D \neq 0)$$

where

$$D = \begin{vmatrix} a_1 & b_1 \\ a_2 & b_2 \end{vmatrix} \qquad D_x = \begin{vmatrix} c_1 & b_1 \\ c_2 & b_2 \end{vmatrix} \qquad D_y = \begin{vmatrix} a_1 & c_1 \\ a_2 & c_2 \end{vmatrix} \qquad (D \neq 0)$$

The determinant D is made up of the coefficients of x and y in the original system. The determinants D_x and D_y are found by replacing the coefficients of x or y by the constant terms in the original system. Notice also that Cramer's rule does not apply if $D = 0$. In this case the equations are either inconsistent or dependent.

EXAMPLE 1   Use Cramer's rule to solve:
$$2x - 3y = 4$$
$$4x + 5y = 3$$

SOLUTION   We begin by calculating the determinants D, D_x, and D_y:

$$D = \begin{vmatrix} 2 & -3 \\ 4 & 5 \end{vmatrix} = 2(5) - 4(-3) = 22$$

$$D_x = \begin{vmatrix} 4 & -3 \\ 3 & 5 \end{vmatrix} = 4(5) - 3(-3) = 29$$

$$D_y = \begin{vmatrix} 2 & 4 \\ 4 & 3 \end{vmatrix} = 2(3) - 4(4) = -10$$

$$x = \frac{D_x}{D} = \frac{29}{22} \qquad \text{and} \qquad y = \frac{D_y}{D} = \frac{-10}{22} = -\frac{5}{11}$$

The solution set for the system is $\{(\frac{29}{22}, -\frac{5}{11})\}$.

Cramer's rule can be applied to systems of linear equations in three variables also.

THEOREM 7.2: ALSO CRAMER'S RULE

The solution set to the system

$$a_1 x + b_1 y + c_1 z = d_1$$
$$a_2 x + b_2 y + c_2 z = d_2$$
$$a_3 x + b_3 y + c_3 z = d_3$$

is given by

$$x = \frac{D_x}{D} \qquad y = \frac{D_y}{D} \qquad z = \frac{D_z}{D} \qquad (D \neq 0)$$

where

$$D = \begin{vmatrix} a_1 & b_1 & c_1 \\ a_2 & b_2 & c_2 \\ a_3 & b_3 & c_3 \end{vmatrix} \qquad (D \neq 0)$$

$$D_x = \begin{vmatrix} d_1 & b_1 & c_1 \\ d_2 & b_2 & c_2 \\ d_3 & b_3 & c_3 \end{vmatrix}$$

$$D_y = \begin{vmatrix} a_1 & d_1 & c_1 \\ a_2 & d_2 & c_2 \\ a_3 & d_3 & c_3 \end{vmatrix}$$

$$D_z = \begin{vmatrix} a_1 & b_1 & d_1 \\ a_2 & b_2 & d_2 \\ a_3 & b_3 & d_3 \end{vmatrix}$$

Again the determinant D consists of the coefficients of x, y, and z in the original system. The determinants D_x, D_y, and D_z are found by replacing the coefficients of x, y, and z, respectively, with the constant terms from the original system. If $D = 0$, there is no unique solution to the system.

EXAMPLE 2   Use Cramer's rule to solve:

$$\begin{aligned} x + y + z &= 6 \\ 2x - y + z &= 3 \\ x + 2y - 3z &= -4 \end{aligned}$$

Recall that there is more than one way to evaluate a 3×3 determinant. Since we have four of these determinants, we can use both Methods 1 and 2 from the previous section.

SOLUTION This is the same system we solved in Example 1 of Section 7.2. We begin by setting up and evaluating D, D_x, D_y, and D_z. We evaluate D using Method 1 from Section 7.3.

$$D = \begin{vmatrix} 1 & 1 & 1 \\ 2 & -1 & 1 \\ 1 & 2 & -3 \end{vmatrix}$$

$$= 3 + 1 + 4 - (-1) - (2) - (-6)$$
$$= 13$$

We evaluate D_x using Method 2 from Section 7.3 and expanding across row 1:

$$D_x = \begin{vmatrix} 6 & 1 & 1 \\ 3 & -1 & 1 \\ -4 & 2 & -3 \end{vmatrix} = 6 \begin{vmatrix} -1 & 1 \\ 2 & -3 \end{vmatrix} - 1 \begin{vmatrix} 3 & 1 \\ -4 & -3 \end{vmatrix} + 1 \begin{vmatrix} 3 & -1 \\ -4 & 2 \end{vmatrix}$$

$$= 6(1) - 1(-5) + 1(2)$$
$$= 13$$

Find D_y by expanding across row 2:

$$D_y = \begin{vmatrix} 1 & 6 & 1 \\ 2 & 3 & 1 \\ 1 & -4 & -3 \end{vmatrix} = -2 \begin{vmatrix} 6 & 1 \\ -4 & -3 \end{vmatrix} + 3 \begin{vmatrix} 1 & 1 \\ 1 & -3 \end{vmatrix} - 1 \begin{vmatrix} 1 & 6 \\ 1 & -4 \end{vmatrix}$$

$$= -2(-14) + 3(-4) - 1(-10)$$
$$= 26$$

Find D_z by expanding down column 1:

$$D_z = \begin{vmatrix} 1 & 1 & 6 \\ 2 & -1 & 3 \\ 1 & 2 & -4 \end{vmatrix} = 1 \begin{vmatrix} -1 & 3 \\ 2 & -4 \end{vmatrix} - 2 \begin{vmatrix} 1 & 6 \\ 2 & -4 \end{vmatrix} + 1 \begin{vmatrix} 1 & 6 \\ -1 & 3 \end{vmatrix}$$

$$= 1(-2) - 2(-16) + 1(9)$$
$$= 39$$

$$x = \frac{D_x}{D} = \frac{13}{13} = 1$$

$$y = \frac{D_y}{D} = \frac{26}{13} = 2$$

$$z = \frac{D_z}{D} = \frac{39}{13} = 3$$

The solution set is $\{(1, 2, 3)\}$.

EXAMPLE 3 Use Cramer's rule to solve:

$$x + y = -1$$
$$2x - z = 3$$
$$y + 2z = -1$$

SOLUTION It is helpful to rewrite the system using 0's for the coefficients of those variables not shown:

$$x + y + 0z = -1$$
$$2x + 0y - z = 3$$
$$0x + y + 2z = -1$$

The four determinants used in Cramer's rule are

$$D = \begin{vmatrix} 1 & 1 & 0 \\ 2 & 0 & -1 \\ 0 & 1 & 2 \end{vmatrix} = -3 \qquad D_x = \begin{vmatrix} -1 & 1 & 0 \\ 3 & 0 & -1 \\ -1 & 1 & 2 \end{vmatrix} = -6$$

$$D_y = \begin{vmatrix} 1 & -1 & 0 \\ 2 & 3 & -1 \\ 0 & -1 & 2 \end{vmatrix} = 9 \qquad D_z = \begin{vmatrix} 1 & 1 & -1 \\ 2 & 0 & 3 \\ 0 & 1 & -1 \end{vmatrix} = -3$$

$$x = \frac{D_x}{D} = \frac{-6}{-3} = 2 \qquad y = \frac{D_y}{D} = \frac{9}{-3} = -3 \qquad z = \frac{D_z}{D} = \frac{-3}{-3} = 1$$

The solution set is $\{(2, -3, 1)\}$.

Finally, we should mention the possible situations that can occur when we are using Cramer's rule and the determinant D is 0.

If $D = 0$ and at least one of the other determinants, D_x or D_y (or D_z in a system of three equations in three variables), is not 0, then the system is inconsistent. In this case, there is no solution to the system.

On the other hand, if $D = 0$ and both D_x and D_y (and D_z in a system of three equations in three variables) are 0, then the system is dependent.

Problem Set

7.4

Solve each of the following systems using Cramer's rule.

1. $2x - 3y = 3$
     $4x - 2y = 10$

2. $3x + y = -2$
     $-3x + 2y = -4$

3. $5x - 2y = 4$
     $-10x + 4y = 1$

4. $-4x + 3y = -11$
     $5x + 4y = 6$

5. $4x - 7y = 3$
     $5x + 2y = -3$

6. $3x - 4y = 7$
     $6x - 2y = 5$

7. $9x - 8y = 4$
$2x + 3y = 6$

8. $4x - 7y = 10$
$-3x + 2y = -9$

9. $x + y + z = 4$
$x - y - z = 2$
$2x + 2y - z = 2$

10. $-x + y + 3z = 6$
$x + y + 2z = 7$
$2x + 3y + z = 4$

11. $x + y - z = 2$
$-x + y + z = 3$
$x + y + z = 4$

12. $-x - y + z = 1$
$x - y + z = 3$
$x + y - z = 4$

13. $3x - y + 2z = 4$
$6x - 2y + 4z = 8$
$x - 5y + 2z = 1$

14. $2x - 3y + z = 1$
$3x - y - z = 4$
$4x - 6y + 2z = 3$

15. $2x - y + 3z = 4$
$x - 5y - 2z = 1$
$-4x - 2y + z = 3$

16. $4x - y + 5z = 1$
$2x + 3y + 4z = 5$
$x + y + 3z = 2$

17. $-x - 7y = 1$
$x + 3z = 11$
$2y + z = 0$

18. $x + y = 2$
$-x + 3z = 0$
$2y + z = 3$

19. $x - y = 2$
$3x + z = 11$
$y - 2z = -3$

20. $4x + 5y = -1$
$2y + 3z = -5$
$x + 2z = -1$

Applying the Concepts

21. Break-Even Point If a company has fixed costs of $100 per week and each item it produces costs $10 to manufacture, then the total cost, y, per week to produce x items is

$$y = 10x + 100$$

If the company sells each item it manufactures for $12, then the total amount of money, y, the company brings in for selling x items is

$$y = 12x$$

Use Cramer's rule to solve the system

$$y = 10x + 100$$
$$y = 12x$$

for x to find the number of items the company must sell per week in order to break even.

22. Break-Even Point Suppose a company has fixed costs of $200 per week, and each item it produces costs $20 to manufacture.
(a) Write an equation that gives the total cost per week, y, to manufacture x items.
(b) If each item sells for $25, write an equation that gives the total amount of money, y, the company brings in for selling x items.
(c) Use Cramer's rule to find the number of items the company must sell each week to break even.

Review Problems

The problems below review material we covered in Section 6.4.

Solve each equation.

23. $x^4 - 2x^2 - 8 = 0$
24. $x^4 - 8x^2 - 9 = 0$
25. $x^{2/3} - 5x^{1/3} + 6 = 0$
26. $x^{2/3} - 3x^{1/3} + 2 = 0$
27. $2x - 5\sqrt{x} + 3 = 0$
28. $3x - 8\sqrt{x} + 4 = 0$
29. $(3x + 1) - 6\sqrt{3x + 1} + 8 = 0$
30. $(2x - 1) - 2\sqrt{2x - 1} - 15 = 0$
31. Solve $kx^2 + 4x - k = 0$ for x.
32. Solve $4x^2 - 4x + k = 0$ for x.

One Step Further

Solve for x and y using Cramer's rule. Your answers will contain the constants a and b.

33. $ax + by = -1$
$bx + ay = 1$

34. $ax + y = b$
$bx + y = a$

35. $a^2x + by = 1$
$b^2x + ay = 1$

36. $ax + by = a$
$bx + ay = a$

37. Name the system of equations for which Cramer's rule yields the following determinants.

$$D = \begin{vmatrix} 1 & 2 \\ 3 & 4 \end{vmatrix} \qquad D_x = \begin{vmatrix} 1 & 2 \\ 0 & 4 \end{vmatrix}$$

38. Name the system of equations for which Cramer's rule yields the following determinants.

$$D = \begin{vmatrix} 1 & 3 & 2 \\ -1 & 0 & 4 \\ 2 & 5 & -1 \end{vmatrix} \qquad D_y = \begin{vmatrix} 1 & 1 & 2 \\ -1 & 3 & 4 \\ 2 & 5 & -1 \end{vmatrix}$$

SECTION

7.5

Applications

Many times word problems involve more than one unknown quantity. If a problem is stated in terms of two unknowns and we represent each unknown quantity with a different variable, then we must write the relationships between the variables with two equations. The two equations written in terms of the two variables form a system of linear equations that we solve using the methods developed in this chapter. If we find a problem that relates three unknown quantities, then we need three different equations in order to form a linear system we can solve.

EXAMPLE 1   One number is 2 more than 3 times another. Their sum is 26. Find the two numbers.

SOLUTION   If we let x and y represent the two numbers, then the translation of the first sentence in the problem into an equation would be

$$y = 3x + 2$$

The second sentence gives us a second equation:

$$x + y = 26$$

The linear system that describes the situation is

$$x + y = 26$$
$$y = 3x + 2$$

Substituting the expression for y from the second equation into the first and solving for x yields

$$x + (3x + 2) = 26$$
$$4x + 2 = 26$$
$$4x = 24$$
$$x = 6$$

Using $x = 6$ in $y = 3x + 2$ gives the second number:

$$y = 3(6) + 2$$
$$y = 20$$

The two numbers are 6 and 20. Their sum is 26, and the second is 2 more than 3 times the first. ◢

EXAMPLE 2   Suppose 850 tickets were sold for a game for a total of $1,100. If adult tickets cost $1.50 and children's tickets cost $1.00, how many of each kind of ticket were sold?

SOLUTION   If we let x = the number of adult tickets and y = the number of children's tickets, then

$$x + y = 850$$

since a total of 850 tickets were sold. Also, since each adult ticket costs $1.50 and each children's ticket costs $1.00, and the total amount of money paid for tickets was $1,100, a second equation is

$$1.50x + 1.00y = 1,100$$

The same information can also be obtained by summarizing the problem with a table, as shown below. Notice that the two equations obtained above are given by the two rows of the table.

	Adult Tickets	Children's Tickets	Total ($)
Number	x	y	850
Value ($)	$1.50x$	$1.00y$	1,100

Whether we use a table to summarize the information in the problem, or just talk our way through the problem, the system of equations that describes the situation is

$$x + y = 850$$
$$1.50x + 1.00y = 1,100$$

If we multiply the second equation by 10 to clear it of decimals, we have the system

$$x + y = 850$$
$$15x + 10y = 11,000$$

Multiplying the first equation by -10 and adding the result to the second equation eliminates the variable y from the system:

$$
\begin{aligned}
-10x - 10y &= -8,500 \\
15x + 10y &= 11,000 \\
\hline
5x &= 2,500 \\
x &= 500
\end{aligned}
$$

The number of adult tickets sold was 500. To find the number of children's tickets, we substitute $x = 500$ into $x + y = 850$ to get

$$500 + y = 850$$
$$y = 350$$

The number of children's tickets sold was 350.

EXAMPLE 3   Suppose a person invests a total of $10,000 in two accounts. One account earns 8% annually and the other earns 9% annually. If the total interest earned from both accounts in a year is $850, how much is invested in each account?

SOLUTION   The form of the solution to this problem is very similar to that of Example 2. We let x equal the amount invested at 9% and y equal the amount invested at 8%. Since the total investment is $10,000, one relationship between x and y can be written as

$$x + y = 10,000$$

The total interest earned from both accounts is $860. The amount of interest earned on x dollars at 9% is $0.09x$, while the amount of interest earned on y dollars at 8% is $0.08y$. This relationship is represented by the equation

$$0.09x + 0.08y = 860$$

The two equations we have just written can also be found by first summarizing the information from the problem in a table. Again, the two rows of the table yield the two equations just written.

	Dollars at 9%	Dollars at 8%	Total ($)
Number	x	y	10,000
Interest ($)	$0.09x$	$0.08y$	860

The system of equations that describes this situation is given by

$$x + \quad y = 10,000$$
$$0.09x + 0.08y = 860$$

Multiplying the second equation by 100 will clear it of decimals. The system that results is

$$x + \quad y = 10,000$$
$$9x + 8y = 86,000$$

We can eliminate y from this system by multiplying the first equation by -8 and adding the result to the second equation:

$$-8x - 8y = -80,000$$
$$\underline{9x + 8y = \quad 86,000}$$
$$x \quad\quad = \quad 6,000$$

The amount of money invested at 9% is $6,000. Since the total investment was $10,000, the amount invested at 8% must be $4,000. ◢

EXAMPLE 4 How much 20% alcohol solution and 50% alcohol solution must be mixed to get 12 gallons of 30% alcohol solution?

SOLUTION To solve this problem we must first understand that a 20% alcohol solution is 20% alcohol and 80% water.

Let x = the number of gallons of 20% alcohol solution needed, and let y = the number of gallons of 50% alcohol solution needed. Since we must end up with a total of 12 gallons of solution, one equation for the system is

$$x + y = 12$$

The amount of alcohol in the x gallons of 20% solution is $0.20x$, while the amount of alcohol in the y gallons of 50% solution is $0.50y$. Since the total amount of alcohol in the 20% and 50% solutions must add up to the amount of alcohol in the 12 gallons of 30% solution, the second equation in our system can be written as

$$0.20x + 0.50y = 0.30(12)$$

Again, let's make a table that summarizes the information we have to this point in the problem:

	20% Solution	50% Solution	Final Solution
Total Number of Gallons	x	y	12
Gallons of Alcohol	$0.20x$	$0.50y$	$0.30(12)$

Our system of equations is

$$x + y = 12$$
$$0.20x + 0.50y = 0.30(12) = 3.6$$

Multiplying the second equation by 10 gives us an equivalent system:

$$x + y = 12$$
$$2x + 5y = 36$$

Multiplying the top equation by -2 to eliminate the x variable, we have

$$
\begin{aligned}
-2x - 2y &= -24 \\
2x + 5y &= 36 \\
\hline
3y &= 12 \\
y &= 4
\end{aligned}
$$

Substituting $y = 4$ into $x + y = 12$, we solve for x:

$$x + 4 = 12$$
$$x = 8$$

It takes 8 gallons of 20% alcohol solution and 4 gallons of 50% alcohol solution to produce 12 gallons of 30% alcohol solution.

EXAMPLE 5   It takes 2 hours for a boat to travel 28 miles downstream (with the current). The same boat can travel 18 miles upstream (against the current) in 3 hours. What is the speed of the boat in still water, and what is the speed of the current of the river?

SOLUTION   Let $x =$ the speed of the boat in still water and let $y =$ the speed of the current. The average speed (rate) of the boat upstream is $x - y$, since it is traveling against the current. The rate of the boat downstream is $x + y$, since the boat is traveling with the current. Putting the information into a table, we have

	d (distance, mi)	r (rate, mi/hr)	t (time, hr)
Upstream	18	$x - y$	3
Downstream	28	$x + y$	2

The formula for the relationship between distance d, rate r, and time t is $d = rt$ (the rate equation). Since $d = r \cdot t$, the system we need to solve the problem is

$$18 = (x - y) \cdot 3$$
$$28 = (x + y) \cdot 2$$

which is equivalent to

$$6 = x - y$$
$$14 = x + y$$

Adding the two equations, we have

$$20 = 2x$$
$$x = 10$$

Substituting $x = 10$ into $14 = x + y$, we see that

$$y = 4$$

The speed of the boat in still water is 10 mi/hr; the speed of the current is 4 mi/hr.

EXAMPLE 6   A coin collection consists of 14 coins with a total value of $1.35. If the coins are nickels, dimes, and quarters, and the number of nickels is 3 less than twice the number of dimes, how many of each coin are in the collection?

SOLUTION   Since we have three types of coins we will have to use three variables. Let x = the number of nickels, y = the number of dimes, and z = the number of quarters. Since the total number of coins is 14, our first equation is

$$x + y + z = 14$$

Since the number of nickels is 3 less than twice the number of dimes, we have a second equation:

$$x = 2y - 3 \qquad \text{which is equivalent to} \qquad x - 2y = -3$$

Our last equation is obtained by considering the value of each coin and the total value of the collection. Let's write the equation in terms of cents, so we won't have to clear it of decimals later:

$$5x + 10y + 25z = 135$$

Here is our system, with the equations numbered for reference:

$$
\begin{aligned}
x + \quad y + \quad z &= 14 \quad (1) \\
x - \quad 2y \qquad\quad &= -3 \quad (2) \\
5x + 10y + 25z &= 135 \quad (3)
\end{aligned}
$$

Let's begin by eliminating x from the first and second equations, and the first and third equations. Adding -1 times the second equation to the first equation gives us an equation in only y and z. We call this equation (4):

$$3y + z = 17 \quad (4)$$

Adding -5 times equation (1) to equation (3) gives us

$$5y + 20z = 65 \quad (5)$$

We can eliminate z from equations (4) and (5) by adding -20 times (4) to (5). Here is the result:

$$
\begin{aligned}
-55y &= -275 \\
y &= 5
\end{aligned}
$$

Substituting $y = 5$ into equation (4) gives us $z = 2$. Substituting $y = 5$ and $z = 2$ into equation (1) gives us $x = 7$. The collection consists of 7 nickels, 5 dimes, and 2 quarters. ◢

Now, let's return to the example discussed in the introduction to this chapter. If you go on to take a chemistry class, you may come across it again.

EXAMPLE 7   In a chemistry lab, students record the temperature of water at room temperature and find that it is 77° on the Fahrenheit temperature scale and 25° on the Celsius temperature scale. The water is then heated until it boils. The temperature of the boiling water is 212°F and 100°C. Assume that the relationship between the two temperature scales is a linear one; then use the given data to find the formula that gives the Celsius temperature C in terms of the Fahrenheit temperature F.

TABLE I	
Corresponding	
Temperatures	
In	In
Degrees	Degrees
Fahrenheit	Celsius
77	25
212	100

SOLUTION The data are summarized in Table 1.

If we assume the relationship is linear, then the formula that relates the two temperature scales can be written in slope-intercept form as

$$C = mF + b$$

Substituting $C = 25$ and $F = 77$ into this formula gives us

$$25 = 77m + b$$

Substituting $C = 100$ and $F = 212$ into the formula yields

$$100 = 212m + b$$

Together, the two equations form a system of equations, which we can solve using the addition method.

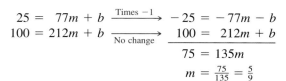

$$75 = 135m$$
$$m = \frac{75}{135} = \frac{5}{9}$$

To find the value of b we substitute $m = \frac{5}{9}$ into $25 = 77m + b$ and solve for b.

$$25 = 77(\tfrac{5}{9}) + b$$
$$25 = \tfrac{385}{9} + b$$
$$b = 25 - \tfrac{385}{9} = \tfrac{225}{9} - \tfrac{385}{9} = -\tfrac{160}{9}$$

The equation that gives C in terms of F is

$$C = \tfrac{5}{9}F - \tfrac{160}{9}$$

77°

Problem Set

7.5

Number Problems

1. One number is 3 more than twice another. The sum of the numbers is 18. Find the two numbers.

2. The sum of two numbers is 32. One of the numbers is 4 less than 5 times the other. Find the two numbers.

3. The difference of two numbers is 6. Twice the smaller is 4 more than the larger. Find the two numbers.

4. The larger of two numbers is 5 more than twice the smaller. If the smaller is subtracted from the larger, the result is 12. Find the two numbers.

5. The sum of three numbers is 8. Twice the smallest is 2 less than the largest, while the sum of the largest and smallest is 5. Use a linear system in three variables to find the three numbers.

6. The sum of three numbers is 14. The largest is 4 times the smallest, while the sum of the smallest and twice the largest is 18. Use a linear system in three variables to find the three numbers.

Ticket and Interest Problems

7. A total of 925 tickets were sold for a game, for a total of $1,150. If adult tickets sold for $2.00 and children's tickets sold for $1.00, how many of each kind of ticket were sold?

8. If tickets for a show cost $2.00 for adults and $1.50 for children, how many of each kind of ticket were sold if a total of 300 tickets were sold for $525?

9. Mr. Jones has $20,000 to invest. He invests part at 6% and the rest at 7%. If he earns $1,280 in interest after 1 year, how much did he invest at each rate?

10. A man invests $17,000 in two accounts. One account earns 5% interest per year and the other earns 6.5%. If his total interest after 1 year is $970, how much does he invest at each rate?

11. Ms. Smith invests twice as much money at 7.5% as she does at 6%. If her total interest after a year is $840, how much does she have invested at each rate?

12. A woman earns $1,350 in interest from two accounts in 1 year. If she has three times as much invested at 7% as she does at 6%, how much does she have in each account?

13. A man invests $2,200 in three accounts that pay 6%, 8%, and 9% in annual interest. He has three times as much invested at 9% as he does at 6%. If his total interest for the year is $178, how much is invested at each rate?

14. A student has money in three accounts that pay 5%, 7%, and 8% in annual interest. She has three times as much invested at 8% as she does

at 5%. If the total amount she has invested is $1,600 and her interest for the year comes to $115, how much money does she have in each account?

Mixture Problems

15. How many gallons of 20% alcohol solution and 50% alcohol solution must be mixed to get 9 gallons of 30% alcohol solution?

16. How many ounces of 30% hydrochloric acid solution and 80% hydrochloric acid solution must be mixed to get 10 ounces of 50% hydrochloric acid solution?

17. A mixture of 16% disinfectant solution is to be made from 20% and 14% disinfectant solutions. How much of each solution should be used if 15 gallons of the 16% solution are needed?

18. How much 25% antifreeze and 50% antifreeze should be combined to give 40 gallons of 30% antifreeze?

Rate Problems

19. It takes a boat 2 hours to travel 24 miles downstream and 3 hours to travel 18 miles upstream. What is the speed of the boat in still water? What is the speed of the current of the river?

20. A boat on a river travels 20 miles downstream in only 2 hours. It takes the same boat 6 hours to travel 12 miles upstream. What are the speed of the boat and the speed of the current?

21. An airplane flying with the wind can cover a certain distance in 2 hours. The return trip against the wind takes $2\frac{1}{2}$ hours. How fast is the plane and what is the speed of the wind, if the distance is 600 miles?

22. An airplane covers a distance of 1,500 miles in 3 hours when it flies with the wind and $3\frac{1}{3}$ hours when it flies against the wind. What is the speed of the plane in still air?

Coin Problems

23. Bob has 20 coins totaling $1.40. If he has only dimes and nickels, how many of each coin does he have?

24. If Amy has 15 coins totaling $2.70, and the coins are quarters and dimes, how many of each coin does she have?

25. A collection of nickels, dimes, and quarters consists of 9 coins with a total value of $1.20. If the number of dimes is equal to the number of nickels, find the number of each type of coin.

26. A coin collection consists of 12 coins with a total value of $1.20. If the collection consists only of nickels, dimes, and quarters, and the number of dimes is 2 more than twice the number of nickels, how many of each type of coin are in the collection?

Finding the Equation

27. A manufacturing company finds that they can sell 300 items if the price per item is $2.00, and 400 items if the price is $1.50 per item. If the relationship between the number of items sold, x, and the price per item, p, is a linear one, find a formula that gives x in terms of p. Then use the formula to find the number of items they will sell if the price per item is $3.00.

28. A company manufactures and sells bracelets. They have found from past experience that they can sell 300 bracelets each week if the price per bracelet is $2.00, but only 150 bracelets are sold if the price is $2.50 per bracelet. If the relationship between the number of bracelets sold, x, and the price per bracelet, p, is a linear one, find a formula that gives x in terms of p. Then use the formula to find the number of bracelets they will sell at $3.00 each.

29. Five Cities Garbage charges a flat monthly fee for their services plus a certain amount for each bag of trash they pick up. A customer notices that the January bill for picking up 5 bags of trash was $25.60, while the February bill for 7 bags of trash was $27.10. Assume the relationship between the total monthly charges, C, and the number of bags, x, of trash picked up is a linear relationship. Use the data to find the formula that gives C in terms of x. Then use the formula to predict the cost for picking up 12 bags of trash in 1 month.

30. A bottled water company charges a flat fee each month for the use of their water dispenser plus a certain amount for each gallon of water delivered. Suppose that the company delivers 10 gallons of water in March, and the March bill is $18. Then, in April, 15 gallons of water are delivered for a total charge of $23.50. Assume the relationship between the total monthly charges, C, and the number of gallons, x, of water delivered is a linear relationship. Use the data to find the formula that gives C in terms of x. Then use the formula to predict the cost if 20 gallons of water are delivered in 1 month.

31. A ball is tossed into the air so that the height after 1, 3, and 5 seconds is as given in the table below:

t (sec)	h (ft)
1	128
3	128
5	0

The relationship between the height of the ball h and the time t is

$$h = at^2 + bt + c$$

Use the information in the table to write a system of three equations in three variables, a, b, and c. Solve the system to find the exact relationship between h and t.

32. A ball is tossed into the air, and its height above the ground after 1, 3, and 4 seconds is recorded as shown in the table:

t (sec)	h (ft)
1	96
3	64
4	0

The relationship between the height of the ball h and the time t can be written as

$$h = at^2 + bt + c$$

Use the information in the table to write a system of three equations in three variables, a, b, and c. Solve the system to find the exact relationship between the variables h and t.

Review Problems

 The problems below review material we covered in Section 6.5.

Find the vertex for each of the following parabolas and then indicate whether it is the highest or lowest point on the graph.

33. $y = 2x^2 + 8x - 15$
34. $y = 3x^2 - 9x - 10$
35. $y = 12x - 4x^2$
36. $y = 18x - 6x^2$

37. An object is projected into the air with an initial upward velocity of 64 feet/second. Its height h (in feet) at any time t (in seconds) is given by the formula $h = 64t - 16t^2$. Find the time at which the object reaches its maximum height. Then find the maximum height.

38. An object is projected into the air with an initial upward velocity of 64 feet/second from the top of a building 40 feet high. If the height h (in feet) of the object t seconds after it is projected into the air is $h = 40 + 64t - 16t^2$, find the time at which the object reaches its maximum height. Then find the maximum height it attains.

REVIEW FOR CHAPTER 7

Examples

1. The solution to the system

$$x + 2y = 4$$
$$x - y = 1$$

is the ordered pair (2, 1). It is the only ordered pair that satisfies both equations.

2. We can eliminate the y variable from the system in Example 1 above by multiplying both sides of the second equation by 2 and adding the result to the first equation:

$$
\begin{array}{ll}
x + 2y = 4 & \xrightarrow{\text{No change}} \quad x + 2y = 4 \\
x - y = 1 & \xrightarrow[\;2\;]{\text{Times}} \quad \dfrac{2x - 2y = 2}{3x \qquad = 6} \\
& \qquad\qquad\qquad x = 2
\end{array}
$$

Substituting $x = 2$ into either of the original two equations gives $y = 1$. The solution is (2, 1).

3. We can apply the substitution method to the system in Example 1 above by first solving the second equation for x to get $x = y + 1$.

Substituting this expression for x into the first equation, we have

$$(y + 1) + 2y = 4$$
$$3y + 1 = 4$$
$$3y = 3$$
$$y = 1$$

Using $y = 1$ in either of the original equations gives $x = 2$.

Chapter 7 Summary

Systems of Linear Equations [7.1, 7.2]

A **system of linear equations** consists of two or more linear equations considered simultaneously. The solution set to a linear system in two variables is the set of ordered pairs that satisfy both equations. The solution set to a linear system in three variables consists of the **ordered triples** that satisfy all three equations in the system.

Strategy for Solving a System by the Addition Method [7.1]

Step 1 Look the system over to decide which variable will be easier to eliminate.

Step 2 Use the multiplication property of equality on each equation separately if necessary to ensure that the coefficients of the variable to be eliminated are opposites.

Step 3 Add the left and right sides of the system produced in Step 2 and solve the resulting equation.

Step 4 Substitute the solution from Step 3 back into any equation with both x and y variables and solve.

Step 5 Check your solution in both equations if necessary.

Strategy for Solving a System by the Substitution Method [7.1]

Step 1 Solve either of the equations for one of the variables (this step is not necessary if one of the equations has the correct form already).

Step 2 Substitute the results of Step 1 into the other equation and solve.

Step 3 Substitute the results of Step 2 into an equation with both x and y variables, and solve. (The equation produced in Step 1 is usually a good one to use.)

Step 4 Check your solution if necessary.

4. If the two lines are parallel, then the system will be inconsistent and the solution is Ø. If the two lines coincide, then the system is dependent.

Inconsistent and Dependent Equations [7.1, 7.2]

Two linear equations that have no points in common are said to be **inconsistent.** Two linear equations that have all their solutions in common are said to be **dependent.**

5. $\begin{vmatrix} 3 & 4 \\ -2 & 5 \end{vmatrix} = 15 - (-8)$
$= 23$

2 × 2 Determinants [7.3]

The value of a **2 × 2 determinant** is

$$\begin{vmatrix} a & c \\ b & d \end{vmatrix} = ad - bc$$

6. Expanding
$$\begin{vmatrix} 1 & 3 & -2 \\ 2 & 0 & 1 \\ 4 & -1 & 1 \end{vmatrix}$$
across the first row gives us

$1 \begin{vmatrix} 0 & 1 \\ -1 & 1 \end{vmatrix} - 3 \begin{vmatrix} 2 & 1 \\ 4 & 1 \end{vmatrix} - 2 \begin{vmatrix} 2 & 0 \\ 4 & -1 \end{vmatrix}$

$= 1(1) - 3(-2) - 2(-2)$
$= 11$

3 × 3 Determinants [7.3]

The value of a **3 × 3 determinant** is given by

$$\begin{vmatrix} a_1 & b_1 & c_1 \\ a_2 & b_2 & c_2 \\ a_3 & b_3 & c_3 \end{vmatrix} = \begin{matrix} a_1b_2c_3 + a_3b_1c_2 + a_2b_3c_1 \\ - a_3b_2c_1 - a_1b_3c_2 - a_2b_1c_3 \end{matrix}$$

There are two methods of finding the six products in the expansion of a 3 × 3 determinant. One method involves a cross-multiplication scheme. The other method involves **expanding** the determinant **by minors.**

7. For the system

$x + y = 6$
$3x - 2y = -2$

we have

$D = \begin{vmatrix} 1 & 1 \\ 3 & -2 \end{vmatrix} = -5$

$D_x = \begin{vmatrix} 6 & 1 \\ -2 & -2 \end{vmatrix} = -10$

$D_y = \begin{vmatrix} 1 & 6 \\ 3 & -2 \end{vmatrix} = -20$

$x = \dfrac{-10}{-5} = 2$

$y = \dfrac{-20}{-5} = 4$

Cramer's Rule for a Linear System in Two Variables [7.4]

The solution to the system

$$a_1x + b_1y = c_1$$
$$a_2x + b_2y = c_2$$

is given by

$$x = \frac{D_x}{D} \qquad y = \frac{D_y}{D} \qquad (D \neq 0)$$

where

$$D = \begin{vmatrix} a_1 & b_1 \\ a_2 & b_2 \end{vmatrix} \qquad D_x = \begin{vmatrix} c_1 & b_1 \\ c_2 & b_2 \end{vmatrix} \qquad D_y = \begin{vmatrix} a_1 & c_1 \\ a_2 & c_2 \end{vmatrix} \qquad (D \neq 0)$$

REVIEW FOR CHAPTER 7

8. For the system

$$x + y = -1$$
$$2x - z = 3$$
$$y + 2z = -1$$

$$D = \begin{vmatrix} 1 & 1 & 0 \\ 2 & 0 & -1 \\ 0 & 1 & 2 \end{vmatrix} = -3$$

$$D_x = \begin{vmatrix} -1 & 1 & 0 \\ 3 & 0 & -1 \\ -1 & 1 & 2 \end{vmatrix} = -6$$

$$D_y = \begin{vmatrix} 1 & -1 & 0 \\ 2 & 3 & -1 \\ 0 & -1 & 2 \end{vmatrix} = 9$$

$$D_z = \begin{vmatrix} 1 & 1 & -1 \\ 2 & 0 & 3 \\ 0 & 1 & -1 \end{vmatrix} = -3$$

$$x = \frac{-6}{-3} = 2$$

$$y = \frac{9}{-3} = -3$$

$$z = \frac{-3}{-3} = 1$$

Cramer's Rule for a Linear System in Three Variables [7.4]

The solution to the system

$$a_1x + b_1y + c_1z = d_1$$
$$a_2x + b_2y + c_2z = d_2$$
$$a_3x + b_3y + c_3z = d_3$$

is given by

$$x = \frac{D_x}{D} \qquad y = \frac{D_y}{D} \qquad z = \frac{D_z}{D} \qquad (D \neq 0)$$

where

$$D = \begin{vmatrix} a_1 & b_1 & c_1 \\ a_2 & b_2 & c_2 \\ a_3 & b_3 & c_3 \end{vmatrix} \qquad D_x = \begin{vmatrix} d_1 & b_1 & c_1 \\ d_2 & b_2 & c_2 \\ d_3 & b_3 & c_3 \end{vmatrix}$$

$$D_y = \begin{vmatrix} a_1 & d_1 & c_1 \\ a_2 & d_2 & c_2 \\ a_3 & d_3 & c_3 \end{vmatrix} \qquad D_z = \begin{vmatrix} a_1 & b_1 & d_1 \\ a_2 & b_2 & d_2 \\ a_3 & b_3 & d_3 \end{vmatrix}$$

Chapter 7 Review Problems

Solve each system using the addition method. [7.1]

1. $x + y = 4$
$2x - y = 14$

2. $3x + y = 2$
$2x + y = 0$

3. $2x - 4y = 5$
$-x + 2y = 3$

4. $3x - y = 2$
$-6x + 2y = -4$

5. $5x - 2y = 7$
$3x + y = 2$

6. $6x - 5y = -5$
$3x + y = 1$

7. $6x + 4y = 8$
$9x + 6y = 12$

8. $4x - 8y = 6$
$6x - 12y = 6$

9. $3x - 7y = 2$
$-4x + 6y = -6$

10. $6x + 5y = 9$
$4x + 3y = 6$

11. $-7x + 4y = -1$
$5x - 3y = 0$

12. $-9x + 3y = 1$
$5x - 2y = -2$

13. $\frac{1}{2}x - \frac{3}{4}y = -4$
$\frac{1}{4}x + \frac{3}{2}y = 13$

14. $\frac{2}{3}x - \frac{1}{6}y = 0$
$\frac{4}{3}x + \frac{5}{6}y = 14$

15. $-\frac{1}{2}x + \frac{1}{3}y = -\frac{13}{6}$
$\frac{4}{5}x + \frac{3}{4}y = \frac{9}{10}$

16. $-\frac{1}{5}x + \frac{4}{3}y = \frac{14}{15}$
$\frac{1}{3}x - \frac{1}{4}y = \frac{5}{12}$

Solve each system by the substitution method. [7.1]

17. $x + y = 2$
$y = x - 1$

18. $x - y = 2$
$y = 3x + 1$

19. $2x - 3y = 5$
$y = 2x - 7$

20. $3x - 2y = 5$
$y = 3x - 7$

21. $x + y = 4$
$2x + 5y = 2$

22. $x + y = 3$
$2x + 5y = -6$

23. $3x + 7y = 6$
$x = -3y + 4$

24. $4x + 7y = -3$
$x = -2y - 2$

25. $5x - y = 4$
$y = 5x - 3$

26. $2x + y = 3$
$y = -2x + 3$

Solve each system. [7.2]

27. $x + y + z = 6$
$x - y - 3z = -8$
$x + y - 2z = -6$

28. $x - 2y + 3z = 4$
$x + 2y - z = 0$
$x + 2y + z = 8$

29. $3x + 2y + z = 4$
$2x - 4y + z = -1$
$x + 6y + 3z = -4$

30. $2x + 3y - 8z = 2$
$3x - y + 2z = 10$
$4x + y + 8z = 16$

31. $5x + 8y - 4z = -7$
$7x + 4y + 2z = -2$
$3x - 2y + 8z = 8$

32. $5x - 3y - 6z = 5$
$4x - 6y - 3z = 4$
$-x + 9y + 9z = 7$

33. $5x - 2y + z = 6$
$-3x + 4y - z = 2$
$6x - 8y + 2z = -4$

34. $4x - 6y + 8z = 4$
$5x + y - 2z = 4$
$6x - 9y + 12z = 6$

35. $2x - y = 5$
$3x - 2z = -2$
$5y + z = -1$

36. $2x + y = 8$
$4y - z = -9$
$3x - 2z = -6$

37. $x - y = 2$
$y - z = -3$
$x - z = -1$

38. $x + y = 4$
$x + z = 1$
$y - 2z = 5$

Evaluate each determinant. [7.3]

39. $\begin{vmatrix} 2 & 3 \\ -5 & 4 \end{vmatrix}$

40. $\begin{vmatrix} 3 & 0 \\ 5 & -1 \end{vmatrix}$

41. $\begin{vmatrix} 1 & 0 \\ -7 & -3 \end{vmatrix}$

42. $\begin{vmatrix} -5 & -2 \\ 0 & -6 \end{vmatrix}$

43. $\begin{vmatrix} 3 & 0 & 2 \\ -1 & 4 & 0 \\ 2 & 0 & 0 \end{vmatrix}$

44. $\begin{vmatrix} 3 & -1 & 0 \\ 0 & 2 & -4 \\ 6 & 0 & 2 \end{vmatrix}$

45. $\begin{vmatrix} -3 & -2 & 0 \\ 0 & -4 & 2 \\ 5 & 1 & 1 \end{vmatrix}$

46. $\begin{vmatrix} 3 & 4 & -1 \\ 0 & 0 & 2 \\ -5 & 1 & 2 \end{vmatrix}$

Solve for x. [7.3]

47. $\begin{vmatrix} 2 & 3x \\ -1 & 2x \end{vmatrix} = 4$

48. $\begin{vmatrix} 6 & 2x \\ -2 & 3x \end{vmatrix} = 5$

Use Cramer's rule to solve each system. [7.4]

49. $3x - 5y = 4$
$7x - 2y = 3$

50. $7x - 5y = 8$
$4x + 3y = 2$

51. $3x - 6y = 9$
$2x - 4y = 6$

52. $5x - 10y = 15$
$-2x + 4y = -6$

53. $2x - y + 3z = 4$
$5x + 2y - z = 3$
$-x - 3y + 2z = 1$

54. $5x - 3y + z = 2$
$3x - y + 4z = 3$
$2x + 3y - z = -1$

Use a system of equations to solve each application problem. In each case be sure to show the system used. [7.5]

55. One number is 5 more than twice another. Their sum is 11. Find the two numbers.

56. The larger of two numbers is 4 more than 5 times the smaller. If the smaller number is subtracted from 3 times the larger, the result is 26. Find the two numbers.

REVIEW FOR CHAPTER 7

57. The sum of three numbers is 2. The smallest number is 11 less than the largest number. If the sum of twice the largest with the other two numbers is 7, find the three numbers.

58. The sum of three numbers is 9. The sum of the two larger ones is 7, while the sum of the two smaller ones is 5. Find the three numbers.

59. Tickets for the show cost $2.00 for adults and $1.50 for children. How many adult tickets and how many children's tickets were sold if a total of 127 tickets were sold for $214?

60. John has 20 coins totaling $3.20. If he has only dimes and quarters, how many of each coin does he have?

61. Ms. Jones invests money in two accounts, one of which pays 12% per year, while the other pays 15% per year. If her total investment is $12,000 and the interest after 1 year is $1,650, how much is invested in each account?

62. A man invests twice as much money in an account that pays 18% per year as he does in an account that pays 13% per year. If the total amount of interest is $1,470, how much is invested at each rate?

63. How many ounces of 30% HCl solution and 70% HCl solution must be mixed to get 15 ounces of 50% HCl solution?

64. How many gallons of 25% alcohol solution and 50% alcohol solution should be mixed to get 20 gallons of 42.5% alcohol solution?

65. It takes a boat on a river 2 hours to travel 28 miles downstream and 3 hours to travel 30 miles upstream. What is the speed of the boat, and what is the speed of the current of the river?

66. A boat travels 36 miles down a river in 3 hours. If it takes the boat 9 hours to travel the same distance going up the river, what is the speed of the boat? What is the speed of the current of the river?

Chapter 7 Test

Solve the following systems by the addition method. [7.1]

1. $2x - 5y = -8$
$3x + y = 5$

2. $4x - 7y = -2$
$-5x + 6y = -3$

3. $\frac{1}{3}x - \frac{1}{6}y = 3$
$-\frac{1}{5}x + \frac{1}{4}y = 0$

Solve the following systems by the substitution method. [7.1]

4. $2x - 5y = 14$
$y = 3x + 8$

5. $6x - 3y = 0$
$x + 2y = 5$

6. Solve the system. [7.2]
$2x - y + z = 9$
$x + y - 3z = -2$
$3x + y - z = 6$

Evaluate each determinant. [7.3]

7. $\begin{vmatrix} 3 & -5 \\ -4 & 2 \end{vmatrix}$

8. $\begin{vmatrix} 1 & 0 & -3 \\ 2 & 1 & 0 \\ 0 & 5 & 4 \end{vmatrix}$

Use Cramer's rule to solve. [7.4]

9. $5x - 4y = 2$
$-2x + y = 3$

10. $2x + 4y = 3$
$-4x - 8y = -6$

11. $2x - y + 3z = 2$
$x - 4y - z = 6$
$3x - 2y + z = 4$

12. $x - 3y = 12$
$2x + 4y = -26$

13. $x + 2y = 4$
$y + 3z = 18$
$2x - 5z = -29$

Solve each application problem. [7.5]

14. A number is 1 less than twice another. Their sum is 14. Find the two numbers.

15. John invests twice as much money at 6% as he does at 5%. If his investments earn a total of $680 in 1 year, how much does he have invested at each rate?

16. There were 750 tickets sold for a basketball game for a total of $1,090. If adult tickets cost $2.00 and children's tickets cost $1.00, how many of each kind were sold?

17. How much 30% alcohol solution and 70% alcohol solution must be mixed to get 16 gallons of 60% solution?

18. A boat can travel 20 miles downstream in 2 hours. The same boat can travel 18 miles upstream in 3 hours. What is the speed of the boat in still water, and what is the speed of the current?

19. A collection of nickels, dimes, and quarters consists of 15 coins with a total value of $1.10. If the number of nickels is 1 less than 4 times the number of dimes, how many of each coin is contained in the collection?

20. For a woman of average height and weight between the ages of 19 and 22, the Food and Nutrition Board of the National Academy of Sciences has determined the Recommended Daily Allowance (RDA) of ascorbic acid to be 45 mg (milligrams). They have also determined the RDA for niacin to be 14 mg for the same woman.

Each ounce of cereal I contains 10 mg of ascorbic acid and 4 mg of niacin, while each ounce of cereal II contains 15 mg of ascorbic acid and 2 mg of niacin. If the cereals are combined, how many ounces of each cereal must the average woman between the ages of 19 and 22 consume in order to have the RDAs for both ascorbic acid and niacin?

The following table is a summary of the information given.

	Cereal I	Cereal II	RDA
Ascorbic Acid	10 mg	15 mg	45 mg
Niacin	4 mg	2 mg	14 mg

8

Exponential and Logarithmic Functions

Contents

Introduction

If you have ever had your thyroid gland tested, then you may have come into contact with radioactive iodine-131. Like all radioactive elements, iodine-131 decays naturally. The half-life of iodine-131 is 8 days, which means that every 8 days a sample of iodine-131 will decrease (decay) to half of its original amount. If you start with 1 gram of iodine-131, then 8 days later one half gram will remain, and 8 days after that only one fourth gram will be left. Table 1 and Figure 1 show what happens to a 1,600 microgram (μg) sample of iodine-131 over time.

TABLE 1
Iodine-131 as a Function of Time

Input	Output
t (days)	A (μg)
0	1,600
8	800
16	400
24	200
32	100

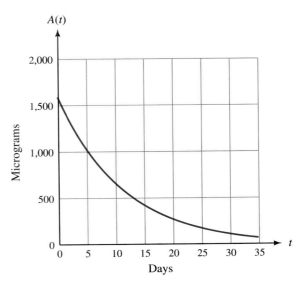

FIGURE 1

The relationship between the quantities shown in the table and graph falls into a general category of relationships called *exponential decay*. The function represented by the information in the table and graph is

$$A(t) = 1,600 \cdot 2^{-t/8}$$

This is an *exponential function*. There are many other relationships in the world around us that can be modeled by exponential functions. They are used to find

453

things as commonplace as the amount of money that accumulates in an interest-bearing savings account, and they are used to make decisions as to where to store the radioactive waste from nuclear power plants.

Overview

This chapter begins with a definition of *exponential functions,* followed by a general definition for the *inverse of a function.* These two topics lead directly into the definition for *logarithms* and *logarithmic functions.* There are many applications of these functions in this chapter. For example, the pH of a liquid is defined in terms of logarithms. (That's the same pH that is given on the labels of many hair conditioners.) The Richter scale for measuring earthquake intensity is a logarithmic scale, as is the decibel scale used for measuring the intensity of sound.

SECTION 8.1

Exponential Functions and the Inverse of a Function

After we have gained some experience with *exponential functions,* we will give a general definition for the *inverse of a function.* Later, in Section 8.2, we will make the connection between exponential functions and *logarithmic functions.*

> **DEFINITION**
>
> An **exponential function** is any function that can be written in the form
> $$f(x) = b^x$$
> where b is a positive real number other than 1.

Each of the following is an exponential function:
$$f(x) = 2^x \qquad y = 3^x \qquad f(x) = (\tfrac{1}{4})^x$$

The first step in becoming familiar with exponential functions is to find some values for specific exponential functions.

EXAMPLE 1   If the exponential functions f and g are defined by
$$f(x) = 2^x \qquad \text{and} \qquad g(x) = 3^x$$
then

$$f(0) = 2^0 = 1 \qquad\qquad g(0) = 3^0 = 1$$
$$f(1) = 2^1 = 2 \qquad\qquad g(1) = 3^1 = 3$$
$$f(2) = 2^2 = 4 \qquad\qquad g(2) = 3^2 = 9$$
$$f(3) = 2^3 = 8 \qquad\qquad g(3) = 3^3 = 27$$
$$f(-2) = 2^{-2} = \frac{1}{2^2} = \frac{1}{4} \qquad\qquad g(-2) = 3^{-2} = \frac{1}{3^2} = \frac{1}{9}$$
$$f(-3) = 2^{-3} = \frac{1}{2^3} = \frac{1}{8} \qquad\qquad g(-3) = 3^{-3} = \frac{1}{3^3} = \frac{1}{27}$$

We will now turn our attention to the graphs of exponential functions. Since the notation y is easier to use when graphing, and $y = f(x)$, for convenience we will write the exponential functions as

$$y = b^x$$

EXAMPLE 2   Sketch the graph of the exponential function $y = 2^x$.

SOLUTION   Using the results of Example 1, we have the table shown in the margin. Graphing the ordered pairs given in the table and connecting them with a smooth curve, we have the graph of $y = 2^x$ shown in Figure 1.

x	y
-3	$\frac{1}{8}$
-2	$\frac{1}{4}$
-1	$\frac{1}{2}$
0	1
1	2
2	4
3	8

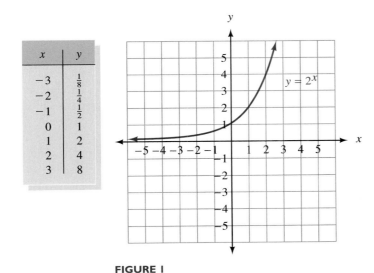

FIGURE 1

Notice that the graph does not cross the x-axis. It *approaches* the x-axis—in fact, we can get it as close to the x-axis as we want without it actually intersecting the x-axis. In order for the graph of $y = 2^x$ to intersect the x-axis, we would have to find a value of x that would make $2^x = 0$. Because no such value of x exists, the graph of $y = 2^x$ cannot intersect the x-axis.

EXAMPLE 3   Sketch the graph of $y = (\frac{1}{3})^x$.

SOLUTION   The table in the margin on the next page gives some ordered pairs that satisfy the equation. Using the ordered pairs from the table, we have the graph shown in Figure 2.

x	y
-3	27
-2	9
-1	3
0	1
1	$\frac{1}{3}$
2	$\frac{1}{9}$
3	$\frac{1}{27}$

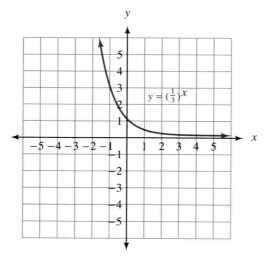

FIGURE 2

The graphs of all exponential functions have two things in common: (1) each crosses the y-axis at (0, 1), since $b^0 = 1$; and (2) none can cross the x-axis, since $b^x = 0$ is impossible because of the restrictions on b.

The Inverse of a Function

Suppose the function f is given by

$$f = \{(1, 4), (2, 5), (3, 6), (4, 7)\}$$

The **inverse** of f is obtained by reversing the order of the coordinates in each ordered pair in f. The inverse of f is the relation given by

$$g = \{(4, 1), (5, 2), (6, 3), (7, 4)\}$$

It is obvious that the domain of f is now the range of g, and the range of f is now the domain of g. Every function (or relation) has an inverse that is obtained from the original function by interchanging the components of each ordered pair.

Suppose a function f is defined with an equation instead of a list of ordered pairs. We can obtain the equation of the inverse of f by interchanging the role of x and y in the equation for f.

EXAMPLE 4 If the function f is defined by $f(x) = 2x - 3$, find the equation that represents the inverse of f.

SOLUTION Since the inverse of f is obtained by interchanging the components of all the ordered pairs belonging to f, and each ordered pair in f satisfies the equation $y = 2x - 3$, we simply exchange x and y in the equation to get the formula for the inverse of f:

$$x = 2y - 3$$

We now solve this equation for y in terms of x:

$$x + 3 = 2y$$

$$y = \frac{x + 3}{2}$$

The last line gives the equation that defines the inverse of f. The graphs of f and its inverse are shown in Figure 3.

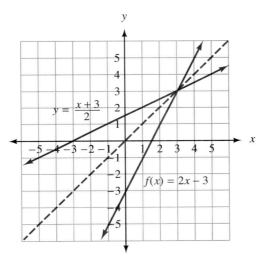

FIGURE 3

The graphs of f and its inverse have symmetry about the line $y = x$ (the dashed line in Figure 3). This is a reasonable result since one function was obtained from the other by interchanging x and y in the equation. The ordered pairs (a, b) and (b, a) always have symmetry about the line $y = x$.

EXAMPLE 5 Graph the function $y = x^2 - 2$ and its inverse. Give the equation for the inverse.

SOLUTION We can obtain the graph of the inverse of $y = x^2 - 2$ by graphing $y = x^2 - 2$ by the usual methods, and then reflecting the graph about the line $y = x$, as shown in Figure 4 (at the top of the next page).

The equation that corresponds to the inverse of $y = x^2 - 2$ is obtained by interchanging x and y to get $x = y^2 - 2$. Then we solve for y:

$$x = y^2 - 2$$

$$x + 2 = y^2$$

$$y = \pm \sqrt{x + 2}$$

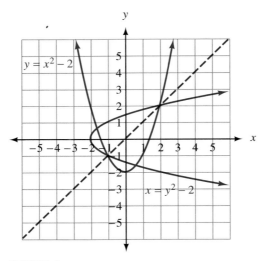

FIGURE 4

Comparing the graphs from Examples 4 and 5, we observe that the inverse of a function is not always a function. In Example 4, both f and its inverse have graphs that are nonvertical straight lines, and therefore both represent functions. In Example 5, the inverse of the function is not a function, since a vertical line crosses it in more than one place.

One-to-One Functions

We can distinguish between those functions with inverses that are also functions and those functions with inverses that are not functions with the following definition.

DEFINITION

A function is a **one-to-one function** if every element in the range comes from exactly one element in the domain.

This definition indicates that a one-to-one function will yield a set of ordered pairs in which no two different ordered pairs have the same second coordinates. For example, the function

$$f = \{(2, 3), (-1, 3), (5, 8)\}$$

is not one-to-one because the element 3 in the range comes from both 2 and -1 in the domain. On the other hand, the function

$$g = \{(5, 7), (3, -1), (4, 2)\}$$

is a one-to-one function because every element in the range comes from only one element in the domain.

Horizontal Line Test

If we have the graph of a function, we can determine whether the function is one-to-one with the **horizontal line test.** If a horizontal line crosses the graph of a function in more than one place, then the function is not a one-to-one function because the points at which the horizontal line crosses the graph will be points with the same y-coordinates but different x-coordinates. Therefore, the function will have an element in the range (the y-coordinate) that comes from more than one element in the domain (the x-coordinates).

Of the functions we have covered previously, all the linear functions and exponential functions are one-to-one functions because no horizontal lines can be found that will cross their graphs in more than one place.

Functions with Inverses That Are Also Functions

Because one-to-one functions do not repeat second coordinates, when we reverse the order of the ordered pairs in a one-to-one function, we obtain a relation in which no two ordered pairs have the same first coordinate—by definition, this relation must be a function. In other words, every one-to-one function has an inverse that is itself a function. Because of this, we can use function notation to represent that inverse.

Inverse Function Notation

If $y = f(x)$ is a one-to-one function, then the inverse of f is also a function and can be denoted by $y = f^{-1}(x).$

To illustrate, in Example 4 we found the inverse of $f(x) = 2x - 3$ was the function $y = \dfrac{x + 3}{2}$. We can write this inverse function with inverse function notation as

$$f^{-1}(x) = \frac{x + 3}{2}$$

On the other hand, the inverse of the function in Example 5 is not itself a function, so we do not use the notation $f^{-1}(x)$ to represent it.

Note: The notation f^{-1} does not represent the reciprocal of f. That is, the $^{-1}$ in this notation is not an exponent. The notation f^{-1} is defined as representing the inverse function for a one-to-one function.

EXAMPLE 6 Find the inverse of $g(x) = \dfrac{x - 4}{x - 2}.$

SOLUTION To find the inverse for g, we begin by replacing $g(x)$ with y to obtain

$$y = \frac{x - 4}{x - 2} \qquad \text{The original function}$$

Section 8.1 Exponential Functions and the Inverse of a Function

To find an equation for the inverse, we exchange x and y:

$$x = \frac{y - 4}{y - 2} \qquad \text{The inverse of the original function}$$

To solve for y, we first multiply each side by $y - 2$ to obtain

$$x(y - 2) = y - 4$$
$$xy - 2x = y - 4 \qquad \text{Distributive property}$$
$$xy - y = 2x - 4 \qquad \begin{array}{l}\text{Collect all terms containing} \\ y \text{ on the left side.}\end{array}$$
$$y(x - 1) = 2x - 4 \qquad \text{Factor } y \text{ from each term on the left side.}$$
$$y = \frac{2x - 4}{x - 1} \qquad \text{Divide each side by } x - 1.$$

Figure 5 shows that the graph of this function passes the horizontal line test. Therefore it is a one-to-one function.

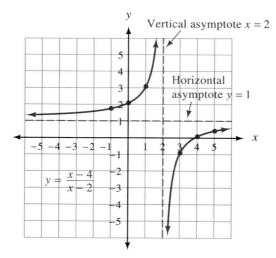

FIGURE 5

Since our original function is one-to-one (see Figure 5), its inverse is also a function. Therefore we can use inverse function notation to write

$$g^{-1}(x) = \frac{2x - 4}{x - 1}$$

EXAMPLE 7   Graph the function $y = 2^x$ and its inverse $x = 2^y$.

SOLUTION   We graphed $y = 2^x$ in Example 2. We simply reflect its graph about the line $y = x$ to obtain the graph of its inverse $x = 2^y$. (See Figure 6.)

As you can see from the graph, $x = 2^y$ is a function. However, we do not have the mathematical tools to solve this equation for y. Therefore, we are unable to use the inverse function notation to represent this function. In the next section we will give a definition that solves this problem. For now, we simply leave the equation as $x = 2^y$.

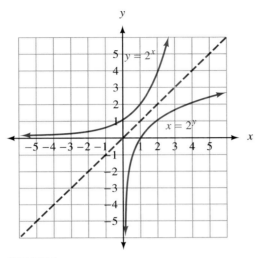

FIGURE 6

Functions, Relations, and Inverses—A Summary

Here is a summary of some of the things we know about functions, relations, and their inverses:

1. Every function is a relation, but not every relation is a function.
2. Every function has an inverse, but only one-to-one functions have inverses that are also functions.
3. The domain of a function is the range of its inverse, and the range of a function is the domain of its inverse.
4. If $y = f(x)$ is a one-to-one function, then we can use the notation $y = f^{-1}(x)$ to represent its inverse function.
5. The graphs of a function and its inverse have symmetry about the line $y = x$.
6. If (a, b) belongs to the function f, then the point (b, a) belongs to its inverse.

Problem Set
8.1

Let $f(x) = 3^x$ and $g(x) = (\frac{1}{2})^x$, *and evaluate each of the following:*

1. $g(0)$
2. $f(0)$
3. $g(-1)$
4. $g(-4)$
5. $f(-3)$
6. $f(-1)$
7. $f(2) + g(-2)$
8. $f(2) - g(-2)$

Graph each of the following functions.

9. $y = 4^x$ **10.** $y = 2^{-x}$

11. $y = 3^{-x}$ **12.** $y = (\frac{1}{3})^{-x}$

13. $y = 2^{x+1}$ **14.** $y = 2^{x-3}$

15. $y = 2^{2x}$ **16.** $y = 3^{2x}$

17. **Bacteria Growth** Suppose it takes 1 day for a certain strain of bacteria to reproduce by dividing in half. If there are 100 bacteria present to begin with, then the total number present after x days will be $f(x) = 100 \cdot 2^x$. Find the total number present after 1 day, 2 days, 3 days, and 4 days. How many days must elapse before there are over 100,000 bacteria present?

18. **Bacteria Growth** Suppose it takes 12 hours for a certain strain of bacteria to reproduce by dividing in half. If there are 50 bacteria present to begin with, then the total number present after x days will be $f(x) = 50 \cdot 4^x$. Find the total number present after 1 day, 2 days, and 3 days.

19. **Value of a Painting** A painting is purchased as an investment for $125. If the painting's value doubles every 5 years, then its value is given by the function

$$V(t) = 125 \cdot 2^{t/5} \qquad \text{for } t \geq 0$$

where t is the number of years since it was purchased, and $V(t)$ is its value (in dollars) at that time. Graph this function.

20. **Value of a Painting** A painting is purchased as an investment for $150. If the painting's

value doubles every 3 years, then its value is given by the function

$$V(t) = 150 \cdot 2^{t/3} \qquad \text{for } t \geq 0$$

where t is the number of years since it was purchased, and $V(t)$ is its value (in dollars) at that time. Graph this function.

For each of the following one-to-one functions, find the equation of the inverse. Write the inverse using the notation $f^{-1}(x)$.

21. $f(x) = 3x - 1$ **22.** $f(x) = 2x - 5$

23. $f(x) = \dfrac{x - 3}{x - 1}$ **24.** $f(x) = \dfrac{x - 2}{x - 3}$

25. $f(x) = \dfrac{x - 3}{4}$ **26.** $f(x) = \dfrac{x + 7}{2}$

27. $f(x) = \frac{1}{2}x - 3$ **28.** $f(x) = \frac{1}{3}x + 1$

29. $f(x) = \dfrac{2x + 1}{3x + 1}$ **30.** $f(x) = \dfrac{3x + 2}{5x + 1}$

For each of the following functions, sketch the graph of the relation and its inverse, and write an equation for the inverse.

31. $y = 2x - 1$ **32.** $y = 3x + 1$

33. $y = x^2 - 3$ **34.** $y = x^2 + 1$

35. $y = x^2 - 2x - 3$ **36.** $y = x^2 + 2x - 3$

37. $y = 3^x$ **38.** $y = (\frac{1}{2})^x$

39. $y = 4$ **40.** $y = -2$

41. $y = \frac{1}{2}x^3$ **42.** $y = x^3 - 2$

43. $y = \frac{1}{2}x + 2$ **44.** $y = \frac{1}{3}x - 1$

45. If $f(x) = 3x - 2$, then $f^{-1}(x) = \dfrac{x + 2}{3}$.

Use these two functions to find:
(a) $f(2)$ (b) $f^{-1}(2)$
(c) $f[f^{-1}(2)]$ (d) $f^{-1}[f(2)]$

46. If $f(x) = \frac{1}{2}x + 5$, then $f^{-1}(x) = 2x - 10$.
Use these two functions to find:
(a) $f(-4)$ (b) $f^{-1}(-4)$
(c) $f[f^{-1}(-4)]$ (d) $f^{-1}[f(-4)]$

47. Let $f(x) = \dfrac{1}{x}$, and find $f^{-1}(x)$.

48. Let $f(x) = \dfrac{a}{x}$, and find $f^{-1}(x)$. (Assume a is a real number constant.)

Review Problems

The problems that follow review material we covered in Section 7.1.

Solve each system by the addition method.

49. $4x + 3y = 10$
     $2x + y = 4$

50. $3x - 5y = -2$
     $2x - 3y = 1$

51. $4x + 5y = 5$
     $\frac{6}{5}x + y = 2$

52. $4x + 2y = -2$
     $\frac{1}{2}x + y = 0$

Solve each system by the substitution method.

53. $x + y = 3$
     $ y = x + 3$

54. $x + y = 6$
     $ y = x - 4$

55. $2x - 3y = -6$
     $ y = 3x - 5$

56. $7x - y = 24$
     $ x = 2y + 9$

One Step Further

57. Drag Racing In Chapter 3 we mentioned the dragster equipped with a computer. Table 1 gives the speed of the dragster every second during one race at the 1993 Winternationals. Figure 7 is a line graph constructed from the data in Table 1.

TABLE I
Speed of a Dragster

Elapsed Time (sec)	Speed (mi/hr)
0	0.0
1	72.7
2	129.9
3	162.8
4	192.2
5	212.4
6	228.1

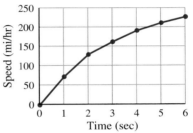

FIGURE 7
Line graph of Table 1 data

The graph of the function below contains the first point and the last point shown in Figure 7. That is, both (0, 0) and (6, 228.1) satisfy the function. Graph the function to see how close it comes to the other points in Figure 7.

$$s(t) = 250(1 - 1.5^{-t})$$

SECTION
8.2

Logarithms Are Exponents

As you know from your work in the previous section, equations of the form

$$y = b^x \qquad b > 0, b \neq 1$$

are called exponential functions. Since the equation of the inverse of a function can be obtained by exchanging x and y in the equation of the original function, the inverse of an exponential function must have the form

$$x = b^y \qquad b > 0, b \neq 1$$

Now, this last equation is actually the equation of a **logarithmic function,** as the definition at the top of the next page indicates.

DEFINITION

The expression $y = \log_b x$ is read "y is the logarithm to the base b of x" and is equivalent to the expression

$$x = b^y \qquad b > 0, b \neq 1$$

In words, we say "y is the number we raise b to in order to get x."

Notation: When an expression is in the form $x = b^y$, it is said to be in *exponential form.* On the other hand, if an expression is in the form $y = \log_b x$, it is said to be in *logarithmic form.*

Here are some equivalent statements written in both forms.

Exponential Form		Logarithmic Form
$8 = 2^3$	$\Leftrightarrow$	$\log_2 8 = 3$
$25 = 5^2$	$\Leftrightarrow$	$\log_5 25 = 2$
$0.1 = 10^{-1}$	$\Leftrightarrow$	$\log_{10} 0.1 = -1$
$\frac{1}{8} = 2^{-3}$	$\Leftrightarrow$	$\log_2 \frac{1}{8} = -3$
$r = z^s$	$\Leftrightarrow$	$\log_z r = s$

EXAMPLE 1 Solve for x: $\log_3 x = -2$.

SOLUTION In exponential form the equation looks like this:

$$x = 3^{-2}$$

or

$$x = \frac{1}{9}$$

The solution is $\frac{1}{9}$.

EXAMPLE 2 Solve: $\log_x 4 = 3$.

SOLUTION Again, we use the definition of logarithms to write the expression in exponential form:

$$4 = x^3$$

Taking the cube root of both sides, we have

$$\sqrt[3]{4} = \sqrt[3]{x^3}$$
$$x = \sqrt[3]{4}$$

The solution set is $\{\sqrt[3]{4}\}$.

EXAMPLE 3   Solve: $\log_8 4 = x$.

SOLUTION   We write the expression in exponential form:

$$4 = 8^x$$

Since both 4 and 8 can be written as powers of 2, we write them in terms of powers of 2:

$$2^2 = (2^3)^x$$
$$2^2 = 2^{3x}$$

The only way the left and right sides of this last equation can be equal is if the exponents are equal — that is, if

$$2 = 3x$$

or
$$x = \tfrac{2}{3}$$

The solution is $\tfrac{2}{3}$.

Check   We check as follows:

$$\log_8 4 = \tfrac{2}{3} \quad \Leftrightarrow \quad 4 = 8^{2/3}$$
$$4 = (\sqrt[3]{8})^2$$
$$4 = 2^2$$
$$4 = 4$$

The solution checks when used in the original equation.

Graphing Logarithmic Functions

Graphing logarithmic functions can be done using the graphs of exponential functions and the fact that the graphs of inverse functions have symmetry about the line $y = x$. Here's an example to illustrate.

EXAMPLE 4   Graph the equation: $y = \log_2 x$.

SOLUTION   The equation $y = \log_2 x$ is, by definition, equivalent to the exponential equation

$$x = 2^y$$

which is the equation of the inverse of the function

$$y = 2^x$$

The graphs of $y = 2^x$ and its inverse, $x = 2^y$, were given in Figure 6 of Section 8.1 and are repeated at the top of the next page in Figure 1. Now, however, we know that we can represent the inverse $x = 2^y$ as $y = \log_2 x$.

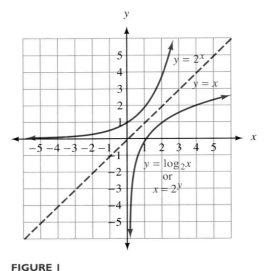

FIGURE I

It is apparent from the graph in Figure 1 that $y = \log_2 x$ is a function, since no vertical line will cross its graph in more than one place. The same is true for all logarithmic equations of the form $y = \log_b x$ where b is a positive number other than 1. Note also that the graph of $y = \log_b x$ will always appear to the right of the y-axis, meaning that x will always be positive in the expression $y = \log_b x$.

Two Special Identities

If b is a positive real number other than 1, then each of the following is a consequence of the definition of a logarithm:

$$(1) \quad b^{\log_b x} = x \qquad \text{and} \qquad (2) \quad \log_b b^x = x$$

The justifications for these identities are similar. Let's consider only the first one. Consider the expression

$$y = \log_b x$$

By definition, it is equivalent to

$$x = b^y$$

Substituting $\log_b x$ for y in the last line gives us

$$x = b^{\log_b x}$$

The following examples show how these identities can be used to simplify expressions involving logarithms.

EXAMPLE 5   Simplify: $\log_2 8$.

SOLUTION   Substitute 2^3 for 8:

$$\log_2 8 = \log_2 2^3$$
$$= 3$$

EXAMPLE 6   Simplify: $\log_{10} 10,000$.

SOLUTION   10,000 can be written as 10^4:

$$\log_{10} 10,000 = \log_{10} 10^4$$
$$= 4$$

EXAMPLE 7   Simplify: $\log_b b$ $(b > 0, b \neq 1)$.

SOLUTION   Since $b^1 = b$, we have

$$\log_b b = \log_b b^1$$
$$= 1$$

EXAMPLE 8   Simplify: $\log_b 1$ $(b > 0, b \neq 1)$.

SOLUTION   Since $1 = b^0$, we have

$$\log_b 1 = \log_b b^0$$
$$= 0$$

EXAMPLE 9   Simplify: $\log_4(\log_5 5)$.

SOLUTION   Since $\log_5 5 = 1$,

$$\log_4(\log_5 5) = \log_4 1$$
$$= 0$$

The Richter Scale

When we talk about the size of a shock wave, we are talking about its amplitude. The amplitude of a wave is half the difference between its highest point and its lowest point.

One application of logarithms is in measuring the magnitude of an earthquake. If an earthquake has a shock wave T times greater than the smallest shock wave that can be measured on a seismograph, then the magnitude M of the earthquake, as measured on the Richter scale, is given by the formula

$$M = \log_{10} T$$

To illustrate the discussion, an earthquake that produces a shock wave that is 10,000 times greater than the smallest shock wave measurable on a seismograph will have a magnitude M on the Richter scale of

$$M = \log_{10} 10,000 = 4$$

EXAMPLE 10

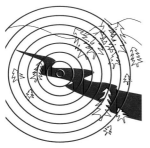

If an earthquake has a magnitude of $M = 5$ on the Richter scale, what can you say about the size of its shock wave?

SOLUTION To answer this question, we put $M = 5$ into the formula $M = \log_{10} T$ to obtain

$$5 = \log_{10} T$$

Writing this expression in exponential form, we have

$$T = 10^5 = 100,000$$

We can say that an earthquake that measures 5 on the Richter scale has a shock wave 100,000 times greater than the smallest shock wave measurable on a seismograph.

From Example 10 and the discussion that preceded it, we find that an earthquake of magnitude 5 has a shock wave that is 10 times greater than that of an earthquake of magnitude 4, because 100,000 is 10 times 10,000.

Problem Set
8.2

Write each of the following expressions in logarithmic form.

1. $2^4 = 16$ **2.** $3^2 = 9$
3. $125 = 5^3$ **4.** $16 = 4^2$
5. $0.01 = 10^{-2}$ **6.** $0.001 = 10^{-3}$
7. $2^{-5} = \frac{1}{32}$ **8.** $4^{-2} = \frac{1}{16}$
9. $(\frac{1}{2})^{-3} = 8$ **10.** $(\frac{1}{3})^{-2} = 9$
11. $27 = 3^3$ **12.** $81 = 3^4$

Write each of the following expressions in exponential form.

13. $\log_{10} 100 = 2$ **14.** $\log_2 8 = 3$
15. $\log_2 64 = 6$ **16.** $\log_2 32 = 5$
17. $\log_8 1 = 0$ **18.** $\log_9 9 = 1$
19. $\log_{10} 0.001 = -3$ **20.** $\log_{10} 0.0001 = -4$
21. $\log_6 36 = 2$ **22.** $\log_7 49 = 2$
23. $\log_5 \frac{1}{25} = -2$ **24.** $\log_3 \frac{1}{81} = -4$

Solve each of the following equations for x.

25. $\log_3 x = 2$ **26.** $\log_4 x = 3$
27. $\log_5 x = -3$ **28.** $\log_2 x = -4$
29. $\log_2 16 = x$ **30.** $\log_3 27 = x$

31. $\log_8 2 = x$ **32.** $\log_{25} 5 = x$
33. $\log_x 4 = 2$ **34.** $\log_x 16 = 4$
35. $\log_x 5 = 3$ **36.** $\log_x 8 = 2$

Sketch the graph of each of the following logarithmic equations.

37. $y = \log_3 x$ **38.** $y = \log_{1/2} x$
39. $y = \log_{1/3} x$ **40.** $y = \log_4 x$
41. $y = \log_5 x$ **42.** $y = \log_{1/5} x$
43. $y = \log_{10} x$ **44.** $y = \log_{1/4} x$

Simplify each of the following.

45. $\log_2 16$ **46.** $\log_3 9$
47. $\log_{25} 125$ **48.** $\log_9 27$
49. $\log_{10} 1,000$ **50.** $\log_{10} 10,000$
51. $\log_3 3$ **52.** $\log_4 4$
53. $\log_5 1$ **54.** $\log_{10} 1$
55. $\log_3(\log_6 6)$ **56.** $\log_5(\log_3 3)$
57. $\log_4[\log_2(\log_2 16)]$ **58.** $\log_4[\log_3(\log_2 8)]$

Applying the Concepts

Measuring Acidity *In chemistry, the pH of a solution is defined in terms of logarithms as* pH $= -\log_{10}[H^+]$, *where* $[H^+]$ *is the concentration of the hydrogen ions in solution. An acid solution has a* pH *below 7, and a basic solution has a* pH *higher than 7.*

59. In distilled water, the concentration of hydrogen ions is $[H^+] = 10^{-7}$. What is the pH?

60. Find the pH of a bottle of vinegar, if the concentration of hydrogen ions is $[H^+] = 10^{-3}$.

61. A hair conditioner has a pH of 6. Find the concentration of hydrogen ions, $[H^+]$, in the conditioner.

62. If a glass of orange juice has a pH of 4, what is the concentration of hydrogen ions, $[H^+]$, in the orange juice?

63. Magnitude of an Earthquake Find the magnitude M of an earthquake with a shock wave that measures $T = 100$ on a seismograph.

64. Magnitude of an Earthquake Find the magnitude M of an earthquake with a shock wave that measures $T = 100,000$ on a seismograph.

65. Shock Wave If an earthquake has a magnitude of 8 on the Richter scale, how many times greater is its shock wave than the smallest shock wave measurable on a seismograph?

66. Shock Wave If an earthquake has a magnitude of 6 on the Richter scale, how many times greater is its shock wave than the smallest shock wave measurable on a seismograph?

Review Problems

 The problems below review material from Section 7.2.

Solve each system.

67. $\begin{aligned} x + y + z &= 6 \\ 2x - y + z &= 3 \\ x + 2y - 3z &= -4 \end{aligned}$

68. $\begin{aligned} x + y + z &= 6 \\ x - y + 2z &= 7 \\ 2x - y - z &= 0 \end{aligned}$

69. $\begin{aligned} 3x + 4y &= 15 \\ 2x - 5z &= -3 \\ 4y - 3z &= 9 \end{aligned}$

70. $\begin{aligned} x + 3y &= 5 \\ 6y + z &= 12 \\ x - 2z &= -10 \end{aligned}$

SECTION 8.3

Properties of Logarithms

For the following three properties, x, y, and b are all positive real numbers, $b \neq 1$, and r is any real number.

▸ **PROPERTY 1**

$$\log_b(xy) = \log_b x + \log_b y$$

In Words: The logarithm of a product is the sum of the logarithms.

PROPERTY 2

$$\log_b\left(\frac{x}{y}\right) = \log_b x - \log_b y$$

In Words: The logarithm of a quotient is the difference of the logarithms.

PROPERTY 3

$$\log_b x^r = r \log_b x$$

In Words: The logarithm of a number raised to a power is the product of the power and the logarithm of the number.

Proof of Property 1

To prove property 1, we simply apply the first identity for logarithms given at the end of the preceding section:

$$b^{\log_b(xy)} = xy = (b^{\log_b x})(b^{\log_b y}) = b^{\log_b x + \log_b y}$$

Since the first and last expressions are equal and the bases are the same, the exponents $\log_b(xy)$ and $\log_b x + \log_b y$ must be equal. Therefore,

$$\log_b(xy) = \log_b x + \log_b y$$

The proofs of properties 2 and 3 proceed in much the same manner, so we will omit them here. The examples that follow show how the three properties can be used.

EXAMPLE 1 Expand, using the properties of logarithms: $\log_5 \dfrac{3xy}{z}$.

SOLUTION Applying property 2, we can write the quotient of $3xy$ and z in terms of a difference:

$$\log_5 \frac{3xy}{z} = \log_5 3xy - \log_5 z$$

Applying property 1 to the product $3xy$, we write it in terms of addition:

$$\log_5 \frac{3xy}{z} = (\log_5 3 + \log_5 x + \log_5 y) - \log_5 z$$

EXAMPLE 2 Expand, using the properties of logarithms: $\log_2 \dfrac{x^4}{\sqrt{y} \cdot z^3}$.

SOLUTION We write $\sqrt{y}$ as $y^{1/2}$ and apply the properties:

$$
\begin{aligned}
\log_2 \frac{x^4}{\sqrt{y} \cdot z^3} &= \log_2 \frac{x^4}{y^{1/2} z^3} && \sqrt{y} = y^{1/2} \\
&= \log_2 x^4 - \log_2(y^{1/2} \cdot z^3) && \text{Property 2} \\
&= \log_2 x^4 - (\log_2 y^{1/2} + \log_2 z^3) && \text{Property 1} \\
&= \log_2 x^4 - \log_2 y^{1/2} - \log_2 z^3 && \text{Remove parentheses.} \\
&= 4 \log_2 x - \tfrac{1}{2} \log_2 y - 3 \log_2 z && \text{Property 3}
\end{aligned}
$$

We can also use the three properties to write an expression in expanded form as just one logarithm.

EXAMPLE 3   Write as a single logarithm:

$$2 \log_{10} a + 3 \log_{10} b - \tfrac{1}{3} \log_{10} c$$

SOLUTION   We begin by applying property 3:

$$
\begin{aligned}
&2 \log_{10} a + 3 \log_{10} b - \tfrac{1}{3} \log_{10} c \\
&= \log_{10} a^2 + \log_{10} b^3 - \log_{10} c^{1/3} && \text{Property 3} \\
&= \log_{10}(a^2 \cdot b^3) - \log_{10} c^{1/3} && \text{Property 1} \\
&= \log_{10} \frac{a^2 b^3}{c^{1/3}} && \text{Property 2} \\
&= \log_{10} \frac{a^2 b^3}{\sqrt[3]{c}} && c^{1/3} = \sqrt[3]{c}
\end{aligned}
$$

The properties of logarithms along with the definition of logarithms are useful in solving equations that involve logarithms.

EXAMPLE 4   Solve for x: $\log_2(x + 2) + \log_2 x = 3$.

SOLUTION   Applying property 1 to the left side of the equation allows us to write it as a single logarithm:

$$
\begin{aligned}
\log_2(x + 2) + \log_2 x &= 3 \\
\log_2[(x + 2)(x)] &= 3
\end{aligned}
$$

The last line can be written in exponential form using the definition of logarithms:

$$(x + 2)(x) = 2^3$$

Solve as usual:

$$
\begin{aligned}
x^2 + 2x &= 8 \\
x^2 + 2x - 8 &= 0 \\
(x + 4)(x - 2) &= 0 \\
x + 4 = 0 \quad &\text{or} \quad x - 2 = 0 \\
x = -4 \quad &\text{or} \quad x = 2
\end{aligned}
$$

In the previous section, we noted the fact that x in the expression $y = \log_b x$ cannot be a negative number. Since substitution of $x = -4$ into the original equation gives

$$\log_2(-2) + \log_2(-4) = 3$$

which contains logarithms of negative numbers, we cannot use -4 as a solution. The solution set is $\{2\}$.

Problem Set

8.3

Use the three properties of logarithms given in this section to expand each expression as much as possible.

1. $\log_3 4x$

2. $\log_2 5x$

3. $\log_6 \dfrac{5}{x}$

4. $\log_3 \dfrac{x}{5}$

5. $\log_2 y^5$

6. $\log_7 y^3$

7. $\log_9 \sqrt[3]{z}$

8. $\log_8 \sqrt{z}$

9. $\log_6 x^2 y^4$

10. $\log_{10} x^2 y^4$

11. $\log_5 \sqrt{x} \cdot y^4$

12. $\log_8 \sqrt[3]{xy^6}$

13. $\log_b \dfrac{xy}{z}$

14. $\log_b \dfrac{3x}{y}$

15. $\log_{10} \dfrac{4}{xy}$

16. $\log_{10} \dfrac{5}{4y}$

17. $\log_{10} \dfrac{x^2 y}{\sqrt{z}}$

18. $\log_{10} \dfrac{\sqrt{x} \cdot y}{z^3}$

19. $\log_{10} \dfrac{x^3 \sqrt{y}}{z^4}$

20. $\log_{10} \dfrac{x^4 \sqrt[3]{y}}{\sqrt{z}}$

21. $\log_b \sqrt[3]{\dfrac{x^2 y}{z^4}}$

22. $\log_b \sqrt[4]{\dfrac{x^4 y^3}{z^5}}$

Write each expression as a single logarithm.

23. $\log_b x + \log_b z$

24. $\log_b x - \log_b z$

25. $2 \log_3 x - 3 \log_3 y$

26. $4 \log_2 x + 5 \log_2 y$

27. $\frac{1}{2} \log_{10} x + \frac{1}{3} \log_{10} y$

28. $\frac{1}{3} \log_{10} x - \frac{1}{4} \log_{10} y$

29. $3 \log_2 x + \frac{1}{2} \log_2 y - \log_2 z$

30. $2 \log_3 x + 3 \log_3 y - \log_3 z$

31. $\frac{1}{2} \log_2 x - 3 \log_2 y - 4 \log_2 z$

32. $3 \log_{10} x - \log_{10} y - \log_{10} z$

33. $\frac{3}{2} \log_{10} x - \frac{3}{4} \log_{10} y - \frac{4}{5} \log_{10} z$

34. $3 \log_{10} x - \frac{4}{3} \log_{10} y - 5 \log_{10} z$

Solve each of the following equations.

35. $\log_2 x + \log_2 3 = 1$

36. $\log_3 x + \log_3 3 = 1$

37. $\log_3 x - \log_3 2 = 2$

38. $\log_3 x + \log_3 2 = 2$

39. $\log_3 x + \log_3(x - 2) = 1$

40. $\log_6 x + \log_6(x - 1) = 1$

41. $\log_3(x + 3) - \log_3(x - 1) = 1$

42. $\log_4(x - 2) - \log_4(x + 1) = 1$

43. $\log_2 x + \log_2(x - 2) = 3$

44. $\log_4 x + \log_4(x + 6) = 2$

45. $\log_8 x + \log_8(x - 3) = \frac{2}{3}$

46. $\log_{27} x + \log_{27}(x + 8) = \frac{2}{3}$

47. $\log_5 \sqrt{x} + \log_5 \sqrt{6x + 5} = 1$

48. $\log_2 \sqrt{x} + \log_2 \sqrt{6x + 5} = 1$

Applying the Concepts

49. **Food Processing** The formula $M = 0.21(\log_{10} a - \log_{10} b)$ is used in the food processing industry to find the number of minutes, M, of heat processing a certain food should undergo at 250°F to reduce the probability of survival of *C. botulinum* spores. The letter a represents the number of spores per can before heating, and b represents the number of spores per can after heating. Find M if $a = 1$ and $b = 10^{-12}$. Then find M using the same values for a and b in the formula

$$M = 0.21 \log_{10} \frac{a}{b}$$

50. **Acoustic Power** The formula $N = \log_{10} \dfrac{P_1}{P_2}$ is used in radio electronics to find the ratio of the acoustic powers of two electric circuits in terms of their electric powers. Find N if P_1 is 100 and P_2 is 1. Then use the same two values of P_1 and P_2 to find N in the formula $N = \log_{10} P_1 - \log_{10} P_2$.

51. Use the properties of logarithms to show that $\log_{10}(8.43 \times 10^2)$ can be written as $2 + \log_{10} 8.43$.

52. Use the properties of logarithms to show that $\log_{10}(2.76 \times 10^3)$ can be written as $3 + \log_{10} 2.76$.

53. Use the properties of logarithms to show that the formula $\log_{10} A = \log_{10}[100(1.06)^t]$ can be written as $\log_{10} A = 2 + t \log_{10} 1.06$.

54. Use the properties of logarithms to show that the formula $\log_{10} A = \log_{10}[3(2)^{t/5,600}]$ can be written as $\log_{10} A = \log_{10} 3 + \dfrac{t}{5,600} \log_{10} 2$.

Evaluate each determinant.

59. $\begin{vmatrix} 3 & 5 \\ -6 & 2 \end{vmatrix}$

60. $\begin{vmatrix} -2 & 0 \\ 0 & -1 \end{vmatrix}$

61. $\begin{vmatrix} 1 & -2 & 3 \\ 0 & 4 & -1 \\ 2 & -4 & 6 \end{vmatrix}$

62. $\begin{vmatrix} 2 & 0 & 0 \\ 0 & -3 & 0 \\ 0 & 0 & 4 \end{vmatrix}$

Review Problems

 NOTE: The problems that follow review material we covered in Section 7.3.

If $A = \begin{bmatrix} 5 & -2 \\ 1 & 0 \end{bmatrix}$ and $B = \begin{bmatrix} -1 & 0 \\ 8 & 3 \end{bmatrix}$, *find:*

55. $2A - B$

56. $-3B$

57. AB

58. BA

SECTION 8.4

Common Logarithms and Natural Logarithms

There are two kinds of logarithms that occur more frequently than any other logarithms. Logarithms with a base of 10 are very common because our number system is a base-10 number system. For this reason, we call base-10 logarithms **common logarithms.** After our discussion of common logarithms, we'll discuss the second kind—**natural logarithms.**

Common Logarithms

> **DEFINITION**
>
> A **common logarithm** is a logarithm with a base of 10. Since common logarithms are used so frequently, it is customary to omit the base when we write them. That is,
>
> $$\log_{10} x = \log x$$
>
> When the base is not shown, it is assumed to be 10.

Common logarithms of powers of 10 are very simple to evaluate. We only need to recognize that $\log 10 = \log_{10} 10 = 1$ and apply the third property of logarithms: $\log_b x^r = r \log_b x$.

$$
\begin{aligned}
\log 1{,}000 &= \log 10^3 &=& \quad 3 \log 10 = && 3(1) = 3 \\
\log 100 &= \log 10^2 &=& \quad 2 \log 10 = && 2(1) = 2 \\
\log 10 &= \log 10^1 &=& \quad 1 \log 10 = && 1(1) = 1 \\
\log 1 &= \log 10^0 &=& \quad 0 \log 10 = && 0(1) = 0 \\
\log 0.1 &= \log 10^{-1} &=& -1 \log 10 = && -1(1) = -1 \\
\log 0.01 &= \log 10^{-2} &=& -2 \log 10 = && -2(1) = -2 \\
\log 0.001 &= \log 10^{-3} &=& -3 \log 10 = && -3(1) = -3
\end{aligned}
$$

To find common logarithms of numbers that are not powers of 10, we use a calculator with a $\boxed{\log}$ key or a table of logarithms. We will assume the use of a calculator for the rest of this chapter.

Check the following logarithms to be sure you know how to use your calculator. (These answers have been rounded to the nearest ten thousandth.)

$$
\begin{array}{ll}
\log 7.02 = 0.8463 & \log 6.00 = 0.7782 \\
\log 1.39 = 0.1430 & \log 9.99 = 0.9996
\end{array}
$$

EXAMPLE 1 Use a calculator to find log 2,760.

SOLUTION $\quad \log 2{,}760 = 3.4409$

To work this problem on a scientific calculator, we simply enter the number 2,760 and press the key labeled $\boxed{\log}$. To work the problem on a graphing calculator, we press the $\boxed{\text{LOG}}$ key first, then enter the number 2,760, and then press the $\boxed{\text{ENTER}}$ key.

$$2760 \ \boxed{\log} \quad \text{or} \quad \boxed{\text{LOG}} \ 2760 \ \boxed{\text{ENTER}}$$

The 3 in the answer is called the **characteristic,** while the decimal part of the logarithm is called the **mantissa.**

EXAMPLE 2 Find log 0.0391.

SOLUTION $\quad \log 0.0391 = -1.4078$

EXAMPLE 3 Find log 0.00523.

SOLUTION $\quad \log 0.00523 = -2.2815$

EXAMPLE 4 Find x if $\log x = 3.8774$.

SOLUTION We are looking for the number whose logarithm is 3.8774. On a scientific calculator, we enter 3.8774 and press the key labeled $\boxed{10^x}$. On a graphing calculator we press $\boxed{10^x}$ first, then enter the number 3.8774, and then press the

ENTER key. The result is 7,540 to four significant digits.

$$\text{If} \quad \log x = 3.8774$$
$$\text{then} \quad x = 10^{3.8774}$$
$$x = 7,540$$

The number 7,540 is called the **antilogarithm**, or just **antilog**, of 3.8774. That is, 7,540 is the number whose logarithm is 3.8774.

EXAMPLE 5 Find x if $\log x = -2.4179$.

SOLUTION Using the $\boxed{10^x}$ key on a calculator, with $x = -2.4179$, the result is 0.00382.

$$\text{If} \quad \log x = -2.4179$$
$$\text{then} \quad x = 10^{-2.4179}$$
$$= 0.00382$$

The antilog of -2.4179 is 0.00382. That is, the logarithm of 0.00382 is -2.4179.

In Section 8.2, we found that the magnitude M of an earthquake that produces a shock wave T times larger than the smallest shock wave that can be measured on a seismograph is given by the formula

$$M = \log_{10} T$$

We can rewrite this formula using our shorthand notation for common logarithms as

$$M = \log T$$

EXAMPLE 6 The San Francisco earthquake of 1906 measured 8.3 on the Richter scale. The San Fernando earthquake of 1971 measured 6.6 on the Richter scale. Find T for each earthquake and then indicate how much stronger the 1906 earthquake was than the 1971 earthquake.

SOLUTION For the 1906 earthquake:

$$\text{If} \quad \log T = 8.3, \quad \text{then} \quad T \approx 2.00 \times 10^8$$

For the 1971 earthquake:

$$\text{If} \quad \log T = 6.6, \quad \text{then} \quad T \approx 3.98 \times 10^6$$

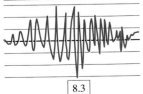

8.3

Dividing the two values of T and rounding our answer to the nearest whole number, we have

$$\frac{2.00 \times 10^8}{3.98 \times 10^6} = 50$$

The shock wave for the 1906 earthquake was approximately 50 times as large as the shock wave for the 1971 earthquake.

Natural Logarithms

If this bothers you, try to think of the last time you saw a precise definition for the number π. Even though you may not know the definition of π, you are still able to work problems that use the number π, simply by knowing that it is an irrational number that is approximately 3.1416.

The second kind of frequently used logarithms are **natural logarithms.** In order to give a definition for natural logarithms, we need to talk about a special number that is denoted by the letter e. The number e is a number like π. It is irrational and occurs in many formulas that describe the world around us. Like π, it can be approximated with a decimal number. Whereas π is approximately 3.1416, e is approximately 2.7183. (If you have a calculator with a key labeled $\boxed{e^x}$, you can use it to find e^1 to see a more accurate approximation to e.) We cannot give a more precise definition of the number e without using some of the topics taught in calculus. But for the work we are going to do with the number e, we only need to know that it is an irrational number that is approximately 2.7183.

Here is our definition for natural logarithms.

> **DEFINITION**
>
> A **natural logarithm** is a logarithm with a base of e. The natural logarithm of x is denoted by **ln x.** That is,
>
> $$\ln x = \log_e x$$

We can assume that all our properties of exponents and logarithms hold for expressions with a base of e, since e is a real number. Here are some examples intended to make you more familiar with the number e and natural logarithms.

EXAMPLE 7 Simplify each of the following expressions:
(a) $e^0 = 1$
(b) $e^1 = e$
(c) $\ln e = 1$ In exponential form, $e^1 = e$.
(d) $\ln 1 = 0$ In exponential form, $e^0 = 1$.
(e) $\ln e^3 = 3$
(f) $\ln e^{-4} = -4$
(g) $\ln e^t = t$

EXAMPLE 8 Use the properties of logarithms to expand the expression $\ln Ae^{5t}$.

SOLUTION Since the properties of logarithms hold for natural logarithms, we have

$$\ln Ae^{5t} = \ln A + \ln e^{5t}$$
$$= \ln A + 5t \ln e$$
$$= \ln A + 5t \qquad \text{Because } \ln e = 1 \text{ (see Example 7c).}$$

EXAMPLE 9 If $\ln 2 = 0.6931$ and $\ln 3 = 1.0986$, find:
(a) $\ln 6$ (b) $\ln 0.5$ (c) $\ln 8$

SOLUTION

(a) Since $6 = 2 \cdot 3$, we have

$$\ln 6 = \ln 2 \cdot 3$$
$$= \ln 2 + \ln 3$$
$$= 0.6931 + 1.0986$$
$$= 1.7917$$

(b) Writing 0.5 as $\frac{1}{2}$ and applying property 2 for logarithms gives us

$$\ln 0.5 = \ln \frac{1}{2}$$
$$= \ln 1 - \ln 2$$
$$= 0 - 0.6931$$
$$= -0.6931$$

(c) Writing 8 as 2^3 and applying property 3 for logarithms, we have

$$\ln 8 = \ln 2^3$$
$$= 3 \ln 2$$
$$= 3(0.6931)$$
$$= 2.0793$$

Problem Set

8.4

Common Logarithms

Find the following logarithms:

1. log 378
2. log 426
3. log 37.8
4. log 42,600
5. log 3,780
6. log 0.4260
7. log 0.0378
8. log 0.0426
9. log 37,800
10. log 4,900
11. log 600
12. log 900
13. log 2,010
14. log 10,200
15. log 0.00971
16. log 0.0312
17. log 0.0314
18. log 0.00052
19. log 0.399
20. log 0.111

Find x in the following equations:

21. $\log x = 2.8802$
22. $\log x = 4.8802$
23. $\log x = -2.1198$
24. $\log x = -3.1198$
25. $\log x = 3.1553$
26. $\log x = 5.5911$
27. $\log x = -5.3497$
28. $\log x = -1.5670$
29. $\log x = -7.0372$
30. $\log x = -4.2000$

31. $\log x = 10$
32. $\log x = -1$
33. $\log x = -10$
34. $\log x = 1$
35. $\log x = 20$
36. $\log x = -20$
37. $\log x = -2$
38. $\log x = 4$
39. $\log x = \log_2 8$
40. $\log x = \log_3 9$
41. $\log x = \log 5$
42. $\log x = \log 10$

Applying the Concepts

Measuring Acidity *In Problem Set 8.2, we indicated that the pH of a solution is defined in terms of logarithms as*

$$pH = -\log[H^+]$$

where $[H^+]$ is the concentration of hydrogen ions in that solution.

43. Find the pH of orange juice if the concentration of hydrogen ions in the juice is $[H^+] = 6.50 \times 10^{-4}$.

44. Find the pH of milk if the concentration of hydrogen ions in the milk is $[H^+] = 1.88 \times 10^{-6}$.

45. Find the concentration of hydrogen ions in a bottle of vinegar if the pH is 4.75.

46. Find the concentration of hydrogen ions in a glass of wine if the pH is 5.75.

The Richter Scale *Find the relative size, T, of the shock wave of an earthquake with each of the following magnitudes, as measured on the Richter scale.*

47. 5.5   **48.** 6.6
49. 8.3   **50.** 8.7

51. Shock Wave How much larger is the shock wave of an earthquake that measures 6.5 on the Richter scale than one that measures 5.5 on the same scale?

52. Shock Wave How much larger is the shock wave of an earthquake that measures 8.5 on the Richter scale than one that measures 5.5 on the same scale?

Depreciation *The annual rate of depreciation, r, on a car that is purchased for P dollars and is worth W dollars t years later can be found from the formula*

$$\log(1 - r) = \frac{1}{t} \log \frac{W}{P}$$

53. Find the annual rate of depreciation on a car that is purchased for $9,000 and sold 5 years later for $4,500.

54. Find the annual rate of depreciation on a car that is purchased for $9,000 and sold 4 years later for $3,000.

55. Find the annual rate of depreciation on a car that is purchased for $7,550 and sold 5 years later for $5,750.

56. Find the annual rate of depreciation on a car that is purchased for $7,550 and sold 3 years later for $5,750.

Natural Logarithms

Simplify each of the following expressions:

57. $\ln e$ **58.** $\ln 1$
59. $\ln e^5$ **60.** $\ln e^{-3}$
61. $\ln e^x$ **62.** $\ln e^y$

Use the properties of logarithms to expand each of the following expressions:

63. $\ln 10e^{3t}$ **64.** $\ln 10e^{4t}$
65. $\ln Ae^{-2t}$ **66.** $\ln Ae^{-3t}$

If $\ln 2 = 0.6931$, $\ln 3 = 1.0986$, and $\ln 5 = 1.6094$, find each of the following:

67. $\ln 15$ **68.** $\ln 10$
69. $\ln \frac{1}{3}$ **70.** $\ln \frac{1}{5}$
71. $\ln 9$ **72.** $\ln 25$
73. $\ln 16$ **74.** $\ln 81$

Review Problems

 The problems below review material we covered in Section 7.4.

Solve each system by using Cramer's rule.

75. $4x - 7y = 3$
         $5x + 2y = -3$

76. $9x - 8y = 4$
         $2x + 3y = 6$

77. $3x + 4y = 15$
         $2x - 5z = -3$
         $4y - 3z = 9$

78. $x + 3y = 5$
         $6y + z = 12$
         $x - 2z = -10$

Exponential Equations and Change of Base

Logarithms are very important in solving equations in which the variable appears as an exponent. The equation

$$5^x = 12$$

is an example of one such equation. Equations of this form are called **exponential equations.** Since the quantities 5^x and 12 are equal, so are their common logarithms. We begin our solution by taking the logarithm of both sides:

$$\log 5^x = \log 12$$

We now apply property 3 for logarithms, $\log x^r = r \log x$, to turn x from an exponent into a coefficient:

$$x \log 5 = \log 12$$

Dividing both sides by $\log 5$ gives us

$$x = \frac{\log 12}{\log 5}$$

If we want a decimal approximation to the solution, we can find $\log 12$ and $\log 5$ on a calculator and divide:

$$x = \frac{1.0792}{0.6990}$$
$$= 1.5439$$

The complete problem looks like this:

$$5^x = 12$$
$$\log 5^x = \log 12$$
$$x \log 5 = \log 12$$
$$x = \frac{\log 12}{\log 5}$$
$$= \frac{1.0792}{0.6990}$$
$$= 1.5439$$

Here is another example of solving an exponential equation using logarithms.

EXAMPLE 1 Solve for x: $25^{2x+1} = 15$.

SOLUTION Taking the logarithm of both sides and then writing the exponent $(2x + 1)$ as a coefficient, we proceed as follows:

$$25^{2x+1} = 15$$

$$\log 25^{2x+1} = \log 15 \qquad \text{Take the log of both sides.}$$

$$(2x + 1)\log 25 = \log 15 \qquad \text{Property 3}$$

$$2x + 1 = \frac{\log 15}{\log 25} \qquad \text{Divide by log 25.}$$

$$2x = \frac{\log 15}{\log 25} - 1 \qquad \text{Add } -1 \text{ to both sides.}$$

$$x = \frac{1}{2}\left(\frac{\log 15}{\log 25} - 1\right) \qquad \text{Multiply both sides by } \tfrac{1}{2}.$$

Using a calculator, we can write a decimal approximation to the answer:

$$x = \frac{1}{2}\left(\frac{1.1761}{1.3979} - 1\right)$$

$$= \frac{1}{2}(0.8413 - 1)$$

$$= \frac{1}{2}(-0.1587)$$

$$= -0.0794$$

The next two examples involve interest. You probably receive interest on money you have in the bank. First, you will need the formula for compound interest.

If you invest P dollars in an account with an annual interest rate r that is compounded n times a year, then t years later the amount of money A in that account will be

$$A = P\left(1 + \frac{r}{n}\right)^{nt}$$

EXAMPLE 2 If $5,000 is placed in an account with an annual interest rate of 12% compounded twice a year, how much money will be in the account 10 years later?

SOLUTION Substituting $P = 5,000$, $r = 0.12$, $n = 2$, and $t = 10$ into the formula above, we have

$$A = 5,000\left(1 + \frac{0.12}{2}\right)^{2\cdot10}$$

$$= 5,000(1.06)^{20}$$

To evaluate this last expression on a calculator, we use the following sequence:

$$1.06 \boxed{y^x} \ 20 \ \boxed{\times} \ 5000 \ \boxed{=}$$

We get 16,035.68 as the result.

We could also use logarithms to solve this problem. With logarithms, we take the common logarithm of each side of our last equation to obtain

$$\log A = \log[5{,}000(1.06)^{20}]$$
$$\log A = \log 5{,}000 + 20 \log 1.06$$
$$\log A = 4.2050873$$
$$A = 16{,}035.68$$

The original amount, \$5,000, will become \$16,035.68 in 10 years if invested at 12% interest compounded twice a year.

EXAMPLE 3 How long does it take for \$5,000 to double if it is deposited in an account that yields 5% interest compounded once a year?

SOLUTION Substituting $P = 5{,}000$, $r = 0.05$, $n = 1$, and $A = 10{,}000$ into our formula, we have

$$10{,}000 = 5{,}000(1 + 0.05)^t$$
$$10{,}000 = 5{,}000(1.05)^t$$
$$2 = (1.05)^t \qquad \text{Divide by 5,000.}$$

This is an exponential equation. We solve by taking the logarithm of both sides:

$$\log 2 = \log(1.05)^t$$
$$\log 2 = t \log 1.05$$

Dividing both sides by log 1.05, we have

$$t = \frac{\log 2}{\log 1.05}$$
$$= 14.2 \qquad \text{To the nearest tenth}$$

It takes a little over 14 years for \$5,000 to double if it earns 5% interest per year, compounded once a year.

There is a fourth property of logarithms we have not yet considered. This last property allows us to change from one base to another and is therefore called the **change-of-base property.**

PROPERTY 4: CHANGE OF BASE

If a and b are both positive numbers other than 1, and if $x > 0$, then

$$\log_a x = \frac{\log_b x}{\log_b a}$$

Base a \qquad Base b

The logarithm on the left side has a base of a, while both logarithms on the right side have a base of b. This allows us to change from base a to any other base b that is a positive number other than 1. Here is a proof of property 4 for logarithms.

Proof
We begin by writing the identity

$$a^{\log_a x} = x$$

Taking the logarithm base b of both sides and writing the exponent $\log_a x$ as a coefficient, we have

$$\log_b a^{\log_a x} = \log_b x$$
$$(\log_a x)\log_b a = \log_b x$$

Dividing both sides by $\log_b a$, we have the desired result:

$$\frac{(\log_a x)\log_b a}{\log_b a} = \frac{\log_b x}{\log_b a}$$
$$\log_a x = \frac{\log_b x}{\log_b a}$$

We can use this property to find logarithms we could not otherwise compute on our calculators — that is, logarithms with bases other than 10 or e. The next example illustrates the use of this property.

EXAMPLE 4 Find $\log_8 24$.

SOLUTION Since we do not have base-8 logarithms on our calculators, we can change this expression to an equivalent expression that contains only base-10 logarithms:

$$\log_8 24 = \frac{\log 24}{\log 8} \qquad \text{Property 4}$$

Don't be confused. We did not just drop the base, we changed to base 10. We could have written the last line like this:

$$\log_8 24 = \frac{\log_{10} 24}{\log_{10} 8}$$

From our calculators, we write

$$\log_8 24 = \frac{1.3802}{0.9031}$$
$$= 1.5283$$

Here is the complete calculator solution:

SCIENTIFIC CALCULATOR

24 $\boxed{\log}$ $\boxed{\div}$ 8 $\boxed{\log}$ $\boxed{=}$

GRAPHING CALCULATOR

$\boxed{\text{LOG}}$ 24 $\boxed{\div}$ $\boxed{\text{LOG}}$ 8 $\boxed{\text{ENTER}}$

EXAMPLE 5

Suppose the population in a small city is 32,000 in the beginning of 1994 and the city council assumes that the population size t years later can be estimated by the equation

$$P = 32,000e^{0.05t}$$

Approximately when will the city have a population of 50,000?

SOLUTION   We substitute 50,000 for P in the equation and solve for t:

$$50,000 = 32,000e^{0.05t}$$
$$1.56 = e^{0.05t} \qquad \frac{50,000}{32,000} \text{ is approximately } 1.56.$$

To solve this equation for t, we can take the natural logarithm of each side:

$$\ln 1.56 = \ln e^{0.05t}$$
$$\ln 1.56 = 0.05t \ln e \qquad \text{Property 3 for logarithms}$$
$$\ln 1.56 = 0.05t \qquad \text{Because } \ln e = 1$$
$$t = \frac{\ln 1.56}{0.05} \qquad \text{Divide each side by 0.05.}$$
$$= \frac{0.4447}{0.05}$$
$$= 8.89 \text{ years}$$

We can estimate that the population will reach 50,000 toward the end of 2002.

Problem Set
8.5

Solve each exponential equation. Use a calculator to write the answer in decimal form.

1. $3^x = 5$

2. $4^x = 3$

3. $5^x = 3$

4. $3^x = 4$

5. $5^{-x} = 12$

6. $7^{-x} = 8$

7. $12^{-x} = 5$

8. $8^{-x} = 7$

9. $8^{x+1} = 4$

10. $9^{x+1} = 3$

11. $4^{x-1} = 4$

12. $3^{x-1} = 9$

13. $3^{2x+1} = 2$

14. $2^{2x+1} = 3$

15. $3^{1-2x} = 2$

16. $2^{1-2x} = 3$

17. $15^{3x-4} = 10$

18. $10^{3x-4} = 15$

19. $6^{5-2x} = 4$

20. $9^{7-3x} = 5$

Applying the Concepts

21. **Compound Interest** If $5,000 is placed in an account with an annual interest rate of 12% compounded once a year, how much money will be in the account 10 years later?

22. **Compound Interest** If $5,000 is placed in an account with an annual interest rate of 12% compounded four times a year, how much money will be in the account 10 years later?

23. **Compound Interest** If $200 is placed in an account with an annual interest rate of 8% compounded twice a year, how much money will be in the account 10 years later?

24. **Compound Interest** If $200 is placed in an account with an annual interest rate of 8% compounded once a year, how much money will be in the account 10 years later?

25. **Compound Interest** How long will it take for $500 to double if it is invested at 6% annual interest compounded twice a year?

26. **Compound Interest** How long will it take for $500 to double if it is invested at 6% annual interest compounded twelve times a year?

27. **Compound Interest** How long will it take for $1,000 to triple if it is invested at 12% annual interest compounded six times a year?

28. **Compound Interest** How long will it take for $1,000 to become $4,000 if it is invested at 12% annual interest compounded six times a year?

Change of Base

Use the change-of-base property and a calculator to find a decimal approximation to each of the following logarithms.

29. $\log_8 16$
30. $\log_9 27$
31. $\log_{16} 8$
32. $\log_{27} 9$
33. $\log_7 15$
34. $\log_3 12$
35. $\log_{15} 7$
36. $\log_{12} 3$
37. $\log_8 240$
38. $\log_6 180$
39. $\log_4 321$
40. $\log_5 462$

Natural Logarithms

Find a decimal approximation to each of the following natural logarithms.

41. $\ln 345$
42. $\ln 3,450$
43. $\ln 0.345$
44. $\ln 0.0345$
45. $\ln 10$
46. $\ln 100$
47. $\ln 45,000$
48. $\ln 450,000$

Applying the Concepts

49. **Population Growth** Suppose the population in a small city is 32,000 at the beginning of 1994 and the city council assumes that the population size t years later can be estimated by the equation

$$P = 32,000e^{0.05t}$$

Approximately when will the city have a population of 64,000?

50. **Population Growth** Suppose the population of a city is given by the equation

$$P = 100,000e^{0.05t}$$

where t is the number of years from the present time. How large is the population now? ("Now" corresponds to a certain value of t. Once you realize what that value of t is, the problem becomes very simple.)

51. **Population Growth** Suppose the population of a city is given by the equation

$$P = 15,000e^{0.04t}$$

where t is the number of years from the present time. How long will it take for the population to reach 45,000?

52. **Population Growth** Suppose the population of a city is given by the equation

$$P = 15,000e^{0.08t}$$

where t is the number of years from the present time. How long will it take for the population to reach 45,000?

53. Solve the formula $A = Pe^{rt}$ for t.
54. Solve the formula $A = Pe^{-rt}$ for t.
55. Solve the formula $A = P2^{-kt}$ for t.
56. Solve the formula $A = P2^{kt}$ for t.
57. Solve the formula $A = P(1 - r)^t$ for t.
58. Solve the formula $A = P(1 + r)^t$ for t.

Review Problems

NOTE: The problems below review material from Section 7.5. They are taken from the book *Algebra for the Practical Man,* written by J. E. Thompson and published by D. Van Nostrand Company in 1931.

59. A man spent $112.80 for 108 geese and ducks, each goose costing 14 dimes and each duck 6 dimes. How many of each did he buy?

60. If 15 lb of tea and 10 lb of coffee together cost $15.50, while 25 lb of tea and 13 lb of coffee at the same prices cost $24.55, find the price per pound of each.

61. A number of oranges at the rate of three for ten cents and apples at fifteen cents a dozen cost, together, $6.80. Five times as many oranges and one-fourth as many apples at the same rates would have cost $25.45. How many of each were bought?

62. An estate is divided among three persons: *A*, *B*, and *C*. *A*'s share is three times that of *B* and *B*'s share is twice that of *C*. If *A* receives $9,000 more than *C*, how much does each receive?

REVIEW FOR CHAPTER 8

Examples

1. For the exponential function $f(x) = 2^x$,

$f(0) = 2^0 = 1$

$f(1) = 2^1 = 2$

$f(2) = 2^2 = 4$

$f(3) = 2^3 = 8$

Chapter 8 Summary

Exponential Functions [8.1]

Any function of the form

$$f(x) = b^x$$

where $b > 0$ and $b \neq 1$, is an **exponential function.**

2. The function $f(x) = x^2$ is not one-to-one because 9, which is in the range, comes from both 3 and -3 in the domain.

One-to-One Functions [8.1]

A function is a **one-to-one function** if every element in the range comes from exactly one element in the domain.

3. The inverse of $f(x) = 2x - 3$ is $f^{-1}(x) = \dfrac{x + 3}{2}$.

Inverse Functions [8.1]

The **inverse** of a function is obtained by reversing the order of the coordinates of the ordered pairs belonging to the function. Only one-to-one functions have inverses that are also functions.

4. The definition allows us to write expressions like

$$y = \log_3 27$$

equivalently in exponential form as

$$3^y = 27$$

which makes it apparent that y is 3.

Definition of Logarithms [8.2]

If b is a positive number not equal to 1, then the expression

$$y = \log_b x$$

is equivalent to $x = b^y$. That is, in the expression $y = \log_b x$, y is the number to which we raise b in order to get x. Expressions written in the form $y = \log_b x$ are said to be in *logarithmic form*. Expressions like $x = b^y$ are in *exponential form*.

5. Examples of the two special identities are

$$5^{\log_5 12} = 12$$

and

$$\log_8 8^3 = 3$$

Two Special Identities [8.2]

For $b > 0$, $b \neq 1$, the following two expressions hold for all positive real numbers x:

(1) $b^{\log_b x} = x$

(2) $\log_b b^x = x$

6. We can rewrite the expression

$$\log_{10} \frac{45^6}{273}$$

using the properties of logarithms, as

$$6 \log_{10} 45 - \log_{10} 273$$

7.
$$\log_{10} 10{,}000 = \log 10{,}000$$
$$= \log 10^4$$
$$= 4$$

8. $\ln e = 1$
   $\ln 1 = 0$

9. $\log_6 475 = \dfrac{\log 475}{\log 6}$

$$= \frac{2.6767}{0.7782}$$

$$= 3.44$$

Properties of Logarithms [8.3]

If x, y, and b are positive real numbers, $b \neq 1$, and r is any real number, then:

1. $\log_b(xy) = \log_b x + \log_b y$

2. $\log_b \left(\dfrac{x}{y} \right) = \log_b x - \log_b y$

3. $\log_b x^r = r \log_b x$

Common Logarithms [8.4]

Common logarithms are logarithms with a base of 10. To save time in writing, we omit the base when working with common logarithms. That is,

$$\log x = \log_{10} x$$

Natural Logarithms [8.4]

Natural logarithms, written **ln x,** are logarithms with a base of e, where the number e is an irrational number (like the number π). A decimal approximation for e is 2.7183. All the properties of exponents and logarithms hold when the base is e.

Change of Base [8.5]

If x, a, and b are positive real numbers, $a \neq 1$ and $b \neq 1$, then

$$\log_a x = \frac{\log_b x}{\log_b a}$$

Common Mistakes

The most common mistakes that occur with logarithms come from trying to apply the three properties of logarithms to situations in which they don't apply. For example, a very common mistake looks like this:

$$\frac{\log 3}{\log 2} = \log 3 - \log 2 \quad \text{Mistake}$$

This is not a property of logarithms. In order to write the expression $\log 3 - \log 2$, we would have to start with

$$\log \frac{3}{2} \quad NOT \quad \frac{\log 3}{\log 2}$$

There is a difference.

REVIEW FOR CHAPTER 8

Chapter 8 Review Problems

Let $f(x) = 2^x$ and $g(x) = (\frac{1}{3})^x$, and find the following. [8.1]

1. $f(4)$
2. $f(-1)$
3. $g(2)$
4. $f(2) - g(-2)$
5. $f(-1) + g(1)$
6. $g(-1) + f(2)$
7. The graph of $y = f(x)$
8. The graph of $y = g(x)$

For each relation that follows, sketch the graph of the relation and its inverse, and write an equation for the inverse. [8.1]

9. $y = 2x + 1$
10. $y = x^2 - 4$

For each function below, find the equation of the inverse. Write the inverse using the notation $f^{-1}(x)$ if the inverse is itself a function. [8.1]

11. $f(x) = 2x + 3$
12. $f(x) = x^2 - 1$
13. $f(x) = \frac{1}{2}x + 2$
14. $f(x) = 4 - 2x^2$

Write each expression in logarithmic form. [8.2]

15. $3^4 = 81$
16. $7^2 = 49$
17. $0.01 = 10^{-2}$
18. $2^{-3} = \frac{1}{8}$

Write each expression in exponential form. [8.2]

19. $\log_2 8 = 3$
20. $\log_3 9 = 2$
21. $\log_4 2 = \frac{1}{2}$
22. $\log_4 4 = 1$

Solve for x. [8.2]

23. $\log_5 x = 2$
24. $\log_3 x = 3$
25. $\log_{16} 8 = x$
26. $\log_9 27 = x$
27. $\log_x 0.01 = -2$
28. $\log_x 0.1 = -1$

Graph each equation. [8.2]

29. $y = \log_2 x$
30. $y = \log_{1/2} x$

Simplify each expression. [8.2]

31. $\log_4 16$
32. $\log_3 81$
33. $\log_{27} 9$
34. $\log_8 16$
35. $\log_4(\log_3 3)$
36. $\log_5[\log_2(\log_3 9)]$

Use the properties of logarithms to expand each expression. [8.3]

37. $\log_2 5x$
38. $\log_3 4x$

39. $\log_{10} \dfrac{2x}{y}$
40. $\log_3 x^2 y^4$

41. $\log_a \dfrac{y^3 \sqrt{x}}{z}$
42. $\log_{10} \dfrac{x^2}{y^3 z^4}$

Write each expression as a single logarithm. [8.3]

43. $\log_2 x + \log_2 y$
44. $\log_2 5 + \log_2 x$
45. $\log_3 x - \log_3 4$
46. $2 \log_{10} x + 3 \log_{10} y$
47. $2 \log_a 5 - \frac{1}{2} \log_a 9$
48. $3 \log_2 x + 2 \log_2 y - 4 \log_2 z$

Solve each equation. [8.3]

49. $\log_2 x + \log_2 4 = 3$
50. $\log_2 x - \log_2 3 = 1$
51. $\log_3 x + \log_3(x - 2) = 1$
52. $\log_2 x + \log_2(x - 2) = 3$
53. $\log_4(x + 1) - \log_4(x - 2) = 1$
54. $\log_3(x - 3) - \log_3(x + 2) = 1$
55. $\log_6(x - 1) + \log_6 x = 1$
56. $\log_4(x - 3) + \log_4 x = 1$

Evaluate each expression. [8.4]

57. $\log 346$
58. $\log 3,460$
59. $\log 0.713$
60. $\log 0.00713$

Find x. [8.4]

61. $\log x = 3.9652$
62. $\log x = 5.9652$
63. $\log x = -1.6003$
64. $\log x = -2.6003$

Simplify. [8.4]

65. $\ln e$
66. $\ln 1$
67. $\ln e^2$
68. $\ln e^{-4}$

If $\ln 3 = 1.0986$ and $\ln 7 = 1.9459$, find each of the following. [8.4]

69. $\ln 21$
70. $\ln 27$
71. $\ln 9$
72. $\ln 49$
73. $\ln 343$
74. $\ln \frac{1}{3}$

Use the formula pH $= -\log[H^+]$ to find the pH of a solution with the given hydrogen ion concentration. [8.4]

75. $[H^+] = 7.9 \times 10^{-3}$
76. $[H^+] = 8.1 \times 10^{-6}$

Find $[H^+]$ for a solution with the given pH. [8.4]

77. pH $= 2.7$ **78.** pH $= 7.5$

Solve each equation. [8.5]

79. $4^x = 8$ **80.** $8^{x+2} = 4$
81. $4^{3x+2} = 5$ **82.** $10^{5x+4} = 20$

Use the change-of-base property and a calculator to evaluate each expression. Round your answers to the nearest hundredth. [8.5]

83. $\log_{16} 8$ **84.** $\log_3 24$
85. $\log_{12} 421$ **86.** $\log_8 320$

Use the formula $A = P\left(1 + \dfrac{r}{n}\right)^{nt}$ to solve each problem below. [8.5]

87. How much money is in an account after 10 years if $5,000 was deposited originally at 12% compounded annually?

88. A $10,000 T-bill earns 14% compounded twice a year. How much is the T-bill worth after 4 years?

89. How long does it take $5,000 to double if it is deposited in an account that pays 16% annual interest compounded once a year?

90. How long does it take $10,000 to triple if it is deposited in an account that pays 12% annual interest compounded 6 times a year?

Chapter 8 Test

Graph each exponential function. [8.1]

1. $f(x) = 2^x$ **2.** $g(x) = 3^{-x}$

Sketch the graph of each function and its inverse. Find $f^{-1}(x)$ for Problem 3. [8.1]

3. $f(x) = 2x - 3$ **4.** $f(x) = x^2 - 4$

Solve for x. [8.2]

5. $\log_4 x = 3$ **6.** $\log_x 5 = 2$

Graph each of the following. [8.2]

7. $y = \log_2 x$ **8.** $y = \log_{1/2} x$

Evaluate each of the following. [8.2, 8.4, 8.5]

9. $\log_8 4$ **10.** $\log_7 21$
11. $\log 23{,}400$ **12.** $\log 0.0123$
13. $\ln 46.2$ **14.** $\ln 0.0462$

Use the properties of logarithms to expand each expression. [8.3]

15. $\log_2 \dfrac{8x^2}{y}$ **16.** $\log \dfrac{\sqrt{x}}{(y^4)\sqrt[5]{z}}$

Write each expression as a single logarithm. [8.3]

17. $2 \log_3 x - \frac{1}{2} \log_3 y$
18. $\frac{1}{3} \log x - \log y - 2 \log z$

Use a calculator to find x. [8.4]

19. $\log x = 4.8476$ **20.** $\log x = -2.6478$

Solve for x. [8.3, 8.5]

21. $5 = 3^x$ **22.** $4^{2x-1} = 8$
23. $\log_5 x - \log_5 3 = 1$
24. $\log_2 x + \log_2(x - 7) = 3$

25. Find the pH of a solution in which $[H^+] = 6.6 \times 10^{-7}$. [8.4]

26. If $400 is deposited in an account that earns 10% annual interest compounded twice a year, how much money will be in the account after 5 years? [8.5]

27. How long will it take $600 to become $1,800 if the $600 is deposited in an account that earns 8% annual interest compounded four times a year? [8.5]

Sequences and Series

Contents

Introduction

Much of what we do in mathematics is concerned with recognizing patterns and classifying groups of numbers that share a common characteristic. The sequence below is called the **Fibonacci sequence** after the mathematician Leonardo Fibonacci.

$$1, 1, 2, 3, 5, 8, \ldots$$

Can you recognize the pattern? Can you predict what the next term will be? If you see that each term is the sum of the two previous terms, then you know that the next term is 13. It may surprise you to know that the Fibonacci sequence can be used to model many of the things we see in the world around us.

For example, the Fibonacci sequence gives us the number of bees in each generation of the family tree of the male honeybee. A male honeybee has one parent, its mother, while a female honeybee has two parents, a mother and a father. (A male honeybee comes from an unfertilized egg; a female honeybee comes from a fertilized egg.) Using these facts, we construct the family tree of a male honeybee using M to represent a male honeybee and F to represent a female honeybee (Figure 1).

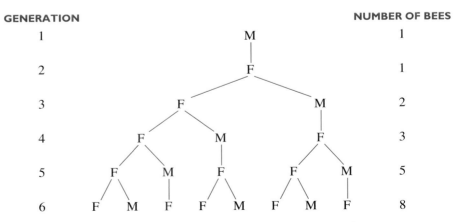

FIGURE I
Family tree of a male honeybee

Looking at the numbers in the right column in Figure 1, the sequence that gives us the number of bees in each generation of the family tree of a male honeybee is

1   1   2   3   5   8

As you can see, this is the Fibonacci sequence. We have taken our original diagram (the family tree of the male honeybee) and reduced it to a *mathematical model* (the Fibonacci sequence). The model can be used in place of the diagram to find the number of bees in any generation.

Overview

We begin this chapter with a look at **sequences** in general. Then we look at what happens when we add the terms of a sequence to give us what is called a **series.** From there we look at two special types of sequences, **arithmetic sequences** and **geometric sequences,** and their associated series. We end the chapter with a look at **binomial expansions,** which give us a way to expand expressions such as $(x + 3)^6$ without actually doing any multiplication.

Sequences

Many of the sequences in this chapter will be familiar to you on an intuitive level because you have worked with them for some time. Here are some of those sequences:

The sequence of odd numbers:

1, 3, 5, 7, . . .

The sequence of even numbers:

2, 4, 6, 8, . . .

The sequence of squares:

$1^2, 2^2, 3^2, 4^2, . . . = 1, 4, 9, 16, . . .$

The numbers in each of these sequences can be found from the formulas that define functions. For example, the sequence of even numbers can be found from the function

$$f(x) = 2x$$

by finding $f(1), f(2), f(3), f(4)$, and so forth. This gives us justification for the formal definition of a sequence.

> **DEFINITION**
>
> A **sequence** is a function whose domain is the set of positive integers $\{1, 2, 3, 4, . . .\}$.

As you can see, sequences are simply functions with a specific domain. If we want to form a sequence from the function $f(x) = 3x + 5$, we simply find $f(1), f(2), f(3)$, and so on. Doing so gives us the sequence

$$8, 11, 14, 17, \ldots$$

because $f(1) = 3(1) + 5 = 8$, $f(2) = 3(2) + 5 = 11$, $f(3) = 3(3) + 5 = 14$, and $f(4) = 3(4) + 5 = 17$.

Notation: Since the domain for a sequence is always the set $\{1, 2, 3, \ldots\}$, we can simplify the notation we use to represent the terms of a sequence. Using the letter a instead of f, and subscripts instead of numbers enclosed by parentheses, we can represent the sequence from the discussion above as follows:

$$a_n = 3n + 5$$

- Instead of $f(1)$, we write a_1 for the *first term* of the sequence.
- Instead of $f(2)$, we write a_2 for the *second term* of the sequence.
- Instead of $f(3)$, we write a_3 for the *third term* of the sequence.
- Instead of $f(4)$, we write a_4 for the *fourth term* of the sequence.
- Instead of $f(n)$, we write a_n for the *nth term* of the sequence.

The **nth term** is also called the **general term** of the sequence. The general term is used to define the other terms of the sequence. That is, if we are given the formula for the general term, a_n, we can find any other term in the sequence. The following examples illustrate.

EXAMPLE 1 Find the first four terms of the sequence whose general term is given by $a_n = 2n - 1$.

SOLUTION The subscript notation a_n works the same way function notation works. To find the first, second, third, and fourth terms of this sequence, we simply substitute 1, 2, 3, and 4 for n in the formula $2n - 1$:

If: The general term is $a_n = 2n - 1$
Then: The first term is $a_1 = 2(1) - 1 = 1$
The second term is $a_2 = 2(2) - 1 = 3$
The third term is $a_3 = 2(3) - 1 = 5$
The fourth term is $a_4 = 2(4) - 1 = 7$

The first four terms of this sequence are the odd numbers 1, 3, 5, and 7. The whole sequence can be written as

$$1, 3, 5, \ldots, 2n - 1, \ldots$$

Since each term in the sequence in Example 1 is larger than the preceding term, we say the sequence is an **increasing sequence.**

EXAMPLE 2 Write the first four terms of the sequence defined by

$$a_n = \frac{1}{n + 1}$$

SOLUTION Replacing n with 1, 2, 3, and 4, we have the first four terms:

$$\text{First term} = a_1 = \frac{1}{1 + 1} = \frac{1}{2}$$

$$\text{Second term} = a_2 = \frac{1}{2 + 1} = \frac{1}{3}$$

$$\text{Third term} = a_3 = \frac{1}{3 + 1} = \frac{1}{4}$$

$$\text{Fourth term} = a_4 = \frac{1}{4 + 1} = \frac{1}{5}$$

The sequence defined by $a_n = \frac{1}{n + 1}$ can be written as

$$\frac{1}{2}, \frac{1}{3}, \frac{1}{4}, \ldots, \frac{1}{n + 1}, \ldots$$

Since each term in the sequence in Example 2 is smaller than the term preceding it, the sequence is said to be a **decreasing sequence.**

EXAMPLE 3 Find the fifth and sixth terms of the sequence whose general term is given by

$$a_n = \frac{(-1)^n}{n^2}$$

SOLUTION For the fifth term, we replace n with 5. For the sixth term, we replace n with 6:

$$\text{Fifth term} = a_5 = \frac{(-1)^5}{5^2} = \frac{-1}{25}$$

$$\text{Sixth term} = a_6 = \frac{(-1)^6}{6^2} = \frac{1}{36}$$

The sequence in Example 3 can be written as

$$-1, \frac{1}{4}, -\frac{1}{9}, \frac{1}{16}, \ldots, \frac{(-1)^n}{n^2}, \ldots$$

Since the terms alternate in sign—if one term is positive, then the next term is negative—we call this an **alternating sequence.**

The first three examples illustrate how we work with a sequence in which we are given a formula for the general term. Now we'll see that there is another way to write the general term.

Recursion Formulas

Let's go back to one of the first sequences we looked at in this section:

$$8, 11, 14, 17, \ldots$$

Each term in the sequence can be found by simply substituting positive integers for n in the formula $a_n = 3n + 5$. However, another way to look at this sequence is to notice that each term can be found by adding 3 to the preceding term; so, we could give all the terms of this sequence by simply saying

Start with 8, then add 3 to each term to get the next term.

The same idea, expressed in symbols, looks like this:

$$a_1 = 8 \qquad \text{and} \qquad a_n = a_{n-1} + 3 \qquad \text{for } n > 1$$

This formula is called a **recursion formula** because each term is written *recursively* using the term (or terms) that precede it.

EXAMPLE 4   Write the first four terms of the sequence given recursively by

$$a_1 = 4 \qquad \text{and} \qquad a_n = 5a_{n-1} \qquad \text{for } n > 1$$

SOLUTION   The formula tells us to start the sequence with the number 4, and then multiply each term by 5 to get the next term. Therefore,

$$a_1 = 4$$
$$a_2 = 5a_1 = 5(4) = 20$$
$$a_3 = 5a_2 = 5(20) = 100$$
$$a_4 = 5a_3 = 5(100) = 500$$

The sequence is 4, 20, 100, 500, ...

Finding the General Term

In the first four examples, we found some terms of a sequence after being given the general term. In the next two examples, we will do the reverse. That is, given some terms of a sequence, we will find the formula for the general term.

EXAMPLE 5   Find a formula for the nth term of the sequence 2, 8, 18, 32, ...

SOLUTION   Solving a problem like this involves some guessing. Looking over the first four terms, we see that each is twice a perfect square:

$$2 = 2(1)$$
$$8 = 2(4)$$
$$18 = 2(9)$$
$$32 = 2(16)$$

If we write each square with an exponent of 2, the formula for the nth term becomes obvious:

$$a_1 = 2 = 2(1)^2$$
$$a_2 = 8 = 2(2)^2$$
$$a_3 = 18 = 2(3)^2$$
$$a_4 = 32 = 2(4)^2$$
$$\cdot$$
$$\cdot$$
$$\cdot$$
$$a_n = \quad 2(n)^2 = 2n^2$$

The general term of the sequence 2, 8, 18, 32, . . . is $a_n = 2n^2$.

EXAMPLE 6   Find the general term for the sequence $2, \frac{3}{8}, \frac{4}{27}, \frac{5}{64}, \ . \ . \ . \ .$

SOLUTION   The first term can be written as $\frac{2}{1}$. The denominators are all perfect cubes. The numerators are all 1 more than the base of the cubes in the denominators:

$$a_1 = \frac{2}{1} = \frac{1 + 1}{1^3}$$

$$a_2 = \frac{3}{8} = \frac{2 + 1}{2^3}$$

$$a_3 = \frac{4}{27} = \frac{3 + 1}{3^3}$$

$$a_4 = \frac{5}{64} = \frac{4 + 1}{4^3}$$

Observing this pattern, we recognize the general term to be

$$a_n = \frac{n + 1}{n^3}$$

Note: Finding the nth term of a sequence from the first few terms is not always automatic. That is, it sometimes takes a while to recognize the pattern. Don't be afraid to guess at the formula for the general term. Many times an incorrect guess leads to the correct formula.

Problem Set

9.1

Write the first five terms of the sequences with the following general terms.

1. $a_n = 3n + 1$

2. $a_n = 2n + 3$

3. $a_n = 4n - 1$

4. $a_n = n + 4$

5. $a_n = n$

6. $a_n = -n$

7. $a_n = n^2 + 3$

8. $a_n = n^3 + 1$

9. $a_n = \dfrac{n}{n + 3}$

10. $a_n = \dfrac{n}{n + 2}$

11. $a_n = \dfrac{n + 1}{n + 2}$

12. $a_n = \dfrac{n + 3}{n + 4}$

13. $a_n = \dfrac{1}{n^2}$

14. $a_n = \dfrac{1}{n^3}$

15. $a_n = 2^n$

16. $a_n = 3^n$

17. $a_n = 3^{-n}$

18. $a_n = 2^{-n}$

19. $a_n = 1 + \dfrac{1}{n}$

20. $a_n = 1 - \dfrac{1}{n}$

21. $a_n = n - \dfrac{1}{n}$

22. $a_n = n + \dfrac{1}{n}$

23. $a_n = (-2)^n$

24. $a_n = (-3)^n$

Write the first five terms of the sequences defined by the following recursion formulas.

25. $a_1 = 3$ $a_n = -3a_{n-1}$ $n > 1$

26. $a_1 = -3$ $a_n = 3a_{n-1}$ $n > 1$

27. $a_1 = 3$ $a_n = a_{n-1} - 3$ $n > 1$

28. $a_1 = -3$ $a_n = a_{n-1} - 3$ $n > 1$

29. $a_1 = 1$ $a_n = 2a_{n-1} + 3$ $n > 1$

30. $a_1 = 1$ $a_n = 3a_{n-1} + 2$ $n > 1$

31. $a_1 = 1$ $a_n = a_{n-1} + n$ $n > 1$

32. $a_1 = 2$ $a_n = a_{n-1} - n$ $n > 1$

Determine the general term for each of the following sequences.

33. $2, 3, 4, 5, \ldots$

34. $3, 6, 9, 12, \ldots$

35. $4, 8, 12, 16, 20, \ldots$

36. $3, 4, 5, 6, \ldots$

37. $7, 10, 13, 16, \ldots$

38. $4, 9, 14, 19, \ldots$

39. $1, 4, 9, 16, \ldots$

40. $1, 8, 27, 64, \ldots$

41. $3, 12, 27, 48, \ldots$

42. $2, 16, 54, 128, \ldots$

43. $4, 8, 16, 32, \ldots$

44. $3, 9, 27, 81, \ldots$

45. $-2, 4, -8, 16, \ldots$

46. $-3, 9, -27, 81, \ldots$

47. $\frac{1}{4}, \frac{1}{8}, \frac{1}{16}, \frac{1}{32}, \ldots$

48. $\frac{1}{3}, \frac{1}{9}, \frac{1}{27}, \frac{1}{81}, \ldots$

49. $\frac{1}{4}, \frac{2}{9}, \frac{3}{16}, \frac{4}{25}, \ldots$

50. $\frac{1}{4}, \frac{2}{10}, \frac{3}{28}, \frac{4}{82}, \ldots$

Review Problems

 The problems that follow review material we covered in Section 8.2.

Write each expression in logarithmic form.

51. $100 = 10^2$

52. $4^{3/2} = 8$

Write each expression in exponential form.

53. $\log_3 81 = 4$

54. $-2 = \log_{10} 0.01$

Find x in each of the following.

55. $\log_9 x = \frac{3}{2}$

56. $\log_x \frac{1}{4} = -2$

Simplify each expression.

57. $\log_2 32$

58. $\log_{10} 10{,}000$

59. $\log_3(\log_2 8)$

60. $\log_5(\log_6 6)$

One Step Further

61. As n increases, the terms in the sequence $a_n = \left(1 + \dfrac{1}{n}\right)^n$ get closer and closer to the number e (that's the same e we used in defining natural logarithms). However, it takes some fairly large values of n before we can see this happening. Use a calculator to find $a_{100}, a_{1,000}, a_{10,000}$, and $a_{100,000}$, and compare the values you get to the decimal approximation we gave for the number e.

498

62. The sequence $a_n = \left(1 + \dfrac{1}{n}\right)^{-n}$ gets close to the number $\dfrac{1}{e}$ as n becomes large. Use a calculator to find approximations for a_{100} and $a_{1,000}$, and then compare them to $\dfrac{1}{2.7183}$.

63. Write the first 10 terms of the sequence defined by the recursion formula

$$a_1 = 1, a_2 = 1, a_n = a_{n-1} + a_{n-2} \qquad n > 2$$

64. Write the first 10 terms of the sequence defined by the recursion formula

$$a_1 = 2, a_2 = 2, a_n = a_{n-1} + a_{n-2} \qquad n > 2$$

65. Simplify each complex fraction in the sequence below and then compare this sequence with the sequence you wrote in Problem 63.

$$1 + \dfrac{1}{1+1}, 1 + \dfrac{1}{1+\dfrac{1}{1+1}}, 1 + \dfrac{1}{1+\dfrac{1}{1+\dfrac{1}{1+1}}}, \ldots$$

66. Write the first five terms of the sequence given by the recursion formula below. Compare this sequence with the sequence in Problem 65.

$$a_1 = \dfrac{3}{2}, a_n = 1 + \dfrac{1}{a_{n-1}} \qquad n > 1$$

Series

There is an interesting relationship between the sequence of odd numbers and the sequence of squares that is found by adding the terms in the sequence of odd numbers:

$$
\begin{aligned}
1 &= 1 \\
1 + 3 &= 4 \\
1 + 3 + 5 &= 9 \\
1 + 3 + 5 + 7 &= 16
\end{aligned}
$$

When we add the terms of a sequence the result is called a **series.**

> **DEFINITION**
>
> The sum of a number of terms in a sequence is called a **series.**

A sequence can be finite or infinite, depending on whether the sequence ends at the nth term. For example,

$$1, 3, 5, 7, 9$$

is a finite sequence, while

$$1, 3, 5, \ldots$$

is an infinite sequence. Associated with each of the above sequences is a series found by adding the terms of the sequence:

$$1 + 3 + 5 + 7 + 9 \quad \text{Finite series}$$

$$1 + 3 + 5 + \cdots \quad \text{Infinite series}$$

In this section we will consider only finite series. We can introduce a new kind of notation here that is a compact way of indicating a finite series. The notation is called **summation notation,** or **sigma notation** (because it is written using a stylized version of the Greek letter sigma). The expression

$$\sum_{i=1}^{4}(8i - 10)$$

is an example of how we use summation notation. The summation notation in this expression indicates the sum of all the expressions with the formula $8i - 10$ from $i = 1$ up to and including $i = 4$. That is,

$$\sum_{i=1}^{4}(8i - 10) = (8 \cdot 1 - 10) + (8 \cdot 2 - 10) + (8 \cdot 3 - 10) + (8 \cdot 4 - 10)$$

$$= -2 + 6 + 14 + 22$$
$$= 40$$

The letter i used here is called the **index of summation,** or just **index** for short.

Here are some examples illustrating the use of summation notation.

EXAMPLE 1 Expand and simplify: $\sum_{i=1}^{5}(i^2 - 1)$.

SOLUTION We replace i in the expression $i^2 - 1$ with all consecutive integers from 1 up to 5, including 1 and 5:

$$\sum_{i=1}^{5}(i^2 - 1) = (1^2 - 1) + (2^2 - 1) + (3^2 - 1) + (4^2 - 1) + (5^2 - 1)$$

$$= 0 + 3 + 8 + 15 + 24$$
$$= 50$$

EXAMPLE 2 Expand and simplify: $\sum_{i=3}^{6}(-2)^i$.

SOLUTION We replace i in the expression $(-2)^i$ with the consecutive integers beginning at 3 and ending at 6:

$$\sum_{i=3}^{6}(-2)^i = (-2)^3 + (-2)^4 + (-2)^5 + (-2)^6$$

$$= -8 + 16 + (-32) + 64$$
$$= 40$$

EXAMPLE 3   Expand: $\displaystyle\sum_{i=2}^{5} (x^i - 3)$.

SOLUTION   We must be careful not to confuse the letter x with i. The index i is the quantity we replace by the consecutive integers from 2 to 5, not x:

$$\sum_{i=2}^{5}(x^i - 3) = (x^2 - 3) + (x^3 - 3) + (x^4 - 3) + (x^5 - 3)$$

In the first three examples, we were given an expression with summation notation and asked to expand it. The next examples illustrate how we can rewrite an expression given in expanded form as an expression using summation notation.

EXAMPLE 4   Write with summation notation: $1 + 3 + 5 + 7 + 9$.

SOLUTION   A formula that gives us the terms of this sum is

$$a_i = 2i - 1$$

where i ranges from 1 up to and including 5. Notice that we are using the subscript i in exactly the same way we used the subscript n in the last section—to indicate the general term. Writing the sum

$$1 + 3 + 5 + 7 + 9$$

with summation notation looks like this:

$$\sum_{i=1}^{5}(2i - 1)$$

EXAMPLE 5   Write with summation notation: $3 + 12 + 27 + 48$.

SOLUTION   We need a formula in terms of i that will give each term in the sum. Writing the sum as

$$3 \cdot 1^2 + 3 \cdot 2^2 + 3 \cdot 3^2 + 3 \cdot 4^2$$

we see the formula

$$a_i = 3 \cdot i^2$$

where i ranges from 1 up to and including 4. Using this formula and summation notation, we can represent the sum

$$3 + 12 + 27 + 48$$

as

$$\sum_{i=1}^{4} 3i^2$$

EXAMPLE 6   Write with summation notation:

$$\frac{x+3}{x^3} + \frac{x+4}{x^4} + \frac{x+5}{x^5} + \frac{x+6}{x^6}$$

SOLUTION   A formula that gives each of these terms is

$$a_i = \frac{x+i}{x^i}$$

where i assumes all integer values between 3 and 6, including 3 and 6. The sum can be written as

$$\sum_{i=3}^{6} \frac{x+i}{x^i}$$

Problem Set

9.2

Expand and simplify each of the following:

1. $\displaystyle\sum_{i=1}^{4}(2i+4)$

2. $\displaystyle\sum_{i=1}^{5}(3i-1)$

3. $\displaystyle\sum_{i=1}^{3}(2i-1)$

4. $\displaystyle\sum_{i=1}^{4}(2i-1)$

5. $\displaystyle\sum_{i=2}^{3}(i^2-1)$

6. $\displaystyle\sum_{i=3}^{6}(i^2+1)$

7. $\displaystyle\sum_{i=1}^{4}\frac{i}{1+i}$

8. $\displaystyle\sum_{i=1}^{4}\frac{i^2}{1+i}$

9. $\displaystyle\sum_{i=1}^{3}\frac{i^2}{2i-1}$

10. $\displaystyle\sum_{i=3}^{5}(i^3+4)$

11. $\displaystyle\sum_{i=1}^{4}(-3)^i$

12. $\displaystyle\sum_{i=1}^{4}\left(-\frac{1}{3}\right)^i$

13. $\displaystyle\sum_{i=3}^{6}(-2)^i$

14. $\displaystyle\sum_{i=4}^{6}\left(-\frac{1}{2}\right)^i$

Expand the following:

15. $\displaystyle\sum_{i=1}^{5}(x+i)$

16. $\displaystyle\sum_{i=3}^{6}(x-i)$

17. $\displaystyle\sum_{i=2}^{7}(x+1)^i$

18. $\displaystyle\sum_{i=1}^{4}(x+3)^i$

19. $\displaystyle\sum_{i=1}^{5}\frac{x+i}{x-1}$

20. $\displaystyle\sum_{i=1}^{6}\frac{x-3i}{x+3i}$

21. $\displaystyle\sum_{i=3}^{8}(x+i)^i$

22. $\displaystyle\sum_{i=4}^{7}(x-2i)^i$

23. $\displaystyle\sum_{i=1}^{5}(x+i)^{i+1}$

24. $\displaystyle\sum_{i=2}^{6}(x+i)^{i-1}$

Write each of the following sums with summation notation.

25. $2 + 4 + 8 + 16$
26. $3 + 5 + 7 + 9 + 11$
27. $4 + 8 + 16 + 32 + 64$
28. $1 + 3 + 5$
29. $5 + 10 + 17 + 26 + 37$
30. $3 + 8 + 15 + 24$
31. $\frac{3}{4} + \frac{4}{5} + \frac{5}{6} + \frac{6}{7} + \frac{7}{8}$
32. $\frac{1}{2} + \frac{2}{3} + \frac{3}{4} + \frac{4}{5}$
33. $\frac{1}{3} + \frac{2}{5} + \frac{3}{7} + \frac{4}{9}$
34. $\frac{3}{1} + \frac{5}{3} + \frac{7}{5} + \frac{9}{7}$
35. $(x - 3) + (x - 4) + (x - 5) + (x - 6)$

36. $x^2 + x^3 + x^4 + x^5 + x^6$

37. $\dfrac{x}{x+3} + \dfrac{x}{x+4} + \dfrac{x}{x+5}$

38. $\dfrac{x-3}{x^3} + \dfrac{x-4}{x^4} + \dfrac{x-5}{x^5} + \dfrac{x-6}{x^6}$

39. $x^2(x+2) + x^3(x+3) + x^4(x+4)$

40. $x(x+2)^2 + x(x+3)^3 + x(x+4)^4$

Applying the Concepts

41. Skydiving A skydiver jumps from a plane and falls 16 feet the first second, 48 feet the second second, and 80 feet the third second. If he continues to fall in the same manner, how far will he fall the seventh second? What is the distance he falls in 7 seconds?

42. Bacteria Growth After 1 day, a colony of 50 bacteria reproduces to become 200 bacteria. After 2 days, they reproduce to become 800 bacteria. If they continue to reproduce at this rate, how many bacteria will be present after 4 days?

Review Problems

 The problems below review material we covered in Section 8.3.

Use the properties of logarithms to expand each of the following expressions:

43. $\log_2 x^3 y$

44. $\log_7 \dfrac{x^2}{y^4}$

45. $\log_{10} \dfrac{\sqrt[3]{x}}{y^2}$

46. $\log_{10} \sqrt[3]{\dfrac{x}{y^2}}$

Write each expression as a single logarithm.

47. $\log_{10} x - \log_{10} y^2$

48. $\log_{10} x^2 + \log_{10} y^2$

49. $2 \log_3 x - 3 \log_3 y - 4 \log_3 z$

50. $\frac{1}{2} \log_6 x + \frac{1}{3} \log_6 y + \frac{1}{4} \log_6 z$

Solve each equation.

51. $\log_4 x - \log_4 5 = 2$

52. $\log_3 6 + \log_3 x = 4$

53. $\log_2 x + \log_2(x-7) = 3$

54. $\log_5(x+1) + \log_5(x-3) = 1$

SECTION
9.3

Arithmetic Sequences

In this and the following section, we will review and extend two major types of sequences, which we have worked with previously—**arithmetic sequences** and **geometric sequences.**

> **DEFINITION**
>
> An **arithmetic sequence** is a sequence of numbers in which each term is obtained from the preceding term by adding the same amount each time. An arithmetic sequence is also called an **arithmetic progression.**

The sequence

$$2, 6, 10, 14, \ldots$$

is an example of an arithmetic sequence, since each term is obtained from the preceding term by adding 4 each time. The amount we add each time—in this case, 4—is called the **common difference,** since it can be obtained by subtracting any two consecutive terms. (When we subtract, the term with the larger subscript must be written first.) The common difference is denoted by d.

EXAMPLE 1 Give the common difference d for the arithmetic sequence 4, 10, 16, 22, ...

SOLUTION Since each term can be obtained from the preceding term by adding 6, the common difference is 6. That is, $d = 6$.

EXAMPLE 2 Give the common difference for 100, 93, 86, 79, ...

SOLUTION The common difference in this case is $d = -7$, since adding -7 to any term always produces the next consecutive term.

EXAMPLE 3 Give the common difference for $\frac{1}{2}$, 1, $\frac{3}{2}$, 2, ...

SOLUTION The common difference is $d = \frac{1}{2}$.

The General Term

The general term, a_n, of an arithmetic progression can always be written in terms of the first term a_1 and the common difference d. Consider the sequence from Example 1:

$$4, 10, 16, 22, \ldots$$

We can write each term in terms of the first term 4 and the common difference 6:

$$4, \quad 4 + (1 \cdot 6), \quad 4 + (2 \cdot 6), \quad 4 + (3 \cdot 6), \ldots$$
$$a_1, \quad\quad a_2, \quad\quad\quad a_3, \quad\quad\quad a_4, \quad\quad \ldots$$

Observing the relationship between the subscript on the terms in the second line and the coefficients of the 6's in the first line, we write the general term for the sequence as

$$a_n = 4 + (n - 1)6$$

We generalize this result to include the general term of any arithmetic sequence.

DEFINITION

The **general term of an arithmetic sequence** with first term a_1 and common difference d is given by

$$a_n = a_1 + (n - 1)d$$

EXAMPLE 4   Find the general term for the sequence 7, 10, 13, 16, ...

SOLUTION   The first term is $a_1 = 7$, and the common difference is $d = 3$. Substituting these numbers into the formula given above, we have

$$a_n = 7 + (n - 1)3$$

which we can simplify, if we choose, to

$$a_n = 7 + 3n - 3$$
$$= 3n + 4$$

EXAMPLE 5   Find the general term of the arithmetic progression whose third term, a_3, is 7 and whose eighth term, a_8, is 17.

SOLUTION   According to the formula for the general term, the third term can be written as $a_3 = a_1 + 2d$, and the eighth term can be written as $a_8 = a_1 + 7d$. Since these terms are also equal to 7 and 17, respectively, we can write

$$a_3 = a_1 + 2d = 7$$
$$a_8 = a_1 + 7d = 17$$

To find a_1 and d, we simply solve the system:

$$a_1 + 2d = 7$$
$$a_1 + 7d = 17$$

We add the opposite of the top equation to the bottom equation. The result is

$$5d = 10$$
$$d = 2$$

To find a_1, we simply substitute 2 for d in either of the original equations and get

$$a_1 = 3$$

The general term for this progression is

$$a_n = 3 + (n - 1)2$$

which we can simplify to

$$a_n = 2n + 1$$

◢

The sum of the first n terms of an arithmetic sequence is denoted by S_n. The following theorem gives the formula for finding S_n, which is sometimes called the **nth partial sum.**

THEOREM 9.1

The sum of the first n terms of an arithmetic sequence whose first term is a_1 and whose nth term is a_n is given by

$$S_n = \frac{n}{2}(a_1 + a_n)$$

Proof

We can write S_n in expanded form as

$$S_n = a_1 + [a_1 + d] + [a_1 + 2d] + \cdots + [a_1 + (n - 1)d]$$

We can arrive at this same series by starting with the last term, a_n, and subtracting d each time. Writing S_n this way, we have

$$S_n = a_n + [a_n - d] + [a_n - 2d] + \cdots + [a_n - (n - 1)d]$$

If we add the preceding two expressions term by term, we have

$$2S_n = (a_1 + a_n) + (a_1 + a_n) + (a_1 + a_n) + \cdots + (a_1 + a_n)$$

$$2S_n = n(a_1 + a_n)$$

$$S_n = \frac{n}{2}(a_1 + a_n)$$

EXAMPLE 6   Find the sum of the first 10 terms of the arithmetic progression 2, 10, 18, 26, ...

SOLUTION   The first term is 2 and the common difference is 8. The tenth term is

$$a_{10} = 2 + 9(8)$$
$$= 2 + 72$$
$$= 74$$

Substituting $n = 10$, $a_1 = 2$, and $a_{10} = 74$ into the formula

$$S_n = \frac{n}{2}(a_1 + a_n)$$

we have

$$S_{10} = \frac{10}{2}(2 + 74)$$

$$= 5(76)$$
$$= 380$$

The sum of the first 10 terms is 380.

◢

506

Problem Set

9.3

Determine which of the following sequences are arithmetic progressions. For those that are arithmetic progressions, identify the common difference d.

1. 1, 2, 3, 4, . . .   **2.** 4, 6, 8, 10, . . .
3. 1, 2, 4, 7, . . .   **4.** 1, 2, 4, 8, . . .
5. 50, 45, 40, . . .   **6.** $1, \frac{1}{2}, \frac{1}{4}, \frac{1}{8}$, . . .
7. 1, 4, 9, 16 . . .   **8.** 5, 7, 9, 11, . . .
9. $\frac{1}{3}, 1, \frac{5}{3}, \frac{7}{3}$, . . .   **10.** 5, 11, 17, . . .

Each of the following problems refers to an arithmetic sequence.

11. If $a_1 = 3$ and $d = 4$, find a_n and a_{24}.

12. If $a_1 = 5$ and $d = 10$, find a_n and a_{100}.

13. If $a_1 = 6$ and $d = -2$, find a_{10} and S_{10}.

14. If $a_1 = 7$ and $d = -1$, find a_{24} and S_{24}.

15. If $a_6 = 17$ and $a_{12} = 29$, find the first term a_1, the common difference d, and then find a_{30}.

16. If $a_5 = 23$ and $a_{10} = 48$, find the first term a_1, the common difference d, and then find a_{40}.

17. If the third term is 16 and the eighth term is 26, find the first term, the common difference, and then find a_{20} and S_{20}.

18. If the third term is 16 and the eighth term is 51, find the first term, the common difference, and then find a_{50} and S_{50}.

19. Find the sum of the first 100 terms of the sequence 5, 9, 13, 17, ...

20. Find the sum of the first 50 terms of the sequence 8, 11, 14, 17, ...

21. Find a_{35} for the sequence 12, 7, 2, −3, ...

22. Find a_{45} for the sequence 25, 20, 15, 10, ...

23. Find the tenth term and the sum of the first 10 terms of the sequence $\frac{1}{2}, 1, \frac{3}{2}, 2$, ...

24. Find the fifteenth term and the sum of the first 15 terms of the sequence $-\frac{1}{3}, 0, \frac{1}{3}, \frac{2}{3}$, ...

Applying the Concepts

25. **Increasing Salary** Suppose a woman earns $28,000 the first year she works and then gets a raise of $850 every year after that. Write a sequence that gives her salary for each of the first 5 years she works. What is the general term of this sequence? At this rate, how much will she be making the tenth year she works?

26. **Increasing Salary** Suppose a school teacher makes $26,500 the first year he works and then gets a $900 raise every year after that. Write a sequence that gives his salary for the first 5 years he works. What is the general term of this sequence? How much will he be making the twentieth year he works?

Review Problems

 The problems that follow review material we covered in Section 8.4.

Find the following common logarithms:

27. log 576   **28.** log 57,600
29. log 0.0576   **30.** log 0.000576

Find x.

31. log x = 2.6484   **32.** log x = 7.9832
33. log x = −7.3516   **34.** log x = −2.0168

Geometric Sequences

This section is concerned with the second major classification of sequences, called **geometric sequences.** The problems in this section are very similar to the problems in the preceding section.

> **DEFINITION**
>
> A sequence of numbers in which each term is obtained from the previous term by multiplying by the same amount each time is called a **geometric sequence.** Geometric sequences are also called **geometric progressions.**

The sequence

$$3, 6, 12, 24, \ldots$$

is an example of a geometric progression. Each term is obtained from the previous term by multiplying by 2. The amount by which we multiply each time—in this case, 2—is called the **common ratio.** The common ratio is denoted by r and can be found by taking the ratio of any two consecutive terms. (The term with the larger subscript must be in the numerator.)

EXAMPLE 1 Find the common ratio for the geometric progression

$$\frac{1}{2}, \frac{1}{4}, \frac{1}{8}, \frac{1}{16}, \ldots$$

SOLUTION Since each term can be obtained from the term before it by multiplying by $\frac{1}{2}$, the common ratio is $\frac{1}{2}$. That is, $r = \frac{1}{2}$.

EXAMPLE 2 Find the common ratio for $\sqrt{3}, 3, 3\sqrt{3}, 9, \ldots$.

SOLUTION If we take the ratio of the third term to the second term, we have

$$\frac{3\sqrt{3}}{3} = \sqrt{3}$$

The common ratio is $r = \sqrt{3}$.

> **DEFINITION**
>
> The **general term of a geometric sequence** with first term a_1 and common ratio r is given by
>
> $$a_n = a_1 r^{n-1}$$

To see how we arrive at this formula, consider the following geometric progression whose common ratio is 3:

$$2, 6, 18, 54, \ldots$$

We can write each term of the sequence in terms of the first term 2 and the common ratio 3:

$$2 \cdot 3^0, \quad 2 \cdot 3^1, \quad 2 \cdot 3^2, \quad 2 \cdot 3^3, \ldots$$
$$a_1, \qquad a_2, \qquad a_3, \qquad a_4, \quad \ldots$$

Observing the relationship between the two lines written above, we find we can write the general term of this progression as

$$a_n = 2 \cdot 3^{n-1}$$

Since the first term can be designated by a_1 and the common ratio by r, the formula

$$a_n = 2 \cdot 3^{n-1}$$

coincides with the formula

$$a_n = a_1 r^{n-1}$$

EXAMPLE 3 Find the general term for the geometric progression

$$5, 10, 20, \ldots$$

SOLUTION The first term is $a_1 = 5$, and the common ratio is $r = 2$. Using these values in the formula

$$a_n = a_1 r^{n-1}$$

we have

$$a_n = 5 \cdot 2^{n-1}$$

EXAMPLE 4 Find the tenth term of the sequence $3, \frac{3}{2}, \frac{3}{4}, \frac{3}{8}, \ldots$.

SOLUTION The sequence is a geometric progression with first term $a_1 = 3$ and common ratio $r = \frac{1}{2}$. The tenth term is

$$a_{10} = 3\left(\frac{1}{2}\right)^9 = \frac{3}{512}$$

EXAMPLE 5 Find the general term for the geometric progression whose fourth term is 16 and whose seventh term is 128.

SOLUTION The fourth term can be written as $a_4 = a_1 r^3$, and the seventh term can be written as $a_7 = a_1 r^6$.

$$a_4 = a_1 r^3 = 16$$
$$a_7 = a_1 r^6 = 128$$

We can solve for r by using the ratio a_7/a_4:

$$\frac{a_7}{a_4} = \frac{a_1 r^6}{a_1 r^3} = \frac{128}{16}$$
$$r^3 = 8$$
$$r = 2$$

The common ratio is 2. To find the first term, we substitute $r = 2$ into either of the original two equations. The result is

$$a_1 = 2$$

The general term for this progression is

$$a_n = 2 \cdot 2^{n-1}$$

which we can simplify by adding exponents, since the bases are equal:

$$a_n = 2^n$$

As was the case in the preceding section, the sum of the first n terms of a geometric progression is denoted by S_n, which is called the **nth partial sum** of the progression.

THEOREM 9.2

The sum of the first n terms of a geometric progression with first term a_1 and common ratio r is given by the formula

$$S_n = \frac{a_1(r^n - 1)}{r - 1}$$

Proof
We can write the sum of the first n terms in expanded form:

$$S_n = a_1 + a_1 r + a_1 r^2 + \cdots + a_1 r^{n-1} \tag{1}$$

Then multiplying both sides by r, we have

$$r S_n = a_1 r + a_1 r^2 + a_1 r^3 + \cdots + a_1 r^n \tag{2}$$

If we subtract the left side of equation (1) from the left side of equation (2) and do the same for the right sides, we end up with

$$r S_n - S_n = a_1 r^n - a_1$$

We factor S_n from both terms on the left side and a_1 from both terms on the right side of this equation:

$$S_n(r - 1) = a_1(r^n - 1)$$

Dividing both sides by $r - 1$ gives the desired result:

$$S_n = \frac{a_1(r^n - 1)}{r - 1}$$

EXAMPLE 6 Find the sum of the first 10 terms of the geometric progression 5, 15, 45, 135, ...

SOLUTION The first term is $a_1 = 5$, and the common ratio is $r = 3$. Substituting these values into the formula for S_{10}, we have the sum of the first 10 terms of the sequence:

$$S_{10} = \frac{5(3^{10} - 1)}{3 - 1}$$

$$= \frac{5(3^{10} - 1)}{2}$$

The answer can be left in this form. A calculator will give the result as 147,620.

Infinite Geometric Series

Suppose the common ratio for a geometric sequence is a number whose absolute value is less than 1—for instance $\frac{1}{2}$. The sum of the first n terms is given by the formula

$$S_n = \frac{a_1[(\frac{1}{2})^n - 1]}{\frac{1}{2} - 1}$$

As n becomes larger and larger, the term $(\frac{1}{2})^n$ will become closer and closer to 0. That is, for $n = 10, 20$, and 30, we have the following approximations:

$$(\tfrac{1}{2})^{10} \approx 0.001$$

$$(\tfrac{1}{2})^{20} \approx 0.000001$$

$$(\tfrac{1}{2})^{30} \approx 0.000000001$$

Therefore, for large values of n, there is very little difference between the expression

$$\frac{a_1(r^n - 1)}{r - 1}$$

and the expression

$$\frac{a_1(0 - 1)}{r - 1} = \frac{-a_1}{r - 1} = \frac{a_1}{1 - r} \qquad \text{if } |r| < 1$$

In fact, the sum of the terms of a geometric sequence in which $|r| < 1$ actually becomes the expression

$$\frac{a_1}{1 - r}$$

as n approaches infinity. To summarize, we have the following:

THE SUM OF AN INFINITE GEOMETRIC SERIES

If a geometric sequence has first term a_1 and common ratio r such that $|r| < 1$, then the following is called an **infinite geometric series:**

$$S = \sum_{i=0}^{\infty} a_1 r^i = a_1 + a_1 r + a_1 r^2 + a_1 r^3 + \cdots$$

Its sum is given by the formula

$$S = \frac{a_1}{1 - r}$$

EXAMPLE 7   Find the sum of the infinite geometric series

$$\frac{1}{5} + \frac{1}{10} + \frac{1}{20} + \frac{1}{40} + \cdots$$

SOLUTION   The first term is $a_1 = \frac{1}{5}$ and the common ratio is $r = \frac{1}{2}$, which has an absolute value less than 1. Therefore, the sum of this series is

$$S = \frac{a_1}{1 - r} = \frac{\frac{1}{5}}{1 - \frac{1}{2}} = \frac{\frac{1}{5}}{\frac{1}{2}} = \frac{2}{5}$$

EXAMPLE 8   Show that $0.999\ldots$ is equal to 1.

SOLUTION   We begin by writing $0.999\ldots$ as an infinite geometric series:

$$0.999\ldots = 0.9 + 0.09 + 0.009 + 0.0009 + \cdots$$

$$= \frac{9}{10} + \frac{9}{100} + \frac{9}{1,000} + \frac{9}{10,000} + \cdots$$

$$= \frac{9}{10} + \frac{9}{10}\left(\frac{1}{10}\right) + \frac{9}{10}\left(\frac{1}{10}\right)^2 + \frac{9}{10}\left(\frac{1}{10}\right)^3 + \cdots$$

As the last line indicates, we have an infinite geometric series with $a_1 = \frac{9}{10}$ and $r = \frac{1}{10}$. The sum of this series is given by

$$S = \frac{a_1}{1 - r} = \frac{\frac{9}{10}}{1 - \frac{1}{10}} = \frac{\frac{9}{10}}{\frac{9}{10}} = 1$$

Problem Set 9.4

Determine which of the following sequences are geometric progressions. For those that are geometric, give the common ratio r.

1. 1, 5, 25, 125, . . .

2. 6, 12, 24, 48, . . .

3. $\frac{1}{2}, \frac{1}{6}, \frac{1}{18}, \frac{1}{54}, \ldots$

4. 5, 10, 15, 20, . . .

5. 4, 9, 16, 25, . . .

6. $-1, \frac{1}{3}, -\frac{1}{9}, \frac{1}{27}, \ldots$

7. $-2, 4, -8, 16, \ldots$

8. 1, 8, 27, 64, . . .

9. 4, 6, 8, 10, . . .

10. 1, -3, 9, -27, . . .

Each of the following problems gives some information about a specific geometric progression.

11. If $a_1 = 4$ and $r = 3$, find a_n.

12. If $a_1 = 5$ and $r = 2$, find a_n.

13. If $a_1 = -2$ and $r = -\frac{1}{2}$, find a_6.

14. If $a_1 = 25$ and $r = -\frac{1}{5}$, find a_6.

15. If $a_1 = 3$ and $r = -1$, find a_{20}.

16. If $a_1 = -3$ and $r = -1$, find a_{20}.

17. If $a_1 = 10$ and $r = 2$, find S_{10}.

18. If $a_1 = 8$ and $r = 3$, find S_5.

19. If $a_1 = 1$ and $r = -1$, find S_{20}.

20. If $a_1 = 1$ and $r = -1$, find S_{21}.

21. Find a_8 for $\frac{1}{5}, \frac{1}{10}, \frac{1}{20}, \ldots$

22. Find a_8 for $\frac{1}{2}, \frac{1}{10}, \frac{1}{50}, \ldots$

23. Find S_5 for $-\frac{1}{2}, -\frac{1}{4}, -\frac{1}{8}, \ldots$

24. Find S_6 for $-\frac{1}{2}, 1, -2, \ldots$

25. Find a_{10} and S_{10} for $\sqrt{2}, 2, 2\sqrt{2}, \ldots$

26. Find a_8 and S_8 for $\sqrt{3}, 3, 3\sqrt{3}, \ldots$

27. Find a_6 and S_6 for 100, 10, 1, ...

28. Find a_6 and S_6 for 100, -10, 1, ...

29. If $a_4 = 40$ and $a_6 = 160$, find r.

30. If $a_5 = \frac{1}{8}$ and $a_8 = \frac{1}{64}$, find r.

Find the sum of each geometric series.

31. $\frac{1}{2} + \frac{1}{4} + \frac{1}{8} + \cdots$

32. $\frac{1}{3} + \frac{1}{9} + \frac{1}{27} + \cdots$

33. $4 + 2 + 1 + \cdots$

34. $8 + 4 + 2 + \cdots$

35. $\frac{2}{5} + \frac{4}{25} + \frac{8}{125} + \cdots$

36. $\frac{3}{4} + \frac{9}{16} + \frac{27}{64} + \cdots$

37. $\frac{3}{4} + \frac{1}{4} + \frac{1}{12} + \cdots$

38. $\frac{5}{3} + \frac{1}{3} + \frac{1}{15} + \cdots$

39. Show that 0.444. . . is the same as $\frac{4}{9}$.

40. Show that 0.333. . . is the same as $\frac{1}{3}$.

41. Show that 0.272727. . . is the same as $\frac{3}{11}$.

42. Show that 0.545454. . . is the same as $\frac{6}{11}$.

Review Problems

NOTE: The problems below review material we covered in Section 1.5. Reviewing these problems will help you understand the next section.

Expand and multiply.

43. $(x + 5)^2$

44. $(x + y)^2$

45. $(x + y)^3$

46. $(x - 2)^3$

47. $(x + y)^4$

48. $(x - 1)^4$

One Step Further

Use a calculator to find the given term or partial sum.

49. Find a_{20} if $a_1 = 100$ and $r = 2$.

50. Find a_{20} if $a_1 = 81$ and $r = \frac{1}{3}$.

51. Find S_{18} if $a_1 = 100$ and $r = \frac{1}{2}$.

52. Find S_{22} if $a_1 = 64$ and $r = -\frac{1}{2}$.

53. Find the sum for

$$a + \frac{a}{2} + \frac{a}{4} + \cdots \qquad \text{if } a > 1$$

54. Find the sum for

$$\frac{1}{a} + \frac{1}{a^2} + \frac{1}{a^3} + \cdots \qquad \text{if } a > 1$$

55. Find the sum for

$$\frac{a}{b} + \frac{a^2}{b^2} + \frac{a^3}{b^3} + \cdots \qquad \text{if } \left|\frac{a}{b}\right| < 1$$

56. Sierpinski Triangle In the sequence that follows, the figures are moving toward what is known as the *Sierpinski triangle*. To construct the figure in Stage 2, we remove the triangle formed from the midpoints of the sides of the shaded region in Stage 1. Likewise, the figure in Stage 3 is found by removing the triangles formed by connecting the midpoints of the sides of the shaded regions in Stage 2. If we repeat this process infinitely many times, we arrive at the Sierpinski triangle.

(a) If the shaded region in Stage 1 has an area of 1, find the areas of the shaded regions in Stages 2–4.

(b) Do the areas you find in part (a) form an arithmetic sequence or a geometric sequence?

(c) The Sierpinski triangle is the triangle that is formed after the process shown in Stages 1–4 is repeated infinitely many times. What do you think the area of the shaded region of the Sierpinski triangle will be?

(d) Suppose the perimeter of the shaded region of the triangle in Stage 1 is 1. If we were to find the perimeters of the shaded regions in the other stages, would we have an increasing sequence or a decreasing sequence?

| Stage 1 | Stage 2 | Stage 3 | Stage 4 |

The Binomial Expansion

The purpose of this section is to write and apply the formula for the expansion of expressions of the form $(x + y)^n$, where n is any positive integer. In order to write the formula, we must generalize the information in the following chart:

$$
\begin{aligned}
(x + y)^0 &= 1 \\
(x + y)^1 &= x + y \\
(x + y)^2 &= x^2 + 2xy + y^2 \\
(x + y)^3 &= x^3 + 3x^2y + 3xy^2 + y^3 \\
(x + y)^4 &= x^4 + 4x^3y + 6x^2y^2 + 4xy^3 + y^4 \\
(x + y)^5 &= x^5 + 5x^4y + 10x^3y^2 + 10x^2y^3 + 5xy^4 + y^5
\end{aligned}
$$

The polynomials to the right have been found by expanding the binomials on the left—we just haven't shown the work.

There are a number of similarities to notice among the polynomials on the right. Here is a list:

1. In each polynomial, the sequence of exponents on the variable x decreases to 0 from the exponent on the binomial at the left. (The exponent 0 is not shown, since $x^0 = 1$.)
2. In each polynomial, the exponents on the variable y increase from 0 to the exponent on the binomial at the left. (Since $y^0 = 1$, it is not shown in the first term.)
3. The sum of the exponents on the variables in any single term is equal to the exponent on the binomial at the left.

The pattern in the coefficients of the polynomials on the right can best be seen by writing the right side again without the variables. It looks like this:

row 0						1					
row 1					1		1				
row 2				1		2		1			
row 3			1		3		3		1		
row 4		1		4		6		4		1	
row 5	1		5		10		10		5		1

This triangle-shaped array of coefficients is called **Pascal's triangle.** Each entry in the triangular array is obtained by adding the two numbers above it. Each row begins and ends with the number 1. If we were to continue Pascal's triangle, the next two rows would be

row 6		1	6	15	20	15	6	1	
row 7	1	7	21	35	35	21	7	1	

The coefficients for the terms in the expansion of $(x + y)^n$ are given in the nth row of Pascal's triangle.

There is an alternative method of finding these coefficients that does not involve Pascal's triangle. The alternative method involves **factorial notation.**

DEFINITION

The expression $n!$ is read "n **factorial**" and is the product of all the consecutive integers from n down to 1. For example,

$$1! = 1$$
$$2! = 2 \cdot 1 = 2$$
$$3! = 3 \cdot 2 \cdot 1 = 6$$
$$4! = 4 \cdot 3 \cdot 2 \cdot 1 = 24$$
$$5! = 5 \cdot 4 \cdot 3 \cdot 2 \cdot 1 = 120$$

The expression 0! is defined to be 1. We use factorial notation to define **binomial coefficients** as follows:

DEFINITION

The expression $\dbinom{n}{r}$ is called a **binomial coefficient** and is defined by

$$\binom{n}{r} = \frac{n!}{r!(n-r)!}$$

EXAMPLE 1 Calculate the following binomial coefficients:

$$\binom{7}{5}, \quad \binom{6}{2}, \quad \binom{3}{0}$$

SOLUTION We simply apply the definition for binomial coefficients:

$$\binom{7}{5} = \frac{7!}{5!(7-5)!} \qquad\qquad \binom{6}{2} = \frac{6!}{2!(6-2)!}$$

$$= \frac{7!}{5! \cdot 2!} \qquad\qquad\qquad = \frac{6!}{2! \cdot 4!}$$

$$= \frac{7 \cdot 6 \cdot 5 \cdot 4 \cdot 3 \cdot 2 \cdot 1}{(5 \cdot 4 \cdot 3 \cdot 2 \cdot 1)(2 \cdot 1)} \qquad = \frac{6 \cdot 5 \cdot 4 \cdot 3 \cdot 2 \cdot 1}{(2 \cdot 1)(4 \cdot 3 \cdot 2 \cdot 1)}$$

$$= \frac{42}{2} \qquad\qquad\qquad\qquad = \frac{30}{2}$$

$$= 21 \qquad\qquad\qquad\qquad = 15$$

$$\binom{3}{0} = \frac{3!}{0!(3-0)!}$$

$$= \frac{3!}{0! \cdot 3!}$$

$$= \frac{3 \cdot 2 \cdot 1}{(1)(3 \cdot 2 \cdot 1)}$$

$$= 1$$

If we were to calculate all the binomial coefficients in the following array, we would find they match exactly with the numbers in Pascal's triangle. That is why they are called binomial coefficients—because they are the coefficients of the expansion of $(x + y)^n$.

$$\binom{0}{0}$$

$$\binom{1}{0} \qquad \binom{1}{1}$$

$$\binom{2}{0} \qquad \binom{2}{1} \qquad \binom{2}{2}$$

$$\binom{3}{0} \qquad \binom{3}{1} \qquad \binom{3}{2} \qquad \binom{3}{3}$$

$$\binom{4}{0} \qquad \binom{4}{1} \qquad \binom{4}{2} \qquad \binom{4}{3} \qquad \binom{4}{4}$$

$$\binom{5}{0} \qquad \binom{5}{1} \qquad \binom{5}{2} \qquad \binom{5}{3} \qquad \binom{5}{4} \qquad \binom{5}{5}$$

Using the new notation to represent the entries in Pascal's triangle, we can summarize everything we have noticed about the expansion of binomial powers of the form $(x + y)^n$.

THE BINOMIAL EXPANSION

If x and y represent real numbers and n is a positive integer, then the following formula is known as the **binomial expansion** or **binomial formula:**

$$(x + y)^n = \binom{n}{0}x^n y^0 + \binom{n}{1}x^{n-1}y^1 + \binom{n}{2}x^{n-2}y^2 + \cdots + \binom{n}{n}x^0 y^n$$

It does not make any difference, when expanding binomial powers of the form $(x + y)^n$, whether we use Pascal's triangle or the formula

$$\binom{n}{r} = \frac{n!}{r!(n-r)!}$$

to calculate the coefficients. We will show examples of both methods.

EXAMPLE 2 Expand: $(x - 2)^3$.

SOLUTION Applying the binomial formula, we have

$$(x - 2)^3 = \binom{3}{0}x^3(-2)^0 + \binom{3}{1}x^2(-2)^1 + \binom{3}{2}x^1(-2)^2 + \binom{3}{3}x^0(-2)^3$$

The coefficients $\binom{3}{0}$, $\binom{3}{1}$, $\binom{3}{2}$, and $\binom{3}{3}$ can be found in the third row of Pascal's triangle. They are 1, 3, 3, and 1:

$$(x - 2)^3 = 1x^3(-2)^0 + 3x^2(-2)^1 + 3x^1(-2)^2 + 1x^0(-2)^3$$
$$= x^3 - 6x^2 + 12x - 8$$

EXAMPLE 3 Expand: $(3x + 2y)^4$.

    SOLUTION The coefficients can be found in the fourth row of Pascal's triangle. They are

$$1, 4, 6, 4, 1$$

Here is the expansion of $(3x + 2y)^4$:

$$(3x + 2y)^4 = 1(3x)^4 + 4(3x)^3(2y) + 6(3x)^2(2y)^2 + 4(3x)(2y)^3 + 1(2y)^4$$
$$= 81x^4 + 216x^3y + 216x^2y^2 + 96xy^3 + 16y^4$$

EXAMPLE 4 Write the first three terms in the expansion of $(x + 5)^9$.

    SOLUTION The coefficients of the first three terms are $\binom{9}{0}$, $\binom{9}{1}$, and $\binom{9}{2}$, which we calculate as follows:

$$\binom{9}{0} = \frac{9!}{0! \cdot 9!} = \frac{9 \cdot 8 \cdot 7 \cdot 6 \cdot 5 \cdot 4 \cdot 3 \cdot 2 \cdot 1}{(1)(9 \cdot 8 \cdot 7 \cdot 6 \cdot 5 \cdot 4 \cdot 3 \cdot 2 \cdot 1)} = \frac{1}{1} = 1$$

$$\binom{9}{1} = \frac{9!}{1! \cdot 8!} = \frac{9 \cdot 8 \cdot 7 \cdot 6 \cdot 5 \cdot 4 \cdot 3 \cdot 2 \cdot 1}{(1)(8 \cdot 7 \cdot 6 \cdot 5 \cdot 4 \cdot 3 \cdot 2 \cdot 1)} = \frac{9}{1} = 9$$

$$\binom{9}{2} = \frac{9!}{2! \cdot 7!} = \frac{9 \cdot 8 \cdot 7 \cdot 6 \cdot 5 \cdot 4 \cdot 3 \cdot 2 \cdot 1}{(2 \cdot 1)(7 \cdot 6 \cdot 5 \cdot 4 \cdot 3 \cdot 2 \cdot 1)} = \frac{72}{2} = 36$$

From the binomial formula, we write the first three terms:

$$(x + 5)^9 = 1 \cdot x^9 + 9 \cdot x^8(5) + 36x^7(5)^2 + \cdots$$
$$= x^9 + 45x^8 + 900x^7 + \cdots$$

The kth Term of a Binomial Expansion

If we look at each term in the expansion of $(x + y)^n$ as a term in a sequence a_1, $a_2, a_3, \ldots$, we can write

$$a_1 = \binom{n}{0}x^n y^0$$

$$a_2 = \binom{n}{1}x^{n-1}y^1$$

$$a_3 = \binom{n}{2}x^{n-2}y^2$$

$$a_4 = \binom{n}{3}x^{n-3}y^3$$

and so on

To write the formula for the general term, we simply notice that the exponent on y and the number below n in the coefficient are both 1 less than the term number. This observation allows us to write the following:

THE GENERAL TERM OF A BINOMIAL EXPANSION

The kth term in the expansion of $(x + y)^n$ is

$$a_k = \binom{n}{k-1}x^{n-(k-1)}y^{k-1}$$

EXAMPLE 5 Find the fifth term in the expansion of $(2x + 3y)^{12}$.

SOLUTION Applying the formula above, we have

$$a_5 = \binom{12}{4}(2x)^8(3y)^4$$

$$= \frac{12!}{4! \cdot 8!}(2x)^8(3y)^4$$

Notice that once we have one of the exponents, the other exponents and the denominator of the coefficient are determined: The two exponents add to 12 and match the denominator of the coefficient.

Making the calculations from the formula above, we have

$$a_5 = 495(256x^8)(81y^4)$$
$$= 10,264,320x^8y^4$$

Problem Set 9.5

Use the binomial formula to expand each of the following:

1. $(x + 2)^4$
2. $(x - 2)^5$
3. $(x + y)^6$
4. $(x - 1)^6$
5. $(2x + 1)^5$
6. $(2x - 1)^4$
7. $(x - 2y)^5$
8. $(2x + y)^5$
9. $(3x - 2)^4$
10. $(2x - 3)^4$
11. $(4x - 3y)^3$
12. $(3x - 4y)^3$
13. $(x^2 + 2)^4$
14. $(x^2 - 3)^3$
15. $(x^2 + y^2)^3$
16. $(x^2 - 3y)^4$

17. $\left(\frac{x}{2} - 4\right)^3$
18. $\left(\frac{x}{3} + 6\right)^3$
19. $\left(\frac{x}{3} + \frac{y}{2}\right)^4$
20. $\left(\frac{x}{2} - \frac{y}{3}\right)^4$

Write the first four terms in the expansion of the following:

21. $(x + 2)^9$
22. $(x - 2)^9$
23. $(x - y)^{10}$
24. $(x + y)^{10}$
25. $(x + 2y)^{10}$
26. $(x - 2y)^{10}$

Write the first three terms in the expansion of each of the following:

27. $(x + 1)^{15}$
28. $(x - 1)^{15}$
29. $(x - y)^{12}$
30. $(x + y)^{12}$
31. $(x + 2)^{20}$
32. $(x - 2)^{20}$

Write the first two terms in the expansion of each of the following:

33. $(x + 2)^{100}$
34. $(x - 2)^{50}$
35. $(x + y)^{50}$
36. $(x - y)^{100}$

37. Find the ninth term in the expansion of $(2x + 3y)^{12}$.

38. Find the sixth term in the expansion of $(2x + 3y)^{12}$.

39. Find the fifth term of $(x - 2)^{10}$.

40. Find the fifth term of $(2x - 1)^{10}$.

41. Find the fourth term of $(x + 3)^9$.

42. Find the fifth term of $(x + 3)^9$.

43. Write the formula for the twelfth term of $(2x + 5y)^{20}$. Do not simplify.

44. Write the formula for the eighth term of $(2x + 5y)^{20}$. Do not simplify.

Applying the Concepts

45. **Probability** The third term in the expansion of $(\frac{1}{2} + \frac{1}{2})^7$ will give the probability that in a family with 7 children, 5 will be boys and 2 will be girls. Find the third term.

46. **Probability** The fourth term in the expansion of $(\frac{1}{2} + \frac{1}{2})^8$ will give the probability that in a family with 8 children, 3 will be boys and 5 will be girls. Find the fourth term.

Review Problems

 NOTE: The problems that follow review material we covered in Section 8.5.

Solve each equation. Write your answers to the nearest hundredth.

47. $5^x = 7$
48. $10^x = 15$
49. $8^{2x+1} = 16$
50. $9^{3x-1} = 27$

51. **Compound Interest** How long will it take $400 to double if it is invested in an account with an annual interest rate of 10% compounded four times a year?

52. **Compound Interest** How long will it take $200 to become $800 if it is invested in an account with an annual interest rate of 8% compounded four times a year?

Find each of the following to the nearest hundredth:

53. $\log_4 20$
54. $\log_7 21$
55. $\ln 576$
56. $\ln 5{,}760$

57. Solve the formula $A = 10e^{5t}$ for t.
58. Solve the formula $A = P2^{-5t}$ for t.

One Step Further

59. Calculate both $\binom{8}{5}$ and $\binom{8}{3}$ to show that they are equal.

60. Calculate both $\binom{10}{8}$ and $\binom{10}{2}$ to show that they are equal.

61. Simplify $\binom{20}{12}$ and $\binom{20}{8}$.

62. Simplify $\binom{15}{10}$ and $\binom{15}{5}$.

63. Show that $\binom{n}{r}$ and $\binom{n}{n-r}$ are equal.

64. **Pascal's Triangle** Copy the first eight rows of Pascal's triangle into the eight rows of the triangular array below. (Each number in Pascal's triangle will go into one of the hexagons in the array.)

Next, color in each hexagon that contains an odd number. What pattern begins to emerge from this coloring process?

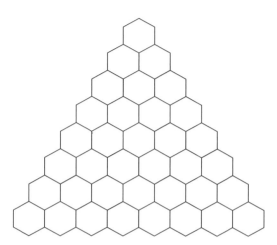

Chapter 9 Summary

Examples

1. In the sequence

$$1, 3, 5, \ldots,$$
$$2n - 1, \ldots$$

$a_1 = 1, a_2 = 3, a_3 = 5,$
and $a_n = 2n - 1.$

Sequences [9.1]

A **sequence** is a function whose domain is the set of positive integers. The terms of a sequence are denoted by

$$a_1, a_2, a_3, \ldots, a_n, \ldots$$

where a_1 (read "a sub 1") is the first term, a_2 is the second term, and a_n is the **nth term,** or **general term.**

2. $\displaystyle\sum_{i=3}^{6} (-2)^i$

$= (-2)^3 + (-2)^4$
$\qquad + (-2)^5 + (-2)^6$
$= -8 + 16$
$\qquad + (-32) + 64$
$= 40$

Summation Notation; Series [9.2]

The notation

$$\sum_{i=1}^{n} a_i = a_1 + a_2 + a_3 + \cdots + a_n$$

is called **summation notation** or **sigma notation.** The letter i is called the **index of summation,** or just the **index.** This notation can be used to represent a finite **series** compactly.

3. For the sequence

$$3, 7, 11, 15, \ldots$$

$a_1 = 3$ and $d = 4.$ The general term is

$a_n = 3 + (n - 1)4$
$\quad = 4n - 1$

Using this formula to find the tenth term, we have

$a_{10} = 4(10) - 1 = 39$

The sum of the first 10 terms is

$S_{10} = \frac{10}{2} (3 + 39) = 210$

Arithmetic Sequences [9.3]

An **arithmetic sequence,** or **arithmetic progression,** is a sequence in which each term is obtained from the preceding term by adding a constant amount each time. If the first term of an arithmetic sequence is a_1 and the amount we add each time (called the **common difference**) is d, then the nth term of the progression is given by

$$a_n = a_1 + (n - 1)d$$

The sum of the first n terms of an arithmetic sequence is

$$S_n = \frac{n}{2} (a_1 + a_n)$$

S_n is called the **nth partial sum.**

REVIEW FOR CHAPTER 9

4. For the geometric progression

$$3, 6, 12, 24, \ldots$$

$a_1 = 3$ and $r = 2$. The general term is

$$a_n = 3 \cdot 2^{n-1}$$

The sum of the first 10 terms is

$$S_{10} = \frac{3(2^{10} - 1)}{2 - 1} = 3{,}069$$

Geometric Sequences [9.4]

A **geometric sequence,** or **geometric progression,** is a sequence of numbers in which each term is obtained from the previous term by multiplying by a constant amount each time. The constant by which we multiply each term to get the next term is called the **common ratio.** If the first term of a geometric sequence is a_1 and the common ratio is r, then the formula that gives the general term, a_n, is

$$a_n = a_1 r^{n-1}$$

The sum of the first n terms of a geometric sequence is given by the formula

$$S_n = \frac{a_1(r^n - 1)}{r - 1}$$

5. The sum of the series

$$\frac{1}{3} + \frac{1}{6} + \frac{1}{12} + \cdots$$

is

$$S = \frac{\frac{1}{3}}{1 - \frac{1}{2}} = \frac{\frac{1}{3}}{\frac{1}{2}} = \frac{2}{3}$$

The Sum of an Infinite Geometric Series [9.4]

If a geometric sequence has first term a_1 and common ratio r such that $|r| < 1$, then the following is called an **infinite geometric series:**

$$S = \sum_{i=0}^{\infty} a_1 r^i = a_1 + a_1 r + a_1 r^2 + a_1 r^3 + \cdots$$

Its sum is given by the formula

$$S = \frac{a_1}{1 - r}$$

Factorials [9.5]

The notation $n!$ is called n **factorial** and is defined to be the product of each consecutive integer from n down to 1. That is,

$$0! = 1 \qquad \text{By definition}$$
$$1! = 1$$
$$2! = 2 \cdot 1$$
$$3! = 3 \cdot 2 \cdot 1$$
$$4! = 4 \cdot 3 \cdot 2 \cdot 1$$

And so on

6. $\binom{7}{3}$

$= \dfrac{7!}{3!(7-3)!}$

$= \dfrac{7!}{3! \cdot 4!}$

$= \dfrac{7 \cdot 6 \cdot 5 \cdot \cancel{4 \cdot 3 \cdot 2 \cdot 1}}{(3 \cdot 2 \cdot 1)(\cancel{4 \cdot 3 \cdot 2 \cdot 1})}$

$= 35$

Binomial Coefficients [9.5]

The notation $\binom{n}{r}$ is called a **binomial coefficient** and is defined by

$$\binom{n}{r} = \dfrac{n!}{r!(n-r)!}$$

Binomial coefficients can be found by using the formula above or by using **Pascal's triangle:**

```
            1
          1   1
        1   2   1
      1   3   3   1
    1   4   6   4   1
  1   5  10  10   5   1
```
And so on

7.
$(x+2)^4$
$= x^4 + 4x^3 \cdot 2 + 6x^2 \cdot 2^2$
$\quad + 4x \cdot 2^3 + 2^4$
$= x^4 + 8x^3 + 24x^2$
$\quad + 32x + 16$

Binomial Expansion [9.5]

If n is a positive integer, then the formula for expanding $(x+y)^n$ is given by

$$(x+y)^n = \binom{n}{0}x^n y^0 + \binom{n}{1}x^{n-1}y^1 + \binom{n}{2}x^{n-2}y^2 + \cdots + \binom{n}{n}x^0 y^n$$

Chapter 9 Review Problems

Write the first four terms of the sequences with the following general terms. [9.1]

1. $a_n = 2n + 5$ **2.** $a_n = 3n - 2$

3. $a_n = n^2 - 1$ **4.** $a_n = \dfrac{n+3}{n+2}$

5. $a_1 = 4, a_n = 4a_{n-1}, n > 1$
6. $a_1 = \frac{1}{4}, a_n = \frac{1}{4}a_{n-1}, n > 1$

Determine the general term for each of the following sequences. [9.1]

7. $2, 5, 8, 11, \ldots$
8. $-3, -1, 1, 3, 5, \ldots$
9. $1, 16, 81, 256, \ldots$ **10.** $2, 5, 10, 17, \ldots$
11. $\frac{1}{2}, \frac{1}{4}, \frac{1}{8}, \frac{1}{16}, \ldots$ **12.** $2, \frac{3}{4}, \frac{4}{9}, \frac{5}{16}, \frac{6}{25}, \ldots$

Expand and simplify each of the following. [9.2]

13. $\sum_{i=1}^{4} (2i + 3)$ **14.** $\sum_{i=1}^{3} (2i^2 - 1)$

15. $\sum_{i=2}^{3} \dfrac{i^2}{i+2}$ **16.** $\sum_{i=1}^{4} (-2)^{i-1}$

17. $\sum_{i=3}^{5} (4i + i^2)$ **18.** $\sum_{i=4}^{6} \dfrac{i+2}{i}$

Write each of the following sums with summation notation. [9.2]

19. $3 + 6 + 9 + 12$
20. $3 + 7 + 11 + 15$
21. $5 + 7 + 9 + 11 + 13$

22. $4 + 9 + 16$

23. $\frac{1}{3} + \frac{1}{4} + \frac{1}{5} + \frac{1}{6}$

24. $\frac{1}{3} + \frac{2}{9} + \frac{3}{27} + \frac{4}{81} + \frac{5}{243}$

25. $(x - 2) + (x - 4) + (x - 6)$

26. $\dfrac{x}{x + 1} + \dfrac{x}{x + 2} + \dfrac{x}{x + 3} + \dfrac{x}{x + 4}$

Determine which of the following sequences are arithmetic progressions, geometric progressions, or neither. [9.3, 9.4]

27. $1, -3, 9, -27, \ldots$ **28.** $7, 9, 11, 13, \ldots$

29. $5, 11, 17, 23, \ldots$ **30.** $\frac{1}{2}, \frac{1}{3}, \frac{1}{4}, \frac{1}{5}, \ldots$

31. $4, 8, 16, 32, \ldots$ **32.** $\frac{1}{2}, \frac{1}{4}, \frac{1}{8}, \frac{1}{16}, \ldots$

33. $12, 9, 6, 3, \ldots$ **34.** $2, 5, 9, 14, \ldots$

Each of the following problems refers to an arithmetic progression. [9.3]

35. If $a_1 = 2$ and $d = 3$, find a_n and a_{20}.

36. If $a_1 = 5$ and $d = -3$, find a_n and a_{16}.

37. If $a_1 = -2$ and $d = 4$, find a_{10} and S_{10}.

38. If $a_1 = 3$ and $d = 5$, find a_{16} and S_{16}.

39. If $a_5 = 21$ and $a_8 = 33$, find the first term a_1, the common difference d, and then find a_{10}.

40. If $a_3 = 14$ and $a_7 = 26$, find the first term a_1, the common difference d, and then find a_9 and S_9.

41. If $a_4 = -10$ and $a_8 = -18$, find the first term a_1, the common difference d, and then find a_{20} and S_{20}.

42. Find the sum of the first 100 terms of the sequence $3, 7, 11, 15, 19, \ldots$.

43. Find a_{40} for the sequence $100, 95, 90, 85, 80, \ldots$.

Each of the following problems refers to an infinite geometric progression. [9.4]

44. If $a_1 = 3$ and $r = 2$, find a_n and a_{20}.

45. If $a_1 = 5$ and $r = -2$, find a_n and a_{16}.

46. If $a_1 = 4$ and $r = \frac{1}{2}$, find a_n and a_{10}.

47. If $a_1 = -2$ and $r = \frac{1}{3}$, find the sum.

48. If $a_1 = 4$ and $r = \frac{1}{2}$, find the sum.

49. If $a_3 = 12$ and $a_4 = 24$, find the first term a_1, the common ratio r, and then find a_6.

50. Find the tenth term of the sequence $3, 3\sqrt{3}, 9, 9\sqrt{3}, \ldots$.

Evaluate each of the following. [9.5]

51. $\dbinom{8}{2}$ **52.** $\dbinom{7}{4}$

53. $\dbinom{6}{3}$ **54.** $\dbinom{9}{2}$

55. $\dbinom{10}{8}$ **56.** $\dbinom{100}{3}$

Use the binomial formula to expand each of the following. [9.5]

57. $(x - 2)^4$ **58.** $(2x + 3)^4$

59. $(3x + 2y)^3$ **60.** $(x^2 - 2)^5$

61. $\left(\dfrac{x}{2} + 3\right)^4$ **62.** $\left(\dfrac{x}{3} - \dfrac{y}{2}\right)^3$

Use the binomial formula to write the first three terms in the expansion of the following. [9.5]

63. $(x + 3y)^{10}$ **64.** $(x - 3y)^9$

65. $(x + y)^{11}$ **66.** $(x - 2y)^{12}$

Use the binomial formula to write the first two terms in the expansion of the following. [9.5]

67. $(x - 2y)^{16}$ **68.** $(x + 2y)^{32}$

69. $(x - 1)^{50}$ **70.** $(x + y)^{150}$

71. Find the sixth term in $(x - 3)^{10}$. [9.5]

72. Find the fourth term in $(2x + 1)^9$. [9.5]

Chapter 9 Test

Write the first five terms of the sequences with the following general terms. [9.1]

1. $a_n = 3n - 5$
2. $a_1 = 3, a_n = a_{n-1} + 4, n > 1$
3. $a_n = n^2 + 1$
4. $a_n = 2n^3$
5. $a_n = \dfrac{n + 1}{n^2}$
6. $a_1 = 4, a_n = -2a_{n-1}, n > 1$

Give the general term for each sequence. [9.1]

7. $6, 10, 14, 18, \ldots$
8. $1, 2, 4, 8, \ldots$
9. $\frac{1}{2}, \frac{1}{4}, \frac{1}{8}, \frac{1}{16}, \ldots$
10. $-3, 9, -27, 81, \ldots$

11. Expand and simplify each of the following. [9.2]

    (a) $\displaystyle\sum_{i=1}^{5}(5i + 3)$ (b) $\displaystyle\sum_{i=3}^{5}(2^i - 1)$

    (c) $\displaystyle\sum_{i=2}^{6}(i^2 + 2i)$

12. Find the first term of an arithmetic progression if $a_5 = 11$ and $a_9 = 19$. [9.3]
13. Find the second term of a geometric progression for which $a_3 = 18$ and $a_5 = 162$. [9.4]

Find the sum of the first 10 terms of the following arithmetic progressions. [9.3]

14. $5, 11, 17, \ldots$ 15. $25, 20, 15, \ldots$

16. Write a formula for the sum of the first 50 terms of the geometric progression $3, 6, 12, \ldots$. [9.4]
17. Find the sum of $\frac{1}{2} + \frac{1}{6} + \frac{1}{18} + \frac{1}{54} + \ldots$. [9.4]

Use the binomial formula to expand each of the following. [9.5]

18. $(x - 3)^4$ 19. $(2x - 1)^5$

20. Find the first 3 terms in the expansion of $(x - 1)^{20}$. [9.5]
21. Find the sixth term in $(2x - 3y)^8$. [9.5]

Conic Sections

Contents

Introduction

If you tack down the two ends of a piece of string, then place a pencil anywhere on the string and trace around the two tacks, the shape you will trace is called an **ellipse** (Figure 1).

Pencil tracing out an ellipse from a string anchored by two tacks.

FIGURE 1

An ellipse is one of a general category of shapes called **conic sections.** Conic sections include ellipses, circles, hyperbolas, and parabolas. They are called conic sections because each can be found by slicing a cone with a plane as shown in Figure 2.

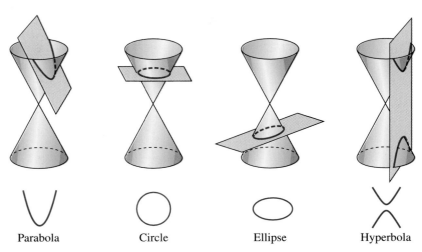

Parabola   Circle   Ellipse   Hyperbola

FIGURE 2

There are many applications associated with conic sections. The planets orbit the sun in elliptical orbits. Many of the comets that come in contact with the gravitational field surrounding the earth travel in parabolic or hyperbolic paths. Flashlight and searchlight mirrors have elliptical or parabolic shapes because of the way those surfaces reflect light.

Overview

You have already been introduced to parabolas in Chapters 3 and 6. We begin this chapter with a study of circles. Then we look at ellipses and hyperbolas. Our approach to the three new conic sections will be to investigate the equations that produce the conic sections. We end the chapter with a look at systems of equations in which one or both of the equations represent conic sections.

The Circle

Before we find the general equation of a circle, we must first derive what is known as the **distance formula.**

Suppose (x_1, y_1) and (x_2, y_2) are any two points in the first quadrant. (Actually, we could choose the two points to be anywhere on the coordinate plane. It is just more convenient to have them in the first quadrant.) We can name the points P_1 and P_2, respectively, and draw the diagram shown in Figure 1.

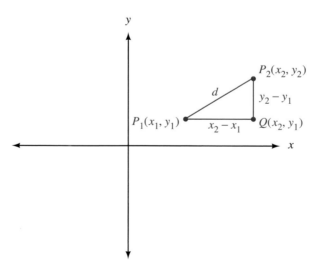

FIGURE 1

The notation with a bar over two points, such as $\overline{P_1P_2}$, refers to the line segment formed by connecting the two points—in this case, points P_1 and P_2.

Notice the coordinates of point Q. The x-coordinate is x_2 since Q is directly below point P_2. The y-coordinate of Q is y_1 since Q is directly across from point P_1. It is evident from the diagram that the length of $\overline{P_2Q}$ is $y_2 - y_1$ and the length of $\overline{P_1Q}$ is $x_2 - x_1$. Using the Pythagorean Theorem, we have

$$(\overline{P_1P_2})^2 = (\overline{P_1Q})^2 + (\overline{P_2Q})^2$$

or

$$d^2 = (x_2 - x_1)^2 + (y_2 - y_1)^2$$

Taking the square root of both sides, we have

$$d = \sqrt{(x_2 - x_1)^2 + (y_2 - y_1)^2}$$

We know this is the positive square root, since d is the distance from P_1 to P_2 and must therefore be positive. This formula is called the **distance formula.**

EXAMPLE 1   Find the distance between $(3, 5)$ and $(2, -1)$.

SOLUTION   If we let $(3, 5)$ be (x_1, y_1) and $(2, -1)$ be (x_2, y_2) and apply the distance formula, we have

$$\begin{aligned} d &= \sqrt{(2 - 3)^2 + (-1 - 5)^2} \\ &= \sqrt{(-1)^2 + (-6)^2} \\ &= \sqrt{1 + 36} \\ &= \sqrt{37} \end{aligned}$$

EXAMPLE 2   Find x if the distance from $(x, 5)$ to $(3, 4)$ is $\sqrt{2}$.

SOLUTION   Using the distance formula, we have

$$\begin{aligned} \sqrt{2} &= \sqrt{(x - 3)^2 + (5 - 4)^2} \\ 2 &= (x - 3)^2 + 1^2 \\ 2 &= x^2 - 6x + 9 + 1 \\ 0 &= x^2 - 6x + 8 \\ 0 &= (x - 4)(x - 2) \\ x &= 4 \quad \text{or} \quad x = 2 \end{aligned}$$

The two solutions are 4 and 2, which indicates there are two points, $(4, 5)$ and $(2, 5)$, which are $\sqrt{2}$ units from $(3, 4)$.

We can use the distance formula to derive the equation of a circle.

THEOREM 10.1

The equation of the **circle** with center at (a, b) and radius r is given by

$$(x - a)^2 + (y - b)^2 = r^2$$

530

Proof

By definition, all points on the circle are a distance r from the center (a, b). If we let (x, y) represent any point on the circle, then (x, y) is r units from (a, b). Applying the distance formula, we have

$$r = \sqrt{(x - a)^2 + (y - b)^2}$$

Squaring both sides of this equation gives the equation of the circle:

$$(x - a)^2 + (y - b)^2 = r^2$$

We can use Theorem 10.1 to find the equation of a circle given its center and radius, or to find its center and radius given the equation.

EXAMPLE 3   Find the equation of the circle with center at $(-3, 2)$ having a radius of 5.

SOLUTION   We have $(a, b) = (-3, 2)$ and $r = 5$. Applying Theorem 10.1 yields

$$[x - (-3)]^2 + (y - 2)^2 = 5^2$$
$$(x + 3)^2 + (y - 2)^2 = 25$$

EXAMPLE 4   Give the equation of the circle with radius 3 whose center is at the origin.

SOLUTION   The coordinates of the center are $(0, 0)$, and the radius is 3. The equation must be

$$(x - 0)^2 + (y - 0)^2 = 3^2$$
$$x^2 + y^2 = 9$$

We can see from Example 4 that the equation of any circle with its center at the origin and radius r will be $x^2 + y^2 = r^2$.

EXAMPLE 5   Find the center and radius of the circle whose equation is

$$(x - 1)^2 + (y + 3)^2 = 4$$

Sketch the graph of the circle.

SOLUTION   Writing the equation in the form

$$(x - a)^2 + (y - b)^2 = r^2$$

we have

$$(x - 1)^2 + [y - (-3)]^2 = 2^2$$

The center is at $(1, -3)$ and the radius is 2. (See Figure 2.)

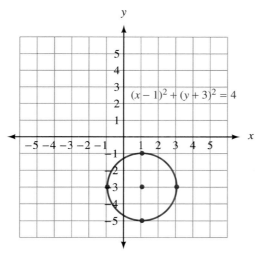

FIGURE 2

EXAMPLE 6 Sketch the graph of $x^2 + y^2 = 9$.

SOLUTION Since the equation can be written in the form

$$(x - 0)^2 + (y - 0)^2 = 3^2$$

it must have its center at $(0, 0)$ and a radius of 3. (See Figure 3.)

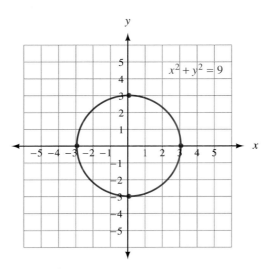

FIGURE 3

EXAMPLE 7   Sketch the graph of $x^2 + y^2 + 6x - 4y - 12 = 0$.

SOLUTION   To sketch the graph we must find the center and radius. The center and radius can be identified if the equation has the form

$$(x - a)^2 + (y - b)^2 = r^2$$

The original equation can be written in this form by completing the squares on x and y:

$$x^2 + y^2 + 6x - 4y - 12 = 0$$
$$x^2 + 6x + y^2 - 4y = 12$$
$$x^2 + 6x + \mathbf{9} + y^2 - 4y + \mathbf{4} = 12 + \mathbf{9} + \mathbf{4}$$
$$(x + 3)^2 + (y - 2)^2 = 25$$
$$[x - (-3)]^2 + (y - 2)^2 = 5^2$$

From the last line it is apparent that the center is at $(-3, 2)$ and the radius is 5. (See Figure 4.)

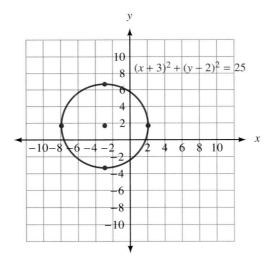

FIGURE 4

$\mathcal{P}roblem\ \mathcal{S}et$

10.1

Find the distance between the following points.

1. $(3, 7)$ and $(6, 3)$
2. $(4, 7)$ and $(8, 1)$
3. $(0, 9)$ and $(5, 0)$
4. $(-3, 0)$ and $(0, 4)$
5. $(3, -5)$ and $(-2, 1)$
6. $(-8, 9)$ and $(-3, -2)$
7. $(-1, -2)$ and $(-10, 5)$

8. $(-3, -8)$ and $(-1, 6)$

9. Find x so the distance between $(x, 2)$ and $(1, 5)$ is $\sqrt{13}$.

10. Find x so the distance between $(-2, 3)$ and $(x, 1)$ is 3.

11. Find y so the distance between $(7, y)$ and $(8, 3)$ is 1.

12. Find y so the distance between $(3, -5)$ and $(3, y)$ is 9.

Write the equation of the circle with the given center and radius.

13. Center $(2, 3)$; $r = 4$

14. Center $(3, -1)$; $r = 5$

15. Center $(3, -2)$; $r = 3$

16. Center $(-2, 4)$; $r = 1$

17. Center $(-5, -1)$; $r = \sqrt{5}$

18. Center $(-7, -6)$; $r = \sqrt{3}$

19. Center $(0, -5)$; $r = 1$

20. Center $(0, -1)$; $r = 7$

21. Center $(0, 0)$; $r = 2$

22. Center $(0, 0)$; $r = 5$

Give the center and radius, and sketch the graph of each of the following circles.

23. $x^2 + y^2 = 4$
24. $x^2 + y^2 = 16$
25. $(x - 1)^2 + (y - 3)^2 = 25$
26. $(x - 4)^2 + (y - 1)^2 = 36$
27. $(x + 2)^2 + (y - 4)^2 = 8$
28. $(x - 3)^2 + (y + 1)^2 = 12$
29. $(x + 1)^2 + (y + 1)^2 = 1$
30. $(x + 3)^2 + (y + 2)^2 = 9$
31. $x^2 + y^2 - 6y = 7$
32. $x^2 + y^2 - 4y = 5$
33. $x^2 + y^2 + 2x = 1$
34. $x^2 + y^2 + 10x = 0$
35. $x^2 + y^2 - 4x - 6y = -4$
36. $x^2 + y^2 - 4x + 2y = 4$
37. $x^2 + y^2 + 2x + y = \frac{11}{4}$
38. $x^2 + y^2 - 6x - y = -\frac{1}{4}$

39. Find the equation of the circle with center at the origin that contains the point $(3, 4)$.

40. Find the equation of the circle with center at the origin that contains the point $(-5, 12)$.

41. Find the equation of the circle with center at the origin and x-intercepts 3 and -3.

42. Find the equation of the circle with y-intercepts 4 and -4, and center at the origin.

43. A circle with center at $(-1, 3)$ passes through the point $(4, 3)$. Find the equation.

44. A circle with center at $(2, 5)$ passes through the point $(-1, 4)$. Find the equation.

45. If we solved the equation $x^2 + y^2 = 9$ for y, we would obtain the equation $y = \pm\sqrt{9 - x^2}$. This last equation is equivalent to the two equations $y = \sqrt{9 - x^2}$, in which y is always positive, and $y = -\sqrt{9 - x^2}$, in which y is always negative. Look at the graph of $x^2 + y^2 = 9$ in Example 6 of this section and indicate what part of the graph each of the two equations corresponds to.

46. Solve the equation $x^2 + y^2 = 9$ for x, and then indicate what part of the graph in Example 6 each of the resulting equations corresponds to.

47. The formula for the circumference of a circle is $C = 2\pi r$. If the units of the coordinate system used in Problem 25 are in meters, what is the circumference of that circle?

48. The formula for the area of a circle is $A = \pi r^2$. What is the area of the circle mentioned in Problem 47?

Review Problems

 The problems below review material we covered in Sections 9.1 and 9.2.

Find the general term of each sequence. [9.1]

49. 5, 9, 13, 17, . . .
50. 3, 8, 15, 24, . . .

Expand and simplify each series. [9.2]

51. $\displaystyle\sum_{i=2}^{5}\left(\frac{1}{2}\right)^{i}$

52. $\displaystyle\sum_{i=3}^{6}(i^{2}-5)$

Write using summation notation. [9.2]

53. $1 + 3 + 5 + 7 + 9$
54. $\frac{2}{3} + \frac{3}{4} + \frac{4}{5} + \frac{5}{6}$

One Step Further

*A circle is **tangent** to a line if it touches, but does not cross, the line.*

55. Find the equation of the circle with center at $(2, 3)$ if the circle is tangent to the y-axis.

56. Find the equation of the circle with center at $(3, 2)$ if the circle is tangent to the x-axis.

57. Find the equation of the circle with center at $(2, 3)$ if the circle is tangent to the vertical line $x = 4$.

58. Find the equation of the circle with center at $(3, 2)$ if the circle is tangent to the horizontal line $y = 6$.

Find the distance from the origin to the center of each circle given below.

59. $x^{2} + y^{2} - 6x + 8y = 144$
60. $x^{2} + y^{2} - 8x + 6y = 144$
61. $x^{2} + y^{2} - 6x - 8y = 144$
62. $x^{2} + y^{2} + 8x + 6y = 144$

SECTION
10.2

Ellipses and Hyperbolas

This section is concerned with the graphs of ellipses and hyperbolas. To simplify matters somewhat we will begin by considering those graphs that are centered about the origin.

Suppose we want to graph the equation

$$\frac{x^{2}}{25} + \frac{y^{2}}{9} = 1$$

We can find the y-intercepts by letting $x = 0$, and the x-intercepts by letting $y = 0$:

When $x = 0$:	When $y = 0$:
$\dfrac{0^{2}}{25} + \dfrac{y^{2}}{9} = 1$	$\dfrac{x^{2}}{25} + \dfrac{0^{2}}{9} = 1$
$y^{2} = 9$	$x^{2} = 25$
$y = \pm 3$	$x = \pm 5$

The graph crosses the y-axis at $(0, 3)$ and $(0, -3)$ and the x-axis at $(5, 0)$ and $(-5, 0)$. Graphing these points and then connecting them with a smooth curve gives the graph shown in Figure 1.

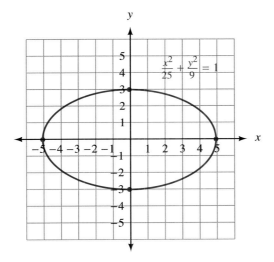

FIGURE I

We can find other ordered pairs on the graph by substituting values for x (or y) and then solving for y (or x). For example, if we let $x = 3$, then

$$\frac{3^2}{25} + \frac{y^2}{9} = 1$$

$$\frac{9}{25} + \frac{y^2}{9} = 1$$

$$0.36 + \frac{y^2}{9} = 1$$

$$\frac{y^2}{9} = 0.64$$

$$y^2 = 5.76$$

$$y = \pm 2.4$$

This would give us the two ordered pairs $(3, -2.4)$ and $(3, 2.4)$.

A graph of the type shown in Figure 1 is called an **ellipse.** If we were to find some other ordered pairs that satisfy our original equation, we would find that their graphs lie on the ellipse. Also, the coordinates of any point on the ellipse will satisfy the equation. We can generalize these results as follows.

ELLIPSES CENTERED AT THE ORIGIN

The graph of any equation of the form

$$\frac{x^2}{a^2} + \frac{y^2}{b^2} = 1 \qquad \textbf{Standard form}$$

(Continued)

will be an **ellipse centered at the origin.** The ellipse will cross the x-axis at $(a, 0)$ and $(-a, 0)$. It will cross the y-axis at $(0, b)$ and $(0, -b)$. When a and b are equal, the ellipse will be a circle. Each of the points $(a, 0)$, $(-a, 0)$, $(0, b)$, and $(0, -b)$ is a **vertex** of the graph.

The most convenient way to graph an ellipse centered at the origin is to locate the intercepts.

EXAMPLE 1 Sketch the graph of $4x^2 + 9y^2 = 36$.

SOLUTION To write the equation in the form

$$\frac{x^2}{a^2} + \frac{y^2}{b^2} = 1$$

we must divide both sides by 36:

$$\frac{4x^2}{36} + \frac{9y^2}{36} = \frac{36}{36}$$

$$\frac{x^2}{9} + \frac{y^2}{4} = 1$$

The graph crosses the x-axis at $(3, 0)$, $(-3, 0)$ and the y-axis at $(0, 2)$, $(0, -2)$. (See Figure 2.)

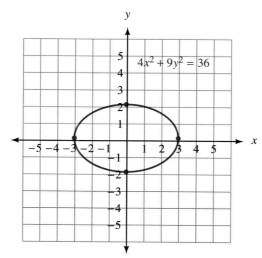

FIGURE 2

Consider the equation

$$\frac{x^2}{9} - \frac{y^2}{4} = 1$$

If we were to find a number of ordered pairs that are solutions to the equation and connect their graphs with a smooth curve, we would have Figure 3.

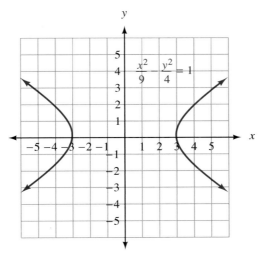

FIGURE 3

This graph is an example of a **hyperbola.** Notice that the graph has x-intercepts at $(3, 0)$ and $(-3, 0)$. The graph has no y-intercepts and hence does not cross the y-axis, since substituting $x = 0$ into the equation yields

$$\frac{0^2}{9} - \frac{y^2}{4} = 1$$
$$-y^2 = 4$$
$$y^2 = -4 \quad \text{No real solution}$$

We can, however, use the denominator of the y^2 term to help sketch the graph. If we draw a rectangle with sides parallel to the x- and y-axes and passing through the x-intercepts and the points on the y-axis corresponding to the square roots of the denominator of the y^2 term ($+2$ and -2), we get the rectangle shown in Figure 4 (p. 538). The lines that connect opposite corners of the rectangle are called **asymptotes.** The graph of the hyperbola

$$\frac{x^2}{9} - \frac{y^2}{4} = 1$$

will approach these lines. Figure 4 shows the graph.

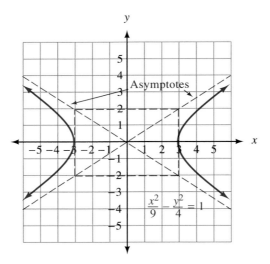

FIGURE 4

EXAMPLE 2   Graph the equation: $\dfrac{y^2}{9} - \dfrac{x^2}{16} = 1$.

SOLUTION   In this case the y-intercepts are 3 and -3, and the x-intercepts do not exist. We can use the square roots of the denominator of the x^2 term, however, to find the asymptotes associated with the graph. The sides of the rectangle used to draw the asymptotes must pass through 3 and -3 on the y-axis, and 4 and -4 on the x-axis. (See Figure 5.)

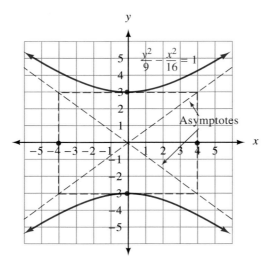

FIGURE 5

Here is a summary of what we know about hyperbolas.

◤ HYPERBOLAS CENTERED AT THE ORIGIN

The graph of the equation

$$\frac{x^2}{a^2} - \frac{y^2}{b^2} = 1$$

will be a **hyperbola centered at the origin.** The graph will have x-intercepts (vertices) at $-a$ and a.

The graph of the equation

$$\frac{y^2}{a^2} - \frac{x^2}{b^2} = 1$$

will be a **hyperbola centered at the origin.** The graph will have y-intercepts (vertices) at $-a$ and a.

As an aid in sketching either of these equations, the asymptotes can be found by drawing lines through opposite corners of the rectangle whose sides pass through $-a$, a, $-b$, and b on the axes.

Ellipses and Hyperbolas Not Centered at the Origin

The equation below is the equation of an ellipse with its center at the point $(4, 1)$:

$$\frac{(x-4)^2}{9} + \frac{(y-1)^2}{4} = 1$$

To see why the center is at $(4, 1)$ we substitute x' (read "x prime") for $x - 4$ and y' for $y - 1$ in the equation. That is:

$$\text{If} \qquad x' = x - 4$$
$$\text{and} \qquad y' = y - 1$$
$$\text{the equation} \quad \frac{(x-4)^2}{9} + \frac{(y-1)^2}{4} = 1$$
$$\text{becomes} \qquad \frac{(x')^2}{9} + \frac{(y')^2}{4} = 1$$

This is the equation of an ellipse in a coordinate system with an x'-axis and a y'-axis. We call this new coordinate system the **$x'y'$-coordinate system.** The center of our ellipse is at the origin in the $x'y'$-coordinate system. The question is this: What are the coordinates of the center of this ellipse in the original xy-coordinate system? To answer this question we go back to our original substitutions:

$$x' = x - 4$$
$$y' = y - 1$$

In the $x'y'$-coordinate system, the center of our ellipse is at $x' = 0$, $y' = 0$ (the origin of the $x'y'$ system). Substituting these numbers for x' and y', we have

$$0 = x - 4$$
$$0 = y - 1$$

Solving these equations for x and y will give us the coordinates of the center of our ellipse in the xy-coordinate system. As you can see, the solutions are $x = 4$ and $y = 1$. Therefore, in the xy-coordinate system, the center of our ellipse is at the point $(4, 1)$. Figure 6 shows the graph.

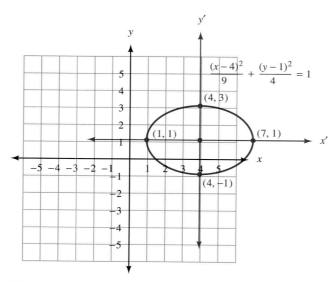

FIGURE 6

The coordinates of all points labeled in Figure 6 are given with respect to the xy-coordinate system. The x' and y' axes are shown simply for reference in our discussion. Note that the horizontal distance from the center to the vertices is 3—the square root of the denominator of the $(x - 4)^2$ term. Likewise, the vertical distance from the center to the other vertices is 2—the square root of the denominator of the $(y - 1)^2$ term.

We summarize the information above with the following:

◗ **AN ELLIPSE WITH CENTER AT (h, k)**

The graph of the equation

$$\frac{(x - h)^2}{a^2} + \frac{(y - k)^2}{b^2} = 1$$

will be an **ellipse with center at (h, k).** The vertices of the ellipse will be at the points $(h + a, k)$, $(h - a, k)$, $(h, k + b)$, and $(h, k - b)$.

EXAMPLE 3 Graph the ellipse: $x^2 + 9y^2 + 4x - 54y + 76 = 0$.

SOLUTION In order to identify the coordinates of the center, we must complete the square on x and also on y. To begin, we rearrange the terms so that the terms containing x are together, the terms containing y are together, and the constant term is on the other side of the equal sign. Doing so gives us the following equation:

$$x^2 + 4x + 9y^2 - 54y = -76$$

Before we can complete the square on y we must factor 9 from each term containing y:

$$x^2 + 4x + 9(y^2 - 6y) = -76$$

To complete the square on x, we add 4 to each side of the equation. To complete the square on y, we add 9 inside the parentheses. This increases the left side of the equation by 81 since each term within the parentheses is multiplied by 9. Therefore, we must add 81 to the right side of the equation also.

$$x^2 + 4x + \mathbf{4} + 9(y^2 - 6y + \mathbf{9}) = -76 + \mathbf{4} + \mathbf{81}$$
$$(x + 2)^2 + 9(y - 3)^2 = 9$$

To identify the distances to the vertices, we divide each term on both sides by 9:

$$\frac{(x + 2)^2}{9} + \frac{9(y - 3)^2}{9} = \frac{9}{9}$$
$$\frac{(x + 2)^2}{9} + \frac{(y - 3)^2}{1} = 1$$

The graph is an ellipse with center at $(-2, 3)$, as shown in Figure 7.

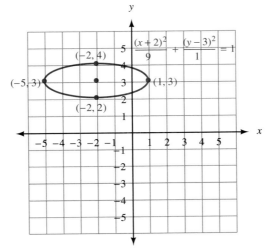

FIGURE 7

The ideas associated with graphing hyperbolas whose centers are not at the origin parallel the ideas just presented about graphing ellipses whose centers have been moved off the origin. Without showing the justification for doing so, we state the following guidelines for graphing hyperbolas:

HYPERBOLAS WITH CENTERS AT (h, k)

The graphs of the equations

$$\frac{(x - h)^2}{a^2} - \frac{(y - k)^2}{b^2} = 1 \quad \text{and} \quad \frac{(y - k)^2}{b^2} - \frac{(x - h)^2}{a^2} = 1$$

will be hyperbolas with their centers at (h, k). The vertices of the graph of the first equation will be at the points ($h + a, k$) and ($h - a, k$), while the vertices for the graph of the second equation will be at ($h, k + b$) and ($h, k - b$). In either case, the asymptotes can be found by connecting opposite corners of the rectangle that contains the four points ($h + a, k$), ($h - a, k$), ($h, k + b$), and ($h, k - b$).

EXAMPLE 4   Graph the hyperbola: $4x^2 - y^2 + 4y - 20 = 0$.

SOLUTION   In order to identify the coordinates of the center of the hyperbola, we need to complete the square on y. (Since there is no linear term in x, we do not need to complete the square on x. The x-coordinate of the center will be $x = 0$.)

$$4x^2 - y^2 + 4y - 20 = 0$$
$$4x^2 - y^2 + 4y = 20 \quad \text{Add 20 to each side.}$$
$$4x^2 - 1(y^2 - 4y) = 20 \quad \text{Factor } -1 \text{ from each term containing } y.$$

To complete the square on y, we add 4 to the terms inside the parentheses. Doing so adds -4 to the left side of the equation since everything inside the parentheses is multiplied by -1. To keep from changing the equation we must add -4 to the right side also.

$$4x^2 - 1(y^2 - 4y + \mathbf{4}) = 20 - \mathbf{4}$$
$$4x^2 - 1(y - 2)^2 = 16$$
$$\frac{4x^2}{16} - \frac{(y - 2)^2}{16} = \frac{16}{16}$$
$$\frac{x^2}{4} - \frac{(y - 2)^2}{16} = 1$$

This is the equation of a hyperbola with center at (0, 2). The graph opens to the right and left as shown in Figure 8.

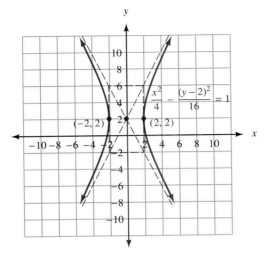

FIGURE 8

Problem Set

10.2

Graph each of the following. Be sure to label both the x- and y-intercepts.

1. $\dfrac{x^2}{9} + \dfrac{y^2}{16} = 1$

2. $\dfrac{x^2}{25} + \dfrac{y^2}{4} = 1$

3. $\dfrac{x^2}{16} + \dfrac{y^2}{9} = 1$

4. $\dfrac{x^2}{4} + \dfrac{y^2}{25} = 1$

5. $\dfrac{x^2}{3} + \dfrac{y^2}{4} = 1$

6. $\dfrac{x^2}{4} + \dfrac{y^2}{3} = 1$

7. $4x^2 + 25y^2 = 100$
8. $4x^2 + 9y^2 = 36$
9. $x^2 + 8y^2 = 16$
10. $12x^2 + y^2 = 36$

Graph each of the following. Show the intercepts and the asymptotes in each case.

11. $\dfrac{x^2}{9} - \dfrac{y^2}{16} = 1$

12. $\dfrac{x^2}{25} - \dfrac{y^2}{4} = 1$

13. $\dfrac{x^2}{16} - \dfrac{y^2}{9} = 1$

14. $\dfrac{x^2}{4} - \dfrac{y^2}{25} = 1$

15. $\dfrac{y^2}{9} - \dfrac{x^2}{16} = 1$

16. $\dfrac{y^2}{25} - \dfrac{x^2}{4} = 1$

17. $\dfrac{y^2}{36} - \dfrac{x^2}{4} = 1$

18. $\dfrac{y^2}{4} - \dfrac{x^2}{36} = 1$

19. $x^2 - 4y^2 = 4$
20. $y^2 - 4x^2 = 4$
21. $16y^2 - 9x^2 = 144$
22. $4y^2 - 25x^2 = 100$

Find the x- and y-intercepts, if they exist, for each of the following. Do not graph.

23. $0.4x^2 + 0.9y^2 = 3.6$
24. $1.6x^2 + 0.9y^2 = 14.4$

25. $\dfrac{x^2}{0.04} - \dfrac{y^2}{0.09} = 1$

26. $\dfrac{y^2}{0.16} - \dfrac{x^2}{0.25} = 1$

27. $\dfrac{25x^2}{9} + \dfrac{25y^2}{4} = 1$

28. $\dfrac{16x^2}{9} + \dfrac{16y^2}{25} = 1$

Graph each of the following ellipses. In each case, label the coordinates of the center and the vertices.

29. $\dfrac{(x-4)^2}{4} + \dfrac{(y-2)^2}{9} = 1$

30. $\dfrac{(x-2)^2}{4} + \dfrac{(y-4)^2}{9} = 1$

31. $4x^2 + y^2 - 4y - 12 = 0$

32. $4x^2 + y^2 - 24x - 4y + 36 = 0$

33. $x^2 + 9y^2 + 4x - 54y + 76 = 0$

34. $4x^2 + y^2 - 16x + 2y + 13 = 0$

Graph each of the following hyperbolas. In each case, label the coordinates of the center and the vertices and show the asymptotes.

35. $\dfrac{(x-2)^2}{16} - \dfrac{y^2}{4} = 1$

36. $\dfrac{(y-2)^2}{16} - \dfrac{x^2}{4} = 1$

37. $9y^2 - x^2 - 4x + 54y + 68 = 0$

38. $4x^2 - y^2 - 24x + 4y + 28 = 0$

39. $4y^2 - 9x^2 - 16y + 72x - 164 = 0$

40. $4x^2 - y^2 - 16x - 2y + 11 = 0$

41. Find y when x is 4 in the equation $\dfrac{x^2}{25} + \dfrac{y^2}{9} = 1$.

42. Find x when y is 3 in the equation $\dfrac{x^2}{4} + \dfrac{y^2}{25} = 1$.

43. Find y when x is 1.8 in $16x^2 + 9y^2 = 144$.

44. Find y when x is 1.6 in $49x^2 + 4y^2 = 196$.

45. Give the equations of the two asymptotes in the graph you found in Problem 15.

46. Give the equations of the two asymptotes in the graph you found in Problem 16.

47. The longer line segment connecting opposite vertices of an ellipse is called the **major axis** of the ellipse. Give the length of the major axis of the ellipse you graphed in Problem 3.

48. The shorter line segment connecting opposite vertices of an ellipse is called the **minor axis** of the ellipse. Give the length of the minor axis of the ellipse you graphed in Problem 3.

Review Problems

 The problems below review material we covered in Sections 9.3 and 9.4.

Find the general term of each sequence.

49. 5, 11, 17, 23, . . .

50. $-3, 9, -27, 81, \ldots$

51. An arithmetic sequence has a first term of $a_1 = 4$ and a common difference of $d = 5$. Find the sum of the first 20 terms, S_{20}.

52. An arithmetic sequence is such that $a_4 = 23$ and $a_9 = 48$. Find a_{40}.

53. A geometric sequence has a first term of $a_1 = 8$ and a common ratio of $r = \frac{1}{2}$. Find the sum of the first 6 terms.

54. Find the sum: $1 + \frac{1}{2} + \frac{1}{4} + \frac{1}{8} + \cdots$

Second-Degree Inequalities and Nonlinear Systems

In Section 3.4 we graphed linear inequalities by first graphing the boundary and then choosing a test point not on the boundary to indicate the region used for the solution set. The problems in this section are very similar. We will use the same general methods for graphing the inequalities in this section that we used in Section 3.4.

EXAMPLE 1   Graph: $x^2 + y^2 < 16$.

SOLUTION   The boundary is $x^2 + y^2 = 16$, which is a circle with center at the origin and a radius of 4. Since the inequality sign is $<$, the boundary is not included in the solution set and must therefore be represented with a broken line. The graph of the boundary is shown in Figure 1.

The solution set for the inequality $x^2 + y^2 < 16$ is either the region inside the circle or the region outside the circle. To see which region represents the solution set, we choose a convenient point not on the boundary and test it in the original inequality. The origin $(0, 0)$ is a convenient point. Since the ori-

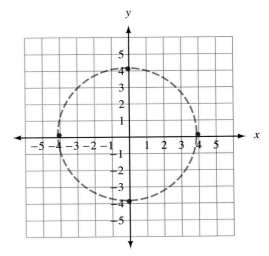

FIGURE 1

gin satisfies the inequality $x^2 + y^2 < 16$, all points in the same region will also satisfy the inequality. The graph of the solution set is shown in Figure 2.

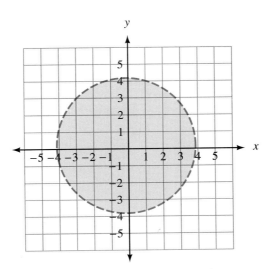

FIGURE 2

EXAMPLE 2   Graph the inequality: $y \le x^2 - 2$.

SOLUTION   The parabola $y = x^2 - 2$ is the boundary and is included in the solution set. Using $(0, 0)$ as the test point, we see that $0 \le 0^2 - 2$ is a false statement, which means that the region containing $(0, 0)$ is not in the solution set. (See Figure 3, at the top of the next page.)

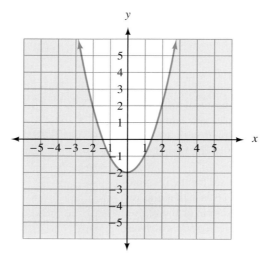

FIGURE 3

EXAMPLE 3 Graph: $4y^2 - 9x^2 < 36$.

SOLUTION The boundary is the hyperbola $4y^2 - 9x^2 = 36$ and is not included in the solution set. Testing $(0, 0)$ in the original inequality yields a true statement, which means that the region containing the origin is the solution set. (See Figure 4.)

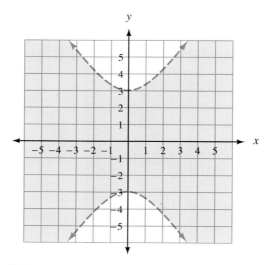

FIGURE 4

Next, we solve systems of equations that contain at least one second-degree equation. The most convenient method of solving a system that contains one or two second-degree equations is by substitution, although the addition method can be used at times.

EXAMPLE 4 Solve the system:

$$x^2 + y^2 = 4$$
$$x - 2y = 4$$

SOLUTION In this case the substitution method is the most convenient. Solving the second equation for x in terms of y, we have

$$x - 2y = 4$$
$$x = 2y + 4$$

We now substitute $2y + 4$ for x in the first equation in our original system and proceed to solve for y:

$$(2y + 4)^2 + y^2 = 4$$
$$4y^2 + 16y + 16 + y^2 = 4$$
$$5y^2 + 16y + 12 = 0$$
$$(5y + 6)(y + 2) = 0$$
$$5y + 6 = 0 \quad \text{or} \quad y + 2 = 0$$
$$y = -\tfrac{6}{5} \quad \text{or} \qquad y = -2$$

These are the y-coordinates of the two solutions to the system. Substituting $y = -\tfrac{6}{5}$ into $x - 2y = 4$ and solving for x gives us $x = \tfrac{8}{5}$. Using $y = -2$ in the same equation yields $x = 0$. The two solutions to our system are $(\tfrac{8}{5}, -\tfrac{6}{5})$ and $(0, -2)$. Although graphing the system is not necessary, it does help us visualize the situation. (See Figure 5.)

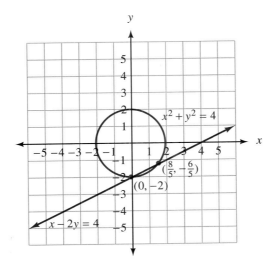

FIGURE 5

EXAMPLE 5 Solve the system:

$$16x^2 - 4y^2 = 64$$
$$x^2 + y^2 = 9$$

SOLUTION Since both equations are second-degree in x and y, it is easier to solve this system by eliminating one of the variables by addition. To eliminate y we multiply the bottom equation by 4 and add the result to the top equation:

$$16x^2 - 4y^2 = 64$$
$$\underline{4x^2 + 4y^2 = 36}$$
$$20x^2 \qquad = 100$$
$$x^2 = 5$$
$$x = \pm\sqrt{5}$$

The x-coordinates of the points of intersection are $\sqrt{5}$ and $-\sqrt{5}$. We substitute each back into the second equation in the original system and solve for y:

When $x = \sqrt{5}$: | When $x = -\sqrt{5}$:

$$(\sqrt{5})^2 + y^2 = 9 \qquad\qquad (-\sqrt{5})^2 + y^2 = 9$$
$$5 + y^2 = 9 \qquad\qquad\qquad 5 + y^2 = 9$$
$$y^2 = 4 \qquad\qquad\qquad\qquad y^2 = 4$$
$$y = \pm 2 \qquad\qquad\qquad\qquad y = \pm 2$$

The four points of intersection are $(\sqrt{5}, 2)$, $(\sqrt{5}, -2)$, $(-\sqrt{5}, 2)$, and $(-\sqrt{5}, -2)$. Graphically the situation is as shown in Figure 6.

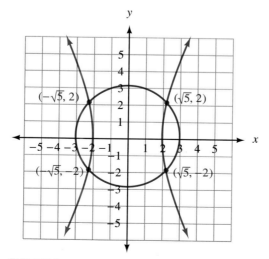

FIGURE 6

EXAMPLE 6 Solve the system:

$$x^2 - 2y = 2$$
$$y = x^2 - 3$$

SOLUTION We can solve this system using the substitution method. Replacing y in the first equation with $x^2 - 3$ from the second equation, we have

$$x^2 - 2(x^2 - 3) = 2$$
$$-x^2 + 6 = 2$$
$$x^2 = 4$$
$$x = \pm 2$$

Using either $+2$ or -2 in the equation $y = x^2 - 3$ gives us $y = 1$. The system has two solutions: $(2, 1)$ and $(-2, 1)$.

EXAMPLE 7 The sum of the squares of two numbers is 34. The difference of their squares is 16. Find the two numbers.

SOLUTION Let x and y be the two numbers. Then the sum of their squares is $x^2 + y^2$, and the difference of their squares is $x^2 - y^2$. (We can assume here that x^2 is the larger number.) The system of equations that describes the situation is

$$x^2 + y^2 = 34$$
$$x^2 - y^2 = 16$$

We can eliminate y by simply adding the two equations. The result is

$$2x^2 = 50$$
$$x^2 = 25$$
$$x = \pm 5$$

Substituting $x = 5$ into either equation in the system gives $y = \pm 3$. Using $x = -5$ gives the same results, $y = \pm 3$. The four pairs of numbers that are solutions to the original problem are $(5, 3)$, $(-5, 3)$, $(5, -3)$, and $(-5, -3)$.

We now turn our attention to systems of inequalities. To solve a system of inequalities by graphing, we simply graph each inequality on the same set of axes. The solution set for the system is the region common to both graphs—the intersection of the individual solution sets.

EXAMPLE 8 Graph the solution set for the system:

$$x^2 + y^2 \leq 9$$
$$\frac{x^2}{4} + \frac{y^2}{25} \geq 1$$

SOLUTION The boundary for the top inequality is a circle with center at the origin and a radius of 3. The solution set lies inside the boundary. The boundary for the second inequality is an ellipse. In this case the solution set lies outside the boundary. For both graphs, the boundary is also part of the solution set. (See Figure 7, p. 550.)

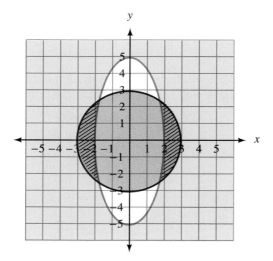

FIGURE 7

The solution set is the intersection of the two individual solution sets, indicated in Figure 7 by the black diagonal lines.

EXAMPLE 9 Graph the solution set for the system:

$$x - 2y \leq 4$$
$$x + y \leq 4$$
$$x \geq -1$$

SOLUTION We have three linear inequalities, representing three sections of the coordinate plane. The graph of the solution set for this system will be the intersection of these three sections. The graph of $x - 2y \leq 4$ is the section above and including $x - 2y = 4$ (the boundary). The graph of $x + y \leq 4$ is the section below and including the boundary line $x + y = 4$. The graph of $x \geq -1$ is all the points to the right of, and including, the vertical line $x = -1$. The intersection of these three graphs is shown in Figure 8.

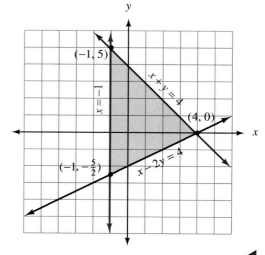

FIGURE 8

Problem Set

10.3

Graph each of the following inequalities:

1. $x^2 + y^2 \leq 49$
2. $x^2 + y^2 < 49$
3. $(x - 2)^2 + (y + 3)^2 < 16$
4. $(x + 3)^2 + (y - 2)^2 \geq 25$
5. $y < x^2 - 6x + 7$
6. $y \geq x^2 + 2x - 8$

7. $\dfrac{x^2}{25} - \dfrac{y^2}{9} \geq 1$

8. $\dfrac{x^2}{25} - \dfrac{y^2}{9} \leq 1$

9. $4x^2 + 25y^2 \leq 100$
10. $25x^2 - 4y^2 > 100$

Graph the solution sets to the following systems:

11. $x^2 + y^2 < 9$
    $y \geq x^2 - 1$

12. $x^2 + y^2 \leq 16$
    $y < x^2 + 2$

13. $\dfrac{x^2}{9} + \dfrac{y^2}{25} \leq 1$
    $\dfrac{x^2}{4} - \dfrac{y^2}{9} > 1$

14. $\dfrac{x^2}{4} + \dfrac{y^2}{16} \geq 1$
    $\dfrac{x^2}{9} - \dfrac{y^2}{25} < 1$

15. $4x^2 + 9y^2 \leq 36$
    $y > x^2 + 2$

16. $9x^2 + 4y^2 \geq 36$
    $y < x^2 + 1$

17. $x + y \leq 3$
    $x - 3y \leq 3$
    $x \geq -2$

18. $x - y \leq 4$
    $x + 2y \leq 4$
    $x \geq -1$

19. $x + y \leq 2$
    $-x + y \leq 2$
    $y \geq -2$

20. $x - y \leq 3$
    $-x - y \leq 3$
    $y \leq -1$

21. $x + y \leq 4$
    $x \geq 0$
    $y \geq 0$

22. $x - y \leq 2$
    $x \geq 0$
    $y \leq 0$

Solve each of the following systems of equations:

23. $x^2 + y^2 = 9$
    $2x + y = 3$

24. $x^2 + y^2 = 9$
    $x + 2y = 3$

25. $x^2 + y^2 = 16$
    $x + 2y = 8$

26. $x^2 + y^2 = 16$
    $x - 2y = 8$

27. $x^2 + y^2 = 25$
    $x^2 - y^2 = 25$

28. $x^2 + y^2 = 4$
    $2x^2 - y^2 = 5$

29. $x^2 + y^2 = 9$
    $y = x^2 - 3$

30. $x^2 + y^2 = 4$
    $y = x^2 - 2$

31. $x^2 + y^2 = 16$
    $y = x^2 - 4$

32. $x^2 + y^2 = 1$
    $y = x^2 - 1$

33. $3x + 2y = 10$
    $y = x^2 - 5$

34. $4x + 2y = 10$
    $y = x^2 - 10$

35. $y = x^2 + 2x - 3$
    $y = -x + 1$

36. $y = -x^2 - 2x + 3$
    $y = x - 1$

37. $y = x^2 - 6x + 5$
    $y = x - 5$

38. $y = x^2 - 2x - 4$
    $y = x - 4$

39. $4x^2 - 9y^2 = 36$
    $4x^2 + 9y^2 = 36$

40. $4x^2 + 25y^2 = 100$
    $4x^2 - 25y^2 = 100$

41. $x - y = 4$
    $x^2 + y^2 = 16$

42. $x + y = 2$
    $x^2 - y^2 = 4$

Applying the Concepts

43. The sum of the squares of two numbers is 89. The difference of their squares is 39. Find the numbers.

44. The difference of the squares of two numbers is 35. The sum of their squares is 37. Find the numbers.

45. One number is 3 less than the square of another. Their sum is 9. Find the numbers.

46. The square of one number is 2 less than twice the square of another. The sum of the squares of the two numbers is 25. Find the numbers.

Review Problems

 The problems below review material we covered in Section 9.5.

Expand and simplify.

47. $(x + 2)^4$

48. $(x - 2)^4$

49. $(2x + y)^3$

50. $(x - 2y)^3$

51. Find the first two terms in the expansion of $(x + 3)^{50}$.

52. Find the first two terms in the expansion of $(x - y)^{75}$.

Examples

Chapter 10 Summary

1. The distance between
(5, 2) and $(-1, 1)$ is

$$d = \sqrt{(5 + 1)^2 + (2 - 1)^2}$$
$$= \sqrt{37}$$

Distance Formula [10.1]

The distance between the two points (x_1, y_1) and (x_2, y_2) is given by the formula

$$d = \sqrt{(x_2 - x_1)^2 + (y_2 - y_1)^2}$$

2. The graph of the circle

$(x - 3)^2 + (y + 2)^2 = 25$

will have its center at
$(3, -2)$, and the radius
will be 5.

The Circle [10.1]

The graph of any equation of the form

$$(x - a)^2 + (y - b)^2 = r^2$$

will be a circle having its center at (a, b) and a radius of r.

3.

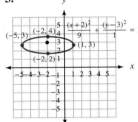

An Ellipse with Center at (h, k) [10.2]

The graph of the equation

$$\frac{(x - h)^2}{a^2} + \frac{(y - k)^2}{b^2} = 1$$

will be an ellipse with center at (h, k). The vertices of the ellipse will be at the points $(h + a, k)$, $(h - a, k)$, $(h, k + b)$, and $(h, k - b)$.

4.

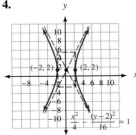

Hyperbolas with Centers at (h, k) [10.2]

The graphs of the equations

$$\frac{(x - h)^2}{a^2} - \frac{(y - k)^2}{b^2} = 1 \quad \text{and} \quad \frac{(y - k)^2}{b^2} - \frac{(x - h)^2}{a^2} = 1$$

will be hyperbolas with their centers at (h, k). The vertices of the graph of the first equation will be at the points $(h + a, k)$ and $(h - a, k)$, while the vertices for the graph of the second equation will be at $(h, k + b)$ and $(h, k - b)$. In either case, the asymptotes can be found by connecting opposite corners of the rectangle that contains the points $(h + a, k)$, $(h - a, k)$, $(h, k + b)$, and $(h, k - b)$.

5. The graph of the inequality

$$x^2 + y^2 < 9$$

is all points inside the circle with center at the origin and radius 3. The circle itself is not part of the solution and is therefore shown with a broken curve.

6. We can solve the system

$$x^2 + y^2 = 4$$
$$x = 2y + 4$$

by substituting $2y + 4$ from the second equation for x in the first equation:

$$(2y + 4)^2 + y^2 = 4$$
$$4y^2 + 16y + 16 + y^2 = 4$$
$$5y^2 + 16y + 12 = 0$$
$$(5y + 6)(y + 2) = 0$$
$$y = -\tfrac{6}{5} \quad \text{or} \quad y = -2$$

Substituting these values of y into the second equation in our system gives $x = \tfrac{8}{5}$ and $x = 0$. The solutions are $(\tfrac{8}{5}, -\tfrac{6}{5})$ and $(0, -2)$.

Second-Degree Inequalities in Two Variables [10.3]

We graph second-degree inequalities in two variables in much the same way that we graphed linear inequalities. That is, we begin by graphing the boundary, using a solid curve if the boundary is included in the solution (this happens when the inequality symbol is $\geq$ or $\leq$), or a broken curve if the boundary is not included in the solution (when the inequality symbol is $>$ or $<$). After we have graphed the boundary, we choose a test point that is not on the boundary and try it in the original inequality. A true statement indicates we are in the region of the solution. A false statement indicates we are not in the region of the solution.

Systems of Nonlinear Equations [10.3]

A system of nonlinear equations is two equations, at least one of which is not linear, considered at the same time. The solution set for the system consists of all ordered pairs that satisfy both equations. In most cases, we use the substitution method to solve these systems; however, the addition method can be used if like variables are raised to the same power in both equations. It is sometimes helpful to graph each equation in the system on the same set of axes in order to anticipate the number and approximate positions of the solutions.

Chapter 10 Review Problems

Find the distance between the following points.
[10.1]

1. $(2, 6), (-1, 5)$

2. $(3, -4), (1, -1)$

3. $(0, 3), (-4, 0)$

4. $(-3, 7), (-3, -2)$

5. Find x so that the distance between $(x, -1)$ and $(2, -4)$ is 5. [10.1]

6. Find y so that the distance between $(3, -4)$ and $(-3, y)$ is 10. [10.1]

Write the equation of the circle with the given center and radius. [10.1]

7. Center $(3, 1)$, $r = 2$

8. Center $(3, -1)$, $r = 4$

9. Center $(-5, 0)$, $r = 3$

10. Center $(-3, 4)$, $r = 3\sqrt{2}$

Find the equation of each circle. [10.1]

11. Center at the origin, x-intercepts ± 5

12. Center at the origin, y-intercepts ± 3

13. Center at $(-2, 3)$ and passing through the point $(2, 0)$

14. Center at $(-6, 8)$ and passing through the origin

Give the center and radius of each circle and then sketch the graph. [10.1]

15. $x^2 + y^2 = 4$
16. $(x - 3)^2 + (y + 1)^2 = 16$
17. $x^2 + y^2 - 6x + 4y = -4$
18. $x^2 + y^2 + 4x - 2y = 4$

Graph each of the following. Label the x- and y-intercepts. [10.2]

19. $\dfrac{x^2}{4} + \dfrac{y^2}{9} = 1$ 20. $4x^2 + y^2 = 16$

Graph the following. Show the asymptotes. [10.2]

21. $\dfrac{x^2}{4} - \dfrac{y^2}{9} = 1$ 22. $4x^2 - y^2 = 16$

Graph each equation. [10.2]

23. $\dfrac{(x + 2)^2}{9} + \dfrac{(y - 3)^2}{1} = 1$

24. $\dfrac{(x - 2)^2}{16} - \dfrac{y^2}{4} = 1$

25. $9y^2 - x^2 - 4x + 54y + 68 = 0$
26. $9x^2 + 4y^2 - 72x - 16y + 124 = 0$

Graph each of the following inequalities. [10.3]

27. $x^2 + y^2 < 9$
28. $(x + 2)^2 + (y - 1)^2 \leq 4$
29. $y \geq x^2 - 1$
30. $9x^2 + 4y^2 \leq 36$

Graph the solution set for each system. [10.3]

31. $x^2 + y^2 < 16$
    $y > x^2 - 4$

32. $x + y \leq 2$
    $-x + y \leq 2$
    $y \geq -2$

Solve each system of equations. [10.3]

33. $x^2 + y^2 = 16$
    $2x + y = 4$

34. $x^2 + y^2 = 4$
    $y = x^2 - 2$

35. $9x^2 - 4y^2 = 36$
    $9x^2 + 4y^2 = 36$

36. $2x^2 - 4y^2 = 8$
    $x^2 + 2y^2 = 10$

Chapter 10 Test

1. Find x so that $(x, 2)$ is $2\sqrt{5}$ units from $(-1, 4)$. [10.1]

2. Give the equation of the circle with center at $(-2, 4)$ and radius 3. [10.1]

3. Give the equation of the circle with center at the origin that contains the point $(-3, -4)$. [10.1]

4. Find the center and radius of the circle $x^2 + y^2 - 10x + 6y = 5$. [10.1]

Graph each of the following. [10.2, 10.3]

5. $4x^2 - y^2 = 16$

6. $\dfrac{x^2}{25} + \dfrac{y^2}{4} = 1$

7. $(x - 2)^2 + (y + 1)^2 \leq 9$
8. $9x^2 + 4y^2 - 72x - 16y + 124 = 0$

Solve the following systems. [10.3]

9. $x^2 + y^2 = 25$
   $2x + y = 5$

10. $x^2 + y^2 = 16$
    $y = x^2 - 4$

CHAPTER I   PROBLEM SET I.I

1. $x + 5 = 2$ **3.** $6 - x = y$ **5.** $2t < y$ **7.** $x + y < x - y$
9. (a) $54 + 18 = 72$ cm^2 (b) $6(12) = 72$ cm^2 **11.** 42   **13.** 50   **15.** 16   **17.** 12   **19.** 18   **21.** 64
23. 34   **25.** 64   **27.** 33   **29.** 33   **31.** 5,431   **33.** 32   **35.** 24   **37.** 41   **39.** 95   **41.** 138
43. 78   **45.** 152   **47.** \$7,471
49. It is a measure of the middle, or center, of the numbers in the group of numbers. It's an average.
51. \$2,638   **53.** \$6,609   **55.** {0, 1, 2, 3, 4, 5, 6}   **57.** {2, 4}   **59.** {1, 3, 5}   **61.** {0, 1, 2, 3, 4, 5, 6}
63. {0, 2}   **65.** {0, 6}   **67.** {0, 1, 2, 3, 4, 5, 6, 7}   **69.** 1, 2   **71.** $-6, -5.2, 0, 1, 2, 2.3, \frac{9}{2}$
73. $-\sqrt{7}, -\pi, \sqrt{17}$ **75.** False   **77.** True   **79.** True   **81.** True   **83.** $2 \cdot 7 \cdot 19$ **85.** $3 \cdot 37$
87. $3^2 \cdot 41$ **89.** $\frac{3}{7}$ **91.** $\frac{11}{21}$ **93.** $\frac{3}{5}$ **95.** 120   **97.** $6! = 6 \cdot 5 \cdot 4 \cdot 3 \cdot 2 \cdot 1 = 6 \cdot 5!$

PROBLEM SET 1.2

1. $6 + x$ **3.** $a + 8$ **5.** $15y$ **7.** x **9.** a **11.** x **13.** $3x + 18$ **15.** $12x + 8$ **17.** $15a + 10b$
19. $\frac{4}{3}x + 2$ **21.** $2 + y$ **23.** $15x + 10$ **25.** $15t + 9$ **27.** $\frac{7}{15}$ **29.** $\frac{29}{35}$ **31.** $\frac{35}{144}$ **33.** $\frac{949}{1,260}$
35. $14a + 7$ **37.** $6y + 6$ **39.** $12x + 2$ **41.** $8y + 11$ **43.** $24a + 15$ **45.** $11x + 20$ **47.** $20t + 5$
49. $7(x + 2) = 7x + 14; 7x + 7(2) = 7x + 14$ **51.** $x(y + 4) = xy + 4x; xy + 4x$ **53.** Commutative
55. Commutative   **57.** Additive inverse   **59.** Commutative   **61.** Associative and commutative
63. Commutative and associative   **65.** Distributive   **67.** 2   **69.** $\frac{3}{4}$ **71.** π **73.** -4 **75.** -2 **77.** $\frac{21}{40}$
79. 2   **81.** $\frac{8}{27}$ **83.** $\frac{1}{10,000}$ **85.** $\frac{72}{385}$ **87.** 1   **89.** $1, -1$ **91.** 0   **93.** $5(4) - 5 = 15; 15 \neq 4$
95. $15 - (8 - 2) = 15 - 6 = 9; (15 - 8) - 2 = 7 - 2 = 5$

PROBLEM SET 1.3

1. 4   **3.** -4 **5.** $-\frac{13}{18}$ **7.** -10 **9.** -4 **11.** $\frac{19}{12}$ **13.** $-\frac{32}{105}$ **15.** 19   **17.** $-\frac{7}{3}$ **19.** -4
21. -1 **23.** -8 **25.** -12 **27.** $-7x$ **29.** 13   **31.** -14 **33.** $6a$ **35.** -15 **37.** 15
39. -24 **41.** -12 **43.** -24 **45.** $-10x$ **47.** x **49.** y **51.** $-8x + 6$ **53.** $12t - 28$
55. $-3a + 4$ **57.** $\frac{3}{2}x + 2$ **59.** -14 **61.** 18   **63.** 16   **65.** 52   **67.** 30   **69.** -19 **71.** 50
73. 20   **75.** -2 **77.** 1   **79.** 80   **81.** -30 **83.** 18   **85.** 277   **87.** -73 **89.** $14x + 12$
91. $7m - 15$ **93.** $-2x + 9$ **95.** $7y + 10$ **97.** $-20x + 5$ **99.** $-11x + 10$ **101.** -2 **103.** 2
105. Undefined   **107.** 0   **109.** $-\frac{2}{3}$ **111.** 32   **113.** 64   **115.** $-\frac{1}{18}$ **117.** 4   **119.** $\frac{5}{3}$ **121.** 11
123. 12   **125.** -3 **127.** -11 **129.** $2x$

PROBLEM SET 1.4

1. 16   **3.** -16 **5.** -0.027 **7.** 32   **9.** $\frac{1}{8}$ **11.** $\frac{25}{36}$ **13.** x^9 **15.** 64   **17.** $-\frac{8}{27}x^6$ **19.** $-6a^6$

21. $\frac{1}{9}$ **23.** $-\frac{1}{32}$ **25.** $\frac{16}{9}$ **27.** 17   **29.** x^3 **31.** $\frac{a^6}{b^{15}}$ **33.** $\frac{8}{125y^{18}}$ **35.** $\frac{1}{5}$ **37.** $\frac{24a^{12}c^6}{b^3}$ **39.** $\frac{8x^{22}}{81y^{23}}$

41. $\frac{1}{x^{10}}$ **43.** a^{10} **45.** $\frac{1}{t^6}$ **47.** x^{12} **49.** x^{18} **51.** $\frac{1}{x^{22}}$ **53.** $\frac{a^3b^7}{4}$ **55.** $\frac{y^{38}}{x^{16}}$ **57.** $\frac{16y^{16}}{x^8}$ **59.** x^4y^6

61. $\frac{b^3}{a^4c^3}$ **63.** 3.78×10^5 **65.** 4.9×10^3 **67.** 3.7×10^{-4} **69.** 4.95×10^{-3} **71.** 5,340

73. 7,800,000   **75.** 0.00344   **77.** 0.49   **79.** 8×10^4 **81.** 2×10^9 **83.** 2.5×10^{-6} **85.** 1.8×10^{-7}

87. 2.37×10^6 **89.** 6.3×10^8 **91.** 22   **93.** 1.003×10^{19} **95.** $\frac{1}{x^3}$ **97.** y^3 **99.** x^5

PROBLEM SET 1.5

1. Trinomial; 2; 5 **3.** Binomial; 1; 3 **5.** Trinomial; 2; 8 **7.** Polynomial; 3; 4 **9.** Monomial; 0; $-\frac{3}{4}$
11. Trinomial; 3; 6 **13.** $7x + 1$ **15.** $2x^2 + 7x - 15$ **17.** $12a^2 - 7ab - 10b^2$ **19.** $x^2 - 13x + 3$
21. $\frac{1}{4}x^2 - \frac{7}{12}x - \frac{1}{4}$ **23.** $-3x$ **25.** $10x - 5$ **27.** $9x - 35$ **29.** -2 **31.** 208 **33.** -15
35. $12x^3 - 10x^2 + 8x$ **37.** $2a^5b - 2a^3b^2 + 2a^2b^4$ **39.** $6a^4 + a^3 - 12a^2 + 5a$ **41.** $a^3 - b^3$
43. $8x^3 + y^3$ **45.** $6a^2 + 13a + 6$ **47.** $20 - 2t - 6t^2$ **49.** $x^6 - 2x^3 - 15$ **51.** $18t^2 - \frac{2}{9}$
53. $b^2 - a^2b - 12a^4$ **55.** $4a^2 - 12a + 9$ **57.** $25x^2 + 20xy + 4y^2$ **59.** $25 - 30t^3 + 9t^6$ **61.** $4a^2 - 9b^2$
63. $9r^4 - 49s^2$ **65.** $\frac{1}{9}x^2 - \frac{4}{25}$ **67.** $x^3 - 6x^2 + 12x - 8$ **69.** $8x^3 - 12x^2 + 6x - 1$
71. $a^2b^2 + b^2 + 8a^2 + 8$ **73.** $3xy^2 + 4x - 6y^2 - 8$ **75.** $3x^2 + 12x + 14$ **77.** $24x$ **79.** $-6x^2 - 2$
81. $P = -300 + 40x - 0.5x^2$; \$300 **83.** $P = -800 + 3.5x - 0.002x^2$; \$700 **85.** Both 240 ft
87. $A = 100 + 400r + 600r^2 + 400r^3 + 100r^4$ **89.** $x^{2n} - 5x^n + 6$ **91.** $x^{4n} - 9$ **93.** $10x^{2n} + 13x^n - 3$
95. $x^{2n} + 10x^n + 25$

PROBLEM SET 1.6

1. $5x^2(2x - 3)$ **3.** $9y^3(y^3 + 2)$ **5.** $3ab(3a - 2b)$ **7.** $7xy^2(3y^2 + x)$ **9.** $3(a^2 - 7a + 10)$
11. $4x(x^2 - 4x - 5)$ **13.** $(a - 2b)(5x - 3y)$ **15.** $3(x + y)^2(x^2 - 2y^2)$ **17.** $(x + 5)(2x^2 + 7x + 6)$
19. $(x + 1)(3y + 2a)$ **21.** $(x + 3)(xy + 1)$ **23.** $(x - a)(x - b)$ **25.** $(b + 5)(a - 1)$ **27.** $(x + 3)(x + 4)$
29. $(x + 3)(x - 4)$ **31.** $(y + 3)(y - 2)$ **33.** $(2 - x)(8 + x)$ **35.** $(2 + x)(6 + x)$ **37.** $(x + 2y)(x + y)$
39. $(a + 6b)(a - 3b)$ **41.** $(x - 8a)(x + 6a)$ **43.** $(x - 6b)^2$ **45.** $(2x - 3)(x + 5)$ **47.** $(2x - 5)(x + 3)$
49. $(2x - 3)(x - 5)$ **51.** $(2x - 5)(x - 3)$ **53.** Does not factor **55.** $(2 + 3a)(1 + 2a)$
57. $15(4y + 3)(y - 1)$ **59.** $x^2(3x - 2)(2x + 1)$ **61.** $10r(2r - 3)^2$ **63.** $(4x + y)(x - 3y)$
65. $(2x - 3a)(5x + 6a)$ **67.** $(3a + 4b)(6a - 7b)$ **69.** $200(1 + 2t)(3 - 2t)$ **71.** $y^2(3y - 2)(3y + 5)$
73. $2a^2(3 + 2a)(4 - 3a)$ **75.** $2x^2y^2(4x + 3y)(x - y)$ **77.** $100(3x^2 + 1)(x^2 + 3)$ **79.** $(x + 5)(2x + 3)(x + 2)$
81. $(2x + 3)(x + 5)(x + 2)$ **83.** $a + 250$
85. $P(1 + r) + P(1 + r)r = (1 + r)(P + Pr) = (1 + r)P(1 + r) = P(1 + r)^2$

PROBLEM SET 1.7

1. $(x - 3)^2$ **3.** $(a - 6)^2$ **5.** $(5 - t)^2$ **7.** $(2y^2 - 3)^2$ **9.** $(4a + 5b)^2$ **11.** $(\frac{1}{5} + \frac{1}{4}t^2)^2$
13. $(x + 2 + 3)^2 = (x + 5)^2$ **15.** $(7x - 8y)(7x + 8y)$ **17.** $(2a - \frac{1}{2})(2a + \frac{1}{2})$ **19.** $(x - \frac{3}{5})(x + \frac{3}{5})$
21. $(5 - t)(5 + t)$ **23.** $(4a^2 + 9)(2a - 3)(2a + 3)$ **25.** $(x - 5 + y)(x - 5 - y)$
27. $(a + 4 + b)(a + 4 - b)$ **29.** $(x + 2)(x + 5)(x - 5)$ **31.** $(2x + 3)(x + 2)(x - 2)$
33. $(x + 3)(2x + 3)(2x - 3)$ **35.** $(x - y)(x^2 + xy + y^2)$ **37.** $(a + 2)(a^2 - 2a + 4)$ **39.** $(y - 1)(y^2 + y + 1)$
41. $10(r - 5)(r^2 + 5r + 25)$ **43.** $(4 + 3a)(16 - 12a + 9a^2)$ **45.** $(t + \frac{1}{3})(t^2 - \frac{1}{3}t + \frac{1}{9})$ **47.** $(x + 9)(x - 9)$
49. $(x - 3)(x + 5)$ **51.** $(x^2 + 2)(y^2 + 1)$ **53.** $2ab(a^2 + 3a + 1)$ **55.** Does not factor
57. $3(2a + 5)(2a - 5)$ **59.** $(5 - t)^2$ **61.** $4x(x^2 + 4y^2)$ **63.** $(x + 5)(x + 3)(x - 3)$ **65.** Does not factor
67. $(x - 3)(x - 7)^2$ **69.** $(2 - 5x)(4 + 3x)$ **71.** $(r + \frac{1}{5})(r - \frac{1}{5})$ **73.** Does not factor **75.** $100(x - 3)(x + 2)$
77. $(3x^2 + 1)(x^2 - 5)$ **79.** $3a^2b(2a - 1)(4a^2 + 2a + 1)$ **81.** $(4 - r)(16 + 4r + r^2)$
83. $5x^2(2x + 3)(2x - 3)$ **85.** $2x^3(4x - 5)(2x - 3)$ **87.** $(y + 1)(y - 1)(y^2 - y + 1)(y^2 + y + 1)$
89. $2(5 + a)(5 - a)$ **91.** $(x - 2 + y)(x - 2 - y)$ **93.** $(x^n - y^n)(x^n + y^n)$ **95.** $(x^n - 2)(x^{2n} + 2x^n + 4)$
97. $(x^n - y^n)(x^{2n} + x^ny^n + y^{2n})$

CHAPTER I REVIEW PROBLEMS

1. $x + 2$ **2.** $x - 2$ **3.** $\dfrac{x}{2}$ **4.** $2(x + y)$ **5.** 17 **6.** 16 **7.** 9 **8.** 30 **9.** $\{1, 2, 3, 4, 5, 6\}$
10. $\{5\}$ **11.** $(2, 4)$ **12.** $\varnothing$ **13.** $\{0, 1, 2\}$ **14.** $\{1, 2, 3, 4, 5, 6\}$ **15.** $3 \cdot 7 \cdot 11$ **16.** $2^2 \cdot 3^2 \cdot 11^2$
17. $\frac{21}{25}$ **18.** $\frac{11}{13}$ **19.** 0, 5 **20.** $-7, 0, 5$ **21.** $-7, -4.2, 0, \frac{3}{4}, 5$ **22.** $-\sqrt{3}, \pi$ **23.** (a) **24.** (c)
25. (a) **26.** (b), (d) **27.** (a), (c) **28.** (f) **29.** 3 **30.** -5 **31.** 4 **32.** 6 **33.** 1 **34.** $\frac{27}{64}$
35. 2 **36.** 1 **37.** 2 **38.** $-\frac{1}{6}$ **39.** 3 **40.** 2 **41.** -42 **42.** 30 **43.** $21x$ **44.** $-6x$
45. $-6x + 10$ **46.** $-6x + 21$ **47.** $-x + 3$ **48.** $-15x + 3$ **49.** $-\frac{5}{6}$ **50.** -36 **51.** $\frac{1}{10}$ **52.** $-\frac{2}{7}$
53. 0 **54.** -36 **55.** 16 **56.** 2 **57.** 13 **58.** 16 **59.** $6x - 3$ **60.** $-2y + 9$ **61.** $-18x - 14$

62. $5a - 22$ **63.** 25 **64.** -25 **65.** $\frac{9}{16}$ **66.** 1 **67.** 16 **68.** x^{10} **69.** $25x^6$ **70.** $-32x^{18}y^8$
71. $\frac{1}{8}$ **72.** $-\frac{1}{8}$ **73.** $\frac{9}{4}$ **74.** $\frac{1}{2}$ **75.** 3.45×10^7 **76.** 5.29×10^{-5} **77.** 44,500 **78.** 0.000445
79. $\dfrac{1}{a^9}$ **80.** $2x^2$ **81.** 8 **82.** $-2x^{15}$ **83.** $\dfrac{x^{12}}{4}$ **84.** x^2 **85.** 8×10^{-2} **86.** 4×10^{-10} **87.** -6
88. $2x^2 - 5x + 7$ **89.** $-x^2 - 7xy + y^2$ **90.** $2x^3 - 2x^2 - 2x - 4$ **91.** $2x - 3$ **92.** $x^2 - 2x - 3$
93. $30x + 12$ **94.** $-35x + 120$ **95.** 15 **96.** -21 **97.** $12x^3 - 6x^2 + 3x$
98. $2a^4b^3 + 4a^3b^4 + 2a^2b^5$ **99.** $18 - 9y + y^2$ **100.** $6x^4 + 5x^2 - 4$ **101.** $2t^3 - 4t^2 - 6t$
102. $x^3 + 27$ **103.** $8x^3 - 27$ **104.** $x^2 + 6x + 9$ **105.** $a^4 - 4a^2 + 4$ **106.** $9x^2 + 30x + 25$
107. $4a^2 + 12ab + 9b^2$ **108.** $x^2 - \frac{1}{9}$ **109.** $x^3 - 3x^2 + 3x - 1$ **110.** $x^{2m} - 4$
111. $3xy(2x^3 - 3y^3 + 6x^2y^2)$ **112.** $4(x + y)^2(x^2 - 2y^2)$ **113.** $x^2(y^3 + 2)(x + 5)$ **114.** $(b + x)(a - x)$
115. $(x - 2)(x - 3)$ **116.** $(x - 3)(x + 2)$ **117.** $2x(x + 5)(x - 3)$ **118.** $(5a - 4b)(4a - 5b)$
119. $x^2(3x + 2)(2x - 5)$ **120.** $(4a + 5)(5a + 3)$ **121.** $3y(4x + 5)(2x - 3)$ **122.** $y^2(2y - 5)(3y + 2)$
123. $(x - 5)^2$ **124.** $(3y - 7)(3y + 7)$ **125.** $(x^2 + 4)(x + 2)(x - 2)$ **126.** $3(a^2 + 3)^2$
127. $(a - 2)(a^2 + 2a + 4)$ **128.** $5x(x + 3y)^2$ **129.** $3ab(a - 3b)(a + 3b)$ **130.** $(x - 5 + y)(x - 5 - y)$
131. $(x + 3)(x - 3)(x + 4)$ **132.** $(x + 2)(x - 2)(x + 5)$

CHAPTER 1 TEST

1. $2a - 3b < 2a + 3b$ **2.** $\{1, 2, 3, 4, 6\}$ **3.** $2 \cdot 5 \cdot 7 \cdot 11$ **4.** Commutative property of addition
5. Multiplicative identity property **6.** -19 **7.** -149 **8.** 2 **9.** $\frac{59}{72}$ **10.** $-4x$ **11.** $-5x - 8$ **12.** $-\frac{5}{2}$
13. x^8 **14.** $\frac{1}{32}$ **15.** a^2 **16.** $\dfrac{2a^{12}}{b^{15}}$ **17.** 6.53×10^6 **18.** 3×10^8 **19.** $\frac{3}{4}x^3 - \frac{5}{4}x^2 - 2x - 1$
20. $4x + 75$ **21.** $6y^2 + y - 35$ **22.** $2x^3 + 3x^2 - 26x + 15$ **23.** $64 - 48t^3 + 9t^6$ **24.** $1 - 36y^2$
25. $2(3x^2 - 1)(2x^2 + 5)$ **26.** $(4a^2 + 9y^2)(2a + 3y)(2a - 3y)$ **27.** $(7a - b^2)(x^2 - 2y)$
28. $(t + \frac{1}{2})(t^2 - \frac{1}{2}t + \frac{1}{4})$ **29.** $(x - 5 + b)(x - 5 - b)$ **30.** $(9 + x^2)(3 + x)(3 - x)$

CHAPTER 2 PROBLEM SET 2.1

1. 5 **3.** $-\frac{9}{2}$ **5.** $-\frac{4}{3}$ **7.** -10 **9.** -4 **11.** -2 **13.** $\frac{3}{4}$ **15.** 3 **17.** 2
19. -3 **21.** 4 **23.** 2 **25.** -3 **27.** 6 **29.** 3 **31.** 6,000 **33.** 1 **35.** $-1, 6$ **37.** 0, 2, 3
39. $\frac{2}{3}, \frac{3}{2}$ **41.** $-3, 0, 7$ **43.** $-4, \frac{5}{2}$ **45.** $0, \frac{4}{3}$ **47.** $-\frac{1}{5}, \frac{1}{3}$ **49.** $-10, 0$ **51.** $-5, 1$ **53.** $-2, 3$
55. $-3, -2, 2$ **57.** $-2, -\frac{3}{2}, 2$ **59.** $\frac{4}{5}$ **61.** $-5, 5$ **63.** 9 **65.** $-\frac{4}{3}$ **67.** 1, 2 **69.** $-2, -5, 5$
71. 1 **73.** $-2, \frac{1}{4}$ **75.** 5,000 **77.** $-\frac{4}{3}, 0, \frac{4}{3}$ **79.** $\frac{3}{2}$ or 1.5 **81.** $-3, -\frac{3}{2}, \frac{3}{2}$
83. Any method of solution results in a false statement.
85. Every attempt at solving the equation results in a true statement. **87.** No solution
89. All real numbers are solutions. **91.** No solution **93.** -6 **95.** 66 **97.** -13 **99.** -4 **101.** 2
103. $-\frac{3}{2}$

PROBLEM SET 2.2

1. -3 **3.** 0 **5.** $\frac{3}{2}$ **7.** 4 **9.** \$5.00 **11.** \$10.00 **13.** 2 cm **15.** 2 in.
17. 5 ft **19.** 1 sec and 2 sec **21.** 0 sec and $\frac{3}{2}$ sec **23.** 2 in. **25.** \$4 or \$8 **27.** \$7 or \$10 **29.** $l = \dfrac{A}{w}$
31. $t = \dfrac{I}{pr}$ **33.** $r = \dfrac{A - P}{Pt}$ **35.** $F = \frac{9}{5}C + 32$ **37.** $v = \dfrac{h - 16t^2}{t}$ **39.** $d = \dfrac{A - a}{n - 1}$ **41.** $y = -\frac{2}{3}x + 2$
43. $y = \frac{3}{5}x + 3$ **45.** $y = 3x - 2$ **47.** $y = \frac{1}{3}x + 2$ **49.** $x = \dfrac{5}{a - b}$ **51.** $P = \dfrac{A}{1 + rt}$ **53.** $x = \dfrac{d - b}{a - c}$
55. 20.52 **57.** 500 **59.** 25% **61.** 925 **63.** $2(x + 3)$ **65.** $2(x + 3) = 16$ **67.** $5(x - 3)$
69. $3x + 2 = x - 4$ **71.** Commutative property of multiplication **73.** Associative property of addition

75. Commutative and associative properties of addition   **77.** Additive identity property   **79.** $x = -\dfrac{a}{b}y + a$

81. $a = \dfrac{bc}{b - c}$ **83.** $R = \dfrac{abc}{bc + ac + ab}$ **85.** \$673.68   **87.** \$674.92   **89.** \$760.05

PROBLEM SET 2.3

 Along with the answers to the odd-numbered problems in this problem set, we are including most of the equations used to solve each problem. Be sure that you try the problems on your own before looking here to see what the correct equations are.

1. The width is x, the length is $2x$; $2x + 4x = 60$; 10 ft by 20 ft
3. The length of a side is x; $4x = 28$; 7 ft
5. The shortest side is x, the medium side is $x + 3$, the longest side is $2x$; $x + (x + 3) + 2x = 23$; 5 in.
7. The width is x, the length is $2x - 3$; $2(2x - 3) + 2x = 18$; 4 m   **9.** \$92.00   **11.** \$9,339.00   **13.** 860 items
15. 41,667   **17.** \$3,260.66 per month   **19.** 20°, 160°   **21.** (a) 20.4°, 69.6°   (b) 38.4°, 141.6°
23. 27°, 72°, 81°   **25.** 102°, 44°, 34°   **27.** 43°, 43°, 94°   **29.** 6, 8, 10   **31.** 2 ft, 8 ft   **33.** 18 in., 4 in.
35. 30 fathers, 45 sons   **37.** \$54   **39.** 44 minutes

45. $-4, 0, 2, 3$ **47.** $-4, -\frac{2}{5}, 0, 2, 3$
49. $\{3, 4, 5, 6, 7, 9\}$ **51.** $\{7, 9\}$

41.

Width (ft)	Length (ft)	Area (ft²)
2	22	44
4	20	80
6	18	108
8	16	128
10	14	140
12	12	144

43.

Time (sec)	Height (ft)
1	112
2	192
3	240
4	256
5	240
6	192

53.

Width	L = 6 − W	Length	Area
0	L = 6 − 0	6	0
0.5	L = 6 − 0.5	5.5	2.75
1	L = 6 − 1	5	5
1.5	L = 6 − 1.5	4.5	6.75
2	L = 6 − 2	4	8
2.5	L = 6 − 2.5	3.5	8.75
3	L = 6 − 3	3	9
3.5	L = 6 − 3.5	2.5	8.75
4	L = 6 − 4	2	8
4.5	L = 6 − 4.5	1.5	6.75
5	L = 6 − 5	1	5
5.5	L = 6 − 5.5	0.5	2.75
6	L = 6 − 6	0	0

Largest area is 9 in.².

55.

Time in Months	Amount in Account	Time in Months	Amount in Account
0	\$100.00	13	\$106.70
1	\$100.50	14	\$107.23
2	\$101.00	15	\$107.77
3	\$101.51	16	\$108.31
4	\$102.02	17	\$108.85
5	\$102.53	18	\$109.39
6	\$103.04	19	\$109.94
7	\$103.55	20	\$110.49
8	\$104.07	21	\$111.04
9	\$104.59	22	\$111.60
10	\$105.11	23	\$112.16
11	\$105.64	24	\$112.72
12	\$106.17		

PROBLEM SET 2.4

1. $x \leq \frac{3}{2}$

3. $x > 4$

5. $x \geq -5$

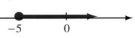

7. $x < 4$

9. $x \geq -6$

11. $x \geq 4$

13. $x < -3$

15. $m \geq -1$

17. $x \geq -3$

19. $y \leq \frac{7}{2}$

21. $x < 6$

23. $y \geq -52$

25. $(-\infty, -2]$　　**27.** $[1, \infty)$　　**29.** $(-\infty, 3)$　　**31.** $(-\infty, -1]$　　**33.** $[-17, \infty)$　　**35.** $(-\infty, -5)$
37. $[3, 7]$　　　　　　　　　　　**39.** $(-4, 2)$　　　　　　　　　　**41.** $[4, 6]$

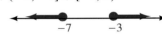

43. $(-4, 2)$

45. $(-3, 3)$

47. $(-\infty, -7] \cup [-3, \infty)$

49. $(-\infty, -1] \cup [\frac{3}{5}, \infty)$

51. $(-\infty, -10) \cup (6, \infty)$

53. $x \geq 5$　　**55.** $x \leq -3$

57. $x \leq 4$　　**59.** $-4 < x < 4$　　**61.** $-4 \leq x \leq 4$　　**63.** $p \leq 2$; set the price at \$2.00 or less per pad.
65. $p > 1.25$; charge more than \$1.25 per pad.　　**67.** $y < -\frac{3}{2}x + 3$　　**69.** $y \leq \frac{4}{5}x - 4$
71. $35°$ to $45°$ Celsius; $35° \leq C \leq 45°$　　**73.** $-25°$ to $10°$ Celsius; $-25° \leq C \leq -10°$　　**75.** $-5 < -x < -2$
77. No.　　**79.** $3xy(2x^3 - 3y^3 + 6x^2y^2)$　　**81.** $(x - 2)(x - 3)$　　**83.** $(2x - 5)^2$　　**85.** $(x^2 + 4)(x + 2)(x - 2)$

87. $(x - 6 - y)(x - 6 + y)$　　**89.** Does not factor; prime polynomial　　**91.** $x < \dfrac{c - b}{a}$　　**93.** $\dfrac{-c - b}{a} < x < \dfrac{c - b}{a}$

PROBLEM SET 2.5

1. $-4, 4$　　**3.** $-2, 2$　　**5.** $\varnothing$　　**7.** $-1, 1$　　**9.** $\varnothing$　　**11.** $-6, 6$　　**13.** $-3, 7$　　**15.** $\frac{17}{3}, \frac{7}{3}$　　**17.** $2, 4$
19. $-\frac{5}{2}, \frac{5}{6}$　　**21.** $-1, 5$　　**23.** $\varnothing$　　**25.** $20, -4$　　**27.** $-4, 8$　　**29.** $\frac{2}{3}, -\frac{10}{3}$　　**31.** $\varnothing$　　**33.** $\frac{3}{2}, -1$
35. $5, 25$　　**37.** $-30, 26$　　**39.** $-12, 28$　　**41.** $-\frac{16}{3}, -6$　　**43.** $-3, 4$　　**45.** $-3, -2$　　**47.** $-5, \frac{3}{5}$
49. $1, \frac{1}{9}$　　**51.** $-\frac{1}{2}$　　**53.** 0　　**55.** $-\frac{1}{2}$　　**57.** $-\frac{1}{6}, -\frac{7}{4}$　　**59.** All real numbers　　**61.** All real numbers

63. $|a - b| = |4 - (-7)| = |11| = 11$　and　$|a - b| = |-5 - (-8)| = |3| = 3$　　**65.** $-2, -1, 0, 1, 2, 3, 4$
 $|b - a| = |-7 - 4| = |-11| = 11$　　　　$|b - a| = |-8 - (-5)| = |-3| = 3$

67. $x \geq 2$　　**69.** $10x^5 + 8x^3 - 6x^2$　　**71.** $12a^2 + 11a - 5$　　**73.** $x^4 - 81$　　**75.** $16y^2 - 40y + 25$

77. $12xy - 6x + 28y - 14$　　**79.** $9 - 6t^2 + t^4$　　**81.** $x = a - b$ or $x = a + b$　　**83.** $x = \dfrac{-b - c}{a}$ or $x = \dfrac{-b + c}{a}$

85. $x = -\dfrac{a}{b}y - a$ or $x = -\dfrac{a}{b}y + a$

PROBLEM SET 2.6

1. $-3 < x < 3$

3. $x \leq -2$ or $x \geq 2$

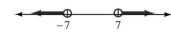

5. $-3 < x < 3$

7. $t < -7$ or $t > 7$

9. $\varnothing$ **11.** All real numbers   **13.** $-4 < x < 10$

15. $a \leq -9$ or $a \geq -1$

17. $\varnothing$ **19.** $-1 < x < 5$

21. $y \leq -5$ or $y \geq -1$

23. $k \leq -5$ or $k \geq 2$

25. $-1 < x < 7$

27. $a \leq -2$ or $a \geq 1$

29. $-6 < x < \frac{8}{3}$

31. $x < 2$ or $x > 8$

33. $x \leq -3$ or $x \geq 12$

35. $x < 2$ or $x > 6$

37. $0.99 < x < 1.01$

39. $x \leq -\frac{3}{5}$ or $x \geq -\frac{2}{5}$ **41.** $-\frac{1}{6} \leq x \leq \frac{3}{2}$ **43.** $-0.05 < x < 0.25$ **45.** $|x| \leq 4$ **47.** $|x - 5| \leq 1$ **49.** $\frac{1}{9}$

51. $\frac{3x^2}{y^2}$ **53.** $\frac{x^7}{y^{12}}$ **55.** 5.4×10^4 **57.** $6,440$ **59.** 1.2×10^4 **61.** $a - b < x < a + b$

63. $x < \dfrac{b - c}{a}$ or $x > \dfrac{b + c}{a}$

CHAPTER 2   REVIEW PROBLEMS

1. 10 **2.** -1 **3.** 2 **4.** 3 **5.** 2 **6.** 8 **7.** -3 **8.** 10 **9.** -3 **10.** 0
11. $\frac{2}{3}$ **12.** -1 **13.** $\frac{10}{13}$ **14.** $\frac{14}{13}$ **15.** $-\frac{5}{11}$ **16.** $\frac{21}{20}$ **17.** $-\frac{4}{9}$ **18.** $\frac{15}{4}$ **19.** $-2, 3$ **20.** $-1, 5$
21. $-\frac{3}{2}, 4$ **22.** $\frac{3}{2}, 4$ **23.** $-\frac{1}{2}, \frac{4}{5}$ **24.** $\frac{1}{2}, \frac{5}{4}$ **25.** $-\frac{5}{3}, \frac{5}{3}$ **26.** $-\frac{3}{4}, \frac{3}{4}$ **27.** $-\frac{1}{9}, \frac{1}{9}$ **28.** $-2, 0$ **29.** 400
30. 600 **31.** $1, 4$ **32.** $2, 6$ **33.** $-4, -3, 3$ **34.** $-2, -\frac{2}{3}, \frac{2}{3}$ **35.** $h = 1$ **36.** $b = 4$ **37.** $h = 17$

38. $b = 40$ **39.** $t = 20$ **40.** $t = 10$ **41.** $n = 5$ **42.** $n = 4$ **43.** $p = \dfrac{I}{rt}$ **44.** $t = \dfrac{I}{pr}$

45. $x = \dfrac{y - b}{m}$ **46.** $m = \dfrac{y - b}{x}$ **47.** $y = \frac{4}{3}x - 4$ **48.** $x = \frac{3}{4}y + 3$ **49.** $v = \dfrac{d - 16t^2}{t}$ **50.** $v = \dfrac{d + 16t^2}{t}$

51. $F = \frac{9}{5}C + 32$ **52.** $C = \frac{5}{9}(F - 32)$ **53.** 2 sec   **54.** $\frac{1}{2}$ sec, 2 sec   **55.** 4 ft by 12 ft   **56.** 5 in. by 11 in.
57. 3 m, 4 m, 5 m   **58.** 6 yd, 8 yd, 10 yd   **59.** $17°, 85°, 78°$ **60.** $80°, 80°, 20°$ **61.** $\$24,875.24$
62. $25°, 65°$ **63.** $3, 4, 5$ **64.** $6, 8, 10$ **65.** $(-\infty, \frac{1}{2})$ **66.** $(-\infty, \frac{1}{3})$ **67.** $(-\infty, 8]$ **68.** $(-\infty, 13]$
69. $(-\infty, 12]$ **70.** $(-\infty, 25]$ **71.** $(-1, \infty)$ **72.** $(-1, \infty)$ **73.** $[2, 6]$ **74.** $[-3, 2]$
75. $(-\infty, -\frac{3}{2}] \cup [3, \infty)$ **76.** $(-\infty, \frac{4}{5}] \cup [2, \infty)$ **77.** $-4, 4$ **78.** $-5, 5$ **79.** $2, 4$ **80.** $-1, 5$
81. $-1, 4$ **82.** $-\frac{5}{3}, 3$ **83.** $-\frac{3}{2}, 3$ **84.** $-\frac{5}{3}, \frac{7}{3}$ **85.** $5, 9$ **86.** $3, 13$ **87.** $-1, 1$ **88.** $-\frac{3}{5}, 5$ **89.** 0
90. 0 **91.** $-5 < x < 5$
92. $a < -2$ or $a > 2$

93. $a \leq -500$ or $a \geq 500$

$-500 \qquad 500$

94. $a \leq -200$ or $a \geq 200$

$-200 \qquad 200$

95. $x = 0$

96. $-1 < y < 5$

$-1 \qquad 5$

97. $y \leq -5.02$ or $y \geq -4.98$

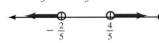

$-5.02 \qquad -4.98$

98. $x < -\frac{2}{5}$ or $x > \frac{4}{5}$

$-\frac{2}{5} \qquad \frac{4}{5}$

99. $-3 < t < 2$

$-3 \qquad 2$

100. $-\frac{7}{2} < t < \frac{5}{2}$

$-\frac{7}{2} \qquad \frac{5}{2}$

101. $\varnothing$ **102.** $\varnothing$ **103.** $\varnothing$

104. All real numbers **105.** $\varnothing$ **106.** $\varnothing$ **107.** All real numbers except 0 **108.** All real numbers **109.** $\varnothing$
110. $\varnothing$ **111.** All real numbers **112.** All real numbers

CHAPTER 2 TEST

1. 28 **2.** -3 **3.** $-\frac{1}{3}, 2$ **4.** 0, 5 **5.** $-\frac{7}{4}$ **6.** 2 **7.** $-5, 2$ **8.** $-4, -2, 4$ **9.** $w = \dfrac{P - 2l}{2}$

10. $B = \dfrac{2A}{h} - b$ **11.** 6 in., 12 in. **12.** 47°, 47°, 86° **13.** \$56.25 **14.** 55°, 125° **15.** 6 in., 8 in., 10 in.

16. 0 sec and 2 sec **17.** $x < 4$ **18.** $y \geq -52$

$0 \qquad 4$ $(-\infty, 4)$

$-52 \qquad 0$ $[-52, \infty)$

19. 2, 6 **20.** $-15, 3$ **21.** $\varnothing$ **22.** $-5, 1$ **23.** $x < -1$ or $x > \frac{4}{3}$

$-1 \qquad \frac{4}{3}$

24. $-\frac{2}{3} \leq x \leq 4$

$-\frac{2}{3} \qquad 4$

25. All real numbers **26.** $\varnothing$

CHAPTER 3 PROBLEM SET 3.1

1.

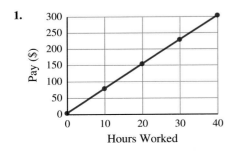

3.

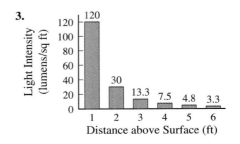

5–15. (odd)

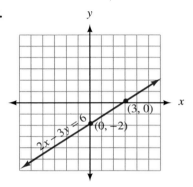

15. $(\frac{1}{2}, 2)$
11. $(0, 2)$ **5.** $(1, 2)$
9. $(5, 0)$
7. $(-1, -2)$
13. $(-5, -5)$

17. $(-\frac{5}{2}, \frac{9}{2})$ **19.** $(-3, \frac{5}{2})$ **21.** $(-2, 0)$ **23.** $(-3, -2)$
25. $(-3, -3)$ **27.** $(3, -4)$

29.

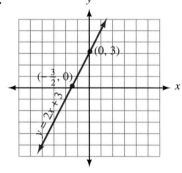

$(3, 0)$
$(0, -2)$
$2x - 3y = 6$

31.

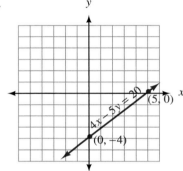

$(5, 0)$
$4x - 5y = 20$
$(0, -4)$

33.

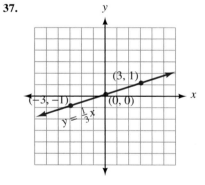

$(0, 3)$
$(-\frac{3}{2}, 0)$
$y = 2x + 3$

35. Table (b)

37.

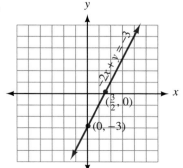

$(3, 1)$
$(-3, -1)$ $(0, 0)$
$y = \frac{1}{3}x$

39.

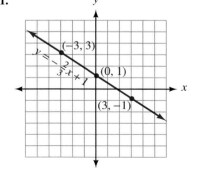

$-2x + y = -3$
$(\frac{3}{2}, 0)$
$(0, -3)$

41.

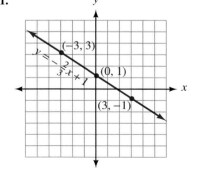

$(-3, 3)$
$y = -\frac{2}{3}x + 1$
$(0, 1)$
$(3, -1)$

43.

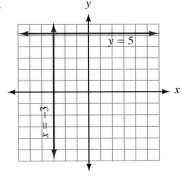

45. Equation (b)

47.

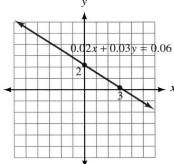

49.

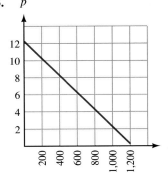

51.

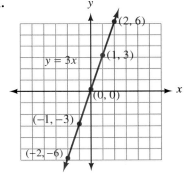

53.

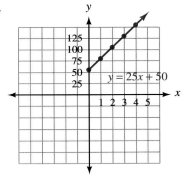

Note that the graph appears in QI only since x and y represent positive numbers.

55. p

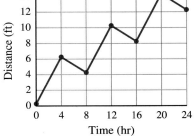

57.

Time (hr)	Distance (ft)
0	0
4	6
8	4
12	10
16	8
20	14
24	12

59. (a)

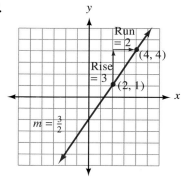

(b)

x	y
1982	16.60
1985	22.67
1989	30.76
1993	38.85
1997	46.93

61.

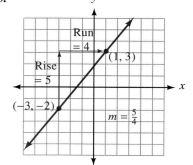

63.

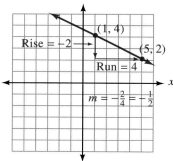

65. 2   **67.** $-\frac{3}{2}, 4$ **69.** -2 **71.** $-3, 0$ **73.** x-intercept $= \dfrac{c}{a}$, y-intercept $= \dfrac{c}{b}$

75. x-intercept $= a$, y-intercept $= b$

PROBLEM SET 3.2

1. $\frac{3}{2}$ **3.** No slope   **5.** $\frac{2}{3}$ **7.**

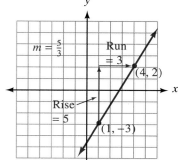

9.

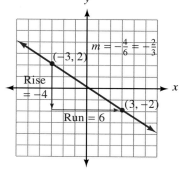

11.

13.

15.

17.

x	y
0	2
3	0

Slope $= -\frac{2}{3}$

19.

x	y
0	-5
3	-3

Slope $= \frac{2}{3}$

21. 5   **23.** -1 **25.**

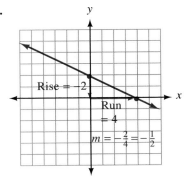

27. $\frac{1}{5}$ **29.** 0   **31.** 8   **33.** 24 ft   **35.** 10 min   **37.** 20; °C per min   **39.** 1st min
41. Slope $= -1{,}250$; dollars per year   **43.** 2 to 3 years   **45.**

47.

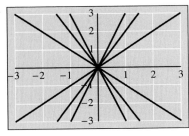

49.

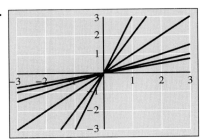

51. 0
53. $y = -\frac{3}{2}x + 6$

55. $t = \dfrac{A - P}{Pr}$

PROBLEM SET 3.3

1. $y = 2x + 3$
3. $y = x - 5$
5. $y = \frac{1}{2}x + \frac{3}{2}$
7. $y = 4$

9. Slope $= 3$
y-intercept $= -2$
Perpendicular slope $= -\frac{1}{3}$

11. Slope $= \frac{2}{3}$
y-intercept $= -4$
Perpendicular slope $= -\frac{3}{2}$

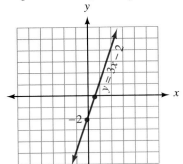

13. Slope $= -\frac{4}{5}$
y-intercept $= 4$
Perpendicular slope $= \frac{5}{4}$

15. Slope $= \frac{1}{2}$, y-intercept $= -4$; $y = \frac{1}{2}x - 4$
17. Slope $= -\frac{2}{3}$, y-intercept $= 3$; $y = -\frac{2}{3}x + 3$

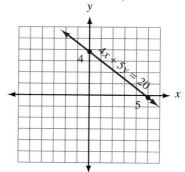

19.

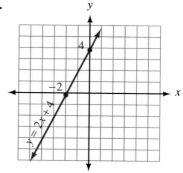

21.

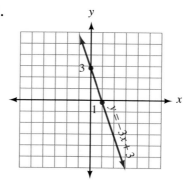

23.

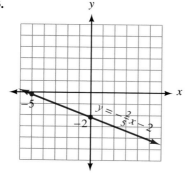

25. $y = 2x - 1$ **27.** $y = -\frac{1}{2}x - 1$ **29.** $y = -3x + 1$ **31.** $x - y = 2$ **33.** $2x - y = 3$
35. $6x - 5y = 3$ **37.** $(0, -4), (2, 0)$; $y = 2x - 4$ **39.** $(-2, 0), (0, 4)$; $y = 2x + 4$

41. Slope $= 0$
y-intercept $= -2$

43. $y = 3x + 7$ **45.** $y = -\frac{5}{2}x - 13$ **47.** $y = \frac{1}{4}x + \frac{1}{4}$
49. $y = -\frac{2}{3}x + 2$ **51.** (b) $86°$ **53.** 5 in., 23 in. **55.** \$46.50

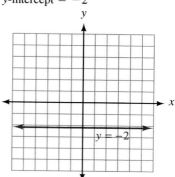

57.

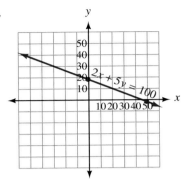

59.

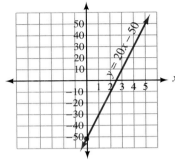

61. $y = -\frac{3}{2}x + 3$: slope $= -\frac{3}{2}$, y-intercept $= 3$, x-intercept $= 2$

63. $y = \frac{3}{2}x + 3$: slope $= \frac{3}{2}$, y-intercept $= 3$, x-intercept $= -2$

65. $y = -\dfrac{b}{a}x + b$: slope $= -\dfrac{b}{a}$, y-intercept $= b$, x-intercept $= a$

PROBLEM SET 3.4

1.

3.

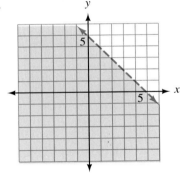

5.

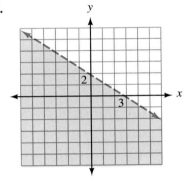

7.

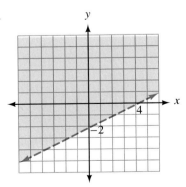

9.

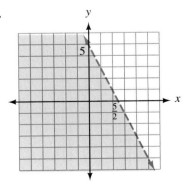

11.

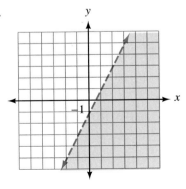

13. $x + y > 4$ **15.** $-x + 2y \le 4$

17.

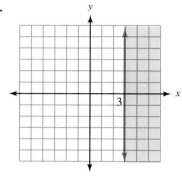

19.

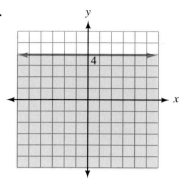

21.

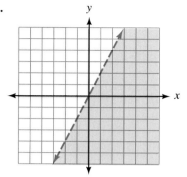

23.

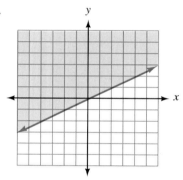

25.

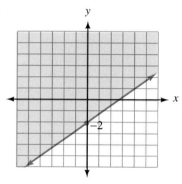

27.

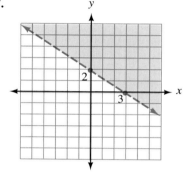

29.

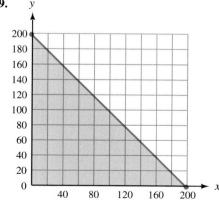

31.

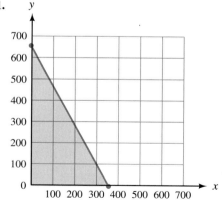

33. $y \le 7$ **35.** $t > -2$ **37.** $-1 < t < 2$

39.

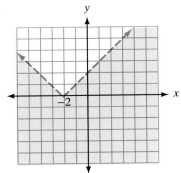

41.

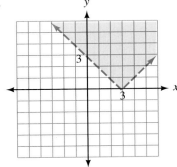

PROBLEM SET 3.5

1. (a) $y = 8.5x$ for $10 \le x \le 40$

(b)

Hours Worked	Function Rule	Gross Pay ($)
x	$y = 8.5x$	y
10	$y = 8.5(10) = 85$	85
20	$y = 8.5(20) = 170$	170
30	$y = 8.5(30) = 255$	255
40	$y = 8.5(40) = 340$	340

(c)

(d) Domain $= \{x \mid 10 \le x \le 40\}$; Range $= \{y \mid 85 \le y \le 340\}$ (e) Minimum $= \$85$; Maximum $= \$340$

3. Domain $= \{1, 2, 4\}$; Range $= \{3, 5, 1\}$; a function **5.** Domain $= \{-1, 1, 2\}$; Range $= \{3, -5\}$; a function

7. Domain $= \{7, 3\}$; Range $= \{-1, 4\}$; not a function **9.** Yes **11.** No **13.** No **15.** Yes **17.** Yes

19. (a)

(b) Domain $= \{t \mid 0 \le t \le 1\}$; Range $= \{h \mid 0 \le h \le 4\}$

Time (sec)	Function Rule	Distance (ft)
t	$h = 16t - 16t^2$	h
0	$h = 16(0) - 16(0)^2$	0
0.1	$h = 16(0.1) - 16(0.1)^2$	1.44
0.2	$h = 16(0.2) - 16(0.2)^2$	2.56
0.3	$h = 16(0.3) - 16(0.3)^2$	3.36
0.4	$h = 16(0.4) - 16(0.4)^2$	3.84
0.5	$h = 16(0.5) - 16(0.5)^2$	4
0.6	$h = 16(0.6) - 16(0.6)^2$	3.84
0.7	$h = 16(0.7) - 16(0.7)^2$	3.36
0.8	$h = 16(0.8) - 16(0.8)^2$	2.56
0.9	$h = 16(0.9) - 16(0.9)^2$	1.44
1	$h = 16(1) - 16(1)^2$	0

(c)

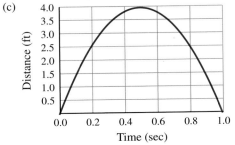

21. Domain = All real numbers
Range = $\{y|y \geq -1\}$
A function

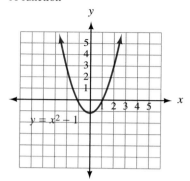

23. Domain = All real numbers
Range = $\{y|y \geq 4\}$
A function

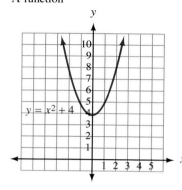

25. Domain = $\{x|x \geq -1\}$
Range = All real numbers
Not a function

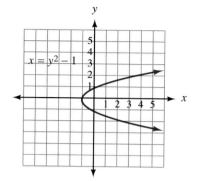

27. Domain = $\{x|x \geq 4\}$
Range = All real numbers
Not a function

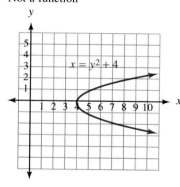

29. Domain = All real numbers
Range = $\{y|y \geq 0\}$
A function

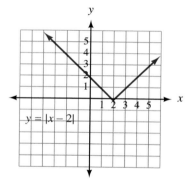

31. Domain = All real numbers
Range = $\{y|y \geq -2\}$
A function

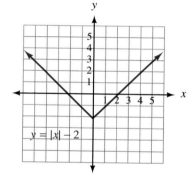

33. Domain = $\{x|x \geq 0\}$
Range = All real numbers
Not a function

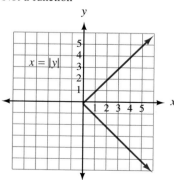

35.

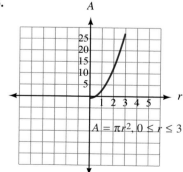

37. x = the width, so $x + 2$ = the length; $P = 2x + 2(x + 2) = 4x + 4$. The variable x must be positive.
39. $A = x(x + 2)$, where $x > 0$
41. (a) Yes   (b) Domain = $\{t|0 \le t \le 6\}$; Range = $\{h|0 \le h \le 60\}$ (c) $t = 3$ (d) $h = 60$ (e) $t = 6$
43. 10   **45.** -14 **47.** 1   **49.** -3 **51.** 130

PROBLEM SET 3.6

1. $V(3) = 300$, the painting is worth \$300 in 3 yr; $V(6) = 600$, the painting is worth \$600 in 6 yr
3. $P(x) = 2x + 2(2x + 3) = 6x + 6$, where $x > 0$
5. $A(2) = 3.14(4) = 12.56$; $A(5) = 3.14(25) = 78.5$; $A(10) = 3.14(100) = 314$ **7.** -1 **9.** -11 **11.** 2
13. 4   **15.** 35   **17.** -13 **19.** 1   **21.** -9 **23.** 8   **25.** 19   **27.** 16   **29.** 0   **31.** $3a^2 - 4a + 1$
33. (a) \$2.49   (b) \$1.53 for a 6 min call   (c) 5 min   **35.** 4   **37.** 0   **39.** 2   **41.** -8 **43.** -1
45. $2a^2 - 8$ **47.** $2b^2 - 8$ **49.** 0   **51.** -2 **53.** -3

55.

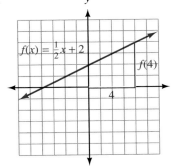

57. $x = 4$ **59.**

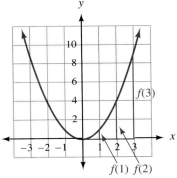

61. $R(p) = 800p - 100p^2$; $R(x) = 8x - 0.01x^2$ **63.** $P(x) = -0.01x^2 + 6x - 200$; $P(40) = \$24$ **65.** $-\frac{2}{3}, 4$
67. $-2, 1$ **69.** $\varnothing$ **71.** $\frac{7}{2}$ **73.** $f(-x) = (-x)^2 - 4 = x^2 - 4 = f(x)$

75. $f(-x) = (-x)^{-2} = \dfrac{1}{(-x)^2} = \dfrac{1}{x^2} = x^{-2} = f(x)$ **77.** $f(-x) = 3(-x) = -3x = -f(x)$

79. $f(-x) = (-x)^3 - (-x) = -x^3 + x = -(x^3 - x) = -f(x)$

PROBLEM SET 3.7

1. 30   **3.** 5   **5.** -6 **7.** $\frac{1}{2}$ **9.** 40   **11.** 225   **13.** $\frac{81}{5}$ **15.** 40.5   **17.** 64   **19.** 8
21. $\frac{50}{3}$ lb   **23.** 12 lb/in.^{2} **25.** $\frac{1,504}{15}$ in.^{2} **27.** 1.5 ohms
29.   **31.**   **33.**

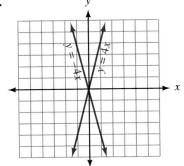

35.

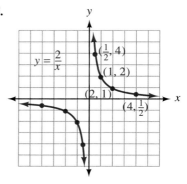

$y = \frac{2}{x}$

$\left(\frac{1}{2}, 4\right)$

$(1, 2)$

$(2, 1)$

$\left(4, \frac{1}{2}\right)$

37.

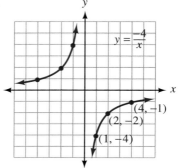

$y = \frac{-4}{x}$

$(4, -1)$

$(2, -2)$

$(1, -4)$

39.

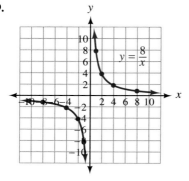

$y = \frac{8}{x}$

41.

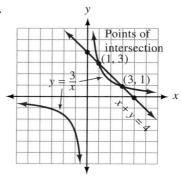

Points of intersection

$(1, 3)$

$y = \frac{3}{x}$

$(3, 1)$

$x + y = 4$

43. $(-1, -2)$, $(1, 2)$

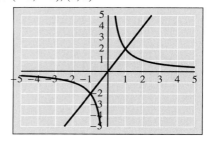

45. $(-1, -0.5)$, $(1, 0.5)$

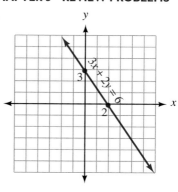

47. $x \le -9$ or $x \ge -1$

$-9 \qquad -1$

49. All real numbers

51. $-6 < y < \frac{8}{3}$

$-6 \qquad \frac{8}{3}$

CHAPTER 3   REVIEW PROBLEMS

1.

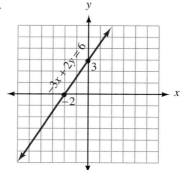

$3x + 2y = 6$

3

2

2.

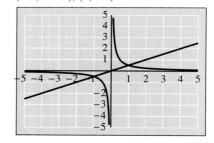

$-3x + 2y = 6$

3

-2

3.

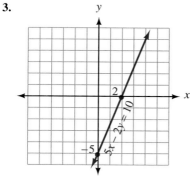

2

-5

$5x - 2y = 10$

4.

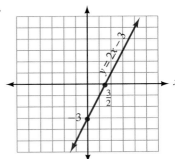

5.

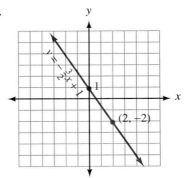

6.

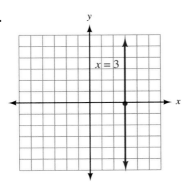

7. -2 **8.** -1 **9.** 0 **10.** No slope   **11.** 3 **12.** 2 **13.** 5 **14.** 3 **15.** -5 **16.** $\frac{4}{3}$ **17.** 6

18. -4 **19.** $y = 3x + 5$ **20.** $y = 5x + 3$ **21.** $y = -2x$ **22.** $y = \frac{1}{3}x - \frac{2}{3}$ **23.** $m = 3, b = -6$

24. $m = \frac{2}{3}, b = -2$ **25.** $m = \frac{2}{3}, b = -3$ **26.** $m = \frac{3}{2}, b = -2$ **27.** $y = 2x$ **28.** $y = 3x - 7$

29. $y = -\frac{1}{3}x$ **30.** $y = -\frac{1}{2}x + \frac{5}{6}$ **31.** $y = 2x + 1$ **32.** $y = 3x + 1$ **33.** $y = 7$ **34.** $x = -2$

35. $y = -\frac{3}{2}x - \frac{17}{2}$ **36.** $y = -\frac{1}{9}x - \frac{26}{9}$ **37.** $y = 2x - 7$ **38.** $y = \frac{3}{2}x - 1$ **39.** $y = \frac{1}{3}x - \frac{2}{3}$

40. $y = -2x - 4$ **41.**

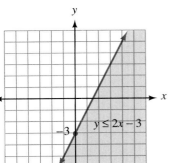

42.

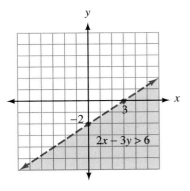

43.

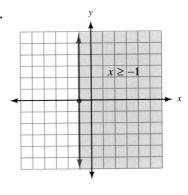

44.

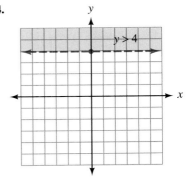

45. Domain $= \{2, 3, 4\}$; Range $= \{4, 3, 2\}$; a function   **46.** Domain $= \{-5, 3\}$; Range $= \{2, 4, -2\}$; not a function
47. Domain $= \{6, -4, -2\}$; Range $= \{3, 0\}$; a function   **48.** Domain $= \{1, 3, -2\}$; Range $= \{-1, 0\}$; a function
49. -2 **50.** 19 **51.** 4 **52.** 12 **53.** 0 **54.** 2 **55.** 1 **56.** 4 **57.** 1 **58.** 7 **59.** $3a + 2$
60. $3a - 4$ **61.** 1 **62.** 53 **63.** 31 **64.** -1 **65.** $P(x) = 6x + 6$, where $x > 0$

66.

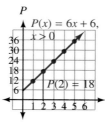

$P(x) = 6x + 6,$
$x > 0$
$P(2) = 18$

67. $V(r) = 2\pi r^3$
68. $V(h) = 4\pi h^3$ **69.** 24
70. 6   **71.** 4   **72.** 25
73. -72 **74.** -108 **75.** 2
76. 1   **77.** 84 lb
78. 270 W   **79.** 16 foot-candles
80. 96 lb

81.

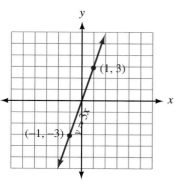

$(1, 3)$
$(-1, -3)$

82.

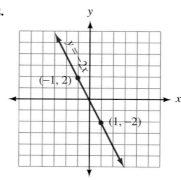

$y = -2x$
$(-1, 2)$
$(1, -2)$

83.

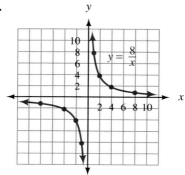

$y = \dfrac{8}{x}$

84.

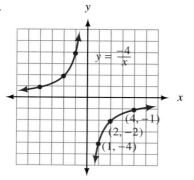

$y = \dfrac{-4}{x}$
$(4, -1)$
$(2, -2)$
$(1, -4)$

CHAPTER 3   TEST

1. x-intercept $= 3$
y-intercept $= 6$
Slope $= -2$

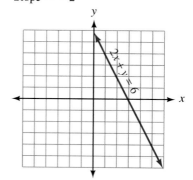

$2x + y = 6$

2. x-intercept $= -\frac{3}{2}$
y-intercept $= -3$
Slope $= -2$

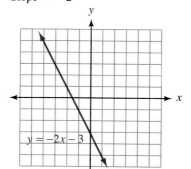

$y = -2x - 3$

3. x-intercept $= -\frac{8}{3}$
y-intercept $= 4$
Slope $= \frac{3}{2}$

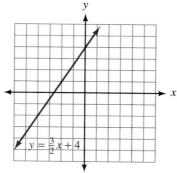

$y = \frac{3}{2}x + 4$

4. x-intercept $= -2$
No y-intercept
No slope

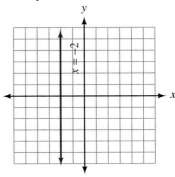

5. $y = 2x + 5$ **6.** $y = -\frac{3}{7}x + \frac{5}{7}$ **7.** $y = \frac{2}{5}x - 5$
8. $y = -\frac{1}{3}x - \frac{7}{3}$ **9.** $x = 4$

10.

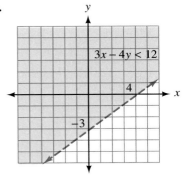

11.

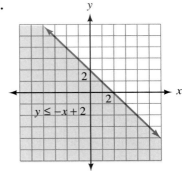

12. Domain $= \{-2, -3\}$; Range $= \{0, 1\}$; not a function
13. Domain $=$ all real numbers; Range $= \{y \mid y \geq -9\}$; a function   **14.** 11   **15.** -4 **16.** 8   **17.** 4
18. $0 < x < 4$ **19.** $V(x) = x(8 - 2x)^2$
20. $V(2) = 32$ in.^{3} is the volume of the box if a square with 2 in. sides is cut from each corner.   **21.** 18   **22.** $\frac{81}{4}$

23. $\frac{2{,}000}{3}$ lb   **24.**

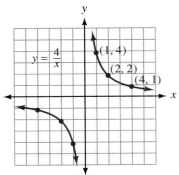

25.

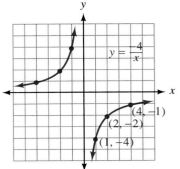

CHAPTER 4   PROBLEM SET 4.1

1. $-\frac{1}{3}$ **3.** $\dfrac{b^3}{2}$ **5.** $-\dfrac{3y^3}{2x}$ **7.** $\dfrac{18c^2}{7a^2}$ **9.** $\dfrac{x-4}{6}$ **11.** $\dfrac{4x-3y}{x(x+y)}$ **13.** $(a^2+9)(a+3)$ **15.** $\dfrac{a-6}{a+6}$

17. $\dfrac{2y+3}{y+1}$ **19.** $\dfrac{5x-2}{x+2}$ **21.** $\dfrac{2-x}{1-x}$ or $\dfrac{x-2}{x-1}$ **23.** $\dfrac{x-3}{x+2}$ **25.** $\dfrac{a^2-ab+b^2}{a-b}$ **27.** $\dfrac{2(x-1)}{x}$

29. $\dfrac{2x+3y}{2x+y}$ **31.** $\dfrac{x+3}{y-4}$ **33.** $\dfrac{x+b}{x-2b}$ **35.** $x+2$ **37.** $\dfrac{1}{x-3}$ **39.** $\dfrac{2x^2-5}{3x-2}$ **41.** -1 **43.** $-(y+6)$

45. $\dfrac{-(3a+1)}{3a-1}$ or $-\dfrac{3a+1}{3a-1}$ **47.** 0   **49.** $4x$

51.

Weeks	Weight (lb)
x	$W(x)$
0	200
1	194
4	184
12	173
24	168

53. $g(0) = -3$, $g(-3) = 0$, $g(3) = 3$, $g(-1) = -1$, $g(1)$ is undefined
55. $h(0) = -3$, $h(-3) = 3$, $h(3) = 0$, $h(-1)$ is undefined, $h(1) = -1$
57. $\{x \,|\, x \neq 1\}$ **59.** $\{x \,|\, x \neq 2\}$ **61.** $\{t \,|\, t \neq 4,\ t \neq -4\}$ **63.** 0.1 mi/min
65. 10.8 mi/gal   **67.** 6.8 ft/sec   **69.** 3,768 in./min; 2,826 in./min   **71.** 2; 2
73. Undefined; 4   **75.** 1; 1   **77.** 3; 3

79. The two graphs differ by the point (2, 4).

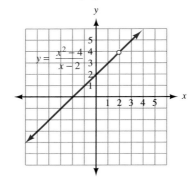

$y = \dfrac{x^2 - 4}{x - 2}$

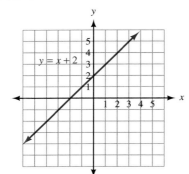

$y = x + 2$

81. (a) Domain $= \{t \,|\, 20 \leq t \leq 50\}$
(b) $r(t)$

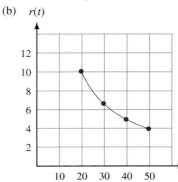

83. (a) Domain $= \{d \,|\, 1 \leq d \leq 6\}$
(b) $I(d)$

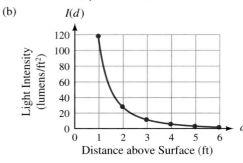

Light Intensity (lumens/ft^2)

Distance above Surface (ft)

85. $3x^2 - 7x + 4$ **87.** 9   **89.** $8x^2$

PROBLEM SET 4.2

1. $2x^2 - 4x + 3$ **3.** $-2x^2 - 3x + 4$ **5.** $2y^2 + \dfrac{5}{2} - \dfrac{3}{2y^2}$ **7.** $4ab^3 + 6a^2b$ **9.** $x + 2$ **11.** $a - 3$

13. $5x + 6y$ **15.** $x^2 + xy + y^2$ **17.** $(y^2 + 4)(y + 2)$ **19.** $(x + 2)(x + 5)$ **21.** 2   **23.** $x + a$

25. $x^2 + xa + a^2$ **27.** $x - 7 + \dfrac{7}{x + 2}$ **29.** $2x + 5 + \dfrac{2}{3x - 4}$ **31.** $2x^2 - 5x + 1 + \dfrac{4}{x + 1}$

33. $y^2 - 3y - 13$ **35.** $x - 3$ **37.** $3y^2 + 6y + 8 + \dfrac{37}{2y - 4}$ **39.** $a^3 + 2a^2 + 4a + 6 + \dfrac{17}{a - 2}$

41. $y^3 + 2y^2 + 4y + 8$ **43.** $x^2 - 2x + 1$ **45.** $(x + 3)(x + 2)(x + 1)$ **47.** $(x + 3)(x + 4)(x - 2)$

49. Yes   **51.** 7   **53.** $\frac{21}{10}$ **55.** $\frac{11}{8}$ **57.** $\frac{1}{18}$ **59.** 32   **61.** x-intercept = 2; y-intercept = $-2\frac{1}{2}$

63. x-intercept = -6; y-intercept = 4   **65.** $4x^3 - x^2 + 3$ **67.** $0.5x^2 - 0.4x + 0.3$ **69.** $\dfrac{3}{2}x - \dfrac{5}{2} + \dfrac{1}{2x + 4}$

71. $\dfrac{2}{3}x + \dfrac{1}{3} + \dfrac{2}{3x - 1}$

PROBLEM SET 4.3

1. $\frac{1}{6}$ **3.** $\frac{9}{4}$ **5.** $\frac{1}{2}$ **7.** $\dfrac{15y}{x^2}$ **9.** $\dfrac{b}{a}$ **11.** $\dfrac{2y^5}{z^3}$ **13.** $\dfrac{x + 3}{x + 2}$ **15.** $y + 1$ **17.** $\dfrac{3(x + 4)}{x - 2}$ **19.** 1

21. $\dfrac{(a - 2)(a + 2)}{a - 5}$ **23.** $\dfrac{9t^2 - 6t + 4}{4t^2 - 2t + 1}$ **25.** $\dfrac{x + 3}{x + 4}$ **27.** $\dfrac{5a - b}{9a^2 + 15ab + 25b^2}$ **29.** 2   **31.** $\dfrac{x(x - 1)}{x^2 + 1}$

33. $\dfrac{(a + 4b)(a - 3b)}{(a - 4b)(a + 5b)}$ **35.** $\dfrac{2y - 1}{2y - 3}$ **37.** $\dfrac{(y - 2)(y + 1)}{(y + 2)(y - 1)}$ **39.** $\dfrac{x - 2}{x + 3}$ **41.** $3x$ **43.** $x - 2$

45. $-(y - 4)$ **47.** $(a - 5)(a + 1)$

49.

Number of Copies	Price per Copy ($)
x	$p(x)$
1	20.33
10	9.33
20	6.40
50	4.00
100	3.05

51. $305.00   **53.** 3   **55.** 5   **57.** $\frac{2}{3}$ **59.** $\frac{2}{5}$ **61.** $\frac{47}{105}$

63. $\frac{1}{35}$ **65.** 3   **67.** $\frac{101}{2}$

PROBLEM SET 4.4

1. $\frac{5}{4}$ **3.** $\frac{1}{3}$ **5.** $\frac{41}{24}$ **7.** $\frac{19}{144}$ **9.** $\frac{31}{24}$ **11.** 1   **13.** -1 **15.** $\dfrac{1}{x + y}$ **17.** 1   **19.** $\dfrac{a^2 + 2a - 3}{a^3}$

21. 1   **23.** $\dfrac{4 - 3t}{2t^2}$ **25.** $\frac{1}{2}$ **27.** $\frac{1}{5}$ **29.** $\dfrac{x + 3}{2(x + 1)}$ **31.** $\dfrac{a - b}{a^2 + ab + b^2}$ **33.** $\dfrac{2y - 3}{4y^2 + 6y + 9}$

35. $\dfrac{2(2x - 3)}{(x - 3)(x - 2)}$ **37.** $\dfrac{1}{2t - 7}$ **39.** $\dfrac{4}{(a - 3)(a + 1)}$ **41.** $\dfrac{-4x^2}{(2x + 1)(2x - 1)(4x^2 + 2x + 1)}$

43. $\dfrac{2}{(2x + 3)(4x + 3)}$ **45.** $\dfrac{a}{(a + 5)(a + 4)}$ **47.** $\dfrac{x + 1}{(x + 3)(x - 2)}$ **49.** $\dfrac{x - 1}{(x + 2)(x + 1)}$ **51.** $\dfrac{1}{(x + 1)(x + 2)}$

53. $\dfrac{1}{(x+2)(x+3)}$ **55.** $\dfrac{4x+5}{2x+1}$ **57.** $\dfrac{22-5t}{4-t}$ **59.** $\dfrac{2x^2+3x-4}{2x+3}$ **61.** $\dfrac{2x-3}{2x}$ **63.** $\frac{1}{2}$ **65.** $\frac{51}{10}=5.1$

67. $(3+4)^{-1}=7^{-1}=\frac{1}{7}$; $3^{-1}+4^{-1}=\frac{1}{3}+\frac{1}{4}=\frac{7}{12}$ **69.** $\frac{4}{3}$ **71.** $\dfrac{x+1}{x}$ **73.** $x+\dfrac{4}{x}=\dfrac{x^2+4}{x}$

75. $\dfrac{1}{x}+\dfrac{1}{x+1}=\dfrac{2x+1}{x(x+1)}$ **77.** Slope $=\frac{2}{3}$; y-intercept $=-2$ **79.** $y=\frac{2}{3}x+6$ **81.** $y=4x-1$

83. 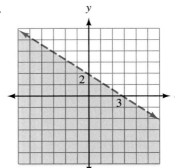   **85.** 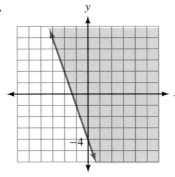   **87.**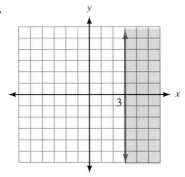

89. $\dfrac{x-1}{x+3}$ **91.** $\dfrac{x-1}{x+50}$ **93.** $\dfrac{x+1}{x-100}$

PROBLEM SET 4.5

1. $\frac{9}{8}$ **3.** $\frac{2}{15}$ **5.** $\frac{119}{20}$ **7.** $\dfrac{1}{x+1}$ **9.** $\dfrac{a+1}{a-1}$ **11.** $\dfrac{y-x}{y+x}$ **13.** $\dfrac{1}{(x-2)(x+5)}$ **15.** $\dfrac{1}{a^2-a+1}$

17. $\dfrac{x+3}{x+2}$ **19.** $\dfrac{a+3}{a-2}$ **21.** $\dfrac{x-3}{x}$ **23.** $\dfrac{x+4}{x+2}$ **25.** $\dfrac{x-3}{x+3}$ **27.** $\dfrac{a-1}{a+1}$ **29.** $-\dfrac{x}{3}$ **31.** $\dfrac{y^2+1}{2y}$

33. $\dfrac{-x^2+x-1}{x-1}$ **35.** $\frac{5}{3}$ **37.** $\dfrac{2x-1}{2x+3}$ **39.** $(a^{-1}+b^{-1})^{-1}=\left(\dfrac{1}{a}+\dfrac{1}{b}\right)^{-1}=\left(\dfrac{a+b}{ab}\right)^{-1}=\dfrac{ab}{a+b}$

41. $\dfrac{1-x^{-1}}{1+x^{-1}}=\dfrac{1-\dfrac{1}{x}}{1+\dfrac{1}{x}}=\dfrac{\dfrac{x-1}{x}}{\dfrac{x+1}{x}}=\dfrac{x-1}{x+1}$ **43.** First three terms: $\frac{3}{2}, \frac{5}{3}, \frac{8}{5}$; Next three terms: $\frac{13}{8}, \frac{21}{13}, \frac{34}{21}$

45. $\frac{4}{3}, \frac{10}{7}, \frac{24}{17}$ **53.** -15 **55.** 5 **57.** 1 **59.** $-3, 4$ **61.** $-2, \frac{5}{3}$ **63.** $2, 3$ **65.** $\dfrac{x+3}{(x+5)(x-3)}$

67. $\dfrac{x-a}{x+a}$

PROBLEM SET 4.6

1. $-\frac{35}{3}$ **3.** $-\frac{18}{5}$ **5.** $\frac{36}{11}$ **7.** 2 **9.** 5 **11.** 2 **13.** $-3, 4$ **15.** $-\frac{4}{3}, 1$
17. Possible solution -1, which does not check; $\varnothing$ **19.** 5 **21.** $-\frac{1}{2}, \frac{5}{3}$ **23.** $\frac{2}{3}$ **25.** 18
27. Possible solution 4, which does not check; $\varnothing$ **29.** Possible solutions 3 and -4; only -4 checks; -4 **31.** -6
33. -5 **35.** $\frac{53}{17}$ **37.** Possible solutions 1 and 2; only 2 checks; 2

39. Possible solution 3, which does not check; $\varnothing$ **41.** $\frac{22}{3}$ **43.** 2   **45.** 1, 5   **47.** $x = \dfrac{ab}{a-b}$

49. $R = \dfrac{R_1 R_2}{R_1 + R_2}$ **51.** $y = \dfrac{x-3}{x-1}$ **53.** $y = \dfrac{1-x}{3x-2}$

55.

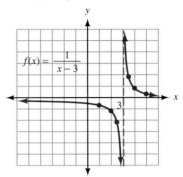

57.

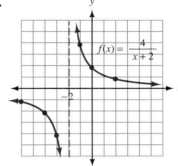

59.

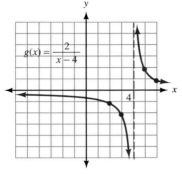

61.

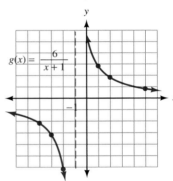

63. $f(0) = -\frac{1}{3}; f(6) = \frac{1}{3}$ **65.** $f(1) = -\frac{1}{2}; f(5) = \frac{1}{2}$

67. Domain $= \{x \mid x \neq 3\}$ **69.** $f(0) = 2; f(-4) = -2$

71. $f(2) = 1; f(-6) = -1$ **73.** Domain $= \{x \mid x \neq -2\}$

75. 147   **77.** ± 10 **79.** 100   **81.** $-\frac{5}{2}, -\frac{2}{3}, \frac{5}{2}$ **83.** $-2, 2, 3$

PROBLEM SET 4.7

NOTE: As you can see, in addition to the answers to these problems we have included some of the equations used to solve them. Remember, you should attempt the problems on your own before looking here to check your answers or equations.

1. $\dfrac{1}{x} + \dfrac{1}{3x} = \dfrac{20}{3}; \dfrac{1}{5}$ and $\dfrac{3}{5}$ **3.** $x + \dfrac{1}{x} = \dfrac{10}{3};$ 3 or $\dfrac{1}{3}$ **5.** $\dfrac{1}{x} + \dfrac{1}{x+1} = \dfrac{7}{12};$ 3, 4   **7.** $\dfrac{7+x}{9+x} = \dfrac{5}{6};$ 3

9. Let $x =$ speed of current; $\dfrac{1.5}{5-x} = \dfrac{3}{5+x}; \dfrac{5}{3}$ mi/hr   **11.** 6 mi/hr   **13.** Train A: 75 mi/hr; Train B: 60 mi/hr

15. 540 mi/hr   **17.** 54 mi/hr   **19.** Let $x =$ time to fill the tank with both open; $\dfrac{1}{8} - \dfrac{1}{16} = \dfrac{1}{x};$ 16 hr   **21.** 15 hr

23. 5.25 min   **25.** 51.1 acres   **27.** 5.9 mi/hr   **29.** 20.7 mi/hr   **31.** 4.6 mi/hr   **33.** 3.6 mi/hr

35.

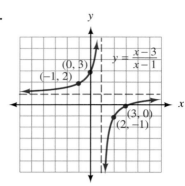

37.

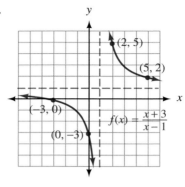

39.

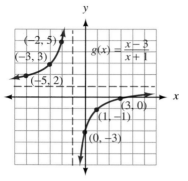

41. $\dfrac{2}{3a}$ **43.** $(x-3)(x+2)$ **45.** 1   **47.** $\dfrac{3-x}{3+x}$ **49.** Possible solution 3, which does not check; $\varnothing$

CHAPTER 4   REVIEW PROBLEMS

1. $\dfrac{25x^2}{7y^3}$ **2.** $\dfrac{2z}{x}$ **3.** $\dfrac{a(a-b)}{4}$ **4.** $\dfrac{3}{4a+3b}$ **5.** $\dfrac{x-5}{x+5}$ **6.** $\dfrac{x-5}{x-3}$ **7.** $\dfrac{a+1}{a-1}$ **8.** $x+3$

9. $-\dfrac{x+3}{x-3}$ **10.** -1 **11.** $3x+2+\dfrac{4}{x}$ **12.** $2x^2-x+3$ **13.** $-9b+5a-7a^2b^2$ **14.** $-2a^2+1-3b^2$

15. $x^{3n}-x^{2n}$ **16.** $2x^{2n}-3x^{3n}$ **17.** $x+2$ **18.** $x-3$ **19.** $5x+6y$ **20.** $x-6y$ **21.** $(y^2+4)(y+2)$

22. $(y^2+9)(y+3)$ **23.** $4x+1-\dfrac{2}{2x-7}$ **24.** $3x+7+\dfrac{10}{3x-4}$ **25.** $y^2-3y-13$ **26.** y^2-5y-1

27. $a^3+2a^2+4a+6+\dfrac{17}{a-2}$ **28.** $a^3-a^2+2a-4+\dfrac{7}{a+2}$ **29.** $\frac{9}{5}$ **30.** 3   **31.** $\dfrac{3x}{4y^2}$ **32.** $\dfrac{6}{y^2}$

33. $\dfrac{x-1}{x^2+1}$ **34.** $x+2$ **35.** 1   **36.** $\dfrac{a-3}{2a-1}$ **37.** $\dfrac{x+2}{x-2}$ **38.** $\dfrac{x+1}{x-1}$ **39.** $(2x-3)(x+3)$

40. $(3x+5)(x+5)$ **41.** $x+3$ **42.** $x-2$ **43.** $\frac{31}{30}$ **44.** $\frac{17}{8}$ **45.** -1 **46.** $-\dfrac{1}{x+y}$ **47.** $\dfrac{x^2+x+1}{x^3}$

48. $\dfrac{x^2-x-1}{x^3}$ **49.** $\dfrac{1}{(y+4)(y+3)}$ **50.** $\dfrac{1}{(y+3)(y+2)}$ **51.** $\dfrac{x-1}{2(x+1)(x+2)}$ **52.** $\dfrac{x-1}{2(x+1)(x+2)}$

53. $\dfrac{15x-2}{5x-2}$ **54.** $\dfrac{49x-23}{7x-3}$ **55.** $\dfrac{1}{(x+2)(x+3)}$ **56.** $\dfrac{1}{(x+3)(x+5)}$ **57.** $\dfrac{12}{(3x+2)(3x-2)}$

58. $\dfrac{24}{(4x+3)(4x-3)}$ **59.** 5   **60.** $\frac{1}{7}$ **61.** $\dfrac{1}{a^2-a+1}$ **62.** $\dfrac{1}{a^2+a+1}$ **63.** $\dfrac{x^2+x+1}{x^2+1}$

64. $\dfrac{x^2+x+1}{x+1}$ **65.** $\dfrac{x+3}{x+2}$ **66.** $\dfrac{2x-1}{2x+1}$ **67.** 6   **68.** $\frac{3}{2}$ **69.** 1   **70.** $-\frac{1}{2}$ **71.** -6 **72.** 5

73. Possible solution -3, which does not check; $\varnothing$ **74.** Possible solution -1, which does not check; $\varnothing$ **75.** $\frac{22}{3}$
76. $-\frac{12}{5}$ **77.** Possible solutions 4 and -5; only 4 checks   **78.** Possible solutions 2 and 3; only 2 checks
79. $-7, -6$ **80.** Possible solutions -3 and -1; only -1 checks   **81.** $-2, 3$ **82.** $\frac{3}{2}, 4$ **83.** 3, 6
84. 2, 8   **85.** $\frac{60}{11}$ min   **86.** 12 min   **87.** 4 or $\frac{1}{2}$ **88.** 3 or 1   **89.** 3 mi/hr
90. car 30 mi/hr, truck 20 mi/hr   **91.** 7.5 mi/hr   **92.** 742 mi/hr

93.

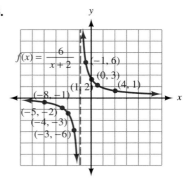

94.

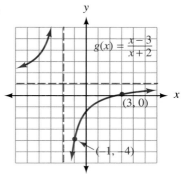

CHAPTER 4   TEST

1. $x + y$ **2.** $\dfrac{x - 1}{x + 1}$ **3.** $6x^2 + 3xy - 4y^2$ **4.** $x^2 - 4x - 2 + \dfrac{8}{2x - 1}$ **5.** $2(a + 4)$ **6.** $4(a + 3)$

7. $x + 3$ **8.** $\dfrac{38}{105}$ **9.** $\dfrac{7}{8}$ **10.** $\dfrac{1}{a - 3}$ **11.** $\dfrac{3(x - 1)}{x(x - 3)}$ **12.** $\dfrac{x}{(x + 4)(x + 5)}$ **13.** $\dfrac{x + 4}{(x + 1)(x + 2)}$

14. $\dfrac{3a + 8}{3a + 10}$ **15.** $\dfrac{x - 3}{x - 2}$ **16.** $-\dfrac{3}{5}$ **17.** Possible solution 3, which does not check; $\varnothing$ **18.** $\dfrac{3}{13}$ **19.** $-2, 3$

20. -7 **21.** 6 mi/hr   **22.** 15 hr   **23.** 2.7 mi   **24.** 1,012 mi/hr

25.

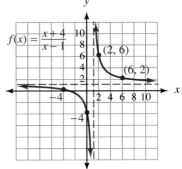

CHAPTER 5   PROBLEM SET 5.1

1. 12   **3.** Not a real number   **5.** -7 **7.** -3 **9.** 2   **11.** Not a real number   **13.** 0.2
15. 0.2   **17.** $6a^4$ **19.** $3a^4$ **21.** xy^2 **23.** $2x^2y$ **25.** $2a^3b^5$ **27.** 6   **29.** -3 **31.** 2   **33.** -2
35. 2   **37.** $\dfrac{9}{5}$ **39.** $\dfrac{4}{5}$ **41.** 9   **43.** 125   **45.** 8   **47.** $\dfrac{1}{3}$ **49.** $\dfrac{1}{27}$ **51.** $\dfrac{6}{5}$ **53.** $\dfrac{8}{27}$ **55.** 7   **57.** $\dfrac{3}{4}$
59. $x^{4/5}$ **61.** a **63.** $\dfrac{1}{x^{2/5}}$ **65.** $x^{1/6}$ **67.** $x^{9/25}y^{1/2}z^{1/5}$ **69.** $\dfrac{b^{7/4}}{a^{1/8}}$ **71.** $y^{3/10}$ **73.** $\dfrac{1}{a^2b^4}$ **75.** $\dfrac{s^{1/2}}{r^{20}}$
77. $10b^3$ **79.** $(9^{1/2} + 4^{1/2})^2 = (3 + 2)^2 = 5^2 = 25 \neq 9 + 4$ **81.** $\sqrt{\sqrt{a}} = (a^{1/2})^{1/2} = a^{1/4} = \sqrt[4]{a}$ **83.** 25 mi/hr
85. 1.618   **87.** $x^6 - x^3$ **89.** $x^2 + 2x - 15$ **91.** $x^4 - 10x^2 + 25$ **93.** $x^3 - 27$ **95.** $3x - 4x^3y$

97. When $x = 2$, $y = 1.7$.
99. When $x = 10$, $y = 5.6$.

101. Graphs intersect at $x = 1$, $y = 1$ and $x = 0$, $y = 0$.

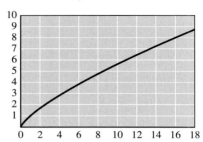

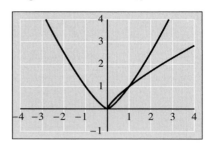

103. (a) 1.62 micrograms   (b) 0.87 microgram   (c) 0.00293 microgram   (d) 0.0000029 microgram

PROBLEM SET 5.2

1. $x + x^2$ **3.** $a^2 - a$ **5.** $6x^3 - 8x^2 + 10x$ **7.** $12x^2 - 36y^2$ **9.** $x^{4/3} - 2x^{2/3} - 8$
11. $a - 10a^{1/2} + 21$ **13.** $20y^{2/3} - 7y^{1/3} - 6$ **15.** $10x^{4/3} + 21x^{2/3}y^{1/2} + 9y$ **17.** $t + 10t^{1/2} + 25$
19. $x^3 + 8x^{3/2} + 16$ **21.** $a - 2a^{1/2}b^{1/2} + b$ **23.** $4x - 12x^{1/2}y^{1/2} + 9y$ **25.** $a - 3$ **27.** $x^3 - y^3$
29. $t - 8$ **31.** $4x^3 - 3$ **33.** $x + y$ **35.** $a - 8$ **37.** $8x + 1$ **39.** $t - 1$ **41.** $2x^{1/2} + 3$
43. $3x^{1/3} - 4y^{1/3}$ **45.** $3a - 2b$ **47.** $3(x - 2)^{1/2}(4x - 11)$ **49.** $5(x - 3)^{7/5}(x - 6)$
51. $3(x + 1)^{1/2}(3x^2 + 3x + 2)$ **53.** $(x^{1/3} - 2)(x^{1/3} - 3)$ **55.** $(a^{1/5} - 4)(a^{1/5} + 2)$ **57.** $(2y^{1/3} + 1)(y^{1/3} - 3)$
59. $(3t^{1/5} + 5)(3t^{1/5} - 5)$ **61.** $(2x^{1/7} + 5)^2$ **63.** $\dfrac{3 + x}{x^{1/2}}$ **65.** $\dfrac{x + 5}{x^{1/3}}$ **67.** $\dfrac{x^3 + 3x^2 + 1}{(x^3 + 1)^{1/2}}$ **69.** $\dfrac{-4}{(x^2 + 4)^{1/2}}$
71. 2   **73.** 27   **75.** 0.871   **77.** 15.8%   **79.** 5.9%   **81.** $\dfrac{1}{x^2 + 9}$ **83.** $3x - 4x^3y$ **85.** $5x - 4$
87. $x^2 + 5x + 25$

PROBLEM SET 5.3

1. $2\sqrt{2}$ **3.** $7\sqrt{2}$ **5.** $12\sqrt{2}$ **7.** $4\sqrt{5}$ **9.** $4\sqrt{3}$ **11.** $15\sqrt{3}$ **13.** $3\sqrt[3]{2}$ **15.** $4\sqrt[3]{2}$ **17.** $6\sqrt[3]{2}$
19. $2\sqrt[5]{2}$ **21.** $3x\sqrt{2x}$ **23.** $2y\sqrt[4]{2y^3}$ **25.** $2xy\sqrt[3]{5xy}$ **27.** $4abc^2\sqrt{3b}$ **29.** $2bc\sqrt[3]{6a^2c}$ **31.** $2xy^2\sqrt[5]{2x^3y^2}$
33. $3xy^2z\sqrt[5]{x^2}$ **35.** $2\sqrt{3}$ **37.** $\sqrt{-20}$, which is not a real number   **39.** $\dfrac{\sqrt{11}}{2}$ **41.** $\dfrac{2\sqrt{3}}{3}$ **43.** $\dfrac{5\sqrt{6}}{6}$
45. $\dfrac{\sqrt{2}}{2}$ **47.** $\dfrac{\sqrt{5}}{5}$ **49.** $2\sqrt[3]{4}$ **51.** $\dfrac{2\sqrt[3]{3}}{3}$ **53.** $\dfrac{\sqrt[4]{24x^2}}{2x}$ **55.** $\dfrac{\sqrt[4]{8y^3}}{y}$ **57.** $\dfrac{\sqrt[3]{36xy^2}}{3y}$ **59.** $\dfrac{\sqrt[3]{6xy^2}}{3y}$ **61.** $\dfrac{\sqrt[4]{2x}}{2x}$
63. $\dfrac{3x\sqrt{15xy}}{5y}$ **65.** $\dfrac{5xy\sqrt{6xz}}{2z}$ **67.** $\dfrac{2ab\sqrt[3]{6ac^2}}{3c}$ **69.** $\dfrac{2xy^2\sqrt[3]{3z^2}}{3z}$ **71.** $5|x|$ **73.** $3|xy|\sqrt{3x}$ **75.** $|x - 5|$
77. $|2x + 3|$ **79.** $2|a(a + 2)|$ **81.** $2|x|\sqrt{x - 2}$ **83.** $\sqrt{9 + 16} = \sqrt{25} = 5$; $\sqrt{9} + \sqrt{16} = 3 + 4 = 7$
85. $5\sqrt{13}$ ft   **89.** $\sqrt{2}, \sqrt{3}, \sqrt{4}, \sqrt{5}, \sqrt{6}, \sqrt{7}, \ldots$; 10th term $= \sqrt{11}$; 100th term $= \sqrt{101}$ **91.** $\dfrac{y^3}{x^2}$ **93.** 1

95. $\dfrac{4x^2 - 6x + 9}{9x^2 - 3x + 1}$ **97.** $12\sqrt[3]{5}$ **99.** $6\sqrt[3]{49}$ **101.** $\dfrac{\sqrt[10]{a^7}}{a}$ **103.** $\dfrac{\sqrt[20]{a^9}}{a}$

105.

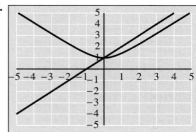

107. About $\frac{3}{4}$ of a unit apart when $x = 2$ **109.** $x = 0$

PROBLEM SET 5.4

1. $7\sqrt{5}$ **3.** $-x\sqrt{7}$ **5.** $\sqrt[3]{10}$ **7.** $9\sqrt[5]{6}$ **9.** 0 **11.** $\sqrt{5}$ **13.** $-32\sqrt{2}$ **15.** $-3x\sqrt{2}$

17. $-2\sqrt[3]{2}$ **19.** $8x\sqrt[3]{xy^2}$ **21.** $3a^2b\sqrt{3ab}$ **23.** $11ab\sqrt[3]{3a^2b}$ **25.** $10xy\sqrt[4]{3y}$ **27.** $\sqrt{2}$ **29.** $\dfrac{8\sqrt{5}}{15}$

31. $\dfrac{(x-1)\sqrt{x}}{x}$ **33.** $\dfrac{3\sqrt{2}}{2}$ **35.** $\dfrac{5\sqrt{6}}{6}$ **37.** $\dfrac{8\sqrt[3]{25}}{5}$ **39.** $\sqrt{12} \approx 3.464;\ 2\sqrt{3} \approx 2(1.732) = 3.464$

41. $\sqrt{8} + \sqrt{18} \approx 2.828 + 4.243 = 7.071;\ \sqrt{50} \approx 7.071;\ \sqrt{26} \approx 5.099$ **43.** $8\sqrt{2x}$ **45.** 5 **53.** 1

55. $\dfrac{13 - 3t}{3 - t}$ **57.** $\dfrac{6}{4x + 3}$ **59.** $\dfrac{x - y}{x^2 + xy + y^2}$ **61.** $11\sqrt{x + 3}$ **63.** 0 **65.** $4x(x + 5)$

PROBLEM SET 5.5

1. $3\sqrt{2}$ **3.** $10\sqrt{21}$ **5.** 720 **7.** 54 **9.** $\sqrt{6} - 9$ **11.** $24 + 6\sqrt[3]{4}$ **13.** $7 + 2\sqrt{6}$
15. $x + 2\sqrt{x} - 15$ **17.** $34 + 20\sqrt{3}$ **19.** $19 + 8\sqrt{3}$ **21.** $x - 6\sqrt{x} + 9$ **23.** $4a - 12\sqrt{ab} + 9b$
25. $x + 4\sqrt{x - 4}$ **27.** $x - 6\sqrt{x - 5} + 4$ **29.** 1 **31.** $a - 49$ **33.** $25 - x$ **35.** $x - 8$

37. $10 + 6\sqrt{3}$ **39.** $\dfrac{\sqrt{3} + 1}{2}$ **41.** $\dfrac{5 - \sqrt{5}}{4}$ **43.** $\dfrac{x + 3\sqrt{x}}{x - 9}$ **45.** $\dfrac{10 + 3\sqrt{5}}{11}$ **47.** $\dfrac{3\sqrt{x} + 3\sqrt{y}}{x - y}$

49. $2 + \sqrt{3}$ **51.** $\dfrac{11 - 4\sqrt{7}}{3}$ **53.** $\dfrac{a + 2\sqrt{ab} + b}{a - b}$ **55.** $\dfrac{x + 4\sqrt{x} + 4}{x - 4}$ **57.** $\dfrac{5 - \sqrt{21}}{4}$ **59.** $\dfrac{\sqrt{x} - 3x + 2}{1 - x}$

63. $10\sqrt{3}$ **65.** $x + 6\sqrt{x} + 9$ **67.** 75 **69.** $\dfrac{5\sqrt{2}}{4}\ \sec;\ \dfrac{5}{2}\ \sec$ **75.** $-\dfrac{1}{8}$ **77.** $\dfrac{y - 2}{y + 2}$ **79.** $\dfrac{2x + 1}{2x - 1}$

81. $\dfrac{x - 4 - 2\sqrt{x - 4}}{x - 8}$ **83.** $\dfrac{x + \sqrt{x^2 - 9}}{3}$ **85.** $\dfrac{\sqrt[3]{x^2} - 2\sqrt[3]{x} + 4}{x + 8}$ **87.** $\dfrac{\sqrt[3]{9} - \sqrt[3]{6} + \sqrt[3]{4}}{5}$

PROBLEM SET 5.6

1. 4 **3.** $\varnothing$ **5.** 5 **7.** $\varnothing$ **9.** $\dfrac{39}{2}$ **11.** $\varnothing$ **13.** 5 **15.** 3 **17.** $-\dfrac{32}{3}$ **19.** $3, 4$
21. $-2, -1$ **23.** -1 **25.** $\varnothing$ **27.** 7 **29.** $0, 3$ **31.** -4 **33.** 8 **35.** 0 **37.** 9 **39.** 0
41. 8 **43.** Possible solution 9, which does not check; $\varnothing$ **45.** Possible solutions 0 and 32; only 0 checks; 0
47. Possible solutions -2 and 6; only 6 checks; 6   **49.** $h = 100 - 16t^2$ **51.** $\dfrac{392}{121} \approx 3.24$ ft

53.

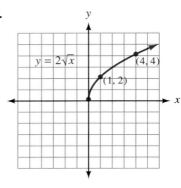

55.

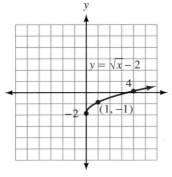

57.

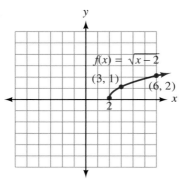

59.

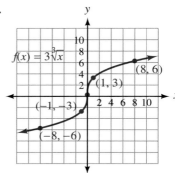

61.

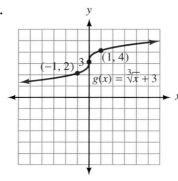

63.

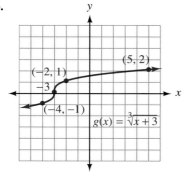

65. $\sqrt{6} - 2$ **67.** $x + 10\sqrt{x} + 25$ **69.** $\dfrac{x - 3\sqrt{x}}{x - 9}$ **71.** Possible solutions 2 and 6; only 6 checks; 6

73. $-3, -1, 1$ **75.** $y = \frac{1}{8} x^2$

77.

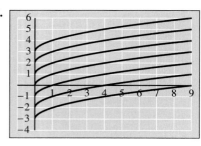

79. The value of b shifts
the curve b units along
the y-axis.

81.

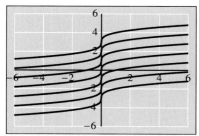

83. The value of b shifts the curve
b units along the y-axis.

85.

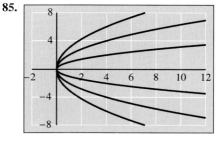

87. The smaller the absolute value
of a, the more slowly the
graph rises or falls. If a is
negative, the graph lies below
the x-axis.

PROBLEM SET 5.7

1. $6i$ **3.** $-5i$ **5.** $6i\sqrt{2}$ **7.** $-2i\sqrt{3}$ **9.** 1 **11.** -1 **13.** $-i$ **15.** $x = 3,\ y = -1$
17. $x = -2,\ y = -\frac{1}{2}$ **19.** $x = -8,\ y = -5$ **21.** $x = 7,\ y = \frac{1}{2}$ **23.** $x = \frac{3}{7},\ y = \frac{2}{5}$ **25.** $5 + 9i$ **27.** $5 - i$
29. $2 - 4i$ **31.** $1 - 6i$ **33.** $2 + 2i$ **35.** $-1 - 7i$ **37.** $6 + 8i$ **39.** $2 - 24i$ **41.** $-15 + 12i$
43. $18 + 24i$ **45.** $10 + 11i$ **47.** $21 + 23i$ **49.** $-2 + 2i$ **51.** $2 - 11i$ **53.** $-21 + 20i$ **55.** $-2i$
57. $-7 - 24i$ **59.** 5 **61.** 40 **63.** 13 **65.** 164 **67.** $-3 - 2i$ **69.** $-2 + 5i$ **71.** $\frac{8}{13} + \frac{12}{13}i$
73. $-\frac{18}{13} - \frac{12}{13}i$ **75.** $-\frac{5}{13} + \frac{12}{13}i$ **77.** $\frac{13}{15} - \frac{2}{5}i$ **79.** $\frac{31}{53} - \frac{24}{53}i$ **81.** $-\frac{3}{2}$ **83.** $-3,\ \frac{1}{2}$ **85.** $\frac{5}{4}$ or $\frac{4}{5}$

CHAPTER 5   REVIEW PROBLEMS

1. 7 **2.** 2 **3.** -3 **4.** 3 **5.** 2 **6.** $\frac{3}{4}$ **7.** 27 **8.** 4 **9.** $2x^3y^2$ **10.** $5x^3y^4$
11. $\frac{1}{16}$ **12.** $\frac{9}{20}$ **13.** x^2 **14.** y **15.** a^2b^4 **16.** $x^{17/12}$ **17.** $a^{7/20}$ **18.** $x^{2/5}y^{3/5}z$ **19.** $a^{5/12}b^{8/3}$
20. $y^{3/2}$ **21.** $12x + 11x^{1/2}y^{1/2} - 15y$ **22.** $20x - 7x^{1/2} - 6$ **23.** $a^{2/3} - 10a^{1/3} + 25$ **24.** $8t - 1$

25. $4x^{1/2} + 2x^{5/6}$ **26.** $3a^{3/7}b^{1/5} - 2a^{1/7}b^{3/5}$ **27.** $2(x - 3)^{1/4}(4x - 13)$ **28.** $(3x^{1/5} + 2)(2x^{1/5} - 5)$ **29.** $\dfrac{x + 5}{x^{1/4}}$

30. $\dfrac{4x^3 + x^2 + 1}{(x^2 + 1)^{1/2}}$ **31.** $2\sqrt{3}$ **32.** $3\sqrt{3}$ **33.** $5\sqrt{2}$ **34.** $2\sqrt{5}$ **35.** $2\sqrt[3]{2}$ **36.** $2\sqrt[3]{4}$ **37.** $3x\sqrt{2}$

38. $6y^2\sqrt{2y}$ **39.** $4ab^2c\sqrt{5a}$ **40.** $3xy\sqrt[3]{x}$ **41.** $2abc\sqrt[4]{2bc^2}$ **42.** $3abc\sqrt[4]{2a^2b}$ **43.** $\dfrac{3\sqrt{2}}{2}$ **44.** $\dfrac{\sqrt{10}}{5}$

45. $3\sqrt[3]{4}$ **46.** $\dfrac{7\sqrt[3]{3}}{3}$ **47.** $\dfrac{4x\sqrt{21xy}}{7y}$ **48.** $\dfrac{5xy\sqrt{6yz}}{2z}$ **49.** $\dfrac{2y\sqrt[3]{45x^2z^2}}{3z}$ **50.** $\dfrac{3xy\sqrt[3]{50xz}}{5z}$ **51.** $-2x\sqrt{6}$

52. $8x\sqrt{7}$ **53.** $3\sqrt{3}$ **54.** $5\sqrt{2}$ **55.** $\dfrac{8\sqrt{5}}{5}$ **56.** $\dfrac{4\sqrt{15}}{5}$ **57.** $7\sqrt{2}$ **58.** $3\sqrt{3}$ **59.** $11a^2b\sqrt{3ab}$

60. $2a^2b\sqrt{2a}$ **61.** $-4xy\sqrt[3]{xz^2}$ **62.** $15xy\sqrt[3]{3x^2y}$ **63.** $\sqrt{6} - 4$ **64.** $40\sqrt{2} - 60$ **65.** $x - 5\sqrt{x} + 6$
66. $18 + 14\sqrt{3}$ **67.** $54 + 3\sqrt{2}$ **68.** $x - 4\sqrt{x} + 4$ **69.** 6 **70.** 44 **71.** $3\sqrt{5} + 6$ **72.** $\sqrt{6} + \sqrt{3}$
73. $6 + \sqrt{35}$ **74.** $\dfrac{x + 2\sqrt{xy} + y}{x - y}$ **75.** $\dfrac{63 + 12\sqrt{7}}{47}$ **76.** $\dfrac{7\sqrt{6} - 12}{5}$ **77.** 0 **78.** $\varnothing$ **79.** 3 **80.** 3

81. 5 **82.** 2 **83.** Possible solution 0, which does not check; $\varnothing$ **84.** 14 **85.** $4, 6$ **86.** $-3, -2$
87.

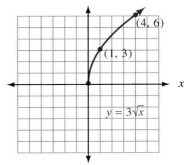

88.

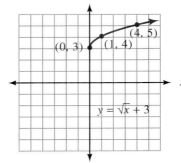

89.

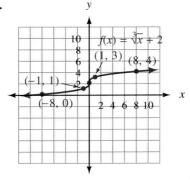

90.

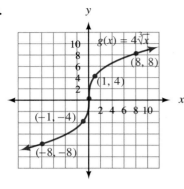

91. $7i$ **92.** $4i\sqrt{5}$ **93.** 1 **94.** $-i$ **95.** $x = -\frac{3}{2},\ y = -\frac{1}{2}$
96. $x = -2,\ y = -4$ **97.** $9 + 3i$ **98.** $-3 + 7i$ **99.** $-7 + 4i$
100. $-2 + 7i$ **101.** $-6 + 12i$ **102.** $-5 - 30i$ **103.** $5 + 14i$
104. $24 + 18i$ **105.** $12 + 16i$ **106.** $2i$ **107.** 25 **108.** 10
109. $1 - 3i$ **110.** $1 + 2i$ **111.** $-\frac{6}{5} + \frac{3}{5}i$ **112.** $\frac{5}{13} + \frac{12}{13}i$
113. $\frac{7}{25} - \frac{24}{25}i$ **114.** $\frac{23}{13} + \frac{11}{13}i$

CHAPTER 5 TEST

1. $\frac{1}{9}$ **2.** $\frac{7}{5}$ **3.** $a^{5/12}$ **4.** $\dfrac{x^{13/12}}{y}$ **5.** $7x^4y^5$ **6.** $2x^2y^4$ **7.** $2a$ **8.** $x^{n^2-n}y^{1-n^3}$ **9.** $6a^2 - 10a$

10. $16a^3 - 40a^{3/2} + 25$ **11.** $(3x^{1/3} - 1)(x^{1/3} + 2)$ **12.** $(3x^{1/3} - 7)(3x^{1/3} + 7)$ **13.** $\dfrac{x+4}{x^{1/2}}$ **14.** $\dfrac{3}{(x^2-3)^{1/2}}$

15. $5xy^2\sqrt{5xy}$ **16.** $2x^2y^2\sqrt[3]{5xy^2}$ **17.** $\dfrac{\sqrt{6}}{3}$ **18.** $\dfrac{2a^2b\sqrt{15bc}}{5c}$ **19.** $-6\sqrt{3}$ **20.** $-ab\sqrt[3]{3}$

21. $x + 3\sqrt{x} - 28$ **22.** $21 - 6\sqrt{6}$ **23.** $\dfrac{5 + 5\sqrt{3}}{2}$ **24.** $\dfrac{x - 2\sqrt{2x} + 2}{x - 2}$

25. Possible solutions 1 and 8; only 8 checks; 8 **26.** -4 **27.** -3
28. **29.**

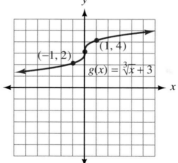

30. $x = \frac{1}{2},\ y = 7$ **31.** $6i$ **32.** $17 - 6i$ **33.** $9 - 40i$ **34.** $-\frac{5}{13} - \frac{12}{13}i$ **35.** $i^{38} = (i^2)^{19} = (-1)^{19} = -1$

CHAPTER 6 PROBLEM SET 6.1

1. ± 5 **3.** $\pm 3i$ **5.** $\pm\dfrac{\sqrt{3}}{2}$ **7.** $\pm 2i\sqrt{3}$ **9.** $\pm\dfrac{3\sqrt{5}}{2}$ **11.** $-2,\ 3$ **13.** $\dfrac{-3 \pm 3i}{2}$ **15.** $\dfrac{-2 \pm 2i\sqrt{2}}{5}$

17. $-4 \pm 3i\sqrt{3}$ **19.** $\dfrac{3 \pm 2i}{2}$ **21.** ± 1 **23.** ± 1 **25.** ± 1 **27.** $x^2 + 12x + 36 = (x + 6)^2$

29. $x^2 - 4x + 4 = (x - 2)^2$ **31.** $a^2 - 10a + 25 = (a - 5)^2$ **33.** $x^2 + 5x + \frac{25}{4} = (x + \frac{5}{2})^2$

35. $y^2 - 7y + \frac{49}{4} = (y - \frac{7}{2})^2$ **37.** $-6,\ 2$ **39.** $-9,\ -3$ **41.** $1 \pm 2i$ **43.** $4 \pm \sqrt{15}$ **45.** $\dfrac{5 \pm \sqrt{37}}{2}$

47. $1 \pm \sqrt{5}$ **49.** $\dfrac{4 \pm \sqrt{13}}{3}$ **51.** $\dfrac{3 \pm i\sqrt{71}}{8}$ **53.** $\dfrac{\sqrt{3}}{2}$ in., 1 in.   **55.** $x\sqrt{3}$, $2x$ **57.** $\sqrt{2}$ in.   **59.** $\dfrac{\sqrt{2}}{2}$ in.

61. $x\sqrt{2}$ **63.** 781 ft   **65.** $\dfrac{1{,}170}{5{,}630} = 0.21$ to the nearest hundredth   **67.** $3\sqrt{5}$ **69.** $3y^2\sqrt{3y}$ **71.** $3x^2y\sqrt[3]{2y^2}$

73. 13   **75.** $\dfrac{3\sqrt{2}}{2}$ **77.** $\sqrt[3]{2}$ **79.** $x = \pm 2a$ **81.** $x = -a$ **83.** $x = 0, -2a$ **85.** $x = \dfrac{-p \pm \sqrt{p^2 - 4q}}{2}$

PROBLEM SET 6.2

1. $-3, -2$ **3.** $2 \pm \sqrt{3}$ **5.** 1, 2   **7.** $\dfrac{2 \pm i\sqrt{14}}{3}$ **9.** 0, 5   **11.** $-\frac{4}{3}, 0$ **13.** $\dfrac{3 \pm \sqrt{5}}{4}$ **15.** $-3 \pm \sqrt{17}$

17. $\dfrac{-1 \pm i\sqrt{5}}{2}$ **19.** 1   **21.** $\dfrac{1 \pm i\sqrt{47}}{6}$ **23.** $4 \pm \sqrt{2}$ **25.** $\frac{1}{2}$, 1   **27.** $-\frac{1}{2}, 3$ **29.** $\dfrac{-1 \pm i\sqrt{7}}{2}$

31. $1 \pm \sqrt{2}$ **33.** $\dfrac{-3 \pm \sqrt{5}}{2}$ **35.** $-5, 3$ **37.** $2, -1 \pm i\sqrt{3}$ **39.** $-\dfrac{3}{2}, \dfrac{3 \pm 3i\sqrt{3}}{4}$ **41.** $\dfrac{1}{5}, \dfrac{-1 \pm i\sqrt{3}}{10}$

43. $0, \dfrac{-1 \pm i\sqrt{5}}{2}$ **45.** $0, 1 \pm i$ **47.** $0, \dfrac{-1 \pm i\sqrt{2}}{3}$ **49.** $\dfrac{-3 - 2i}{5}$ **51.** 2 sec   **53.** $\frac{1}{4}$ sec and 1 sec

55. $40 \pm 20 = 20$ or 60 items   **57.** $\dfrac{3.5 \pm 0.5}{0.004} = 750$ or 1,000 patterns   **59.** $4y + 1 + \dfrac{-2}{2y - 7}$

61. $x^2 + 7x + 12$ **63.** 5   **65.** $\frac{27}{125}$ **67.** $\frac{1}{4}$ **69.** $21x^3y$ **71.** $-2\sqrt{3}, \sqrt{3}$ **73.** $\dfrac{-1 \pm \sqrt{3}}{\sqrt{2}} = \dfrac{-\sqrt{2} \pm \sqrt{6}}{2}$

75. $-2i, i$ **77.** $-i, 4i$

PROBLEM SET 6.3

1. $D = 16$; two rational   **3.** $D = 0$; one rational   **5.** $D = 5$; two irrational   **7.** $D = 17$; two irrational
9. $D = 36$; two rational   **11.** $D = 116$; two irrational   **13.** ± 10 **15.** ± 12 **17.** 9   **19.** -16
21. $\pm 2\sqrt{6}$ **23.** $x^2 - 7x + 10 = 0$ **25.** $t^2 - 3t - 18 = 0$ **27.** $y^3 - 4y^2 - 4y + 16 = 0$
29. $2x^2 - 7x + 3 = 0$ **31.** $4t^2 - 9t - 9 = 0$ **33.** $6x^3 - 5x^2 - 54x + 45 = 0$ **35.** $10a^2 - a - 3 = 0$
37. $9x^3 - 9x^2 - 4x + 4 = 0$ **39.** $x^4 - 13x^2 + 36 = 0$ **41.** $(x - 3)(x + 5)^2 = 0$ or $x^3 + 7x^2 - 5x - 75 = 0$
43. $(x - 3)^2(x + 3)^2 = 0$ or $x^4 - 18x^2 + 81 = 0$ **45.** $-3, -2, -1$ **47.** $-3, -4, 2$ **49.** $1 \pm i$
51. $5, 4 \pm 3i$ **53.** $a^{11/2} - a^{9/2}$ **55.** $x^3 - 6x^{3/2} + 9$ **57.** $6x^{1/2} - 5x$ **59.** $5(x - 3)^{1/2}(2x - 9)$
61. $(2x^{1/3} - 3)(x^{1/3} - 4)$ **63.** $-2, 1, i, -i$ **65.** $x = -a$ is a solution of multiplicity 3   **67.** $-1, 5$
69. $1 + \sqrt{2} \approx 2.41$, $1 - \sqrt{2} \approx -0.41$ **71.** $-1, \frac{1}{2}, 1$ **73.** $\frac{1}{2}, \sqrt{2} \approx 1.41, -\sqrt{2} \approx -1.41$ **75.** $\frac{2}{3}, 1 + i, 1 - i$

PROBLEM SET 6.4

1. 1, 2   **3.** $-8, -\frac{5}{2}$ **5.** $\pm 3, \pm i\sqrt{3}$ **7.** $\pm 2i, \pm i\sqrt{5}$ **9.** $\frac{7}{2}, 4$ **11.** $-\frac{9}{8}, \frac{1}{2}$ **13.** $\pm \dfrac{\sqrt{30}}{6}, \pm i$

15. $\pm \dfrac{\sqrt{21}}{3}, \pm \dfrac{i\sqrt{21}}{3}$ **17.** 4, 25   **19.** Possible solutions 25 and 9; only 25 checks; 25

21. Possible solutions $\frac{25}{9}$ and $\frac{49}{4}$; only $\frac{25}{9}$ checks; $\frac{25}{9}$ **23.** 27, 38   **25.** 4, 12   **27.** $t = \dfrac{1 \pm \sqrt{1 + h}}{4}$

29. $t = \dfrac{v \pm \sqrt{v^2 + 1{,}280}}{32}$ **35.** $t = \dfrac{v \pm \sqrt{v^2 + 64h}}{32}$ **37.** $x = \dfrac{-4 \pm 2\sqrt{4 - k}}{k}$ **39.** $x = -y$ **41.** $3\sqrt{7}$

43. $5\sqrt{2}$ **45.** $39x^2y\sqrt{5x}$ **47.** $-11 + 6\sqrt{5}$ **49.** $x + 4\sqrt{x} + 4$ **51.** $\dfrac{7 + 2\sqrt{7}}{3}$

53. x-intercepts $= -2, 0, 2$; y-intercept $= 0$ **55.** x-intercepts $= -3, -\frac{1}{3}, 3$; y-intercept $= -9$ **57.** $\frac{1}{2}$ and -1

PROBLEM SET 6.5

1. x-intercepts $= -3, 1$
Vertex $= (-1, -4)$

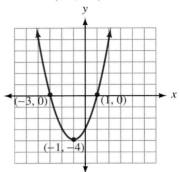

3. x-intercepts $= -5, 1$
Vertex $= (-2, 9)$

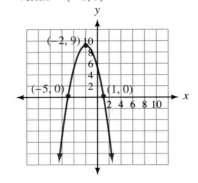

5. x-intercepts $= -1, 1$
Vertex $= (0, -1)$

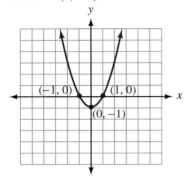

7. x-intercepts $= 3, -3$
Vertex $= (0, 9)$

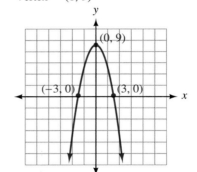

9. x-intercepts $= -1, 3$
Vertex $= (1, -8)$

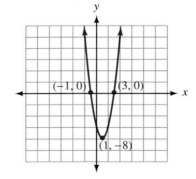

11. x-intercepts $= 1 + \sqrt{5}, 1 - \sqrt{5}$
Vertex $= (1, -5)$

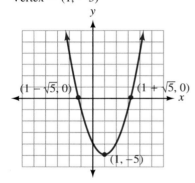

13. Vertex $= (2, -8)$

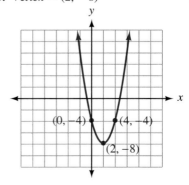

15. Vertex $= (1, -4)$

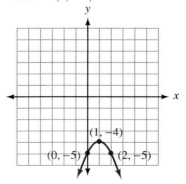

17. Vertex $= (0, 1)$

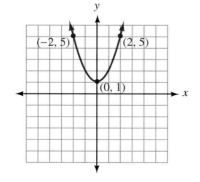

19. Vertex $= (0, -3)$

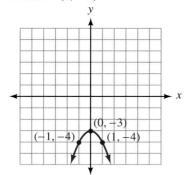

21. Vertex $= (-\frac{2}{3}, -\frac{1}{3})$

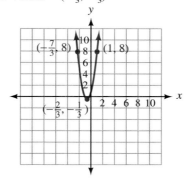

23. $(3, -4)$; lowest
25. $(1, 9)$; highest
27. $(2, 16)$; highest
29. $(-4, 16)$; highest
31. 40 items; maximum profit $500
33. 875 patterns; maximum profit $731.25
35. The ball is in her hand when $h(t) = 0$, which means $t = 0$ or $t = 2$ sec. Maximum height is $h(1) = 16$ ft.
37. 256 ft

39. Maximum $R = \$3,600$ when $p = \$6.00$.

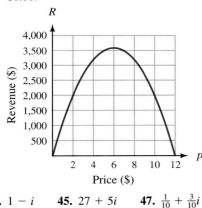

41. Maximum $R = \$7,225$ when $p = \$8.50$.

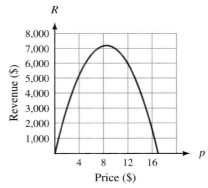

43. $1 - i$ **45.** $27 + 5i$ **47.** $\frac{1}{10} + \frac{3}{10}i$

PROBLEM SET 6.6

1.

3.

5.

7.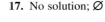

9.

11.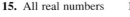

13.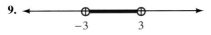

15. All real numbers   **17.** No solution; $\varnothing$

19.

21.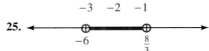

23.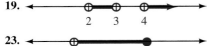

25.

27.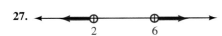

29.

31. 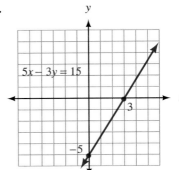 (number line with open circles at 2, 3, 4)

33. (number line with closed circle at 5, open circle at 6)

35. (a) $-2 < x < 2$ (b) $x < -2$ or $x > 2$ (c) $x = -2$ or $x = 2$

37. (a) $-2 < x < 5$ (b) $x < -2$ or $x > 5$ (c) $x = -2$ or $x = 5$

39. (a) $x < -1$ or $1 < x < 3$ (b) $-1 < x < 1$ or $x > 3$ (c) $x = -1$ or $x = 1$ or $x = 3$

41. $x \geq 4$; the width is at least 4 inches.

43. $5 \leq p \leq 8$; charge at least \$5 but no more than \$8 for each radio.

45.

47.
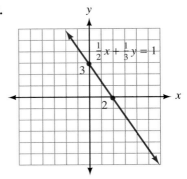

49. $\frac{5}{3}$ **51.** Possible solutions 1 and 6; only 6 checks; 6

53.
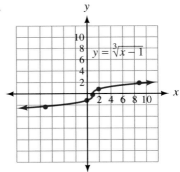

55.
 (number line with open circles at $1 - \sqrt{2}$ and $1 + \sqrt{2}$)

57. (number line with open circles at $4 - \sqrt{3}$ and $4 + \sqrt{3}$)

CHAPTER 6   REVIEW PROBLEMS

1. $0, 5$ **2.** $0, \frac{4}{3}$ **3.** $\dfrac{4 \pm 7i}{3}$ **4.** $\dfrac{-1 \pm 6i}{8}$ **5.** $-3 \pm \sqrt{3}$ **6.** $\dfrac{4 \pm 3\sqrt{2}}{3}$ **7.** $-5, 2$ **8.** $-3, -2$

9. 3 **10.** 2 **11.** $\dfrac{-3 \pm \sqrt{3}}{2}$ **12.** $\dfrac{3 \pm \sqrt{5}}{2}$ **13.** $-5, 2$ **14.** $-\frac{3}{2}, -1$ **15.** $0, \frac{9}{4}$

16. $-\frac{3}{2}, \frac{3}{2}$ **17.** $\dfrac{2 \pm i\sqrt{15}}{2}$ **18.** $\dfrac{1 \pm 2i\sqrt{2}}{2}$ **19.** $1 \pm \sqrt{2}$ **20.** $3 \pm \sqrt{2}$ **21.** $-\frac{4}{3}, \frac{1}{2}$ **22.** $-\frac{1}{2}, \frac{2}{3}$

23. 3 **24.** 1 **25.** $5 \pm \sqrt{7}$ **26.** $5 \pm 3\sqrt{2}$ **27.** $0, \dfrac{5 \pm \sqrt{21}}{2}$ **28.** $0, \dfrac{5 \pm \sqrt{37}}{6}$ **29.** $-1, \frac{3}{5}$

30. $\dfrac{-3 \pm \sqrt{65}}{4}$ **31.** $3, \dfrac{-3 \pm 3i\sqrt{3}}{2}$ **32.** $-\dfrac{3}{2}, \dfrac{3 \pm 3i\sqrt{3}}{4}$ **33.** $\dfrac{1 \pm i\sqrt{2}}{3}$ **34.** $2 \pm \sqrt{2}$ **35.** $\dfrac{3 \pm \sqrt{29}}{2}$

36. $\dfrac{1 \pm \sqrt{29}}{2}$ **37.** 100 or 170 items **38.** 20 or 60 items **39.** $D = 0$; one rational **40.** $D = 0$; one rational

41. $D = 25$; two rational **42.** $D = 361$; two rational **43.** $D = 5$; two irrational **44.** $D = 5$; two irrational
45. $D = -23$; two complex **46.** $D = -87$; two complex **47.** ± 20 **48.** ± 20 **49.** 4 **50.** 4
51. 25 **52.** 49 **53.** $x^2 - 8x + 15 = 0$ **54.** $2y^2 + 7y - 4 = 0$ **55.** $t^3 - 5t^2 - 9t + 45 = 0$
56. $t^3 - 6t^2 - 4t + 24 = 0$ **57.** $-1, 2, 3$ **58.** $-2, 1, 4$ **59.** $-4, 12$ **60.** $-10, 7$ **61.** $-\frac{3}{4}, -\frac{1}{6}$
62. $-\frac{1}{4}, \frac{2}{3}$ **63.** $\pm 2, \pm i\sqrt{3}$ **64.** $\pm 3, \pm i\sqrt{2}$ **65.** Possible solutions 4 and 1; only 4 checks; 4 **66.** 1, 4
67. $\frac{9}{4}, 16$ **68.** Possible solutions $\frac{4}{9}$ and 1; only $\frac{4}{9}$ checks; $\frac{4}{9}$ **69.** 4 **70.** $\frac{9}{4}$ **71.** 4 **72.** 4 **73.** 7

74. 2 **75.** $t = \dfrac{5 \pm \sqrt{25 + 16h}}{16}$ **76.** $t = \dfrac{v \pm \sqrt{v^2 + 640}}{32}$

77.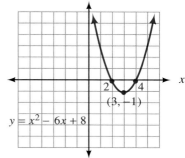

78.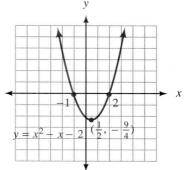

79. 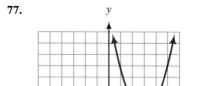 **80.** No solution; $\varnothing$

81.  **82.**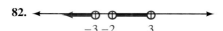

83. **84.**

85. **86.**

CHAPTER 6 TEST

1. $-\frac{9}{2}, \frac{1}{2}$ **2.** $3 \pm i\sqrt{2}$ **3.** $5 \pm 2i$ **4.** $1 \pm i\sqrt{2}$ **5.** $\frac{5}{2}, \dfrac{-5 \pm 5i\sqrt{3}}{4}$ **6.** $-1 \pm i\sqrt{5}$

7. $r = \pm \dfrac{\sqrt{A}}{8} - 1$ **8.** $2 \pm \sqrt{2}$ **9.** $\frac{1}{2}$ sec and $\frac{3}{2}$ sec **10.** 15 cups or 100 cups **11.** 9

12. $D = 81$; two rational **13.** $3x^2 - 13x - 10 = 0$ **14.** $x^3 - 7x^2 - 4x + 28 = 0$ **15.** $2, \dfrac{2 \pm \sqrt{3}}{2}$

16. $\pm \sqrt{2}, \pm \frac{1}{2}i$ **17.** $\frac{1}{2}, 1$ **18.** $\frac{1}{4}, 9$ **19.** $t = \dfrac{7 \pm \sqrt{49 + 16h}}{16}$

20.

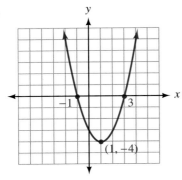

21.

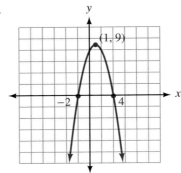

22. Maximum profit = $900 by selling 100 items per week.

23.

24.

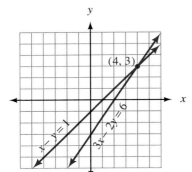

25.

CHAPTER 7   PROBLEM SET 7.1

1. (4, 3)   **3.** (−5, −6)   **5.** (4, 2)

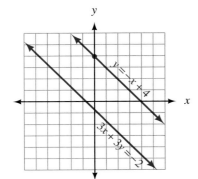

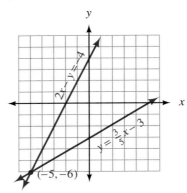

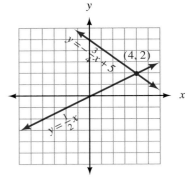

7. Lines are parallel; there is no solution.

9. Lines coincide; any solution to one of the equations is a solution to the other.

11. (2, 3)   **13.** (1, 1)

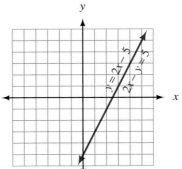

15. Lines coincide; $\{(x, y)|3x - 2y = 6\}$ **17.** $(1, -\frac{1}{2})$ **19.** $(\frac{1}{2}, -3)$ **21.** $(-\frac{8}{3}, 5)$ **23.** $(2, 2)$
25. Lines are parallel; $\varnothing$ **27.** $(12, 30)$ **29.** $(10, 24)$ **31.** $(4, \frac{10}{3})$ **33.** $(3, -3)$ **35.** $(\frac{4}{3}, -2)$ **37.** $(6, 2)$
39. $(2, 4)$ **41.** Lines coincide; $\{(x, y)|2x - y = 5\}$ **43.** Lines coincide; $\{(x, y)|x = \frac{3}{2}y\}$ **45.** $(-\frac{15}{43}, -\frac{27}{43})$
47. $(\frac{60}{43}, \frac{46}{43})$ **49.** $(\frac{9}{41}, -\frac{11}{41})$ **51.** $(6,000, 4,000)$ **53.** 2 **55.** (a) \$3.29 (b) $y = 30x + 45$ (c) 2 min
57. $-2, 3$ **59.** 25 **61.** $5 \pm \sqrt{17}$ **63.** $1 \pm i$ **65.** $a = 3, b = -2$ **67.** $a = 1, b = -2$; vertex $= (1, -1)$

PROBLEM SET 7.2

1. $(1, 2, 1)$ **3.** $(2, 1, 3)$ **5.** $(2, 0, 1)$ **7.** $(\frac{1}{2}, \frac{2}{3}, -\frac{1}{2})$ **9.** No solution; inconsistent system **11.** $(4, -3, -5)$
13. System is dependent. **15.** $(4, -5, -3)$ **17.** System is dependent. **19.** $(\frac{1}{2}, 1, 2)$ **21.** $(\frac{1}{2}, \frac{1}{3}, \frac{1}{4})$
23. $(1, 3, 1)$ **25.** $(-1, 2, -2)$ **27.** $x = 4$ amp, $y = 3$ amp, $z = 1$ amp **29.** $y = \frac{2}{3}x - 2$ **31.** $y = \frac{2}{3}x - \frac{5}{3}$
33. $\dfrac{-2 \pm \sqrt{10}}{2}$ **35.** $\dfrac{2 \pm i\sqrt{3}}{2}$ **37.** $5, \dfrac{-5 \pm 5i\sqrt{3}}{2}$ **39.** $-1, 0, 5$ **41.** $\dfrac{3 \pm \sqrt{29}}{2}$
43. $a = 1, b = -6, c = 5$ **45.** $(1, 2, 3, 4)$

PROBLEM SET 7.3

1. $\begin{bmatrix} 4 & 0 & 0 \\ 3 & 5 & -1 \\ 2 & 0 & 8 \end{bmatrix}$ **3.** $\begin{bmatrix} 2 & -4 & 0 \\ -3 & -3 & 9 \\ 2 & 0 & 2 \end{bmatrix}$ **5.** $\begin{bmatrix} 12 & -8 & 0 \\ 0 & 4 & 16 \\ 8 & 0 & 20 \end{bmatrix}$ **7.** $\begin{bmatrix} 3 & -10 & 0 \\ -9 & -10 & 23 \\ 4 & 0 & 1 \end{bmatrix}$ **9.** 28 **11.** 56

13. $[18 \quad 0 \quad 46]$ **15.** $\begin{bmatrix} -4 & -3 \\ 6 & 5 \end{bmatrix}$ **17.** $\begin{bmatrix} -3 & -2 & 10 \\ 3 & 4 & 7 \\ 2 & 4 & 15 \end{bmatrix}$ **19.** $[32 \quad 20 \quad 44]$ **21.** $\begin{bmatrix} 3 & -5 \\ -1 & 2 \end{bmatrix} = A$

23. $\begin{bmatrix} 1 & 0 \\ 0 & 1 \end{bmatrix} = I$ **25.** Yes **27.** 3 **29.** 5 **31.** -1 **33.** 0 **35.** 10 **37.** 2 **39.** -3

41. -2 **43.** $-2, 5$ **45.** 3 **47.** 0 **49.** 3 **51.** 8 **53.** 6 **55.** -228

57. $\begin{vmatrix} y & x \\ m & 1 \end{vmatrix} = y - mx = b$; $y = mx + b$ **59.** $D = -7$; two complex **61.** $x^2 - 2x - 15 = 0$

63. $3y^2 - 11y + 6 = 0$ **65.** 2, 3 **67.** 4 **69.** 4

PROBLEM SET 7.4

1. $(3, 1)$ **3.** Lines are parallel; $\varnothing$ **5.** $(-\frac{15}{43}, -\frac{27}{43})$ **7.** $(\frac{60}{43}, \frac{46}{43})$ **9.** $(3, -1, 2)$ **11.** $(\frac{1}{2}, \frac{5}{2}, 1)$
13. System is dependent. **15.** $(-\frac{10}{91}, -\frac{9}{13}, \frac{107}{91})$ **17.** $(\frac{71}{13}, -\frac{12}{13}, \frac{24}{13})$ **19.** $(3, 1, 2)$ **21.** $x = 50$ items

23. $\pm 2, \pm i\sqrt{2}$ **25.** 8, 27 **27.** $1, \frac{9}{4}$ **29.** 1, 5 **31.** $\dfrac{-2 \pm \sqrt{4 + k^2}}{k}$

33. $x = -\dfrac{1}{a - b} = \dfrac{1}{b - a}, y = \dfrac{1}{a - b}$ **35.** $x = \dfrac{1}{a^2 + ab + b^2}, y = \dfrac{a + b}{a^2 + ab + b^2}$ **37.** $x + 2y = 1$
$3x + 4y = 0$

PROBLEM SET 7.5

1. $y = 2x + 3$, $x + y = 18$; the two numbers are 5 and 13 **3.** 10, 16 **5.** 1, 3, 4
7. Let $x =$ the number of adult tickets and $y =$ the number of children's tickets:
$x + y = 925$ 225 adult and 700 children's tickets
$2x + y = 1,150$
9. Let $x =$ the amount invested at 6% and $y =$ the amount invested at 7%:
$x + y = 20,000$ He has \$12,000 at 6% and \$8,000 at 7%.
$0.06x + 0.07y = 1,280$

Answers to Odd-Numbered Problems, Chapter Reviews, and Chapter Tests

11. \$4,000 at 6%; \$8,000 at 7.5%   **13.** \$200 at 6%; \$1,400 at 8%; \$600 at 9%
15. 3 gal of 50%; 6 gal of 20%   **17.** 5 gal of 20%; 10 gal of 14%
19. Let x = the speed of the boat and y = the speed of the current:
   $3(x - y) = 18$ The speed of the boat is 9 mi/hr.
   $2(x + y) = 24$ The speed of the current is 3 mi/hr.
21. Airplane, 270 mi/hr; wind, 30 mi/hr   **23.** 12 nickels; 8 dimes   **25.** 3 of each
27. $x = -200p + 700$; when $p =$ \$3, $x = 100$ items   **29.** $C = 0.75x + 21.85$; \$30.85
31. $h = -16t^2 + 64t + 80$ **33.** $(-2, -23)$; lowest point   **35.** $(\frac{3}{2}, 9)$; highest point   **37.** 2 sec; 64 ft

CHAPTER 7 REVIEW PROBLEMS

1. $(6, -2)$ **2.** $(2, -4)$ **3.** Lines are parallel.   **4.** Lines coincide.   **5.** $(1, -1)$ **6.** $(0, 1)$
7. Lines coincide.   **8.** Lines are parallel.   **9.** $(3, 1)$ **10.** $(\frac{3}{2}, 0)$ **11.** $(3, 5)$ **12.** $(\frac{4}{3}, \frac{13}{3})$ **13.** $(4, 8)$
14. $(3, 12)$ **15.** $(3, -2)$ **16.** $(2, 1)$ **17.** $(\frac{3}{2}, \frac{1}{2})$ **18.** $(-\frac{3}{2}, -\frac{7}{2})$ **19.** $(4, 1)$ **20.** $(3, 2)$ **21.** $(6, -2)$
22. $(7, -4)$ **23.** $(-5, 3)$ **24.** $(8, -5)$ **25.** Lines are parallel.   **26.** Lines coincide.   **27.** $(3, -1, 4)$
28. $(-2, 3, 4)$ **29.** $(2, \frac{1}{2}, -3)$ **30.** $(3, 0, \frac{1}{2})$ **31.** $(-1, \frac{1}{2}, \frac{3}{2})$ **32.** $(2, \frac{1}{3}, \frac{2}{3})$ **33.** Dependent system
34. Dependent system   **35.** $(2, -1, 4)$ **36.** $(4, 0, 9)$ **37.** Dependent system   **38.** $(3, 1, -2)$ **39.** 23
40. -3 **41.** -3 **42.** 30   **43.** -16 **44.** 36   **45.** -2 **46.** -46 **47.** $\frac{4}{7}$ **48.** $\frac{5}{22}$
49. $(\frac{7}{29}, -\frac{19}{29})$ **50.** $(\frac{34}{41}, -\frac{18}{41})$ **51.** Lines coincide.   **52.** Lines coincide.   **53.** $(\frac{11}{14}, -\frac{1}{14}, \frac{11}{14})$ **54.** $(\frac{1}{7}, -\frac{18}{77}, \frac{45}{77})$
55. 2, 9   **56.** 1, 9   **57.** $-6, 3, 5$ **58.** 2, 3, 4   **59.** 47 adults; 80 children   **60.** 12 dimes; 8 quarters
61. \$5,000 at 12%; \$7,000 at 15%   **62.** \$6,000 at 18%; \$3,000 at 13%   **63.** 7.5 oz 30%; 7.5 oz 70%
64. 6 gal 25%; 14 gal 50%   **65.** Boat, 12 mi/hr; river, 2 mi/hr   **66.** Boat, 8 mi/hr; river, 4 mi/hr

CHAPTER 7 TEST

1. $(1, 2)$ **2.** $(3, 2)$ **3.** $(15, 12)$ **4.** $(-\frac{54}{13}, -\frac{58}{13})$ **5.** $(1, 2)$ **6.** $(3, -2, 1)$ **7.** -14 **8.** -26
9. $(-\frac{14}{3}, -\frac{19}{3})$ **10.** Lines coincide: $\{(x, y) | 2x + 4y = 3\}$ **11.** $(\frac{5}{11}, -\frac{15}{11}, -\frac{1}{11})$ **12.** $(-3, -5)$
13. $(-2, 3, 5)$ **14.** 5, 9   **15.** \$4,000 at 5%; \$8,000 at 6%   **16.** 340 adults; 410 children
17. 4 gal 30%; 12 gal 70%   **18.** Boat, 8 mi/hr; current, 2 mi/hr   **19.** 11 nickels, 3 dimes, 1 quarter
20. 3 oz of cereal I; 1 oz of cereal II

CHAPTER 8 PROBLEM SET 8.1

1. 1   **3.** 2   **5.** $\frac{1}{27}$ **7.** 13   **9.**

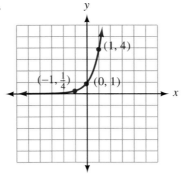

11.

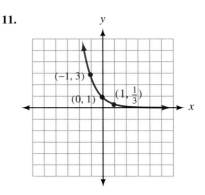

13.

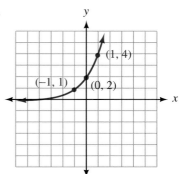

15.

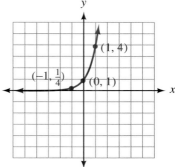

17. 200, 400, 800, 1,600; approximately 10 days

19. $V(t)$

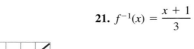

21. $f^{-1}(x) = \dfrac{x+1}{3}$

23. $f^{-1}(x) = \dfrac{x-3}{x-1}$ **25.** $f^{-1}(x) = 4x + 3$ **27.** $f^{-1}(x) = 2(x+3) = 2x + 6$ **29.** $f^{-1}(x) = \dfrac{1-x}{3x-2}$

31.

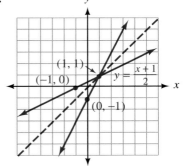

33.

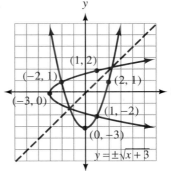

35.

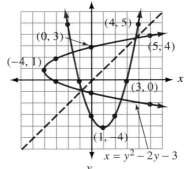

37.

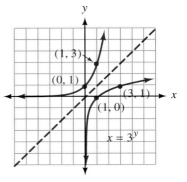

39.

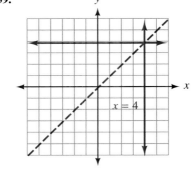

41.

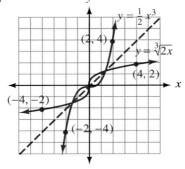

43.

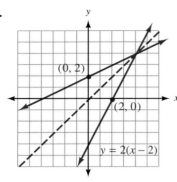

45. (a) 4   (b) $\frac{4}{3}$ (c) 2   (d) 2   **47.** $f^{-1}(x) = \dfrac{1}{x}$ **49.** $(1, 2)$

51. $(\frac{5}{2}, -1)$ **53.** $(0, 3)$ **55.** $(3, 4)$

57. $s(t) = 250(1 - 1.5^{-t})$

t	$s(t)$
0	0
1	83.3
2	138.9
3	175.9
4	200.6
5	217.1
6	228.1

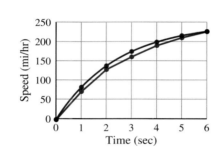

PROBLEM SET 8.2

1. $\log_2 16 = 4$ **3.** $\log_5 125 = 3$ **5.** $\log_{10} 0.01 = -2$ **7.** $\log_2 \frac{1}{32} = -5$ **9.** $\log_{1/2} 8 = -3$
11. $\log_3 27 = 3$ **13.** $10^2 = 100$ **15.** $2^6 = 64$ **17.** $8^0 = 1$ **19.** $10^{-3} = 0.001$ **21.** $6^2 = 36$
23. $5^{-2} = \frac{1}{25}$ **25.** 9   **27.** $\frac{1}{125}$ **29.** 4   **31.** $\frac{1}{3}$ **33.** 2   **35.** $\sqrt[3]{5}$

37.

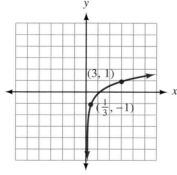

39.

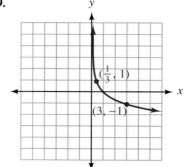

41.

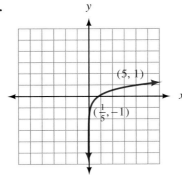

43.

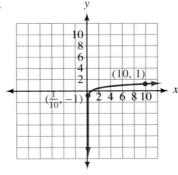

45. 4   **47.** $\frac{3}{2}$ **49.** 3   **51.** 1   **53.** 0   **55.** 0   **57.** $\frac{1}{2}$ **59.** 7   **61.** 10^{-6} **63.** 2
65. 10^8 times as large   **67.** $(1, 2, 3)$ **69.** $(1, 3, 1)$

PROBLEM SET 8.3

1. $\log_3 4 + \log_3 x$ **3.** $\log_6 5 - \log_6 x$ **5.** $5 \log_2 y$ **7.** $\frac{1}{3} \log_9 z$ **9.** $2 \log_6 x + 4 \log_6 y$
11. $\frac{1}{2} \log_5 x + 4 \log_5 y$ **13.** $\log_b x + \log_b y - \log_b z$ **15.** $\log_{10} 4 - \log_{10} x - \log_{10} y$
17. $2 \log_{10} x + \log_{10} y - \frac{1}{2} \log_{10} z$ **19.** $3 \log_{10} x + \frac{1}{2} \log_{10} y - 4 \log_{10} z$ **21.** $\frac{2}{3} \log_b x + \frac{1}{3} \log_b y - \frac{4}{3} \log_b z$

23. $\log_b xz$ **25.** $\log_3 \dfrac{x^2}{y^3}$ **27.** $\log_{10}(\sqrt{x}\,\sqrt[3]{y})$ **29.** $\log_2 \dfrac{x^3 \sqrt{y}}{z}$ **31.** $\log_2 \dfrac{\sqrt{x}}{y^3 z^4}$ **33.** $\log_{10} \dfrac{x^{3/2}}{y^{3/4} z^{4/5}}$ **35.** $\frac{2}{3}$

37. 18   **39.** Possible solutions -1 and 3; only 3 checks; 3   **41.** 3
43. Possible solutions -2 and 4; only 4 checks; 4   **45.** Possible solutions -1 and 4; only 4 checks; 4
47. Possible solutions $-\frac{5}{2}$ and $\frac{5}{3}$; only $\frac{5}{3}$ checks; $\frac{5}{3}$ **49.** 2.52   **55.** $\begin{bmatrix} 11 & -4 \\ -6 & -3 \end{bmatrix}$ **57.** $\begin{bmatrix} -21 & -6 \\ -1 & 0 \end{bmatrix}$ **59.** 36
61. 0

PROBLEM SET 8.4

1. 2.5775   **3.** 1.5775   **5.** 3.5775   **7.** -1.4225 **9.** 4.5775   **11.** 2.7782   **13.** 3.3032   **15.** -2.0128
17. -1.5031 **19.** -0.3990 **21.** 759   **23.** 0.00759   **25.** 1,430   **27.** 0.00000447   **29.** 0.0000000918
31. 10^{10} **33.** 10^{-10} **35.** 10^{20} **37.** $\frac{1}{100}$ **39.** 1,000   **41.** 5   **43.** Approximately 3.19
45. 1.78×10^{-5} **47.** 3.16×10^5 **49.** 2.00×10^8 **51.** 10 times as large   **53.** 12.9%   **55.** 5.3%
57. 1   **59.** 5   **61.** x **63.** $\ln 10 + 3t$ **65.** $\ln A - 2t$ **67.** 2.7080   **69.** -1.0986 **71.** 2.1972
73. 2.7724   **75.** $\left(-\frac{15}{43}, -\frac{27}{43}\right)$ **77.** $(1, 3, 1)$

PROBLEM SET 8.5

1. 1.4650   **3.** 0.6826   **5.** -1.5440 **7.** -0.6477 **9.** -0.3333 **11.** 2.0000   **13.** -0.1845
15. 0.1845   **17.** 1.6168   **19.** 2.1131   **21.** \$15,529.24   **23.** \$438.22   **25.** 11.7 yr   **27.** 9.25 yr
29. 1.3333   **31.** 0.7500   **33.** 1.3917   **35.** 0.7186   **37.** 2.6356   **39.** 4.1632   **41.** 5.8435
43. -1.0642 **45.** 2.3026   **47.** 10.7144   **49.** 13.9 yr later, or toward the end of 2007   **51.** 27.5 yr

53. $t = \dfrac{1}{r} \ln \dfrac{A}{P}$ **55.** $t = \dfrac{1}{k}\left(\dfrac{\log P - \log A}{\log 2} \right)$ **57.** $t = \dfrac{\log A - \log P}{\log(1 - r)}$ **59.** 60 geese; 48 ducks

61. 150 oranges; 144 apples

CHAPTER 8   REVIEW PROBLEMS

1. 16 **2.** $\frac{1}{2}$ **3.** $\frac{1}{9}$ **4.** -5 **5.** $\frac{5}{6}$ **6.** 7

7.

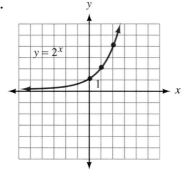

8.

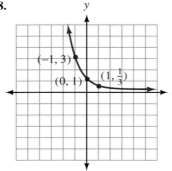

9.

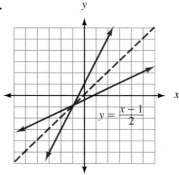

10.

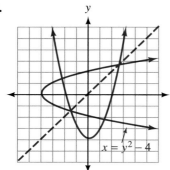

11. $f^{-1}(x) = \dfrac{x-3}{2}$ **12.** $y = \pm\sqrt{x+1}$ **13.** $f^{-1}(x) = 2x - 4$

14. $y = \pm\sqrt{\dfrac{4-x}{2}}$ **15.** $\log_3 81 = 4$ **16.** $\log_7 49 = 2$

17. $\log_{10} 0.01 = -2$ **18.** $\log_2 \frac{1}{8} = -3$ **19.** $2^3 = 8$ **20.** $3^2 = 9$
21. $4^{1/2} = 2$ **22.** $4^1 = 4$ **23.** 25 **24.** 27 **25.** $\frac{3}{4}$ **26.** $\frac{3}{2}$
27. 10 **28.** 10

29.

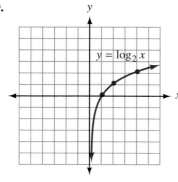

30.

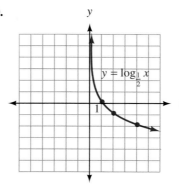

31. 2 **32.** 4 **33.** $\frac{2}{3}$ **34.** $\frac{4}{3}$
35. 0 **36.** 0

37. $\log_2 5 + \log_2 x$ **38.** $\log_3 4 + \log_3 x$ **39.** $\log_{10} 2 + \log_{10} x - \log_{10} y$ **40.** $2\log_3 x + 4\log_3 y$
41. $\frac{1}{2}\log_a x + 3\log_a y - \log_a z$ **42.** $2\log_{10} x - 3\log_{10} y - 4\log_{10} z$ **43.** $\log_2 xy$ **44.** $\log_2 5x$

45. $\log_3 \dfrac{x}{4}$ **46.** $\log_{10} x^2 y^3$ **47.** $\log_a \dfrac{25}{3}$ **48.** $\log_2 \dfrac{x^3 y^2}{z^4}$ **49.** 2 **50.** 6

51. Possible solutions -1 and 3; only 3 checks; 3 **52.** Possible solutions -2 and 4; only 4 checks; 4
53. 3 **54.** Possible solution $-\frac{9}{2}$, which does not check; $\varnothing$ **55.** Possible solutions -2 and 3; only 3 checks; 3

56. Possible solutions -1 and 4; only 4 checks; 4   **57.** 2.5391   **58.** 3.5391   **59.** -0.1469 **60.** -2.1469
61. 9,230   **62.** 923,000   **63.** 0.0251   **64.** 0.00251   **65.** 1   **66.** 0   **67.** 2   **68.** -4 **69.** 3.0445
70. 3.2958   **71.** 2.1972   **72.** 3.8918   **73.** 5.8377   **74.** -1.0986 **75.** 2.1   **76.** 5.1   **77.** 2×10^{-3}

78. 3.2×10^{-8} **79.** $\frac{3}{2}$ **80.** $-\frac{4}{3}$ **81.** $x = \frac{1}{3}\left(\dfrac{\log 5}{\log 4} - 2\right) = -0.28$ **82.** $x = \frac{1}{5}\left(\dfrac{\log 20}{\log 10} - 4\right) = -0.54$

83. 0.75   **84.** 2.89   **85.** 2.43   **86.** 2.77   **87.** \$15,529.24   **88.** \$17,181.86   **89.** About 4.67 yr
90. About 9.25 yr

CHAPTER 8   TEST

1.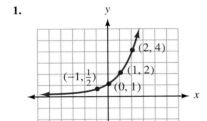

2.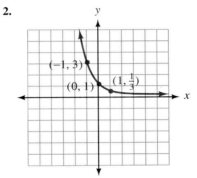

3.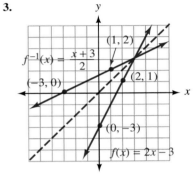

4.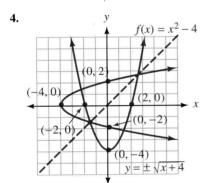

5. 64   **6.** $\sqrt{5}$ **7.**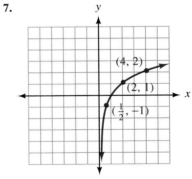

8.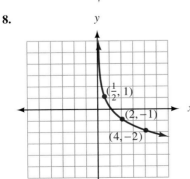

9. $\frac{2}{3}$ **10.** 1.5646   **11.** 4.3692   **12.** -1.9101 **13.** 3.8330
14. -3.0748 **15.** $3 + 2\log_2 x - \log_2 y$

16. $\frac{1}{2}\log x - 4\log y - \frac{1}{5}\log z$ **17.** $\log_3 \dfrac{x^2}{\sqrt{y}}$ **18.** $\log \dfrac{\sqrt[3]{x}}{yz^2}$

19. 7.04×10^4 **20.** 2.25×10^{-3} **21.** 1.46   **22.** $\frac{5}{4}$, or 1.25
23. 15   **24.** Possible solutions -1 and 8; only 8 checks; 8
25. 6.2   **26.** \$651.56   **27.** About 13.9 yr

CHAPTER 9   PROBLEM SET 9.1

1. 4, 7, 10, 13, 16   **3.** 3, 7, 11, 15, 19   **5.** 1, 2, 3, 4, 5   **7.** 4, 7, 12, 19, 28   **9.** $\frac{1}{4}, \frac{2}{5}, \frac{3}{6}, \frac{4}{7}, \frac{5}{8}$ **11.** $\frac{2}{3}, \frac{3}{4}, \frac{4}{5}, \frac{5}{6}, \frac{6}{7}$

13. $1, \frac{1}{4}, \frac{1}{9}, \frac{1}{16}, \frac{1}{25}$ **15.** 2, 4, 8, 16, 32   **17.** $\frac{1}{3}, \frac{1}{9}, \frac{1}{27}, \frac{1}{81}, \frac{1}{243}$ **19.** $2, \frac{3}{2}, \frac{4}{3}, \frac{5}{4}, \frac{6}{5}$ **21.** $0, \frac{3}{2}, \frac{8}{3}, \frac{15}{4}, \frac{24}{5}$

23. $-2, 4, -8, 16, -32$ **25.** $3, -9, 27, -81, 243$ **27.** $3, 0, -3, -6, -9$ **29.** 1, 5, 13, 29, 61

31. 1, 3, 6, 10, 15   **33.** $a_n = n + 1$ **35.** $a_n = 4n$ **37.** $a_n = 3n + 4$, or recursively as $a_1 = 7, a_n = a_{n-1} + 3$

39. $a_n = n^2$ **41.** $a_n = 3n^2$ **43.** $a_n = 2^{n+1}$, or recursively as $a_1 = 4, a_n = 2a_{n-1}$

45. $a_n = (-2)^n$, or recursively as $a_1 = -2, a_n = -2a_{n-1}$ **47.** $a_n = \dfrac{1}{2^{n+1}}$, or recursively as $a_1 = \dfrac{1}{4}, a_n = \dfrac{1}{2}a_{n-1}$

49. $a_n = \dfrac{n}{(n+1)^2}$ **51.** $\log_{10} 100 = 2$ **53.** $3^4 = 81$ **55.** 27   **57.** 5   **59.** 1

61. $a_{100} \approx 2.7048; a_{1,000} \approx 2.7169; a_{10,000} \approx 2.7181; a_{100,000} \approx 2.7183$ **63.** 1, 1, 2, 3, 5, 8, 13, 21, 34, 55

65. $\frac{3}{2}, \frac{5}{3}, \frac{8}{5}$

PROBLEM SET 9.2

1. 36   **3.** 9   **5.** 11   **7.** $\frac{163}{60}$ **9.** $\frac{62}{15}$ **11.** 60   **13.** 40   **15.** $5x + 15$

17. $(x + 1)^2 + (x + 1)^3 + (x + 1)^4 + (x + 1)^5 + (x + 1)^6 + (x + 1)^7$

19. $\dfrac{x + 1}{x - 1} + \dfrac{x + 2}{x - 1} + \dfrac{x + 3}{x - 1} + \dfrac{x + 4}{x - 1} + \dfrac{x + 5}{x - 1} = \dfrac{5x + 15}{x - 1}$

21. $(x + 3)^3 + (x + 4)^4 + (x + 5)^5 + (x + 6)^6 + (x + 7)^7 + (x + 8)^8$

23. $(x + 1)^2 + (x + 2)^3 + (x + 3)^4 + (x + 4)^5 + (x + 5)^6$ **25.** $\sum_{i=1}^{4} 2^i$ **27.** $\sum_{i=2}^{6} 2^i$ **29.** $\sum_{i=2}^{6} (i^2 + 1)$

31. $\sum_{i=3}^{7} \dfrac{i}{i + 1}$ **33.** $\sum_{i=1}^{4} \dfrac{i}{2i + 1}$ **35.** $\sum_{i=3}^{6} (x - i)$ **37.** $\sum_{i=3}^{5} \dfrac{x}{x + i}$ **39.** $\sum_{i=2}^{4} x^i(x + i)$ **41.** 208 ft; 784 ft

43. $3 \log_2 x + \log_2 y$ **45.** $\frac{1}{3} \log_{10} x - 2 \log_{10} y$ **47.** $\log_{10} \dfrac{x}{y^2}$ **49.** $\log_3 \dfrac{x^2}{y^3 z^4}$ **51.** 80

53. Possible solutions -1 and 8; only 8 checks; 8

PROBLEM SET 9.3

1. 1   **3.** Not an arithmetic progression   **5.** -5 **7.** Not an arithmetic progression   **9.** $\frac{2}{3}$

11. $a_n = 4n - 1; a_{24} = 95$ **13.** $a_{10} = -12; S_{10} = -30$ **15.** $a_1 = 7; d = 2; a_{30} = 65$

17. $a_1 = 12; d = 2; a_{20} = 50; S_{20} = 620$ **19.** 20,300   **21.** -158 **23.** $a_{10} = 5; S_{10} = \frac{55}{2}$

25. 28,000, 28,850, 29,700, 30,550, 31,400; $a_n = 27,150 + 850n; a_{10} = 35,650$ **27.** 2.7604   **29.** -1.2396

31. 445   **33.** 4.45×10^{-8}

PROBLEM SET 9.4

1. 5   **3.** $\frac{1}{3}$ **5.** Not geometric   **7.** -2 **9.** Not geometric   **11.** $a_n = 4 \cdot 3^{n-1}$

13. $a_6 = -2\left(-\dfrac{1}{2}\right)^5 = \dfrac{1}{16}$ **15.** $a_{20} = 3(-1)^{19} = -3$ **17.** $S_{10} = \dfrac{10(2^{10} - 1)}{2 - 1} = 10,230$

19. $S_{20} = \dfrac{1[(-1)^{20} - 1]}{-1 - 1} = 0$ **21.** $a_8 = \dfrac{1}{5}\left(\dfrac{1}{2}\right)^7 = \dfrac{1}{640}$ **23.** $S_5 = \dfrac{-\frac{1}{2}[(\frac{1}{2})^5 - 1]}{\frac{1}{2} - 1} = -\dfrac{31}{32}$

25. $a_{10} = \sqrt{2}(\sqrt{2})^9 = (\sqrt{2})^{10} = 32; S_{10} = \dfrac{\sqrt{2}[(\sqrt{2})^{10} - 1]}{\sqrt{2} - 1} = \dfrac{31\sqrt{2}}{\sqrt{2} - 1} = 62 + 31\sqrt{2}$

27. $a_6 = 100\left(\dfrac{1}{10}\right)^5 = \dfrac{1}{10^3} = \dfrac{1}{1,000}; S_6 = \dfrac{100[(\frac{1}{10})^6 - 1]}{\frac{1}{10} - 1} = \dfrac{100(-0.999999)}{-0.9} = 111.111$ **29.** $r = \pm 2$

31. $S = \dfrac{\frac{1}{2}}{1 - \frac{1}{2}} = 1$ **33.** $S = \dfrac{4}{1 - \frac{1}{2}} = 8$ **35.** $S = \dfrac{\frac{2}{5}}{1 - \frac{2}{5}} = \dfrac{2}{3}$ **37.** $S = \dfrac{\frac{3}{4}}{1 - \frac{1}{3}} = \dfrac{9}{8}$ **43.** $x^2 + 10x + 25$

45. $x^3 + 3x^2y + 3xy^2 + y^3$ **47.** $x^4 + 4x^3y + 6x^2y^2 + 4xy^3 + y^4$ **49.** 52,428,800   **51.** 199.99924

53. $S = 2a$ **55.** $S = \dfrac{a}{b - a}$

PROBLEM SET 9.5

1. $x^4 + 8x^3 + 24x^2 + 32x + 16$ **3.** $x^6 + 6x^5y + 15x^4y^2 + 20x^3y^3 + 15x^2y^4 + 6xy^5 + y^6$
5. $32x^5 + 80x^4 + 80x^3 + 40x^2 + 10x + 1$ **7.** $x^5 - 10x^4y + 40x^3y^2 - 80x^2y^3 + 80xy^4 - 32y^5$
9. $81x^4 - 216x^3 + 216x^2 - 96x + 16$ **11.** $64x^3 - 144x^2y + 108xy^2 - 27y^3$ **13.** $x^8 + 8x^6 + 24x^4 + 32x^2 + 16$

15. $x^6 + 3x^4y^2 + 3x^2y^4 + y^6$ **17.** $\dfrac{x^3}{8} - 3x^2 + 24x - 64$ **19.** $\dfrac{x^4}{81} + \dfrac{2x^3y}{27} + \dfrac{x^2y^2}{6} + \dfrac{xy^3}{6} + \dfrac{y^4}{16}$

21. $x^9 + 18x^8 + 144x^7 + 672x^6$ **23.** $x^{10} - 10x^9y + 45x^8y^2 - 120x^7y^3$ **25.** $x^{10} + 20x^9y + 180x^8y^2 + 960x^7y^3$
27. $x^{15} + 15x^{14} + 105x^{13}$ **29.** $x^{12} - 12x^{11}y + 66x^{10}y^2$ **31.** $x^{20} + 40x^{19} + 760x^{18}$ **33.** $x^{100} + 200x^{99}$

35. $x^{50} + 50x^{49}y$ **37.** $a_9 = \dfrac{12!}{8!4!} (2x)^4(3y)^8 = 495(16x^4)(6{,}561y^8) = 51{,}963{,}120x^4y^8$

39. $a_5 = \dfrac{10!}{4!6!} x^6(-2)^4 = 210x^6(16) = 3{,}360x^6$ **41.** $a_4 = \dfrac{9!}{3!6!} x^6(3)^3 = 84x^6(27) = 2{,}268x^6$

43. $a_{12} = \dfrac{20!}{11!9!} (2x)^9(5y)^{11}$ **45.** $\frac{21}{128}$ **47.** $x = \dfrac{\log 7}{\log 5} \approx 1.21$ **49.** $\frac{1}{6}$, or 0.17   **51.** Approximately 7 yr

53. 2.16   **55.** 6.36   **57.** $t = \dfrac{1}{5} \ln \dfrac{A}{10}$ **59.** 56   **61.** 125,970

CHAPTER 9   REVIEW PROBLEMS

1. 7, 9, 11, 13   **2.** 1, 4, 7, 10   **3.** 0, 3, 8, 15   **4.** $\frac{4}{3}, \frac{5}{4}, \frac{6}{5}, \frac{7}{6}$ **5.** 4, 16, 64, 256   **6.** $\frac{1}{4}, \frac{1}{16}, \frac{1}{64}, \frac{1}{256}$

7. $a_n = 3n - 1$ **8.** $a_n = 2n - 5$ **9.** $a_n = n^4$ **10.** $a_n = n^2 + 1$ **11.** $a_n = 2^{-n} = \dfrac{1}{2^n}$ **12.** $a_n = \dfrac{n + 1}{n^2}$

13. 32   **14.** 25   **15.** $\frac{14}{5}$ **16.** -5 **17.** 98   **18.** $\frac{127}{30}$ **19.** $\displaystyle\sum_{i=1}^{4} 3i$ **20.** $\displaystyle\sum_{i=1}^{4} (4i - 1)$ **21.** $\displaystyle\sum_{i=1}^{5} (2i + 3)$

22. $\displaystyle\sum_{i=2}^{4} i^2$ **23.** $\displaystyle\sum_{i=1}^{4} \dfrac{1}{i + 2}$ or $\displaystyle\sum_{i=3}^{6} \dfrac{1}{i}$ **24.** $\displaystyle\sum_{i=1}^{5} \dfrac{i}{3^i}$ **25.** $\displaystyle\sum_{i=1}^{3} (x - 2i)$ **26.** $\displaystyle\sum_{i=1}^{4} \dfrac{x}{x + i}$ **27.** Geometric

28. Arithmetic   **29.** Arithmetic   **30.** Neither   **31.** Geometric   **32.** Geometric   **33.** Arithmetic
34. Neither   **35.** $a_n = 3n - 1$; $a_{20} = 59$ **36.** $a_n = 8 - 3n$; $a_{16} = -40$ **37.** $a_{10} = 34$; $S_{10} = 160$
38. $a_{16} = 78$; $S_{16} = 648$ **39.** $a_1 = 5$; $d = 4$; $a_{10} = 41$ **40.** $a_1 = 8$; $d = 3$; $a_9 = 32$; $S_9 = 180$
41. $a_1 = -4$; $d = -2$; $a_{20} = -42$; $S_{20} = -460$ **42.** 20,100   **43.** -95 **44.** $a_n = 3(2)^{n-1}$; $a_{20} = 3(2)^{19}$
45. $a_n = 5(-2)^{n-1}$; $a_{16} = 5(-2)^{15}$ **46.** $a_n = 4(\frac{1}{2})^{n-1}$; $a_{10} = 4(\frac{1}{2})^9 = \frac{1}{128}$ **47.** -3 **48.** 8
49. $a_1 = 3$; $r = 2$; $a_6 = 96$ **50.** $243\sqrt{3}$ **51.** 28   **52.** 35   **53.** 20   **54.** 36   **55.** 45   **56.** 161,700
57. $x^4 - 8x^3 + 24x^2 - 32x + 16$ **58.** $16x^4 + 96x^3 + 216x^2 + 216x + 81$ **59.** $27x^3 + 54x^2y + 36xy^2 + 8y^3$
60. $x^{10} - 10x^8 + 40x^6 - 80x^4 + 80x^2 - 32$ **61.** $\frac{1}{16}x^4 + \frac{3}{2}x^3 + \frac{27}{2}x^2 + 54x + 81$ **62.** $\frac{1}{27}x^3 - \frac{1}{6}x^2y + \frac{1}{4}xy^2 - \frac{1}{8}y^3$
63. $x^{10} + 30x^9y + 405x^8y^2$ **64.** $x^9 - 27x^8y + 324x^7y^2$ **65.** $x^{11} + 11x^{10}y + 55x^9y^2$
66. $x^{12} - 24x^{11}y + 264x^{10}y^2$ **67.** $x^{16} - 32x^{15}y$ **68.** $x^{32} + 64x^{31}y$ **69.** $x^{50} - 50x^{49}$ **70.** $x^{150} + 150x^{149}y$

71. $\dfrac{10!}{5!5!} x^5(-3)^5 = 252x^5(-243) = -61{,}236x^5$ **72.** $\dfrac{9!}{3!6!} (2x)^6 = 84(64x^6) = 5{,}376x^6$

CHAPTER 9 TEST

1. $-2, 1, 4, 7, 10$ **2.** $3, 7, 11, 15, 19$ **3.** $2, 5, 10, 17, 26$ **4.** $2, 16, 54, 128, 250$ **5.** $2, \frac{3}{4}, \frac{4}{9}, \frac{5}{16}, \frac{6}{25}$

6. $4, -8, 16, -32, 64$ **7.** $a_n = 4n + 2$ **8.** $a_n = 2^{n-1}$ **9.** $a_n = \dfrac{1}{2^n}$ **10.** $a_n = (-3)^n$

11. (a) 90 (b) 53 (c) 130 **12.** 3 **13.** ± 6 **14.** 320 **15.** 25 **16.** $S_{50} = 3(2^{50} - 1)$ **17.** $\frac{3}{4}$
18. $x^4 - 12x^3 + 54x^2 - 108x + 81$ **19.** $32x^5 - 80x^4 + 80x^3 - 40x^2 + 10x - 1$ **20.** $x^{20} - 20x^{19} + 190x^{18}$

21. $\dfrac{8!}{5!3!}(2x)^3(-3y)^5 = 56(8x^3)(-243y^5) = -108{,}864x^3y^5$

CHAPTER 10 PROBLEM SET 10.1

1. 5 **3.** $\sqrt{106}$ **5.** $\sqrt{61}$ **7.** $\sqrt{130}$ **9.** 3 or -1 **11.** 3 **13.** $(x - 2)^2 + (y - 3)^2 = 16$
15. $(x - 3)^2 + (y + 2)^2 = 9$ **17.** $(x + 5)^2 + (y + 1)^2 = 5$ **19.** $x^2 + (y + 5)^2 = 1$ **21.** $x^2 + y^2 = 4$

23. Center $= (0, 0)$
Radius $= 2$

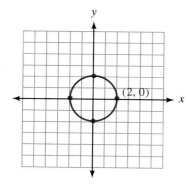

25. Center $= (1, 3)$
Radius $= 5$

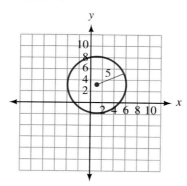

27. Center $= (-2, 4)$
Radius $= 2\sqrt{2}$

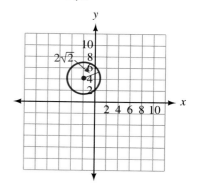

29. Center $= (-1, -1)$
Radius $= 1$

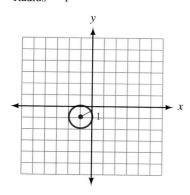

31. Center $= (0, 3)$
Radius $= 4$

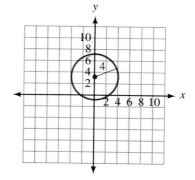

33. Center $= (-1, 0)$
Radius $= \sqrt{2}$

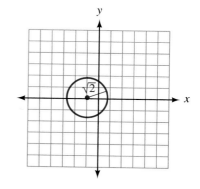

35. Center = (2, 3)
    Radius = 3

37. Center = $(-1, -\frac{1}{2})$
    Radius = 2

39. $x^2 + y^2 = 25$

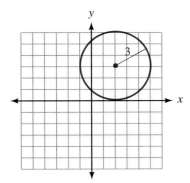

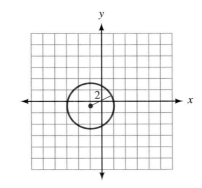

41. $x^2 + y^2 = 9$ **43.** $(x + 1)^2 + (y - 3)^2 = 25$

45. $y = \sqrt{9 - x^2}$ corresponds to the top half; $y = -\sqrt{9 - x^2}$ to the bottom half.   **47.** 10π m   **49.** $a_n = 4n + 1$

51. $\frac{15}{32}$ **53.** $\sum_{i=1}^{5} (2i - 1)$ **55.** $(x - 2)^2 + (y - 3)^2 = 4$ **57.** $(x - 2)^2 + (y - 3)^2 = 4$ **59.** 5   **61.** 5

PROBLEM SET 10.2

1.

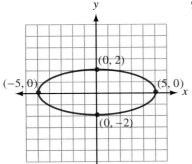

3.

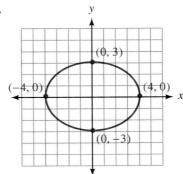

5.

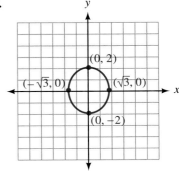

7.

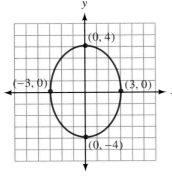

9.

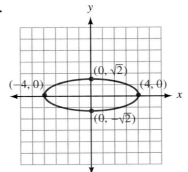

11.

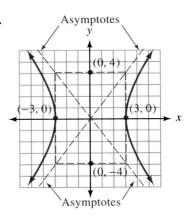

13.

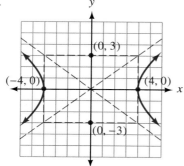

15.

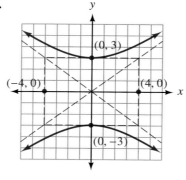

17.

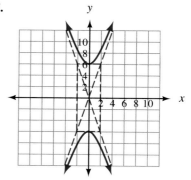

19.

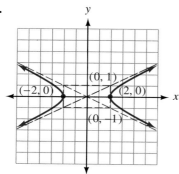

21.

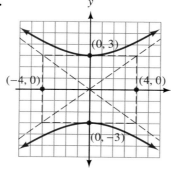

23. x-intercepts $= \pm 3$
y-intercepts $= \pm 2$

25. x-intercepts $= \pm 0.2$
No y-intercepts

27. x-intercepts $= \pm \frac{3}{5}$
y-intercepts $= \pm \frac{2}{5}$

29.

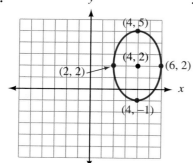

31.

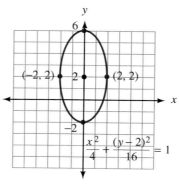

$$\frac{x^2}{4} + \frac{(y-2)^2}{16} = 1$$

33.

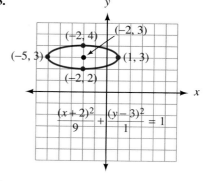

$$\frac{(x+2)^2}{9} + \frac{(y-3)^2}{1} = 1$$

35.

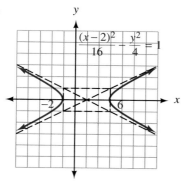

$$\frac{(x-2)^2}{16} - \frac{y^2}{4} = 1$$

37.

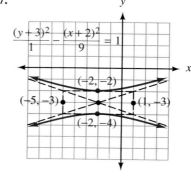

$$\frac{(y+3)^2}{1} - \frac{(x+2)^2}{9} = 1$$

39.

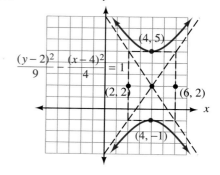

$$\frac{(y-2)^2}{9} - \frac{(x-4)^2}{4} = 1$$

Answers to Odd-Numbered Problems, Chapter Reviews, and Chapter Tests

41. $\pm\frac{9}{5}$ **43.** ± 3.2 **45.** $y = \frac{3}{4}x,\ y = -\frac{3}{4}x$ **47.** 8   **49.** $a_n = 6n - 1$ **51.** 1,030   **53.** $\frac{63}{4}$

PROBLEM SET 10.3

1.

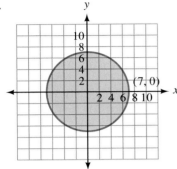

3.

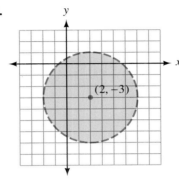

5.

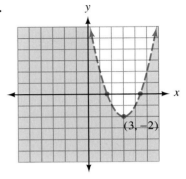

7.

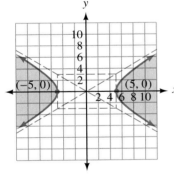

9.

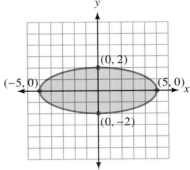

11.

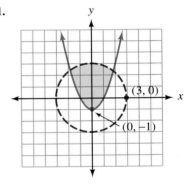

13.

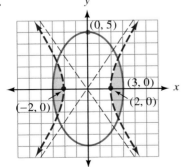

15. No intersection

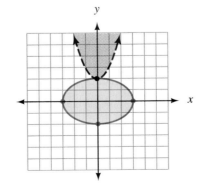

17.

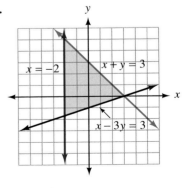

19.

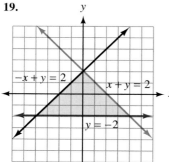

$-x + y = 2$

$x + y = 2$

$y = -2$

21.

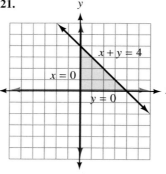

$x + y = 4$

$x = 0$

$y = 0$

23. $(0, 3), (\frac{12}{5}, -\frac{9}{5})$
25. $(0, 4), (\frac{16}{5}, \frac{12}{5})$
27. $(5, 0), (-5, 0)$
29. $(0, -3), (\sqrt{5}, 2), (-\sqrt{5}, 2)$
31. $(0, -4), (\sqrt{7}, 3), (-\sqrt{7}, 3)$
33. $(-4, 11), (\frac{5}{2}, \frac{5}{4})$
35. $(-4, 5), (1, 0)$
37. $(2, -3), (5, 0)$
39. $(3, 0), (-3, 0)$
41. $(4, 0), (0, -4)$
43. 8, 5, or $-8, -5$, or $8, -5$, or $-8, 5$
45. 6, 3 or 13, -4
47. $x^4 + 8x^3 + 24x^2 + 32x + 16$
49. $8x^3 + 12x^2y + 6xy^2 + y^3$
51. $x^{50} + 150x^{49}$

CHAPTER 10   REVIEW PROBLEMS

1. $\sqrt{10}$ **2.** $\sqrt{13}$ **3.** 5   **4.** 9   **5.** $-2, 6$ **6.** $-12, 4$ **7.** $(x - 3)^2 + (y - 1)^2 = 4$
8. $(x - 3)^2 + (y + 1)^2 = 16$ **9.** $(x + 5)^2 + y^2 = 9$ **10.** $(x + 3)^2 + (y - 4)^2 = 18$ **11.** $x^2 + y^2 = 25$
12. $x^2 + y^2 = 9$ **13.** $(x + 2)^2 + (y - 3)^2 = 25$ **14.** $(x + 6)^2 + (y - 8)^2 = 100$

15. $(0, 0); r = 2$

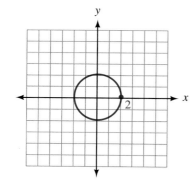

16. $(3, -1); r = 4$

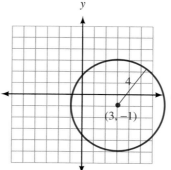

$(3, -1)$

17. $(3, -2); r = 3$

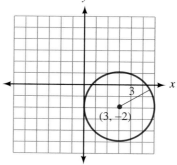

$(3, -2)$

18. $(-2, 1); r = 3$

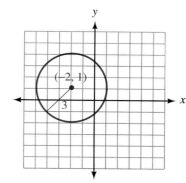

$(-2, 1)$

19.

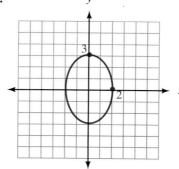

20.

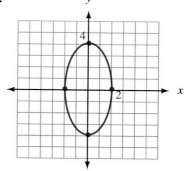

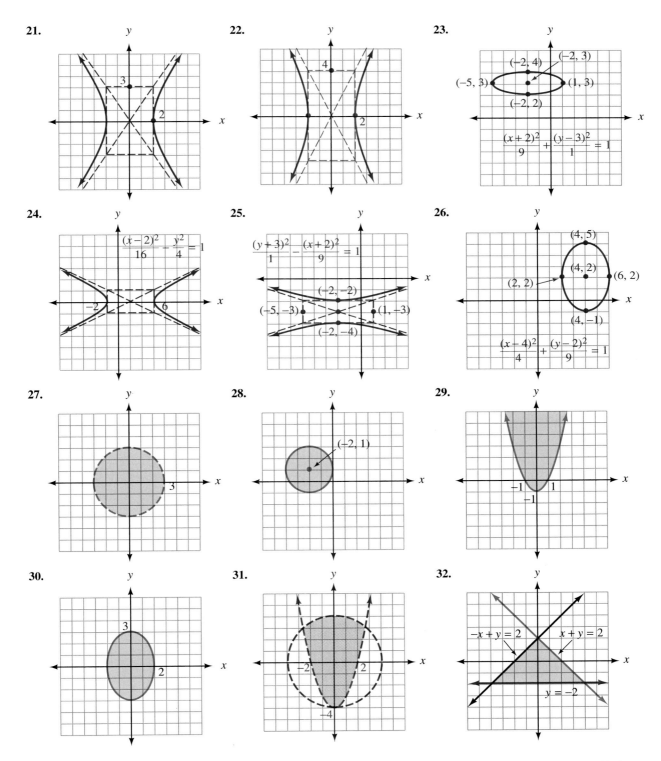

21.

22.

23. $\dfrac{(x+2)^2}{9} + \dfrac{(y-3)^2}{1} = 1$

24. $\dfrac{(x-2)^2}{16} - \dfrac{y^2}{4} = 1$

25. $\dfrac{(y+3)^2}{1} - \dfrac{(x+2)^2}{9} = 1$

26. $\dfrac{(x-4)^2}{4} + \dfrac{(y-2)^2}{9} = 1$

27.

28.

29.

30.

31.

32.

33. $(0, 4), (\frac{16}{5}, -\frac{12}{5})$ **34.** $(0, -2), (\sqrt{3}, 1), (-\sqrt{3}, 1)$ **35.** $(-2, 0), (2, 0)$

36. $\left(-\sqrt{7}, -\frac{\sqrt{6}}{2}\right), \left(-\sqrt{7}, \frac{\sqrt{6}}{2}\right), \left(\sqrt{7}, -\frac{\sqrt{6}}{2}\right), \left(\sqrt{7}, \frac{\sqrt{6}}{2}\right)$

CHAPTER 10   TEST

1. -5 and 3 **2.** $(x + 2)^2 + (y - 4)^2 = 9$ **3.** $x^2 + y^2 = 25$ **4.** $(5, -3); r = \sqrt{39}$

5.

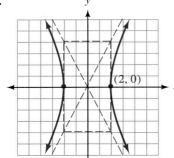

6.

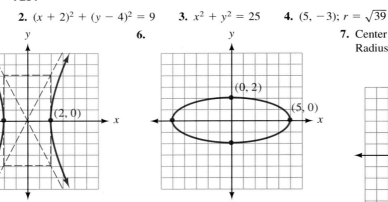

7. Center $= (2, -1)$
Radius $= 3$

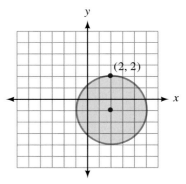

8. $\dfrac{(x - 4)^2}{4} + \dfrac{(y - 2)^2}{9} = 1$

9. $(0, 5), (4, -3)$
10. $(0, -4), (\sqrt{7}, 3), (-\sqrt{7}, 3)$

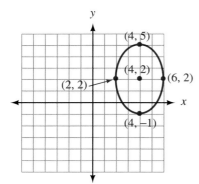